PROGRESS IN BRAIN RESEARCH

VOLUME 176

ATTENTION

Other volumes in PROGRESS IN BRAIN RESEARCH

Volume 140: The Brain's Eye, by J. Hyönä, D.P. Munoz, W. Heide and R. Radach (Eds.) – 2002, ISBN 0-444-51097-4.
Volume 141: Gonadotropin-Releasing Hormone: Molecules and Receptors, by I.S. Parhar (Ed.) – 2002, ISBN 0-444-50979-8.
Volume 142: Neural Control of Space Coding, and Action Production, by C. Prablanc, D. Pélisson and Y. Rossetti (Eds.) – 2003, ISBN 0-444-509771.
Volume 143: Brain Mechanisms for the Integration of Posture and Movement, by S. Mori, D.G. Stuart and M. Wiesendanger (Eds.) – 2004, ISBN 0-444-513892.
Volume 144: The Roots of Visual Awareness, by C.A. Heywood, A.D. Milner and C. Blakemore (Eds.) – 2004, ISBN 0-444-50978-X.
Volume 145: Acetylcholine in the Cerebral Cortex, by L. Descarries, K. Krnjević and M. Steriade (Eds.) – 2004, ISBN 0-444-51125-3.
Volume 146: NGF and Related Molecules in Health and Disease, by L. Aloe and L. Calzà (Eds.) – 2004, ISBN 0-444-51472-4.
Volume 147: Development, Dynamics and Pathology of Neuronal Networks: From Molecules to Functional Circuits, by J. Van Pelt, M. Kamermans, C.N. Levelt, A. Van Ooyen, G.J.A. Ramakers and P.R. Roelfsema (Eds.) – 2005, ISBN 0-444-51663-8.
Volume 148: Creating Coordination in the Cerebellum, by C.I. De Zeeuw and F. Cicirata (Eds.) – 2005, ISBN 0-444-51754-5.
Volume 149: Cortical Function: A View from the Thalamus, by V.A. Casagrande, R.W. Guillery and S.M. Sherman (Eds.) – 2005, ISBN 0-444-51679-4.
Volume 150: The Boundaries of Consciousness: Neurobiology and Neuropathology, by Steven Laureys (Ed.) – 2005, ISBN 0-444-51851-7.
Volume 151: Neuroanatomy of the Oculomotor System, by J.A. Büttner-Ennever (Ed.) – 2006, ISBN 0-444-51696-4.
Volume 152: Autonomic Dysfunction after Spinal Cord Injury, by L.C. Weaver and C. Polosa (Eds.) – 2006, ISBN 0-444-51925-4.
Volume 153: Hypothalamic Integration of Energy Metabolism, by A. Kalsbeek, E. Fliers, M.A. Hofman, D.F. Swaab, E.J.W. Van Someren and R.M. Buijs (Eds.) – 2006, ISBN 978-0-444-52261-0.
Volume 154: Visual Perception, Part 1, Fundamentals of Vision: Low and Mid-Level Processes in Perception, by S. Martinez-Conde, S.L. Macknik, L.M. Martinez, J.M. Alonso and P.U. Tse (Eds.) – 2006, ISBN 978-0-444-52966-4.
Volume 155: Visual Perception, Part 2, Fundamentals of Awareness, Multi-Sensory Integration and High-Order Perception, by S. Martinez-Conde, S.L. Macknik, L.M. Martinez, J.M. Alonso and P.U. Tse (Eds.) – 2006, ISBN 978-0-444-51927-6.
Volume 156: Understanding Emotions, by S. Anders, G. Ende, M. Junghofer, J. Kissler and D. Wildgruber (Eds.) – 2006, ISBN 978-0-444-52182-8.
Volume 157: Reprogramming of the Brain, by A.R. Møller (Ed.) – 2006, ISBN 978-0-444-51602-2.
Volume 158: Functional Genomics and Proteomics in the Clinical Neurosciences, by S.E. Hemby and S. Bahn (Eds.) – 2006, ISBN 978-0-444-51853-8.
Volume 159: Event-Related Dynamics of Brain Oscillations, by C. Neuper and W. Klimesch (Eds.) – 2006, ISBN 978-0-444-52183-5.
Volume 160: GABA and the Basal Ganglia: From Molecules to Systems, by J.M. Tepper, E.D. Abercrombie and J.P. Bolam (Eds.) – 2007, ISBN 978-0-444-52184-2.
Volume 161: Neurotrauma: New Insights into Pathology and Treatment, by J.T. Weber and A.I.R. Maas (Eds.) – 2007, ISBN 978-0-444-53017-2.
Volume 162: Neurobiology of Hyperthermia, by H.S. Sharma (Ed.) – 2007, ISBN 978-0-444-51926-9.
Volume 163: The Dentate Gyrus: A Comprehensive Guide to Structure, Function, and Clinical Implications, by H.E. Scharfman (Ed.) – 2007, ISBN 978-0-444-53015-8.
Volume 164: From Action to Cognition, by C. von Hofsten and K. Rosander (Eds.) – 2007, ISBN 978-0-444-53016-5.
Volume 165: Computational Neuroscience: Theoretical Insights into Brain Function, by P. Cisek, T. Drew and J.F. Kalaska (Eds.) – 2007, ISBN 978 0-444-52823-0.
Volume 166: Tinnitus: Pathophysiology and Treatment, by B. Langguth, G. Hajak, T. Kleinjung, A. Cacace and A.R. Møller (Eds.) – 2007, ISBN 978-0-444-53167-4.
Volume 167: Stress Hormones and Post Traumatic Stress Disorder: Basic Studies and Clinical Perspectives, by E.R. de Kloet, M.S. Oitzl and E. Vermetten (Eds.) – 2008, ISBN 978-0-444-53140-7.
Volume 168: Models of Brain and Mind: Physical, Computational and Psychological Approaches, by R. Banerjee and B.K. Chakrabarti (Eds.) – 2008, ISBN 978-0-444-53050-9.
Volume 169: Essence of Memory, by W.S. Sossin, J.-C. Lacaille, V.F. Castellucci and S. Belleville (Eds.) – 2008, ISBN 978-0-444-53164-3.
Volume 170: Advances in Vasopressin and Oxytocin – From Genes to Behaviour to Disease, by I.D. Neumann and R. Landgraf (Eds.) – 2008, ISBN 978-0-444-53201-5.
Volume 171: Using Eye Movements as an Experimental Probe of Brain Function — A Symposium in Honor of Jean Büttner-Ennever, by Christopher Kennard and R. John Leigh (Eds.) – 2008, ISBN 978-0-444-53163-6.
Volume 172: Serotonin–Dopamine Interaction: Experimental Evidence and Therapeutic Relevance, by Giuseppe Di Giovanni, Vincenzo Di Matteo and Ennio Esposito (Eds.) – 2008, ISBN 978-0-444-53235-0.
Volume 173: Glaucoma: An Open Window to Neurodegeneration and Neuroprotection, by Carlo Nucci, Neville N. Osborne, Giacinto Bagetta and Luciano Cerulli (Eds.) – 2008, ISBN 978-0-444-53256-5.
Volume 174: Mind and Motion: The Bidirectional Link Between Thought and Action, by Markus Raab, Joseph G. Johnson and Hauke R. Heekeren (Eds.) – 2009, ISBN 978-0-444-53356-2.
Volume 175: Neurotherapy: Progress in Restorative Neuroscience and Neurology — Proceedings of the 25th International Summer School of Brain Research, held at the Royal Netherlands Academy of Arts and Sciences, Amsterdam, The Netherlands, August 25–28, 2008, by J. Verhaagen, E.M. Hol, I. Huitinga, J. Wijnholds, A.A. Bergen, G.J. Boer, D.F. Swaab (Eds.) – 2009, ISBN 978-0-12-374511-8.

PROGRESS IN BRAIN RESEARCH

VOLUME 176

ATTENTION

EDITED BY

NARAYANAN SRINIVASAN

Centre of Behavioural and Cognitive Sciences, University of Allahabad, Allahabad, India

ELSEVIER

AMSTERDAM – BOSTON – HEIDELBERG – LONDON – NEW YORK – OXFORD
PARIS – SAN DIEGO – SAN FRANCISCO – SINGAPORE – SYDNEY – TOKYO

Elsevier
360 Park Avenue South, New York, NY 10010-1710
Linacre House, Jordan Hill, Oxford OX2 8DP, UK
Radarweg 29, PO Box 211, 1000 AE Amsterdam, The Netherlands

First edition 2009

British Library Cataloguing in Publication Data
A catalogue record for this book is available from the British Library

Library of Congress Cataloging-in-Publication Data
A catalog record for this book is available from the Library of Congress

ISBN: 978-0-444-53426-2 (this volume)
ISSN: 0079-6123 (Series)

Printed and bound in Great Britain

09 10 11 12 13 10 9 8 7 6 5 4 3 2 1

List of Contributors

H.A. Allen, Behavioural Brain Sciences, School of Psychology, University of Birmingham, Birmingham, UK

S. Baijal, Centre of Behavioural and Cognitive Sciences, University of Allahabad, Allahabad, India

E. Birmingham, Division of Humanities & Social Sciences, California Institute of Technology, Pasadena, CA, USA

J.M. Brown, Department of Psychology, University of Georgia, Athens, GA, USA

M. Carrasco, Department of Psychology & Center for Neural Science, New York University, New York, NY, USA

R. Desimone, McGovern Institute for Brain Research, Massachusetts Institute of Technology (MIT), Cambridge, MA, USA

J.T. Enns, Department of Psychology, University of British Columbia, Vancouver, BC, Canada

S.J. Gotts, Laboratory of Brain and Cognition, National Institute of Mental Health (NIMH), National Institutes of Health, Bethesda, MD, USA

G.G. Gregoriou, Department of Basic Sciences, Medical School, University of Crete, Heraklion, Crete, Greece

R. Gupta, Centre of Behavioural and Cognitive Sciences, University of Allahabad, Allahabad, India

O. Hardt, Department of Psychology, McGill University, Montreal, Quebec, Canada

G.W. Humphreys, Behavioural Brain Sciences, School of Psychology, University of Birmingham, Birmingham, UK

B.R. Kar, Centre of Behavioural and Cognitive Sciences, University of Allahabad, Allahabad, India

J. Kawahara, National Institute of Advanced Industrial Science and Technology, Tsukuba, Japan

R. Kimchi, Department of Psychology & Institute of Information Processing and Decision Making, University of Haifa, Haifa, Israel

A. Kingstone, Department of Psychology, University of British Columbia, Vancouver, BC, Canada

B.R. Levinthal, Department of Psychology, University of Illinois at Urbana-Champaign, Champaign, IL, USA

G. Liu, Department of Psychology, University of British Columbia, Vancouver, BC, Canada

A. Lleras, Department of Psychology, University of Illinois at Urbana-Champaign, Champaign, IL, USA

M. Lohani, Centre of Behavioural and Cognitive Sciences, University of Allahabad, Allahabad, India

E. Mavritsaki, Behavioural Brain Sciences, School of Psychology, University of Birmingham, Birmingham, UK

R.K. Mishra, Centre of Behavioural and Cognitive Sciences, Allahabad University, Allahabad, India

A. Murthy, National Brain Research Centre, Nainwal More, Manesar, Haryana, India

L. Nadel, Department of Psychology, University of Arizona, Tucson, AZ, USA

C. Nakatani, Laboratory for Perceptual Dynamics, RIKEN Brain Science Institute, Wako-shi, Saitama, Japan

M.A. Peterson, Department of Psychology, University of Arizona, Tucson, AZ, USA

A. Raffone, Department of Psychology, "Sapienza" University of Rome, Rome, Italy; Perceptual Dynamics Laboratory, BSI RIKEN, Japan

S.J. Rappaport, Behavioural Brain Sciences, School of Psychology, University of Birmingham, Birmingham, UK
S. Ray, National Brain Research Centre, Nainwal More, Manesar, Haryana, India
J. Raymond, School of Psychology, Bangor University, Bangor, Gwynedd, UK
M.J. Riddoch, Behavioural Brain Sciences, School of Psychology, University of Birmingham, Birmingham, UK
E. Salvagio, Department of Psychology, University of Arizona, Tucson, AZ, USA
K. Shapiro, Wolfson Centre for Clinical and Cognitive Neuroscience, School of Psychology, Bangor University, Bangor, Gwynedd, UK
K.M. Sharika, National Brain Research Centre, Nainwal More, Manesar, Haryana, India
C. Spence, Crossmodal Research Laboratory, Department of Experimental Psychology, University of Oxford, Oxford, UK
N. Srinivasan, Centre of Behavioural and Cognitive Sciences, University of Allahabad, Allahabad, India
P. Srivastava, Centre of Behavioural and Cognitive Sciences, University of Allahabad, Allahabad, India
C. van Leeuwen, Laboratory for Perceptual Dynamics, RIKEN Brain Science Institute, Wako-shi, Saitama, Japan
Y. Yeshurun, Department of Psychology & Institute of Information Processing and Decision Making, University of Haifa, Haifa, Israel
H. Zhou, McGovern Institute for Brain Research, Massachusetts Institute of Technology (MIT), Cambridge, MA, USA

Preface

Attention has been a significant topic of interest for more than 100 years starting from the seminal work of William James in the 19th century. With recent advances in Cognitive Science, we have made tremendous progress in understanding Attention and the way it affects the other mental processes. Attention has been studied at multiple levels and by using different methodologies including behavioral paradigms, eye tracking, single cell studies, local field potentials, EEG, and neural imaging.

The volume arose out of the International Conference on Attention held in December 2008 at the Centre of Behavioural and Cognitive Sciences, University of Allahabad, Allahabad, India. Dr. Anne Triesman gave the inaugural address. Dr. James M. Brown, Dr. James T. Enns, Dr. Glyn Humphreys, Dr. Alan Kingstone, Dr. Alejandro Lleras, Dr. Aditya Murthy, Dr. Lynn Nadel, Dr. Mary Peterson, Dr. Jane Raymond, Dr. Jane Riddoch, and Dr. Kimron Shapiro spoke at the conference and contributed to the volume.

The current volume explores interdisciplinary research on Attention and interaction of Attention with other cognitive processes including perception, learning, and memory. The papers cover major research on attention in Cognitive Neuroscience and Cognitive Psychology. The volume presents recent advances on attention including binding, dynamics of attention, attention and perceptual organization, attention and consciousness, emotion and attention, development of attention, crossmodal attention, computational modeling of attention, control of actions, attention and memory, and meditation. I sincerely believe that the papers in the current volume will add to the growing knowledge on attention and will encourage future scientists to work on attention.

Narayanan Srinivasan
Allahabad

Acknowledgments

I would like to acknowledge the support of the University Grants Commission in generously supporting the Centre and its academic activities. The University of Allahabad has been very supportive of the Centre including the conduct of this International Conference on Attention, none more than Prof. R. G. Harshe, Vice Chancellor, University of Allahabad. The conference would not have been possible without the help of my colleagues Dr. Bhoomika R. Kar and Dr. Ramesh K. Mishra, the office staff, and the students of the Centre. Finally, everything has been made possible due to the tireless work of Prof. Janak Pandey, the then Head of the Centre and the current Vice Chancellor of Central University of Bihar. I thank all of them for their help and encouragement.

The invited speakers to the International Conference on Attention were very supportive and I thank all of them for coming to the conference as well as contributing the chapters. A special thanks to Dr. Marisa Carrasco, Dr. Robert Desimone, Dr. Ruth Kimchi, Dr. Chie Nakatani, Dr. Antonino Raffone, Dr. Cees van Leeuwen, and Dr. Charles Spence for contributing chapters, even though they could not attend the conference. I also want to thank all those who kindly reviewed the chapters. I would like to thank Elsevier for bringing out the volume and everybody at Elsevier who worked on this volume for their support. Finally, I would like to acknowledge the support of my wife Priya Srinivasan.

Narayanan Srinivasan
Allahabad

Contents

See Color Plate Section at the end of this book

N. Srinivasan (Ed.)
Progress in Brain Research, Vol. 176
ISSN 0079-6123

CHAPTER 1

Attention and competition in figure-ground perception

Mary A. Peterson* and Elizabeth Salvagio

Department of Psychology, University of Arizona, Tucson, AZ, USA

Abstract: What are the roles of attention and competition in determining where objects lie in the visual field, a phenomenon known as *figure-ground perception*? In this chapter, we review evidence that attention and other high-level factors such as familiarity affect figure-ground perception, and we discuss models that implement these effects. Next, we consider the *Biased Competition Model of Attention* in which attention is used to resolve the competition for neural representation between two nearby stimuli; in this model the response to the stimulus that loses the competition is suppressed. In the remainder of the chapter we discuss recent behavioral evidence that figure-ground perception entails between-object competition in which the response to the shape of the losing competitor is suppressed. We also describe two experiments testing whether more attention is drawn to resolve greater figure-ground competition, as would be expected if the Biased Competition Model of Attention extends to figure-ground perception. In these experiments we find that responses to targets on the location of a losing strong competitor are slowed, consistent with the idea that the location of the losing competitor is suppressed, but responses to targets on the winning competitor are not speeded, which is inconsistent with the hypothesis that attention is used to resolve figure-ground competition. In closing, we discuss evidence that attention can operate by suppression as well as by facilitation.

Keywords: figure-ground perception; attention; competition; suppression; familiarity; high-level effects

Figure-ground perception and attention: background

This chapter examines the relationship between attention and *figure-ground perception*, a fundamental component of object and scene perception, with a focus on inhibitory competition as a mechanism of figure-ground perception. Figure-ground perception occurs when two contiguous regions share a border; one is often perceived to be an entity (i.e., an object or a *figure*) shaped by the shared border, whereas the other (the *ground*) appears to simply continue behind the figure near their shared border (see Fig. 1). Thus figure-ground perception entails the determination of which regions of the visual field portray near objects and which portray surfaces continuing behind them.

The Gestalt psychologists first introduced figure-ground perception as a topic in perception research. Their position was that figure-ground perception occurred *automatically*, based on innate "configural" cues, image properties that indicated where a configuration (shape/object) lay with

*Corresponding author.
Tel.: +520-621-5365; Fax: +520-621-9306;
E-mail: mapeters@u.arizona.edu

DOI: 10.1016/S0079-6123(09)17601-X

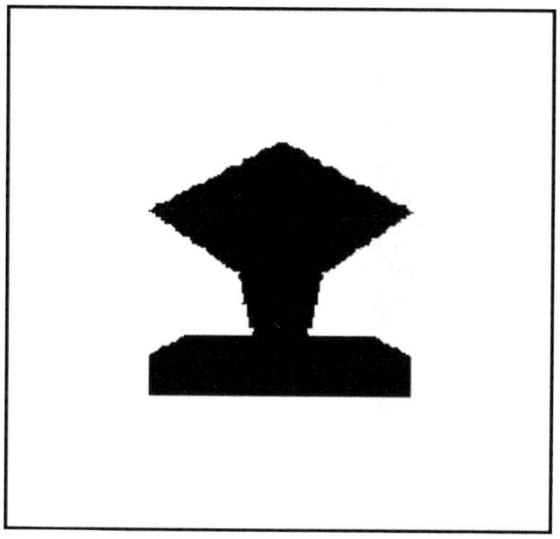

Fig. 1. Here a small, enclosed, symmetric black region shares a border with a larger surrounding white region. The black region is perceived as the figure and white region simply appears to continue behind as its background.

respect to a border shared by two regions. The classic configural cues included image properties such as convexity, closure, small area, and symmetry around a vertical axis. The Gestalt psychologists showed that regions with one or more of the configural properties listed above were more likely to be perceived as figures than contiguous regions with complementary image properties (e.g., regions that were concave, open or surrounding, larger in area, or asymmetric). Figure 2 is an example of the type of display used by the Gestalt psychologists; there multiple black regions with convex parts alternate with multiple white regions with concave parts; the black "convex" regions are more likely to be seen as figure than the white "concave" regions.

The Gestalt psychologists used two-dimensional displays in their experiments, but they assumed that the configural cues operated on three-dimensional displays as well. Inasmuch as figures tend to be nearer to the viewer than grounds, depth cues can affect figure assignment as well (see Peterson and Gibson, 1993; Grossberg, 1994 for tests of how configural and depth cues combine; also see Bertamini et al., 2008; Burge et al., 2005). Modern investigators have identified a number of other image properties that suggest figural status, including lower region (Vecera et al., 2002), base width (Hulleman and Humphreys, 2004), spatial frequency (Klymenko and Weisstein, 1986), extremal edges (Palmer and Ghose, 2008), and edge-induced watercolor fill (Pinna et al., 2003).

For the Gestalt psychologists figures were the substrate on which later processes such as attention and object recognition operated; they held that figure-ground perception per se was unaffected by perceptual experience. According to the traditional Gestalt view, higher-level factors such as experience, knowledge, intention, and/or attention could influence figure *interpretation* but not figure

Fig. 2. Black regions with convex parts alternating with white regions with concave parts. A black frame surrounds the display because it is printed on a white page. In the experiments, no frame was used; displays were presented on a medium gray field.

assignment. That is, high-level factors could operate only after the initial organization was achieved.

Modern research revealing high-level influences on figure-ground perception overturned certain aspects of the traditional view, but not all. For instance, there is some evidence that figure-ground perception can occur pre-attentively (e.g., Kimchi and Peterson, 2008), but this finding does not entail that figure-ground perception always occurs pre-attentively. Other experiments revealed that attention affects figure assignment: Driver and Baylis (1995) showed that regions to which observers allocated attention endogenously (e.g., by following instructions to orient attention to the left or right embodied by arrow cues shown at fixation) were more likely to be perceived as figures than adjacent regions that were unattended. Vecera et al. (2004) extended these findings to regions to which observers' attention was oriented "exogenously" in response to light flashes shown on the right or left side of a display. (Although we note that endogenous attention may underlie Vecera et al.'s effects because their participants may have strategically used the light flashes to accomplish their task.) In addition, Peterson and Gibson (1994b) showed that fixated regions were more likely to be seen as figures than unfixated regions; inasmuch as attention and fixation tend to be coupled, fixation effects may reveal attention effects.

Other experiments showed that, contrary to the traditional Gestalt view, past experience and/or attention can affect figure assignment, per se, and not just figure interpretation. For instance, Peterson et al. (1991) showed that perceptual experience in the form of *familiar configuration*, can influence initial figure assignment: They found that regions portraying portions of familiar objects were more likely to be seen as figures when they portrayed the familiar objects in their typical upright orientation rather than an inverted (relatively unfamiliar) orientation (see Fig. 3). These effects were obtained in both briefly exposed displays (with exposure durations as short as 28 ms) and in reversal experiments where stimulus exposures as long as 40 s were used. (See also Gibson and Peterson, 1994; Peterson and Gibson, 1994a, b; Peterson and Skow-Grant, 2003, for review.) Peterson et al. (1991) and Peterson and Gibson (1994b) also showed that the viewer's perceptual intentions, manipulated via perceptual set instructions, affected which regions they perceived as figures (and not just which regions they reported seeing as figures).

Fig. 3. A familiar configuration of a standing woman depicted by the black region. The standing woman is upright in the display on the left and is inverted in the display on the right. Subjects were more likely to see the black region as figure in the display on the left than the display on the right. The displays above are framed by a black outline. In the experiments, no frame was used; displays were presented on a medium gray field. Adapted from Gibson and Peterson (1994), with permission from the American Psychological Association.

Summing up: In this section, we briefly reviewed the history of figure-ground research, focusing on the cues that affect figure assignment, including both the image properties espoused by the Gestalt psychologists and high-level factors such as attention and familiar configuration identified more recently.

In the next sections, we review both an early, nonbiological, model that shows how attention can affect figure assignment and a contemporary, biologically plausible, model of attention that accounts for competition between objects for neural representation. We then investigate whether the latter model can be applied to figure-ground perception.

Early models of figure-ground perception involving attention and competition

Kienker et al. (1986) and Sejnowski and Hinton (1990) presented a computational model in which attention influenced the determination of which of two contiguous regions was perceived as the figure. Kienker and colleagues proposed this model before empirical research showed that attended regions were more likely to be seen as figures than unattended regions. Accordingly, their model showed that *in principle* attention could affect figure-ground perception.

The Keinker et al. model included "figure" units for every location in the visual field; figure units were essentially feature units. Between pairs of figure/feature units representing adjacent locations in space lay pairs of edge units favoring assigning figural status to one or the other of the paired figure/feature units. (e.g., one of the edge units between two horizontally adjacent figure/feature units would favor assigning figural status to the unit on the left, the other would favor the figure/feature unit on the right.). Edge units facing in opposite directions inhibited each other but engaged in mutual excitation with figure/feature units lying on their preferred side. In this early model, neighboring units responding to the same low-level features (e.g., color, luminance, or texture) engaged in mutual excitation.[1] Kienker et al. (1986) used focused attention as a seed to increase the activity in one set of figure/feature units, which then increased activity in the edge units facing toward them; the activated edge units suppressed the contiguous edge units facing in the opposite direction, which in turn suppressed their associated figure/feature units. The relatively enhanced activity in one set of edge units and their associated figure/feature units was taken to realize figure assignment (see also Grossberg and Mingolla, 1993; Grossberg, 1994). The Kienker et al. model was very simple, including two contiguous equal-area regions, with no distinguishing features. Attention, modeled as a seed that biased the figure/feature units on one side of the edge, was the only cue present.

O'Reilly and Vecera (1998) and Vecera and O'Reilly (2000) extended Kienker et al.'s model to account for Peterson et al.'s (1991; Peterson and Gibson, 1993, 1994a) effects of familiar configuration by using feedback from high-level object representations rather than attention as the seed that increased the activity of the figure/feature units lying on one side of a border. It is important to note that the competitive models proposed by Sejnowski and colleagues and by O'Reilly and Vecera assumed that inhibitory competition between edge units determined the assignment of figure and ground; neither of these models assumed that competition at higher levels, say between object representations, played a direct role in figure-ground perception.

The biased competition model of attention

In this section, we discuss a model of between-object competition that arose in the neurophysiological literature without consideration of figure-ground perception. Later, we will explore the extent to which it applies to figure-ground perception.

Desimone and his colleagues (e.g., see Desimone and Duncan, 1995) showed that objects compete for neural response at many levels of the visual system, including both low and high levels (i.e., V2, V4, TE, IT). In single cell recordings, the competition is evident in reduction of a neuron's response when more than one stimulus is present in its receptive field, even though one of the stimuli is a good stimulus in that it elicits a vigorous response from the neuron when presented alone and the other is a poor stimulus in that it elicits little or no response when presented alone (e.g., Moran and Desimone, 1985; Miller et al., 1993; Rolls and Tovee, 1995). Competition has been demonstrated in both monkeys and humans with a variety of methods (i.e., event-related potentials, and functional magnetic resonance imaging as well as single cell recording).

Desimone and Duncan (1995) showed that the competition can be "biased" toward one stimulus in the neuron's receptive field by contrast (an image property) or by attention (Duncan et al.,

[1]We now know this does not occur; at least at low levels of the visual hierarchy, neurons responding to the same features engage in lateral inhibition.

1997; Reynolds et al., 1999; see Reynolds and Chelazzi, 2004, for a review). For instance, if an animal attends to one of two stimuli within a neuron's receptive field, the neuron's response pattern changes to resemble the pattern obtained when only the attended stimulus is present. Critically, if the animal attends to the poor stimulus, the response to the good stimulus is suppressed (Chelazzi et al., 1993). If, on the other hand, the animal attends to the good stimulus, the response of the neuron is as high as it would be if only the good stimulus were present. Likewise, if one stimulus is higher in contrast than the other, the neuron's response resembles its response to the high-contrast stimulus alone; the response to the other stimulus is suppressed.

The biased competition model has been used primarily to study effects of attention, often in visual search paradigms. As a consequence, it is referred to as the *Biased Competition Model of Attention*. Attention effects have been modeled in terms of contrast units (cf. Carrasco et al., 2000, 2004; Pestilli and Carrasco, 2005; Liu et al., 2009), although there is a debate about whether or not attention can change perceived contrast (cf. Prinzmetal et al., 2008; Schneider, 2006).

Nearby stimuli are more likely than distant stimuli to be represented in the same receptive fields, especially in brain regions lower in the visual hierarchy. Therefore, competition between objects for neural response should increase as between-object distance decreases, and it does; competition-induced suppression is also greater when the stimuli are presented simultaneously rather than sequentially (Moran and Desimone, 1985; Luck et al., 1997; Kastner et al., 1998; Beck and Kastner, 2007; Torralbo and Beck, 2008). These findings regarding proximity and simultaneity in particular led Peterson and Skow (2008) to investigate whether the biased competition model applied to figure-ground perception, as we discuss in the next section.

Biased competition and suppression in figure-ground perception

Peterson and Skow (2008) noted that when two regions in the visual field share a border — the conditions that produce figure-ground perception — the proto-objects that might be seen on opposite sides of the border are highly proximate and therefore highly likely to lie within the same receptive fields and to compete for neural response. This is illustrated by the Rubin vase/faces stimulus shown in Fig. 4A. For the Rubin stimulus, the two objects that compete for figural status are both nameable (a vase/goblet and a face), at least when a large enough set of configured parts is considered. Even for a stimulus like the one in Fig. 4B, portions of object candidates are present on opposite sides of the silhouette's left and right borders, even though neither candidate is familiar/nameable.

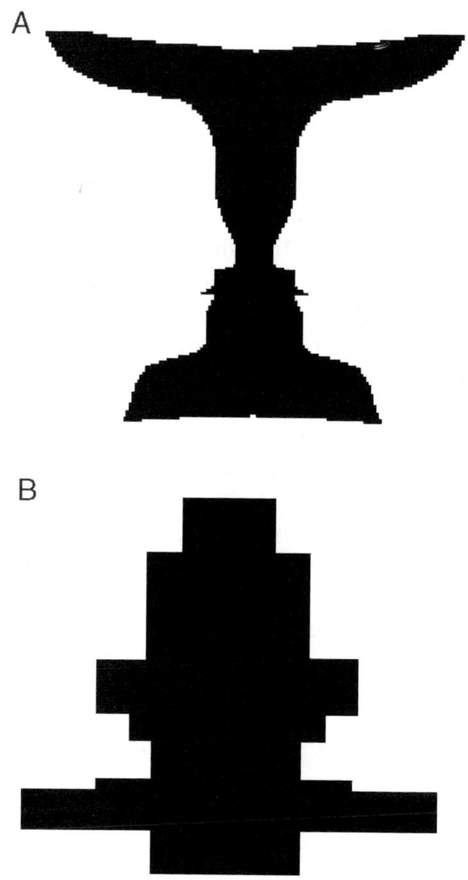

Fig. 4. (A) Rubin's vase/face. (B) Here a small, enclosed, symmetric black region shares a border with a larger surrounding white region. Candidate novel objects are present on the inside and outside of both the left and right vertical edges.

Peterson and Skow (2008) hypothesized that figure-ground perception results from inhibitory competition between portions of candidate objects that might be seen on opposite sides of a border, in addition to (or instead of) competition between lower-level edge units and/or feature units such as those modeled by Kienker et al. (1986), Vecera and O'Reilly (2000), and O'Reilly and Vecera (1998). On this view, the candidate objects can sometimes be novel and at other times can consist of familiar configurations of parts. Note that between-object competition does not necessarily involve whole objects; "familiar configurations" are simply sufficiently large portions of familiar objects to be recognizable. The object candidate that wins the competition at a given border, or portion thereof, is perceived as bounded by the edge locally; in other words it is perceived as the figure. The candidate object, or portion thereof, that loses the competition at a given border is perceived as the ground locally; its shape is not perceived consciously, rather, the response to the losing object is suppressed. On the view that figure-ground perception can involve competition between portions of candidate objects, then suppression should be evident at levels higher than figure and edge units; it should be evident at the level where familiar configurations are represented (at least).

Peterson and Skow (2008) tested for suppression of the response to an object candidate that loses the figure-ground competition using silhouettes like those in Fig. 5. Many cues biased perception toward the interpretation that the figure was located on the inside of the silhouette's border. The insides were closed, symmetric around a vertical axis, and smaller in area than the surrounding region, and they were shown centered on observers' fixation point.

There were two types of silhouettes: "low-competition" (LoC) silhouettes in which few (if any) cues favored perceiving the figure on the outside of the silhouette (see samples in the top row of Fig. 5); and "high-competition" (HiC) silhouettes in which portions of familiar objects were suggested along the outside of the silhouettes' left and right borders; hence familiar configuration favored assigning the figure on the outside and competed with the ensemble of cues favoring the inside as figure. Sample HiC silhouettes are shown in the bottom row of Fig. 5, where portions of boots, butterflies, and bunches of grapes are suggested on the outsides of the silhouettes shown from left to right, respectively. Because the

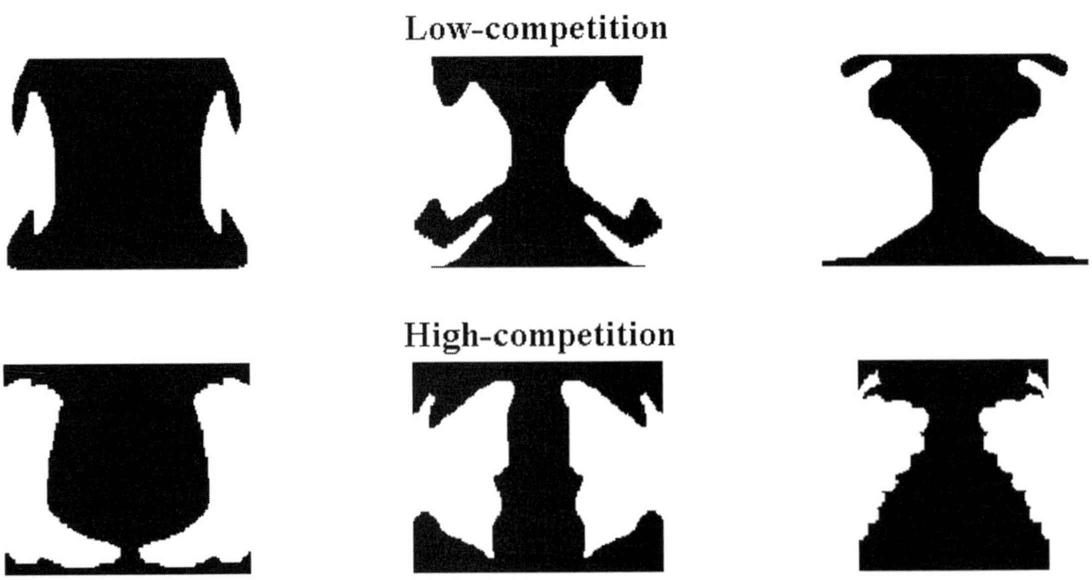

Fig. 5. Top row: Low-competition silhouettes. Bottom row: High-competition silhouettes.

majority of cues favored the inside of the silhouette as figure, and because subjects were naive (unlike anyone who has read the preceding text), Peterson and Skow expected that in the experiments the familiar configuration would lose the competition for figural status in HiC silhouettes and the outside of the silhouette would be seen as a shapeless ground. Indeed, in post-experiment questions, subjects reported that they saw the insides of the silhouettes as figures and did not perceive a familiar object on the outside. The question was whether competition involving suppression of the response to the losing object candidate (the familiar configuration) produced this percept.

Peterson and Skow (2008) presented either a HiC or a LoC silhouette for 50 ms on each trial (see row 1, Fig. 6). Shortly after the silhouette disappeared (33 ms), they presented a line drawing of either a familiar, real world, object (see row 2, Fig. 6), or a novel object drawn from Kroll and Potter's (1984) set (see row 3, Fig. 6). Subjects made no response to the silhouette; their task was to categorize the line drawing as portraying a novel object or an object they had previously encountered in two- or three-dimensional in the real world. Peterson and Skow were interested in the responses to the line drawings of the real world objects; they included the novel objects only so that subjects had to make a decision before making a response. Their hypothesis was that if the response to the losing familiar configuration on the outside of the HiC silhouettes was suppressed in the course of figure-ground competition, then responses to a line drawing of a real world object, say a flower as in Fig. 6, should be longer when it follows a HiC silhouette with the same basic-level objects suggested – but not consciously perceived – on the outside of the silhouette than when it follows a LoC control silhouette (see the "match condition" in the left half of Fig. 6).

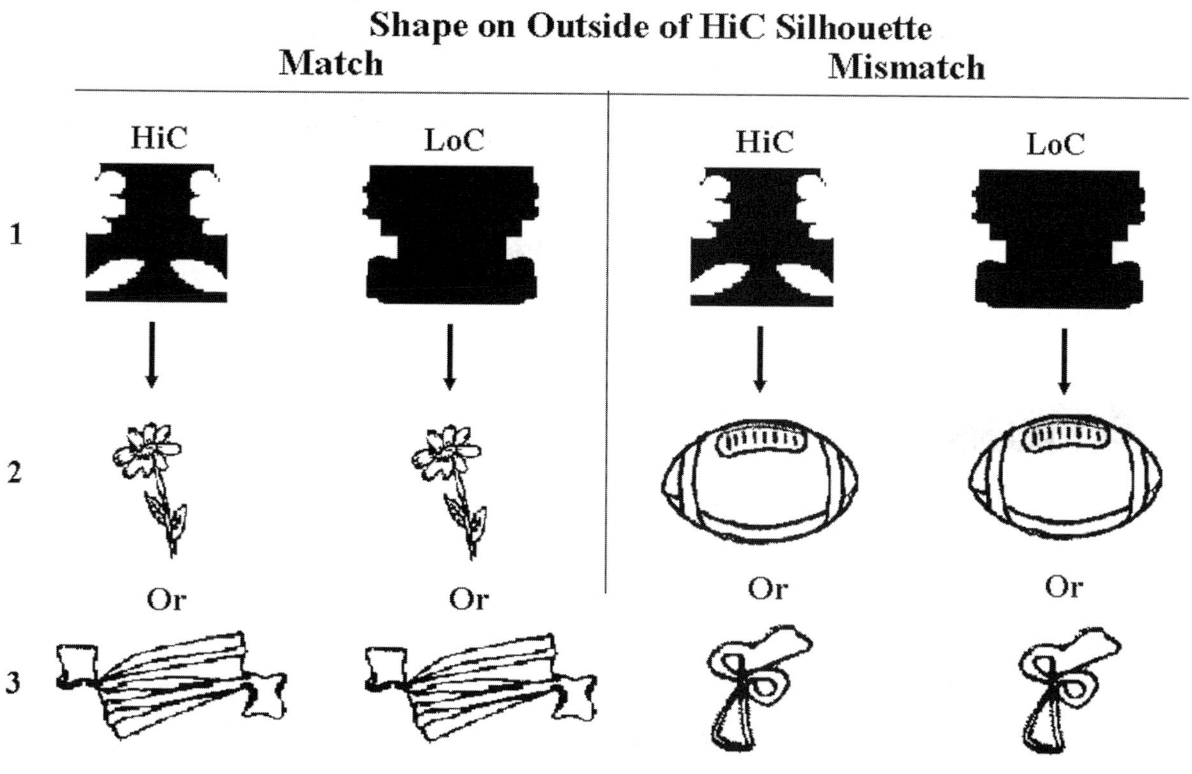

Fig. 6. A schematic of Peterson and Skow's (2008) design.

To be certain that any HiC–LoC RT differences observed in the match condition reflected suppression of the response to the object candidate that lost the competition in HiC silhouettes rather than simply residue of greater competition in HiC than LoC conditions, Peterson and Skow also measured responses to line drawings that portrayed objects from different superordinate categories than the losing object candidate on the outside of the HiC silhouettes preceding it (e.g., a football following a HiC silhouette with a flower suggested on the outside; see the mismatch condition in the right half of Fig. 6). They reasoned that any HiC–LoC RT differences in this mismatch condition did not reflect suppression of the losing competitor. Therefore, only if HiC–LoC RT differences found in the match condition were statistically larger than those found in the mismatch condition could they be taken as evidence for suppression of the response to the familiar configuration that lost the figure-ground competition in the HiC silhouettes.

Peterson and Skow's (2008) results supported the suppression hypothesis, as shown in Fig. 7. The difference between correct object decision RTs in the HiC versus LoC conditions was greater in the match condition than in mismatch condition, $p < 0.01$. This RT difference was short-lived once the silhouette disappeared: it was evident only in subjects' fast responses; and only when the interval between the disappearance of the silhouette and the appearance of the line drawing was short (33 ms, but not 60 ms). Further, consistent with the suppression hypothesis, Peterson and Skow observed greater HiC–LoC RT differences in the match than the mismatch condition only when the familiar configuration was suggested on the ground side of the silhouette edge. Indeed the pattern of results was reversed when the silhouettes were altered such that the familiar objects lay on the figure side of the edge rather than the ground side: Subjects were now faster in the match condition.

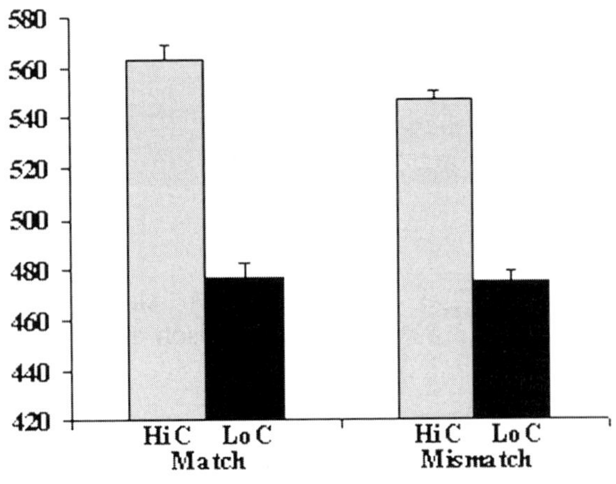

Fig. 7. Fast reaction times measured by Peterson and Skow (2008) for correct "familiar" object decisions following high- and low-competition silhouettes in the match and mismatch conditions. HiC, high competition; LoC, low competition. (Note that the HiC–LoC difference in the mismatch condition does not necessarily index competition time; it reflects regression to the mean due to our method of defining fast responses. See Peterson and Skow, 2008.)

Critically, the borders of the line drawings were always different from those of the silhouettes (even those of same-category line drawings); hence the HiC–LoC RT differences measure suppression at the categorical shape level at least; they cannot be attributed to *edge suppression* alone. These results demanded that extant competitive models of figure-ground competition (Sejnowski and Hinton, 1990; O'Reilly and Vecera, 1998; Vecera and O'Reilly, 2000; Roelfsema et al., 2002) be extended to account for competition between high-level object candidates as well as between edge units and/or feature units.

Because the two figure candidates on opposite sides of a border are so close in space, Peterson and Skow (2008) appealed to the Biased Competition Model of Attention to predict that competition for figural status would occur at the shape level as well as at the lower levels postulated by previous investigators. As evidence for competition, they showed that responses to objects from the same basic-level category as the object that lost the competition for figural status in HiC silhouettes were suppressed, at least for a short time after the silhouette disappeared. In the next section, we describe two recent experiments showing that responses to targets shown in the same location as the losing familiar configuration in HiC silhouettes are slowed; thus suppression of the losing competitor extends to levels lower than shape. These experiments also investigate whether attention is involved in resolving figure-ground competition.

Is attention involved in resolving figure-ground competition?

To continue our investigation of whether the Biased Competition Model of Attention extends to figure-ground competition, we examined whether the amount of attention recruited to resolve figure-ground competition varied with the amount of competition. Torralbo and Beck (2008) recently found that more attention is recruited by objects that are located close to each other rather than at a greater distance. Presumably, more attention is recruited to resolve the greater competition for neural response that occurs for nearby rather than distant objects. In our figure-ground displays, the competing objects on opposite sides of a border are equally nearby in HiC and LoC silhouettes. Yet, by hypothesis, there is more competition in the former than the latter type of silhouette. We next describe two experiments we recently conducted to determine whether more attention was drawn to help resolve the greater competition in HiC than LoC silhouettes.

We tested whether more attention is drawn to the insides of the HiC versus LoC silhouettes, tilted bar targets were displayed at locations just inside or outside the silhouettes' vertical edges. We instructed subjects to report as quickly and as accurately as possible whether the bars were tilted right or left. In discrimination tasks like these, RTs are typically shorter for targets shown in attended than unattended locations (Kim and Cave, 1995, 2001; Cepeda et al., 1998).

Figure 8 illustrates the target tilt discrimination task as used in Experiment 1. Subjects maintained central fixation. On each trial, either a HiC or LoC silhouette ($\sim 3^\circ$ wide) was exposed for 80 ms centered on fixation (a tone sounded during the last 20 ms).[2] The silhouette disappeared and was followed immediately by a 100-ms medium-gray tilted target in a location corresponding to one that was either just inside or just outside the boundary of the previously exposed silhouette. (The target was positioned 0.3° from the location

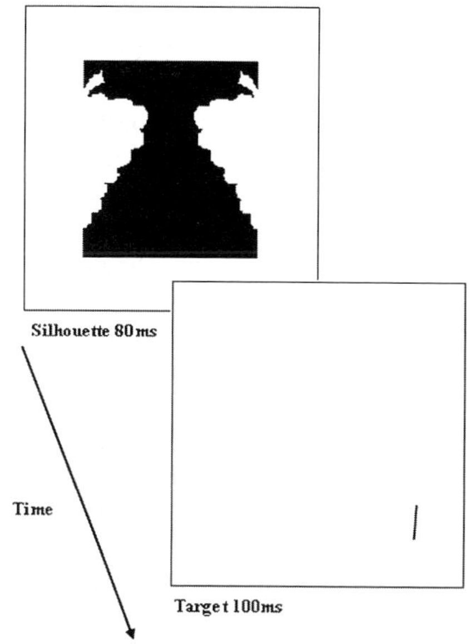

Fig. 8. Schematic of displays used in our target discrimination task; sequential presentation condition. The silhouette shown is a HiC silhouette with a portion of a bunch of grapes suggested on the outside.

previously occupied by one of the silhouette's vertical edges.) Inside and outside targets in HiC and LoC silhouettes were matched for proximity to, and enclosure by, the preceding silhouette's edge. Subjects pressed one of two buttons to report whether the target was tilted left or right.

We had two reasons to expect that RTs would be shorter for targets on locations corresponding to those that were inside versus outside the silhouette that was shown just previously: (1) inside targets were closer to central fixation, hence higher in resolution; and (2) it has been claimed that attention is drawn to figures (Nelson and Palmer, 2007); if attention was drawn to inside locations in the previously viewed silhouette, discrimination RTs should be faster to targets shown in locations corresponding to inside locations.

In addition to these effects, our use of both HiC and LoC silhouettes allowed us to test two hypotheses central to our investigation of whether the Biased Competition Model of Attention can be applied to figure-ground perception. First, is more attention drawn to the inside of HiC than LoC silhouettes to resolve the greater competition

[2]A small number of silhouettes of familiar objects were shown as well; results obtained with the familiar silhouettes are not discussed here.

Table 1. Means and Standard Errors for targets shown on inside and outside locations in High competition and Low competition silhouettes

HiC		LoC	
Inside	Outside	Inside	Outside
A. Experiment 1: Sequential Presentation			
530.54	543.90	515.96	529.38
10.21	10.88	8.63	10.28
B. Experiment 2: Simultaneous Presentation			
542.39	566.70	539.71	539.32
16.35	16.44	16.53	13.10

Note: HiC = High competition; LoC = Low Competition.

from object candidates on the outside in the former than the latter? If so, then RTs should be shorter for targets presented on locations corresponding to the inside of HiC than LoC silhouettes. Second, does suppression of the losing object candidate extend to responses to features at lower levels than familiar configuration, for instance to the location of the losing familiar configuration? If so, then RTs will be longer to report the orientation of targets shown on the outside of HiC than LoC silhouettes.

Table 1A shows the results obtained in Experiment 1 when targets followed the disappearance of the silhouette. RTs were longer for outside than inside targets, $p<0.01$. Contrary to what would be expected if more attention were drawn to the inside of the HiC than the LoC silhouettes to resolve the greater competition in the former than the latter, RTs were *longer* rather than shorter for targets on the inside of the HiC versus LoC silhouettes, $p<0.05$. Consistent with the hypothesis that responses to the location of the losing familiar configuration would be suppressed as well as responses to its shape, RTs were longer for targets on the outside of HiC versus LoC silhouettes, $p<0.05$. We hesitate to take this third finding as evidence for suppression of the location of the familiar configuration that lost the competition in HiC silhouettes, however, because RTs were longer for both inside and outside targets shown after HiC silhouettes, and the inside location is not expected to be suppressed.

The pattern of data we obtained in Experiment 1 could be explained if suppression intended for the outside location of the losing object competitor in HiC silhouettes spread to nearby locations at silhouette offset and affected responses to targets shown on locations corresponding to the inside of the silhouette. (Recall that the inside and outside locations were separated by only 0.6° of visual angle.) Accordingly, in Experiment 2, we examined whether a different pattern of results would be obtained when the silhouettes remained on the screen while the tilted targets were presented. Inasmuch as the borders of the silhouettes might restrict coarsely localized feedback to the outside (Roelfsema et al., 2002), we may be more likely to observe evidence for suppression of the location of the familiar configuration when the silhouettes remain on the screen while the targets are presented. Similarly, given that competition and suppression occur while the silhouette is displayed, and may dissipate quickly after the silhouette is removed, the use of simultaneous rather than successive presentation of the silhouette and the target may allow a more sensitive test of whether more attention is applied to overcome the greater competition in HiC than LoC silhouettes.

The results are shown in Table 1B. With simultaneous presentation, RTs for outside targets were longer for HiC than LoC silhouettes, $p<0.01$, whereas RTs for inside targets were approximately the same in both types of silhouettes. Thus, with simultaneous presentation of the silhouettes and the target we again failed to find evidence that more attention is drawn to the inside of HiC than LoC silhouettes to resolve the greater competition in the former than the latter. Thus, at least as measured by target tilt discrimination responses, our results fail to support this prediction derived from the Biased Competition Model of Attention. Note that Torralbo and Beck (2008) manipulated high versus low competition by varying the proximity of competing objects, whereas we manipulated the amount of competition by manipulating the familiarity of the object candidate on the outside of an edge; in both HiC and LoC silhouettes the competing candidate objects lay on opposite sides of the silhouette edges; hence the proximity of the competing objects was held constant.

The use of simultaneous presentation conditions did allow us to observe evidence for suppression of the location of the familiar configuration that loses the competition in HiC silhouettes. As seen in Table 1B, RTs for outside targets were longer for HiC than for LoC silhouettes.

Taken together the results of Experiments 1 and 2 suggest that (1) response to the location, as well as the categorical shape of the losing familiar configuration in HiC silhouettes is suppressed; (2) suppression is mediated by coarse feedback from higher levels (perhaps shape levels); and (3) the contrast between the features filling the silhouettes versus their background prevented the spread of suppression.

The evidence for greater suppression of responses to the outside location in HiC than LoC silhouettes leaves open the possibility that attention operates to resolve the figure-ground competition via suppression of the losing competitor and its location rather than via facilitation of the winning competitor. Luck (1995) summarized experiments identifying multiple electrophysiologically defined mechanisms of attention. Of particular relevance to the present results is a late operating mechanism that filters (suppresses) distractors in visual search experiments. According to Luck, some of the suppressive effects he observed in visual search reflect feedback from high levels generated as part of a winner-take-all competition engaged when a target is sought in a field of distractors. Given Luck's findings, our behavioral evidence for more suppression of both the shape and the location of the familiar configuration that lost the figure-ground competition could constitute evidence that more attention is drawn to filter out the distracting losing object competitor in HiC than LoC silhouettes. Furthermore, the target discrimination results of Experiments 1 and 2 are consistent with the hypothesis that feedback from higher levels mediates our location suppression effects.

Summary

In this chapter, we reviewed the evidence that attention and other high-level factors such as familiarity affect figure assignment. We discussed early computational accounts in which figure assignment was modeled as inhibitory competition between low-level edge and figure/feature units, with inputs from high levels simply serving to seed the resolution at the lower levels.

We then discussed recent research by Peterson and Skow (2008) showing that competition occurs at high levels at which familiar configurations are represented. As evidence that competition occurs at high levels, Peterson and Skow (2008) showed that the response to a familiar configuration suggested on the ground side of an edge was suppressed when it lost the figure-ground competition. Next, we described two experiments we recently conducted to examine (a) whether responses to the location of the losing familiar configuration were suppressed as well; and (b) whether more attention was recruited to resolve the greater competition that occurs when a familiar configuration is suggested on the outside of a small, closed, symmetric, fixated silhouette.

We found that responses to the location of the familiar configuration that loses the competition for figural status in high-competition silhouettes were suppressed, but we found clear evidence for location-specific suppression only when we presented the targets and silhouettes simultaneously. We hypothesized that suppression is mediated by coarse feedback that is confined to the outside locations by the silhouette edges, but otherwise spreads to nearby locations. These results show that suppression in figure assignment can be measured at multiple levels — at least shape and location.

We found no evidence that responses to the location of the figure were facilitated in high- compared to low-competition silhouettes, as might be expected if more attention had been drawn to resolve the greater competition in the former than the latter. We discuss the possibility that in figure-ground competition, attention may act via suppression (Luck, 1995) rather than via facilitation. In that case, evidence for greater suppression of the location of the losing familiar configuration in HiC versus LoC silhouettes may in fact show that more attention is drawn to resolve the greater competition in the former than the latter. We are pursuing these questions in ongoing research.

Abbreviations

HiC high competition
LoC low competition

Acknowledgments

Mary A. Peterson is grateful to the National Science Foundation (BCS 0425650 & 0418179), for their generous support of the research described in this chapter and to the members of the Centre of Behavioural and Cognitive Sciences (CBCS) at the University of Allahabad, India for their hospitality, intellectual curiosity, and support during the International Conference on Attention, December 7–10, 2008.

References

Beck, D. M., & Kastner, S. (2007). Stimulus similarity modulates competitive interactions in human visual cortex. *Journal of Vision, 7*, 1–12.

Bertamini, M., Martinovic, J., & Wuerger, S. M. (2008). Integration of ordinal and metric cues in depth processing. *Journal of Vision, 8*, 1–12.

Burge, J., Peterson, M. A., & Palmer, S. E. (2005). Ordinal configural cues combine with metric disparity in depth perception. *Journal of Vision, 5*, 534–542.

Carrasco, M., Ling, S., & Read, S. (2004). Attention alters appearance. *Nature Neuroscience, 7*, 308–313.

Carrasco, M., Penpeci-Talgar, C., & Eckstein, M. (2000). Spatial attention increases contrast sensitivity across the CSF: Support for signal enhancement. *Vision Research, 40*, 10–12.

Cepeda, N. J., Cave, K. R., Bichot, N. P., & Kim, M.-S. (1998). Spatial selection via feature-driven inhibition of distractor locations. *Perception & Psychophysics, 60*, 727–746.

Chelazzi, L., Miller, E. K., Duncan, J., & Desimone, R. (1993). A neural basis for visual search in inferior temporal cortex. *Nature, 363*, 345–347.

Desimone, R., & Duncan, J. (1995). Neural mechanisms of selective visual attention. *Annual Reviews of Neurosciences, 18*, 193–222.

Driver, J., & Baylis, G. C. (1995). One-sided edge assignment in vision: 1. Figure-ground segmentation and attention to objects. *Current Directions in Psychologcial Science, 4*, 140–146.

Duncan, J., Humphreys, G., & Ward, R. (1997). Competitive brain activity in visual attention. *Cognitive Neurosciences*, 255–261.

Gibson, B. S., & Peterson, M. A. (1994). Does orientation-independent object recognition precede orientation-dependent recognition? Evidence from a cueing paradigm. *Journal of Experimental Psychology: Human Perception and Performance, 20*, 299–316.

Grossberg, S. (1994). 3-D vision and figure-ground separation by visual cortex. *Perception & Psychophysics, 55*, 48–121.

Grossberg, S., & Mingolla, E. (1993). Neural dynamics of motion perception: direction fields, apertures, and resonant grouping. *Perception & Psychophysics, 53*, 243–278.

Hulleman, J., & Humphreys, G. W. (2004). A new cue to figure-ground coding: Top-bottom polarity. *Vision Research, 44*, 2779–2791.

Kastner, S., de Weerd, P., Desimone, R., & Ungerleider, L. G. (1998). Mechanisms of directed attention to human extrastriate cortex as revealed by functional MRI. *Science, 282*, 108–111.

Kienker, P. K., Sejnowski, T. J., Hinton, G. E., & Schumacher, L. E. (1986). Separating figure from ground with a parallel network. *Perception, 15*, 197–216.

Kim, M. S., & Cave, K. R. (1995). Spatial attention in visual search for features and feature conjunctions. *Psychological Science, 6*, 376–380.

Kim, M. S., & Cave, K. R. (2001). Perceptual grouping via spatial selection in a focused-attention task. *Vision Research, 41*, 611–624.

Kimchi, R., & Peterson, M. A. (2008). Figure-ground segmentation can occur without attention. *Psychological Science, 19*, 660–668.

Klymenko, V., & Weisstein, N. (1986). Spatial frequency differences can determine figure-ground organization. *Journal of Experimental Psychology: Human Perception and Performance, 12*(3), 324–330.

Kroll, J. F., & Potter, M. C. (1984). Recognizing words, pictures, and concepts: A comparison of lexical, object, and reality decisions. *Journal of Verbal Learning and Verbal Behavior, 23*, 39.

Liu, T., Abrams, J., & Carrasco, M. (2009). Voluntary attention enhances contrast appearance. *Psychological Science, 20*, 354–362.

Luck, S. J. (1995). Multiple mechanisms of visual-spatial attention: Recent evidence from human electrophysiology. *Behavioral Brain Research, 71*, 113–123.

Luck, S. J., Chelazzi, L., Hillyard, S. A., & Desimone, R. (1997). Neural mechanisms of spatial selective attention in areas V1, V2, and V4 of macaque visual cortex. *The Journal of Neurophysiology, 77*, 24–42.

Miller, E. K., Gochin, P. M., & Gross, C. G. (1993). Suppression of visual responses of neurons in inferior temporal cortex of the awake macaque by addition of a second stimulus. *Brain Research, 616*, 25–29.

Moran, J., & Desimone, R. (1985). Selective attention gates visual processing in the extrastriate cortex. *Science, 229*, 782–784.

Nelson, R. A., & Palmer, S. E. (2007). Familiar shapes attract attention in figure-ground displays. *Perception & Psychophysics, 69*, 382–392.

O'Reilly, R. C., & Vecera, S. P. (1998). Figure-ground organization and object recognition processes: An interactive account. *Journal of Experimental Psychology: Human Perception and Performance*, *24*, 441–462.

Palmer, S. E., & Ghose, T. (2008). Extremal edges: A powerful cue to depth perception and figure-ground organization. *Psychological Science*, *19*, 77–84.

Pestilli, F., & Carrasco, M. (2005). Attention enhances contrast sensitivity at cued and impairs it at uncued locations. *Vision Research*, *45*, 1867–1875.

Peterson, M. A., & Gibson, B. S. (1993). Shape recognition contributions to figure-ground organization in three-dimensional displays. *Cognitive Psychology*, *25*, 383–429.

Peterson, M. A., & Gibson, B. S. (1994a). Must figure-ground organization precede object recognition? An assumption in peril. *Psychological Science*, *5*, 253–259.

Peterson, M. A., & Gibson, B. S. (1994b). Object recognition contributions to figure-ground organization: Operations on outlines and subjective contours. *Perception & Psychophysics*, *56*, 551–564.

Peterson, M. A., Harvey, E. H., & Weidenbacher, H. L. (1991). Shape recognition inputs to figure-ground organization: Which route counts? *Journal of Experimental Psychology: Human Perception and Performance*, *17*, 1075–1089.

Peterson, M. A., & Skow, E. (2008). Inhibitory competition between shape properties in figure-ground perception. *Journal of Experimental Psychology: Human Perception and Performance*, *34*, 251–267.

Peterson, M. A., & Skow-Grant, E. (2003). Memory and learning in figure-ground perception. In: B. Ross & D. Irwin (Eds.), *Cognitive vision: Psychology of learning and motivation* (Vol. 42, pp. 1–34). New York: Academic Press.

Pinna, B., Werner, J. S., & Spillman, L. (2003). The watercolor effect: A new principle of grouping and figure-ground organization. *Vision Research*, *43*, 43–52.

Prinzmetal, W., Long, V., & Leonhardt, J. (2008). Involuntary attention and brightness contrast. *Perception & Psychophysics*, *70*, 1139–1150.

Reynolds, J. H., & Chelazzi, L. (2004). Attentional modulation of visual processing. *Annual Reviews of Neuroscience*, *27*, 611–647.

Reynolds, J. H., Chelazzi, L., & Desimone, R. (1999). Competitive mechanisms subserve attention in macaque area V2 and V4. *The Journal of Neuroscience*, *19*, 1736–1753.

Roelfsema, P. R., Lamme, V. A. F., Spekreijse, H., & Bosch, H. (2002). Figure-ground segmentation in a recurrent network architecture. *Journal of Cognitive Neuroscience*, *14*, 525–537.

Rolls, E. T., & Tovee, M. J. (1995). Sparseness of the neuronal representation of stimuli in the primate temporal visual cortex. *Journal of Neurophysiology*, *73*, 713–726.

Schneider, K. A. (2006). Does attention alter appearance? *Perception & Psychophysics*, *68*, 800–814.

Sejnowski, T. J., & Hinton, G. E. (1990). Separating figure from ground with a Boltzmann machine. In M. A. Arbib & A. R. Hanson (Eds.), *Vision, brain and cooperative computation* (pp. 703–724). Cambridge: MIT Press.

Torralbo, A., & Beck, D. M. (2008). Perceptual load-induced selection as a result of local competitive interactions in visual cortex. *Psychological Science*, *19*, 1045–1050.

Vecera, S. P., Flevaris, A. V., & Filapek, J. C. (2004). Exogenous spatial attention influences figure-ground assignment. *Psychological Science*, *15*, 20–26.

Vecera, S. P., & O'Reilly, R. C. (2000). Graded effects in hierarchical figure-ground organization: Reply to Peterson (1999). *Journal of Experimental Psychology: Human Perception and Performance*, *26*, 1221–1231.

Vecera, S. P., Vogel, E. K., & Woodman, G. F. (2002). Lower region: A new cue for figure-ground assignment. *Journal of Experimental Psychology: General*, *13*, 194–1205.

N. Srinivasan (Ed.)
Progress in Brain Research, Vol. 176
ISSN 0079-6123

CHAPTER 2

Perceptual organization and visual attention

Ruth Kimchi*

Department of Psychology & Institute of Information Processing and Decision Making, University of Haifa, Haifa, Israel

Abstract: Perceptual organization — the processes structuring visual information into coherent units — and visual attention — the processes by which some visual information in a scene is selected — are crucial for the perception of our visual environment and to visuomotor behavior. Recent research points to important relations between attentional and organizational processes. Several studies demonstrated that perceptual organization constrains attentional selectivity, and other studies suggest that attention can also constrain perceptual organization. In this chapter I focus on two aspects of the relationship between perceptual organization and attention. The first addresses the question of whether or not perceptual organization can take place without attention. I present findings demonstrating that some forms of grouping and figure-ground segmentation can occur without attention, whereas others require controlled attentional processing, depending on the processes involved and the conditions prevailing for each process. These findings challenge the traditional view, which assumes that perceptual organization is a unitary entity that operates preattentively. The second issue addresses the question of whether perceptual organization can affect the automatic deployment of attention. I present findings showing that the mere organization of some elements in the visual field by Gestalt factors into a coherent perceptual unit (an "object"), with no abrupt onset or any other unique transient, can capture attention automatically in a stimulus-driven manner. Taken together, the findings discussed in this chapter demonstrate the multifaceted, interactive relations between perceptual organization and visual attention.

Keywords: perceptual organization; visual attention; grouping; figure-ground segmentation; attentional capture; inattention

Introduction

Perceptual organization and visual attention are crucial for the perception of our visual environment and to visuomotor behavior. Perceptual organization refers to the visual processes structuring the bits and pieces of visual information into coherent units that we eventually experience as environmental objects. The Gestalt psychologists, who were the first to study perceptual organization, suggested that organization is composed of grouping and segregation processes (Koffka, 1935), and identified several stimulus factors that determine organization. These include grouping factors such as proximity, similarity, good continuation, common fate, and closure (Wertheimer, 1955/1923), and factors that govern figure-ground organization, such as size, contrast, convexity, and symmetry (Rubin, 1921). Recently, researchers have identified additional factors that support grouping — common region

*Corresponding author.
Tel.: +972-4-8249746; Fax: +972-4-8249431;
E-mail: rkimchi@research.haifa.ac.il

DOI: 10.1016/S0079-6123(09)17602-1

(Palmer, 1992) and element connectedness (Palmer and Rock, 1994) – and figure-ground assignment – familiarity (Peterson and Gibson, 1994), lower region (Vecera et al., 2002), spatial frequency (Klymenko and Weisstein, 1986), base width (Hulleman and Humphreys, 2004), and extremal edges (Palmer and Ghose, 2008).

Visual attention refers to the processes by which some visual information in a scene is selected, in particular, information that is most relevant to ongoing behavior. Deployment of attention can be goal-directed, based on current behavioral goals of the observer (e.g., Desimone and Duncan, 1995; Posner, 1980). If we know, for example, where is the most probable target location we can use this information to voluntarily (endogenously) direct our attention to this location. Deployment of attention can also be stimulus-driven. In this case, attention is captured involuntarily (exogenously) by certain stimulus events, such as an abrupt onset of a new perceptual object and some types of simple luminance and motion transients (e.g., Abrams and Christ, 2003; Jonides, 1981; Yantis and Hillstrom, 1994), or a salient singleton (e.g., Theeuwes et al., 2003, but see Folk et al., 1992).

Recent research has demonstrated a close interplay between attentional and perceptual organization processes (e.g., Driver et al., 2001; Scholl, 2001). Several studies demonstrated that perceptual organization constrains attentional selectivity. For example, interference from distractor stimuli in selective attention tasks is greater when the target and distractors are strongly grouped by Gestalt cues such as color similarity, good continuation, closure, or common fate (e.g., Baylis and Driver, 1992; Driver and Baylis, 1989; Kahneman and Henik, 1981; Kramer and Jacobson, 1991), and responding to two features is easier when they belong to the same object than when they belong to two separate objects (e.g., Behrmann et al., 1998; Duncan, 1984; Lavie and Driver, 1996; Vecera and Farah, 1994). Also, the cost incurred during target detection when attention is initially cued to a non-target location is smaller for targets that appear in the same object as the cue than for targets appearing in a different object, despite their equivalent distance from the cued location (e.g., Egly et al., 1994; Moore et al., 1998). In addition, neurophysiological studies have found that attended stimuli and unattended stimuli belonging to the same object elicited a very similar spatiotemporal pattern of enhanced neural activity in the visual cortex, even when the objects were defined by illusory boundaries (Martinez et al., 2006, 2007).

Other studies suggest that attention can also constrain perceptual organization. For example, Freeman et al. (2001, 2004) provided evidence for influence of attention on flanker-target integration, demonstrating that detection of a central Gabor target was improved by the presence of collinear flankers when the collinear flankers were attended, but not when the collinear flankers were ignored in favor of flankers with orthogonal orientation. Attention can also influence figure-ground organization (e.g., Peterson and Gibson, 1994; Vecera et al., 2004). For example, Vecera and colleagues demonstrated that when spatial attention is directed to one of the regions of an ambiguous figure-ground stimulus, the attended region is perceived as figure and the shared contour is assigned to the attended region.

These various findings suggest that perceptual organization and visual attention mutually constrain one another.

In this chapter I focus on two issues concerning the relationships between visual attention and perceptual organization. The first focuses on the question of whether or not perceptual organization can be accomplished without attention. The second issue concerns the question of whether perceptual organization can affect the automatic deployment of attention.

Can perceptual organization occur without attention?

Traditional theories of perception assumed that perceptual organization, including grouping and figure-ground segmentation, occurs preattentively, at an early stage of processing and in a bottom-up fashion, to deliver the units for which attention can be allocated for further, more elaborated processing (e.g., Julesz, 1981; Marr, 1982; Neisser, 1967; Treisman, 1982, 1988). Thus, for example,

Treisman (1982, p. 195) noted that "the theories all agree that perceptual grouping occurs automatically and in parallel, without attention." This assumption was based on logical considerations and supported by some empirical findings. Prima facie, if attention is to select candidate objects, then some organization of the visual scene into these objects must occur prior to selection. Empirical findings that were interpreted as supporting this view came from texture segregation and visual search studies showing that certain texture boundaries and certain items "pop-out" under very brief exposures and without effort and scrutiny (e.g., Beck, 1982; Julesz, 1981; Treisman, 1982, 1985), from dual-task studies demonstrating successful texture segregation even though visual attention is engaged with a demanding primary task (e.g., Braun and Sagi, 1990, 1991), and from studies showing that segmentation of the visual field into perceptual groups on the basis of Gestalt principles constrains attentional selectivity (e.g., Baylis and Driver, 1992; Driver and Baylis, 1989; Duncan, 1984; Vecera and Farah, 1994).

An alternative view suggests that no, or very little, perceptual organization can take place without attention (Ben Av et al., 1992; Mack et al., 1992; Mack and Rock, 1998; Palmer and Rock, 1994; Rock et al., 1992). For example, Ben Av et al. (1992) showed that when participants performed a demanding central form identification task and also had to report whether background elements grouped into horizontal or vertical pattern on the basis of proximity or similarity, grouping performance was severely reduced (relative to single-task situation), suggesting that perceptual grouping requires visual attention.

The main support for this view came from the work of Mack and Rock, and their colleagues (Mack et al., 1992; Mack and Rock, 1998; Rock et al., 1992). Mack and Rock argued, and rightfully so, that none of the findings taken as evidence for preattentive perceptual organization were obtained under conditions in which information was truly unattended. Rather, these findings pertain to diffuse or divided attention conditions, in which participants are aware of the potential relevance of the information in the visual scene, including information outside the focus of attention. For example, the secondary-task information in a dual-task procedure is task relevant, and in visual search participants actively search for a predefined target while ignoring distracting information. Similarly, in all the studies examining object-based attentional selection, at least part of the relevant object is attended, and this may cause other parts of the object also to be attended. In contrast, the *inattention method* developed by Mack and Rock attempted to tap processing of unattended stimuli under conditions in which participants are engaged in a highly demanding visual task, and the unattended stimuli are completely irrelevant to the task at hand, so that participants have no reason whatsoever to attend to them.

Grouping under inattention

Mack et al. (1992) used the inattention method to examine whether perceptual grouping can take place under inattention. Participants performed a demanding discrimination task – determining whether the horizontal or vertical line of a centrally, briefly presented cross is longer. In the first few trials the cross was surrounded by ungrouped small elements. On the fourth, *inattention trial*, the surrounding elements were grouped into rows or columns by proximity or lightness similarity, and the participants were asked, after completing the length judgment, about the background organization. Participants were "inattentionally blind" to the grouping of the background elements – they could not report whether the background organization was vertical or horizontal. In a subsequent attention trial, in which participants attended to the background elements, these patterns were easily reported. These kinds of findings led Mack and Rock (Mack et al., 1992; Mack and Rock, 1998) to the conclusion that no Gestalt grouping takes place without attention.

However, Mack and Rock's work was criticized on the ground that poor knowledge of the background organization may reflect poor explicit memory, rather than indicating that no grouping took place when the unattended stimuli were presented. To circumvent the issue of explicit memory, Moore and Egeth (1997) used the inattention paradigm but devised indirect online measures of unattended processing by examining the influence of the unattended information on responses to the attended information. Participants were required to determine which of two horizontal lines is longer. On the inattention trial the elements in the background were grouped by luminance into inducers biasing the length of the horizontal lines by creating Muller-Lyer or Ponzo illusion. Participants were unable to report the background organization, but their line length judgments were influenced by the illusions. These findings suggest that grouping by similarity in luminance occurs under conditions of inattention, albeit without participants' awareness (see also, Lamy et al., 2006). Similar results were found when background elements were grouped by similarity in size (Chan and Chua, 2003).

The method developed by Russell and Driver (2005; originally described in Driver et al., 2001) also provides indirect online measures of unattended processing. On each trial, two successive displays were presented, each of which included a small, centrally located matrix (made up of random black and white pixels) surrounded by task-irrelevant background elements grouped by color similarity into rows or columns, or randomly organized. The task was to judge whether the matrices in the two successive displays were the same or different. When the matrices differed, only one pixel changed its location, rendering the task sufficiently demanding to absorb attention. The background organization stayed the same or changed across the two displays, independently of whether or not the target matrix changed. The results showed that grouping of the background stimuli — whether it stayed the same or changed across successive displays — influenced the detection of changes in the target matrix, even though when probed with surprise questions, participants reported no or little awareness of the background grouping or its changes. These findings suggest that the unattended background elements were perceptually grouped.

Recently, Shomstein et al. (2009) reported similar results in a situation in which the definition of "unattended" did not rely on participants self-report of lack of awareness of the background grouping. They adapted Russell and Driver's (2005) method to test individuals with hemispatial neglect. In their study, patients (and matched controls) performed the target change-detection task on a matrix presented entirely to their intact side of space, and the task-irrelevant grouped elements (columns and rows by color similarity) appeared simultaneously on the unattended side. Changes in the grouping of the neglected task-irrelevant elements produced congruency effects on the target change judgments to the same extent as in the control participants even in patients with severe attentional deficits, suggesting that the grouping was accomplished in the absence of attention.

Figure-ground segmentation under inattention

The view that figure-ground segmentation operates preattentively has been widely accepted, but the evidence is scant (e.g., Driver et al., 1992) and open to alternative interpretations, particularly in light of recent research indicating that exogenous attention can influence figure-ground assignment (Vecera et al., 2004), and that figural cues per se can possibly attract attention (Nelson and Palmer, 2007).

To examine whether figure-ground segmentation can take place without attention, Mary Peterson and I (Kimchi and Peterson, 2008) adapted Russell and Driver's (2005) inattention method. In our study, the target matrix appeared on a task-irrelevant scene of alternating regions organized into figures and grounds by convexity (see Figs. 1a–d). The backdrop region on which the matrix appeared could be convex (figure) or concave (ground). On each trial two successive displays were briefly presented and the task was to judge whether the central matrices are the same or different. The figure-ground organization of the scene backdrop stayed the same or changed across the two successive displays, independently of whether or not the target matrix changed. The edges in the backdrop always changed from the first to the second display regardless of whether or not the figure-ground organization changed, to control for the possibility that a change in backdrop organization could be detected from local changes in edges per se. An example of the display sequence in a single experimental trial is

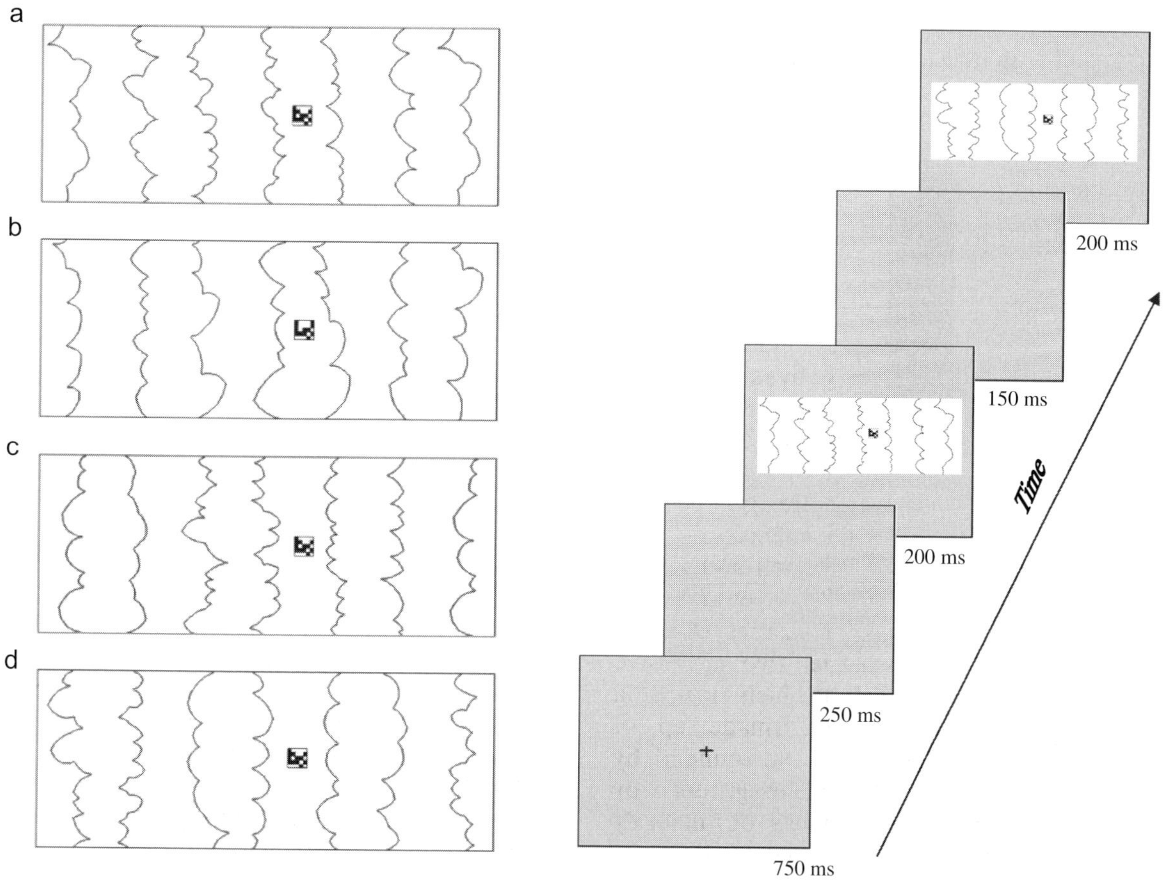

Fig. 1. Left panel: Examples of the displays used by Kimchi and Peterson (2008). In the experiments, displays were presented on a gray field, and no frame was used. The target matrix always appeared on the backdrop region to the right of the central edge (i.e., the fifth region from the left). This region could be convex (figure, F) or concave (ground, G), and the number of parts in this region could be small or large. The examples illustrate (a) the F type with a large part number, (b) the F type with a small part number, (c) the G type with a large part number, and (d) the G type with a small part number. The matrices in (a) and (b) depict an example of a change in matrix (a change in the location of one small black square). Right panel: Sequence of events in a trial. Two successive displays were presented on each trial. The target matrix in successive displays could stay the same or change. The backdrop organization across successive displays could stay the same (FF or GG) or change (FG or GF), independently of whether the target matrix changed or remained the same. The edges in the backdrop always changed from the first to the second display (a backdrop with a small number of parts was paired with a backdrop with a large number of parts). The illustration depicts a same-target trial (matrix is unchanged) on a backdrop that changes from figure to ground. Adapted with permission from Kimchi and Peterson (2008).

presented in Fig. 1, right panel. We examined whether the figure-ground organization of the scene backdrop influenced performance on the matrix-change task. After the last experimental trial, observers were probed with surprise questions asking whether the region on which the target was presented in the preceding display appeared to be figure or ground and whether the figure-ground status of that region had changed between the two displays on that trial.

The main results are presented in Fig. 2. Changes in the scene backdrop's figure-ground organization produced reliable congruency effects on target-change judgments: Target-different judgments were more efficient when backdrop organization changed across the two displays than when it remained the same, and target-same judgments were more efficient when backdrop organization stayed the same than when it changed. These results could not be due to the backdrop's changes in

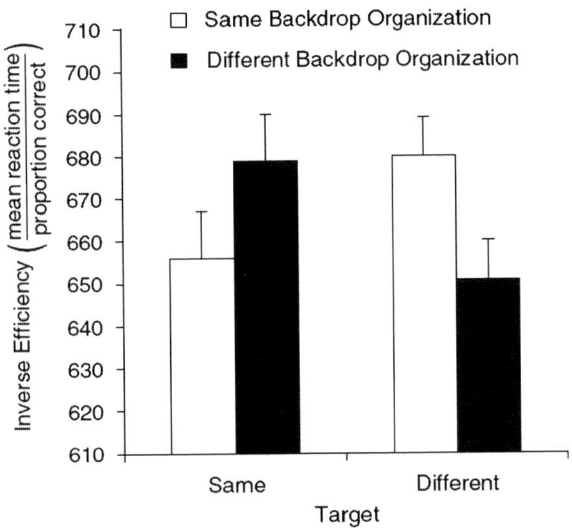

Fig. 2. Results from Kimchi and Peterson (2008). Inverse-efficiency scores for same and different targets as a function of the backdrop's organization (same, different). Error bars indicate standard errors of the means. Adapted with permission from Kimchi and Peterson (2008).

convexity/concavity per se. Performance was less efficient on trials where the backdrop region on which the matrix appeared changed from ground (concave) to figure (convex) — a new figure (a "new object") appeared in the target's backdrop region — than on trials where the backdrop region changed from figure to ground (no new figure in this region). Presumably, implicit processing of a new figure on the former produced less efficient responses to the target. Changes in convexity/concavity per se would not predict a difference between these two types of trials, because in both types convex and concave regions changed their location across successive displays. The congruency effects produced by changes in the backdrop figure-ground organization arose even though, when probed with surprise questions, participants could report neither the figure-ground status of the region on which the matrix appeared nor any change in that status. When attending to this region, participants reported its figure-ground status and changes to it highly accurately. These results strongly suggest that some figure-ground segmentation can occur without attention.

Taken together, the findings reviewed in the last two sections suggest that some forms of perceptual grouping and figure-ground segmentation take place under inattention. In the following section I present findings suggesting that perceptual organization processes vary in their attentional demands.

Perceptual organization and attention: not all organizations are equal

Implicit in traditional theories of perception is the assumption that perceptual organization is a unitary entity. A growing body of research, however, has challenged this monolithic view (e.g., Behrmann and Kimchi, 2003; Ben Av and Sagi, 1995; Hadad and Kimchi, 2006; Han, 2004; Kimchi, 1988, 2000; Kimchi et al., 2005; Kimchi and Razpurker Apfeld, 2004; Kovacs et al., 1999; Kurylo, 1997; Quinn and Bhatt, 2006; Razpurker Apfeld and Kimchi, 2007). For example, several studies showed that groupings guided by different Gestalt principles vary in their time course and developmental trajectory. Experiments with adults showed that grouping by proximity is achieved faster than grouping by similarity in luminance or in shape (Ben Av and Sagi, 1995; Han, 2004) and faster than grouping by good continuation (Kurylo, 1997). Infant studies showed that grouping by common lightness is evident in 3-month-olds (Quinn et al., 1993, 2002), but only 6- to 7-month-olds readily use grouping by shape similarity (Quinn et al., 2002; Quinn and Bhatt, 2006). Sensitivity to good continuation has been documented in 3- to 4-month-old infants (Quinn and Bhatt, 2005), but the ability to group line segments by good continuation appears to be highly constrained by proximity between the segments even at 5 years of age (Hadad and Kimchi, 2006; Kovacs et al., 1999). Also, Kimchi (1998) showed that the global configuration of many small elements was primed at brief exposures and accessible to rapid search, suggesting rapid and effortless grouping, whereas the global configuration of a few relatively large elements was primed at longer exposures and searched inefficiently, suggesting time-consuming and attention-demanding grouping. The former grouping is mature by age 5, whereas the latter improves with age, primarily between ages 5 and 10 (Kimchi et al., 2005).

In addition to noting that grouping involves various principles that may differ from each other, it has been suggested that grouping itself may not be a single process, but rather involves two distinct processes: a process of unit formation or clustering that determines which elements belong together and are segregated from other elements, and a process of shape formation or configuring that determines how the grouped elements appear as a whole based on the interrelations of the elements (Koffka, 1935; Rock, 1986; Trick and Enns, 1997). Trick and Enns (1997) found that enumeration of hierarchical figures — presumably requiring just clustering of local elements — was identical to that of connected figures with both exhibiting equal subitizing, but when the figures were enumerated among distractors — thus involving shape discrimination — only the connected figures were subitized. Trick and Enns interpreted these results as indicating that shape formation requires attention whereas clustering does not. Other studies provide some hints for a continuum of attentional demands rather than a dichotomy (e.g., Behrmann and Kimchi, 2003; Han and Humphreys, 1999; Han et al., 1999). For example, Behrmann and Kimchi (2003) studied perceptual organization in two patients suffering from integrative agnosia. Both patients had no problem grouping elements into columns/rows by proximity or by luminance similarity, but they exhibited different degrees of difficulty grouping elements into a global shape.

To directly examine whether different groupings vary in their attentional demands, Irene Razpuker-Apfeld and I (Kimchi and Razpurker Apfeld, 2004) used Russell and Driver's (2005; Driver et al., 2001) method and manipulated the unattended grouping.

We employed different background organizations (examples of which are presented in Fig. 3), which vary in the processes involved in the grouping. The critical organizations were grouping of columns/rows by color similarity (Fig. 3A), grouping of shape (square/cross or triangle/arrow) by color similarity (Fig. 3B), and grouping of shape (square/cross or triangle/arrow) of homogeneous elements (Fig. 3C). The first two groupings involve elements clustering and segregation (by color similarity) and shape formation. Shape formation, however, may be less demanding for the columns/rows — requiring determination of the orientation (vertical or horizontal) of the grouped pattern, than for the shape by color similarity — requiring the formation of a distinctive shape (Rock, 1986). The third grouping involves clustering and shape formation but no elements segregation; therefore it may be less demanding than the grouping of shape by color similarity. (Additional organizations were connected triangle/arrow and square/cross made of disconnected lines.) On each trial two successive displays were briefly presented and the task was to judge whether the central matrices are the same or different. The background stayed the same or changed across successive displays independently of any change in the target matrix. After the last experimental trial, observers were probed with surprise questions about the immediately preceding background displays.

The results for the critical organizations are presented in Fig. 4 (the results for the triangle/arrows paralleled those for the square/cross). Influence of the background organization on the target-change judgments was observed for grouping of columns/rows by color similarity (Fig. 4A): Target-same judgments were faster when the background stayed the same than when it changed, and target-different judgments were faster when the background organization changed than when it stayed the same, and for grouping of shape when no elements segregation was involved (Fig. 4C): Target-same judgments were faster and more accurate when the background stayed the same, and target-different judgments were more accurate when the background organization changed. No influence of the background organization was found for grouping of shape by color similarity (Fig. 4B). For all three conditions, participants were unable to report the background organization of the immediately preceding background displays.

The difference between the results for the columns/rows and for the shape by common color is of particular interest because both groupings were guided by the same principle of similarity in color, but nevertheless the former took place

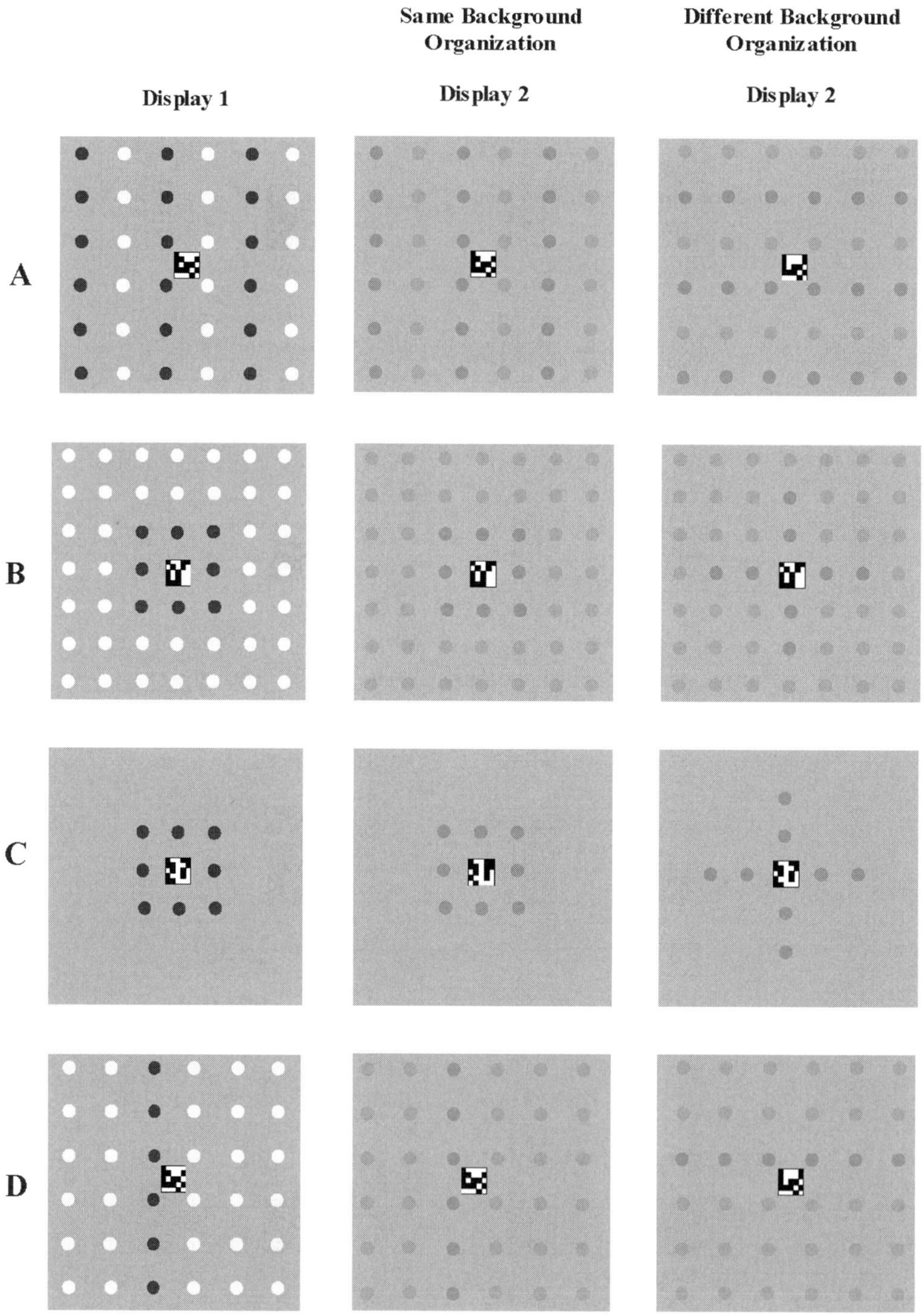

Fig. 3. Examples of the stimulus displays used by Kimchi and Razpurker Apfeld (2004). Two successive displays were presented on each trial. The central target matrix in Displays 1 and 2 were either the same or different. The surrounding colored elements were grouped into (A) columns/rows by color similarity, (B) a square/cross by color similarity, (C) a square/cross, (D) a vertical/horizontal line by color similarity. This background organization either stayed the same across Displays 1 and 2 or changed, independently of whether the target matrix changed or remained the same. The colors of the background elements always changed between Displays 1 and 2. All colors were equiluminant in the experiment. Adapted with permission from Kimchi and Razpurker Apfeld (2004). (See Color Plate 2.3 in color plate section.)

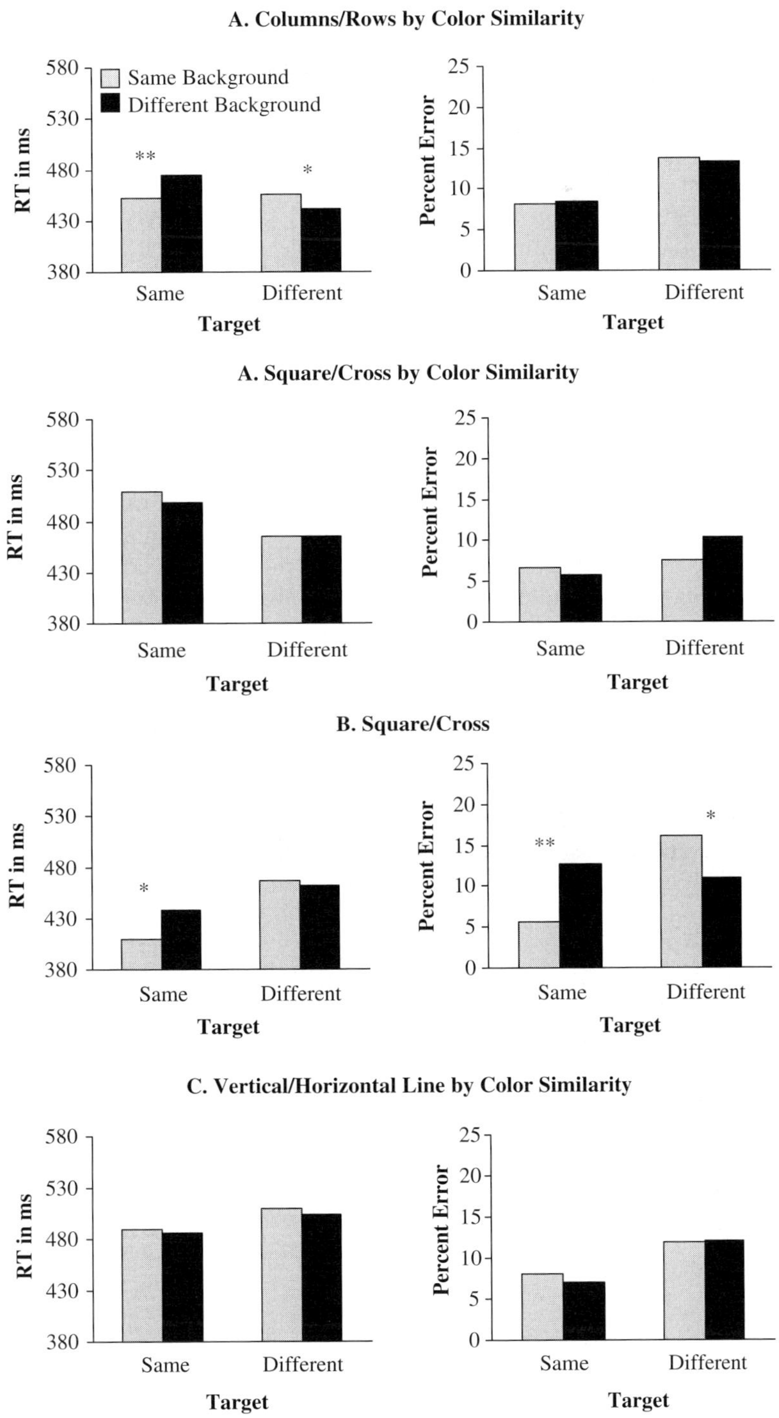

Fig. 4. Results from Kimchi and Razpurker Apfeld (2004). Mean correct reaction times (RTs) (left panel) and error rates (right panel) for target-same and target-different judgments as a function of background similarity (same or different) for each background condition ($^{*}p<0.05$; $^{**}p<0.01$). Adapted with permission from Kimchi and Razpurker Apfeld (2004).

under inattention, whereas the latter did not. Complexity of shape formation per se — forming a shape (e.g., a square or a cross) versus forming lines (columns or rows) — cannot account for this difference because grouping of shape occurred under inattention when no elements segregation was involved. Rather, it is grouping that involves both segregation and shape formation that appeared to require attention. We hypothesized that in this case there was a need to resolve figure-ground relations between groups — designating one of the groups as "figure." In the columns/rows condition, on the other hand, there was no such need because all segmented groups contribute to the global orientation of the pattern (vertical or horizontal). To examine this conjecture, we employed the condition depicted in Fig. 3D — vertical/horizontal line by color similarity. Shape formation for this grouping is as simple (if not simpler) as for the columns/rows (requiring only determination of the orientation of the grouped elements), but unlike the columns/rows, it also requires resolving figure-ground relations, as in the square/cross by color similarity. No influence of the background was observed for the vertical/horizontal line condition (Fig. 4D), suggesting that resolving figure-ground relation may demand attention (see Peterson and Salvagio, this volume).

These results indicate that both clustering and shape formation can take place without attention and thus are incompatible with the view of a dichotomy between these processes in terms of attentional demands (Trick and Enns, 1997).

Rather, these results suggest that a continuum of attentional demands exists as a function of the processes involved in organization and the conditions prevailing for each process. Grouping of columns/row by color similarity can occur under inattention (see also Russell and Driver, 2005; Shomstein et al., 2009). Grouping of shape can also take place without attention when no elements segregation is involved, but grouping of shape that involves elements segregation cannot, presumably because it requires resolving figure-ground relations between groups. Note that according to this view, it is possible, for example, that grouping into columns/rows could have demanded attention were it based on certain shape similarity instead of color similarity (e.g., arrows vs. crosses; see Han and Humphreys, 1999), or if the patterns were not easily resolved, as apparently was the case in Ben Av et al. (1992).

Similarly, figure-ground segmentation can occur without attention under certain conditions but not under others. Thus, in Kimchi and Peterson's (2008) study, figure-ground segmentation was based solely on convexity, which is a powerful cue for figural assignment in multiregion displays (e.g., Hoffman and Singh, 1997; Kanizsa and Gerbino, 1976; Peterson and Salvagio, 2008). It is possible, however, that when other, perhaps less potent, figural cues are involved, segmentation requires the scrutiny of focal attention. Also, resolution of cross-edge competition, which is required for figure-ground assignment when multiple competing cues are involved, may demand focal attention (see Peterson and Salvagio, this volume). Evidence that spatial attention can act as a cue for figure-ground assignment (Peterson and Gibson, 1994; Vecera et al., 2004) also casts serious doubt on the assumption that figure-ground segmentation must necessarily be completed prior to the deployment of focal attention.

Summary

The findings reviewed in the first part of this chapter provide evidence that some perceptual organization, such as some forms of grouping (e.g., grouping of columns/rows by color similarity, or grouping of shape when no elements segregation is involved) and figure-ground segmentation (e.g., figure-ground segmentation by convexity) can occur under inattention. Moore et al. (2003) showed that surface completion can also take place under inattention. Other organizations, however, appear to require focused attention (e.g., grouping of shape that involves elements segregation). Taken together, these findings suggest that perceptual organization is a multiplicity of processes that vary in their attentional demands. Regardless of attentional demands, the products of organization are not available to awareness without attention.

Can perceptual organization affect the automatic deployment of attention?

The critical role of perceptual organization in designating potential objects raises an important issue concerning the relations between perceptual organization and attention: When some elements in the visual scene are organized by Gestalt factors into a coherent perceptual unit (an object),[1] is visual attention automatically deployed to the object? Presumably, favoring a coherent perceptual unit that conforms to Gestalt factors is a desirable characteristic for a system whose one of its important goals is object identification and recognition, because these units are likely to imply objects in the environment.

In this part of the chapter I describe a series of experiments that my colleagues and I have conducted, as a part of an ongoing research, to examine whether the mere organization of some elements into an object, with no abrupt onset or any other unique transient, can capture attention automatically in a stimulus-driven manner, much as exogenous cues capture spatial attention automatically.

As noted earlier, several studies have demonstrated that perceptual organization can constrain attentional selectivity, supporting object-based theories of visual attention. None of these studies, however, show unequivocally that the object per se was the factor that attracted attention, because there were always other factors that directed attention to a part or an attribute of the object, either exogenously or endogenously. Thus, some studies employed a brief flicker presented in one end of the relevant object to exogenously summon attention (e.g., Egly et al., 1994; Moore et al., 1998), and other studies used central cues, instructions, or task-related factors to encourage observers to direct their attention to one of the objects or to its attributes (e.g., Behrmann et al., 1998; Duncan, 1984; Kramer and Jacobson, 1991).

[1]The question of what constitutes a perceptual object is a difficult one and yet to be answered (e.g., Scholl, 2001). I use the term object to refer to "elements in the visual scene organized by Gestalt factors into a coherent unit."

Perceptual objects capture attention

To examine whether an object by itself captures attention, it is crucial that the object has no abrupt onset or any other unique transient, and that the object is irrelevant to the task at hand so there is no incentive for the observer to deliberately attend to the object. To this end, my colleagues and I (Kimchi et al., 2007) modified a paradigm developed by Logan (1995) by substituting the O elements in Logan's original display with L elements in various orientations, and manipulating the organization in the display as described below.

Participants viewed a display composed of nine red and green L elements rotated at different angles and forming the vertices of four adjacent quadrants making up a global diamond (Fig. 5, top panel). The participants' task was to report the color of one of the elements as indicated by an asterisk presented in the center of one of the quadrants and an instruction word — "above," "below," "right," or "left" — that preceded the elements display and specified the position of the target relative to the asterisk. For example, if the word was "above," observers had to identify the color of the element above the asterisk. Each trial began with one of the instruction words, then the display appeared, and 150 ms after the display onset the asterisk appeared in the center of one of the quadrants (Fig. 5, bottom panel). Thus, performing the task required locating the asterisk, locating the target relative to the asterisk, and analyzing the target's color. On half of the trials, the four Ls of one of the quadrants were rotated so as to conform to the Gestalt factors of collinearity, closure, and symmetry, forming a diamond-like object. The asterisk appeared in the object quadrant (*Inside-object* condition, Fig. 5a) on 12.5% of all trials, and in a non-object quadrant (*Outside-object* condition, Fig. 5b) on 37.5% of all trials. On 50% of all trials no object was present in the display (*No-object* condition, Fig. 5c). The diamond-like object was task irrelevant (because the task-relevant feature was the color of a single element) and was not predictive of the relevant quadrant or the target. Moreover, no unique onset was associated with the object because it appeared simultaneously with the onset of the entire

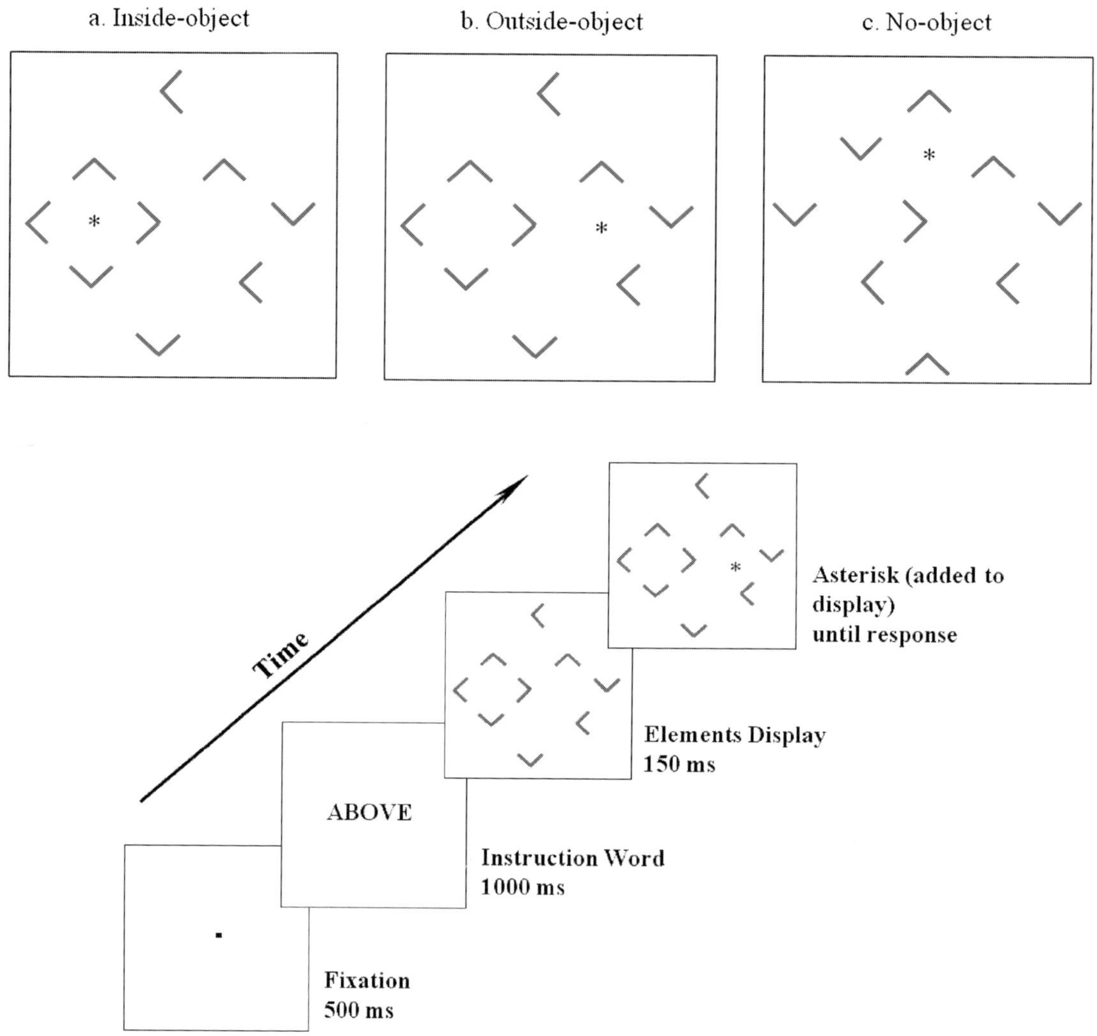

Fig. 5. Top panel: Examples of the displays used by Kimchi et al. (2007). Each display composed of nine red and green elements. (a) Inside-object condition: object present in display and asterisk appearing in center of object quadrant; (b) Outside-object condition: object present in display and asterisk appearing in center of nonobject quadrant; and (c) No-object condition: no object present in display. Fifty percent of the trials were No-object trials, 12.5% were Inside-object trials, and 37.5% were Outside-object trials. Bottom panel: Sequence of events in a trial. The illustration depicts an Outside-object trial with the instruction word *above*. In this trial, the participants had to identify the color of the element above the asterisk (green). Adapted with permission from Kimchi et al. (2007). (See Color Plate 2.5 in color plate section.)

elements display. This is a critical difference from previous research in which attention was captured by the unique appearance of an object defined by discontinuities in luminance, motion, texture, or depth (e.g., Yantis and Hillstrom, 1994; Franconeri et al., 2005). Thus, there was no top-down incentive for the participants to deliberately attend the object, nor was there any previously known stimulus-driven cue, such as feature-singleton, abrupt onset, or any other unique transient, to automatically attract attention to the object quadrant.

We hypothesized that if attention is automatically drawn to the object, then performance will be faster and/or more accurate in the Inside-object condition than in the No-object condition (a benefit) because attention is allocated in advance

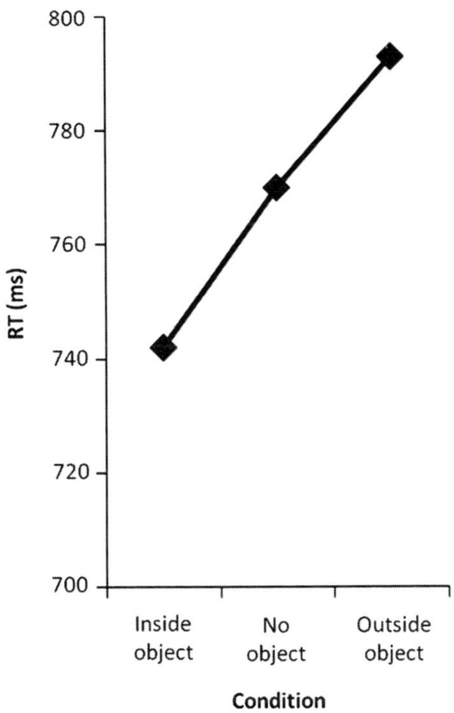

Fig. 6. Data from Kimchi et al. (2007). Mean correct reaction times (RTs) as a function of condition. Adapted with permission from Kimchi et al. (2007).

to the object quadrant, and slower and/or less accurate in the Outside-object condition than in the No-object condition (a cost), because attention has to be redirected from the object quadrant to the actual relevant quadrant. The results (see Fig. 6) showed the expected cost and benefit, demonstrating capture of attention by the irrelevant object.

Kimchi et al.'s (2007) study was the first to show unequivocal evidence for attentional capture by an object. There are, however, two concerns regarding this study. One is the extent to which the observed cost and benefit effects are somehow related to the complexity of the task. The task involved several operations and imposed memory load: Participants had to remember the instruction word, to locate the asterisk, to locate the target relative to the asterisk, and to analyze the target's color. Thus, the observed effects could be, at least partly, a function of task complexity and memory load.

A second concern is the extent to which the observed effects are a consequence of processes that are not necessarily related to attention. This concern arose because of the following observation. In the Outside-object condition, in which the asterisk appeared in a non-object quadrant, the target-element on some of the trials actually "belonged" to the object (i.e., it was one of the four elements forming an object in another quadrant), whereas on the other trials the target-element did not belong to the object. Analysis of the cost for these two types of trials showed costs for both with somewhat higher cost for target-elements that belonged to the object. This finding suggests that some of the observed cost could be attributed to difficulty in "extracting" an element that was already grouped into an object. Thus, the observed effects might be due to a mixture of attentional processes and other processes that are related to the actual processing of the object (e.g., extracting an element from an object).

The experiments described next, conducted in collaboration with Yaffa Yeshurun and Guy Sha'ashoua, addressed these issues by employing a simpler task and a target that is not part of the object.

To examine whether similar results indicating attentional capture by an object emerge with a simpler task that does not impose high memory load, we presented participants with a matrix of 16 black L elements in various orientations (Fig. 7, top panel). One of the Ls changed its color from black to red or orange 150 ms following the onset of the matrix. The task was to identify the color of the changed element. On half of the trials four elements were collinear, forming an object — a square. There were four possible locations where the object could appear (hence there were 12 possible target-elements). The object was present in the display on half of the trials. On 16.6% of all trials the target was an object's element (*Inside-object* condition, Fig. 7a). On 33.4% of all trials the target was a non-object element (*Outside-object* condition, Fig. 7b). On 50% of all trials the elements did not form an object, and the target was one of the twelve possible target-elements (*No-object* condition, Fig. 7c). Note that in the Outside-object condition, the target never belonged to the object. As in Kimchi et al.'s (2007) study, the object was task irrelevant and was not predictive of the target, nor was it associated with any unique transient. The results (Fig. 7,

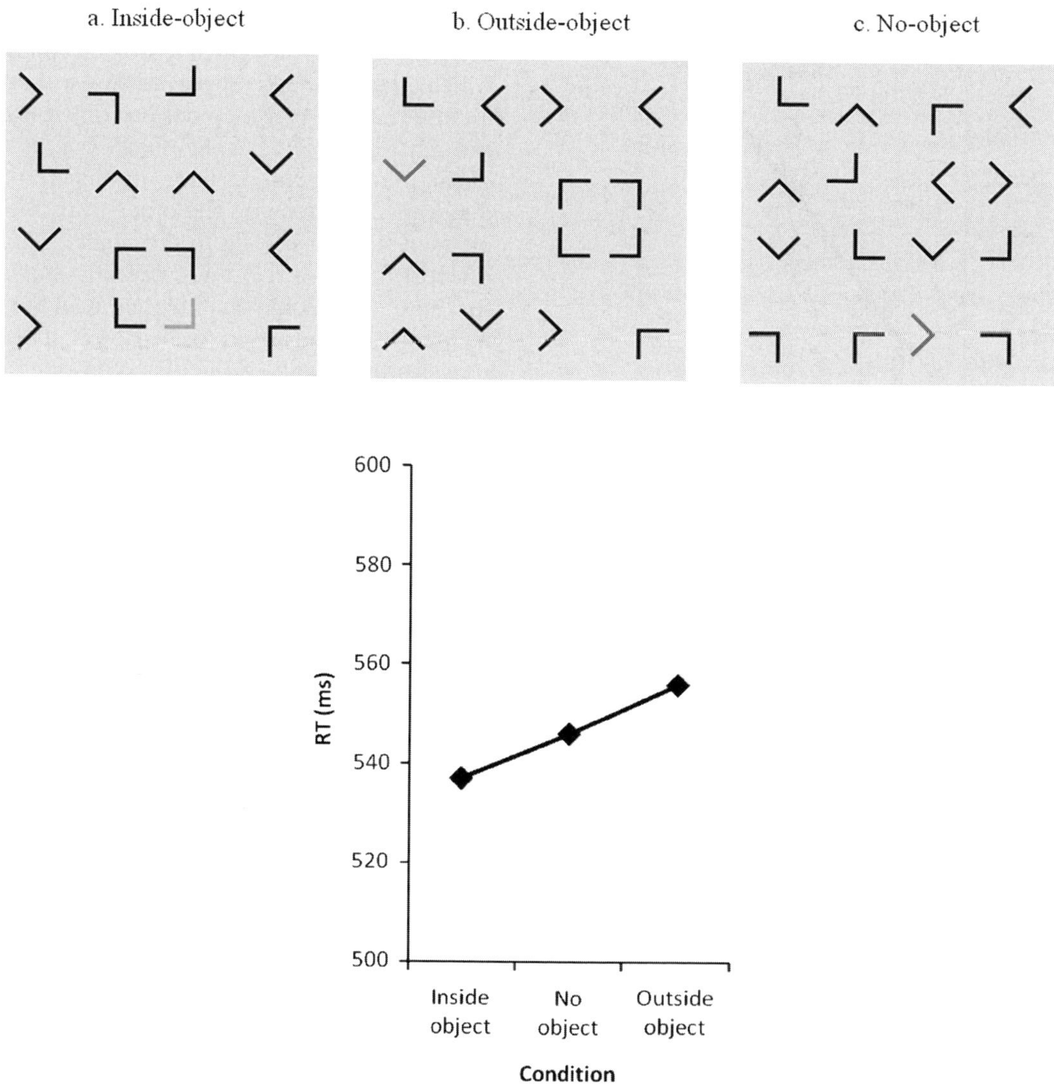

Fig. 7. Top panel: Examples of the displays in the three conditions. (a) Inside-object condition: object present in display and the target is an object element (16.6% of all trials); (b) Outside-object condition: object present in display and the target is a non-object element (33.4% of all trials); and (c) No-object condition: no object present in display (50% of all trials). Bottom panel: Mean correct reaction times (RTs) as a function of condition. (See Color Plate 2.7 in color plate section.)

bottom panel) showed the expected benefit and cost: performance on trials with an object in the display was faster than performance on trials with no object for object-element targets but slower for non-object-element targets, indicating that the object captured attention.

In a second experiment we examined whether a similar automatic attraction of attention by the object can be found with displays in which the target is never a part of the object and has no figural resemblance to the object. The target was a Vernier stimulus composed of two vertical lines with one line appearing above the other and separated by a small horizontal offset. The participants had to discriminate the direction of the offset (right or left). Participants were presented with a matrix of 36 black L elements in various orientations (Fig. 8, top panel). As in the

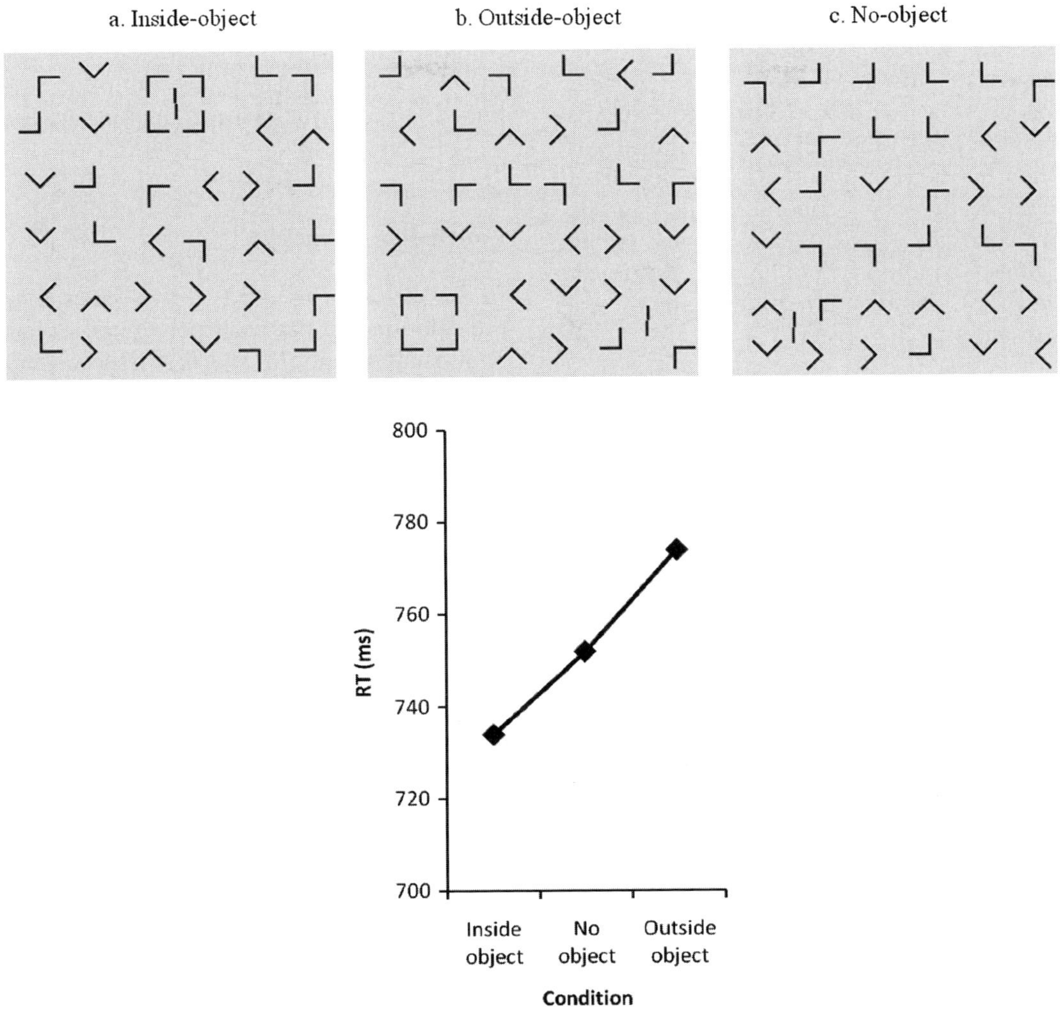

Fig. 8. Top panel: Examples of the displays in the three conditions. (a) Inside-object condition: object present in display and the Vernier target appears at the center of the object (9%of all trials); (b) Outside-object condition: object present in display and the target in another location (64% of all trials); and (c) No-object condition: no object present in display (27% of all trials). Bottom panel: Mean correct reaction times (RTs) as a function of condition.

previous experiment, an object — a square — was formed by four collinear elements. There were eight possible locations in which the object could appear. The Vernier target appeared 150 ms after the onset of the matrix. The Vernier target appeared at the center of the object on 9% of all trials (*Inside-object* condition, Fig. 8a), and outside the object — at one of the other seven possible locations — on 64% of all trials (*Outside-object* condition, Fig. 8b). On 27% of all trials the elements did not form an object, and the target appeared in one of the eight possible locations (*No-object* condition, Fig. 8c).[2] Thus, the matrix

[2]Given the larger number of target and object locations in this experiment, the ratio of Inside-object trials to Outside-object trails is highly in favor of the Outside-object condition. In order to allow for a reasonable number of Inside-object trials while keeping a reasonable number of total trials, we reduced the number of No-object trials. Consequently, the object appeared more frequently, but it was not predictive of target's location.

was completely irrelevant to the task and the object was not predictive of the target location or the direction of offset. Moreover, the Vernier target was never a part of the object. The results (Fig. 8, bottom panel) show that performance was faster when the target appeared in the center of the object (Inside-object condition) than in the No-object condition (benefit), and slower in the Outside-object condition than in the No-object condition (cost), demonstrating automatic attraction of attention to the object.

Summary

The results of the latter two experiments clearly demonstrate that the object-related cost and benefit effects observed in Kimchi et al.'s (2007) study do not depend on high memory load or on the target being a part of the object. These results provide corroborating evidence in support of the hypothesis that attention is automatically attracted to the object.

An automatic, stimulus-driven capture of attention by an object may provide a single account for a variety of "object advantage" effects reported in the literature, demonstrating the special status of objects for our visual system. These include more accurate discrimination of line segments when flashed on the figure than on the ground (Wong and Weisstein, 1982), easier detection of four target lines embedded in distractors when the lines are organized into a face-like pattern than a meaningless cluster (Gorea and Julesz, 1990), higher sensitivity for a target probe when positioned inside a circular contour embedded in a random background rather than outside the circle (Kovacs and Julesz, 1993), better memory for a figure's contour than for ground's contour (Driver and Baylis, 1996), and greater brain activation when the target appears in a region bounded by an object than in an unbounded region (Arrington et al., 2000).

Several outstanding questions await further research. These include uncovering the mechanisms underlying our object effect, examining whether the automatic deployment of attention is exclusively space-based or some combination of object-based and space-based components, and exploring which organization factors (e.g., collinearity, closure, symmetry, etc.) are necessary for an object to capture attention. We are pursuing these questions in ongoing research.

Concluding remarks

In this chapter I focused on two issues concerning the relationship between perceptual organization and visual attention. The first issue concerns the question of whether or not perceptual organization can be accomplished without attention. I reviewed findings demonstrating that some perceptual organization, such as some forms of grouping and figure-ground segmentation can occur without attention, whereas other forms of organization require controlled attentional processing, depending on the processes involved in the organization and the conditions prevailing for each process. These findings challenge the traditional view, which suggests that perceptual organization is a unitary entity that operates preattentively. Nor do they agree with the radical view of Mack and Rock (1998) that no Gestalt grouping can occur without attention. Rather, these findings support the view that perceptual organization is a confluence of multiple processes that vary in attentional demands (Behrmann and Kimchi, 2003; Kimchi, 2003; Kimchi and Razpurker Apfeld, 2004).

The second issue concerns the question of whether perceptual organization can affect the automatic deployment of attention. I presented findings showing that the mere organization of some elements in the visual field by Gestalt factors into a coherent perceptual unit (an object), with no abrupt onset or any other unique transient, can capture attention automatically in a stimulus-driven manner. It is well documented by now that objects play an important role in visual attention (e.g., Scholl, 2001). These findings, however, are the first to demonstrate that an object per se can attract attention automatically.

Taken together, the findings discussed in this chapter (and other findings reported in the literature) demonstrate that the relationship between perceptual organization and visual attention is multifaceted. Thus, a visual scene can be perceptually organized to a degree without attention, yet focused attention may be required to resolve competing organizations; attentional selection can be driven by organization in the visual scene, yet goal-driven attention can affect the organization of a visual scene. These intricate relations between perceptual organization and visual attention suggest a strong interaction between these two important functions of our perceptual system.

Acknowledgments

This chapter was supported in part by the Israel Science Foundation Grant No. 94/06 to the author and in part by Max Wertheimer Minerva Center for Cognitive Processes and Human Performance, University of Haifa.

References

Abrams, R. A., & Christ, S. E. (2003). Motion onset captures attention. *Psychological Science*, *14*(5), 427–432.

Arrington, C. M., Carr, T. H., Mayer, A. R., & Rao, S. M. (2000). Neural mechanisms of visual attention: Object-based selection of a region in space. *Journal of Cognitive Neuroscience*, *12*(Suppl. 2), 106–117.

Baylis, G. C., & Driver, J. (1992). Visual parsing and response competition: The effect of grouping factors. *Perception & Psychophysics*, *51*(2), 145–162.

Beck, J. (1982). Textural segmentation. In J. Beck (Ed.), *Organization and representation in perception* (pp. 285–317). Hillsdale, NJ: Lawrence Erlbaum Associates.

Behrmann, M., & Kimchi, R. (2003). What does visual agnosia tell us about perceptual organization and its relationship to object perception? *Journal of Experimental Psychology: Human Perception and Performance*, *29*(1), 19–42.

Behrmann, M., Zemel, R. S., & Mozer, M. C. (1998). Object-based attention and occlusion: Evidence from normal subjects and a computational model. *Journal of Experimental Psychology: Human Perception and Performance*, *24*(4), 1011–1036.

Ben Av, M. B., & Sagi, D. (1995). Perceptual grouping by similarity and proximity: Experimental results can be predicted by intensity autocorrelations. *Vision Research*, *35*(6), 853–866.

Ben Av, M. B., Sagi, D., & Braun, J. (1992). Visual attention and perceptual grouping. *Perception & Psychophysics*, *52*(3), 277–294.

Braun, J., & Sagi, D. (1990). Vision outside the focus of attention. *Perception & Psychophysics*, *48*(1), 45–58.

Braun, J., & Sagi, D. (1991). Texture-based tasks are little affected by 2Nd tasks requiring peripheral or central attentive fixation. *Perception*, *20*(4), 483–500.

Chan, W. Y., & Chua, F. K. (2003). Grouping with and without attention. *Psychonomic Bulletin and Review*, *10*(4), 932–938.

Desimone, R., & Duncan, J. (1995). Neural mechanisms of selective visual attention. *Annual Review of Neuroscience*, *18*, 193–197.

Driver, J., & Baylis, G. (1996). Edge-assignment and figure-ground segmentation in short-term visual matching. *Cognitive Psychology*, *31*, 248–306.

Driver, J., & Baylis, G. C. (1989). Movement and visual attention: The spotlight metaphor breaks down. *Journal of Experimental Psychology: Human Perception and Performance*, *15*, 448–456.

Driver, J., Baylis, G. C., & Rafal, R. D. (1992). Preserved figure-ground segregation and symmetry perception in visual neglect. *Nature*, *360*, 73–75.

Driver, J., Davis, G., Russell, C., Turatto, M., & Freeman, E. (2001). Segmentation, attention and phenomenal visual objects. *Cognition*, *80*(1–2), 61–95.

Duncan, J. (1984). Selective attention and the organization of visual information. *Journal of Experimental Psychology: General*, *113*(4), 501–517.

Egly, R., Driver, J., & Rafal, R. (1994). Shifting visual attention between objects and locations: Evidence from normal and parietal lesion subjects. *Journal of Experimental Psychology: General*, *123*, 161–177.

Folk, C. L., Remington, R. W., & Johnston, J. C. (1992). Involuntary covert orienting is contingent on attentional control settings. *Journal of Experimental Psychology: Human Perception and Performance*, *18*(4), 1030–1044.

Franconeri, S. L., Hollingworth, A., & Simons, D. J. (2005). Do new objects capture attention? *Psychological Science*, *16*(4), 275–281.

Freeman, E., Sagi, D., & Driver, J. (2001). Lateral interactions between targets and flankers in low-level vision depend on attention to the flankers. *Nature Neuroscience*, *4*(10), 1032–1036.

Freeman, E., Sagi, D., & Driver, J. (2004). Configuration-specific attentional modulation of flanker-target lateral interactions? *Perception*, *33*, 181–194.

Gorea, A., & Julesz, B. (1990). Context superiority in a detection task with line-element stimuli: A low-level effect. *Perception*, *19*(1), 5–16.

Jonides, J. (1981). Voluntary versus automatic control over the mind's eye. In J. Long & A. Baddeley (Eds.), *Attention and performance IX* (pp. 187–203). Hillsdale, NJ: Erlbaum.

Julesz, B. (1981). Textons, the elements of texture-perception, and their interactions. *Nature*, *290*(5802), 91–97.

Hadad, B. S., & Kimchi, R. (2006). Developmental trends in utilizing perceptual closure for grouping of shape: Effects of spatial proximity and collinearity. *Perception & Psychophysics*, *68*(8), 1264–1273.

Han, S., & Humphreys, G. W. (1999). Interactions between perceptual organization based on Gestalt laws and those based on hierarchical processing. *Perception & Psychophysics*, *61*(7), 1287–1298.

Han, S., Humphreys, G. W., & Chen, L. (1999). Parallel and competitive processes in hierarchical analysis: Perceptual grouping and encoding of closure. *Journal of Experimental Psychology: Human Perception and Performance*, *25*(5), 1411–1432.

Han, S. H. (2004). Interactions between proximity and similarity grouping: An event-related brain potential study in humans. *Neuroscience Letters*, *367*(1), 40–43.

Hoffman, D. D., & Singh, M. (1997). Salience of visual parts. *Cognition*, *63*(1), 29–78.

Hulleman, J., & Humphreys, G. W. (2004). A new cue to figure-ground coding: Top-bottom polarity. *Vision Research*, *44*(24), 2779–2791.

Kahneman, D., & Henik, A. (1981). Perceptual organization and attention. In M. Kubovy & J. R. Pomerantz (Eds.), *Perceptual organization* (pp. 181–211). Hillsdale, NJ: Lawrence Erlbaum Associates.

Kanizsa, G., & Gerbino, W. (1976). Convexity and symmetry in figure-ground organization. In M. Henle (Ed.), *Vision and artifact* (pp. 25–32). New York: Springer.

Kimchi, R. (1998). Uniform connectedness and grouping in the perceptual organization of hierarchical patterns. *Journal of Experimental Psychology: Human Perception and Performance*, *24*(4), 1105–1118.

Kimchi, R. (2000). The perceptual organization of visual objects: A microgenetic analysis. *Vision Research*, *40*(10–12), 1333–1347.

Kimchi, R. (2003). Visual perceptual organization: A microgenetic analysis. In R. Kimchi, M. Behrmann, & C. R. Olson (Eds.), *Perceptual organization in vision: Behavioral and neural perspectives* (pp. 117–154). Mahwah, NJ: Lawrence Erlbaum Associates Publishers.

Kimchi, R., Hadad, B., Behrmann, M., & Palmer, S. E. (2005). Microgenesis and ontogenesis of perceptual organization: Evidence from global and local processing of hierarchical patterns. *Psychological Science*, *16*(4), 282–290.

Kimchi, R., & Peterson, M. A. (2008). Figure-ground segmentation can occur without attention. *Psychological Science*, *19*(7), 660–668.

Kimchi, R., & Razpurker Apfeld, I. (2004). Perceptual grouping and attention: Not all groupings are equal. *Psychonomic Bulletin and Review*, *11*(4), 687–696.

Kimchi, R., Yeshurun, Y., & Cohen Savransky, A. (2007). Automatic, stimulus-driven attentional capture by objecthood. *Psychonomic Bulletin and Review*, *14*(1), 166–172.

Klymenko, V., & Weisstein, N. (1986). Spatial frequency differences can determine figure-ground organization. *Journal of Experimental Psychology: Human Perception and Performance*, *12*, 324–330.

Koffka, K. (1935). *Principles of Gestalt psychology*. New York: Harcourt Brace Jovanovich.

Kovacs, I., & Julesz, B. (1993). A closed curve is much more than an incomplete one: Effect of closure in figure-ground segmentation. *Proceedings of the National Academy of Sciences of the United States of America*, *92*, 7495–7497.

Kovacs, I., Kozma, P., Feher, A., & Benedek, G. (1999). Late maturation of visual spatial integration in humans. *Proceedings of the National Academy of Sciences of the United States of America*, *96*(21), 11209–12204.

Kramer, A. F., & Jacobson, A. (1991). Perceptual organization and focused attention: The role of objects and proximity in visual processing. *Perception & Psychophysics*, *50*, 267–284.

Kurylo, D. D. (1997). Time course of perceptual grouping. *Perception & Psychophysics*, *59*(1), 142–147.

Lamy, D., Segal, H., & Ruderman, L. (2006). Grouping does not require attention. *Perception & Psychophysics*, *68*(1), 17–31.

Lavie, N., & Driver, J. (1996). On the spatial extent of attention in object-based selection. *Perception & Psychophysics*, *58*(8), 1238–1251.

Logan, G. D. (1995). Linguistic and conceptual control of visual spatial attention. *Cognitive Psychology*, *28*(2), 103–174.

Mack, A., & Rock, I. (1998). *Inattentional blindness*. Cambridge, MA: MIT Press/Bradford Books series in cognitive psychology, The MIT Press.

Mack, A., Tang, B., Tuma, R., Kahn, S., & Rock, I. (1992). Perceptual organization and attention. *Cognitive Psychology*, *24*, 475–501.

Marr, D. (1982). *Vision*. San Francisco, CA: W. H. Freeman.

Martinez, A., Teder-Salejarvi, W., & Hillyard, S. A. (2007). Spatial attention facilitates selection of illusory objects: Evidence from event-related brain potentials. *Brain Research*, *1139*, 143–152.

Martinez, A., Teder-Salejarvi, W., Vazquez, M., Molholm, S., Foxe, J. J., Javitt, D. C., et al. (2006). Objects are highlighted by spatial attention. *Journal of Cognitive Neuroscience*, *18*(2), 298–310.

Moore, C., & Egeth, H. (1997). Perception without attention: Evidence of grouping under conditions of inattention. *Journal of Experimental Psychology: Human Perception and Performance*, *23*(2), 339–352.

Moore, C., Yantis, S., & Vaughan, B. (1998). Object-based visual selection: Evidence from perceptual completion. *Psychological Science*, *9*(2), 104–110.

Moore, C. M., Grosjean, M., & Lleras, A. E. (2003). Using inattentional blindness as an operational definition of unattended: The case of surface completion. *Visual Cognition*, *10*(3), 299–318.

Neisser, U. (1967). *Cognitive psychology*. New York: Appleton Century Crofts.

Nelson, R. A., & Palmer, S. E. (2007). Familiar shapes attract attention in figure-ground displays. *Perception & Psychophysics*, *69*, 382–392.

Palmer, S., & Rock, I. (1994). Rethinking perceptual organization: The role of uniform connectedness. *Psychonomic Bulletin and Review*, *1*(1), 29–55.

Palmer, S. E. (1992). Common region: A new principle of perceptual grouping. *Cognitive Psychology*, *24*, 436–447.

Palmer, S. E., & Ghose, T. (2008). Extremal edges: A powerful cue to depth perception and figure-ground organization. *Psychological Science*, *19*(1), 77–84.

Peterson, M. A., & Gibson, B. S. (1994). Object recognition contributions to figure-ground organization: Operations on outlines and subjective contours. *Perception & Psychophysics*, *56*(5), 551–564.

Peterson, M. A., & Salvagio, E. (2008). Inhibitory competition in figure-ground perception: Context and convexity. *Journal of Vision*, *8*(16), 1–13.

Posner, M. I. (1980). Orienting of attention. *Quarterly Journal of Experimental Psychology*, *32*, 3–25.

Quinn, P. C., & Bhatt, R. S. (2005). Good continuation affects discrimination of visual pattern information in young infants. *Perception & Psychophysics*, *67*(7), 1171–1176.

Quinn, P. C., & Bhatt, R. S. (2006). Are some Gestalt principles deployed more readily than others during early development? The case of lightness versus form similarity. *Journal of Experimental Psychology: Human Perception and Performance*, *32*(5), 1221–1230.

Quinn, P. C., Bhatt, R. S., Brush, D., Grimes, A., & Sharpnack, H. (2002). Development of form similarity as a Gestalt grouping principle in infancy. *Psychological Science*, *13*(4), 320–328.

Quinn, P. C., Burke, S., & Rush, A. (1993). Part-whole perception in early infancy: Evidence for perceptual grouping produced by lightness similarity. *Infant Behavior and Development*, *16*(1), 19–42.

Razpurker Apfeld, I., & Kimchi, R. (2007). The time course of perceptual grouping: The role of segregation and shape formation. *Perception & Psychophysics*, *69*(5), 732–743.

Rock, I. (1986). The description and analysis of object and event perception. In K. R. Boff, L. Kaufman, & J. P. Thomas (Eds.), *Handbook of perception and human performance* Vol. 33, (pp. 1–71). New York: Wiley.

Rock, I., Linnet, C. M., Grant, P., & Mack, A. (1992). Perception without attention: Results of a new method. *Cognitive Psychology*, *5*, 504–534.

Rubin, E. (1921). *Visuell Wahrgenommene Figuren*. Kobenhaven: Glydenalske boghandel.

Russell, C., & Driver, J. (2005). New indirect measures of "inattentive" visual grouping in a change-detection task. *Perception & Psychophysics*, *67*(4), 606–623.

Scholl, B. J. (2001). Objects and attention: The state of the art. *Cognition*, *80*(1–2), 1–46.

Shomstein, S., Kimchi, R., Hammer, M., & Behrmann, M. (2009). *Perceptual grouping operates independently of attentional selection: Evidence from hemispatial neglect*. Manuscript submitted for publication.

Theeuwes, J., De Vries, G. J., & Godjin, R. (2003). Attentional and oculomotor capture with static singletons. *Perception & Psychophysics*, *65*(5), 735–746.

Treisman, A. (1982). Perceptual grouping and attention in visual search for features and for objects. *Journal of Experimental Psychology: Human Perception and Performance*, *8*, 194–214.

Treisman, A. (1985). Preattentive processing in vision. *Computer Vision, Graphics, and Image Processing*, *31*, 156–177.

Treisman, A. (1988). Features and objects: The fourteenth Bartlett memorial lecture. *Quarterly Journal of Experimental Psychology*, *40A*(2), 201–237.

Trick, L. M., & Enns, J. M. (1997). Clusters precede shapes in perceptual organization. *Psychological Science*, *8*, 124–129.

Vecera, S., & Farah, M. J. (1994). Does visual attention select objects or locations? *Journal of Experimental Psychology: Human Perception and Performance*, *123*(2), 1–14.

Vecera, S. P., Flevaris, A. V., & Filapek, J. C. (2004). Exogenous spatial attention influences figure-ground assignment. *Psychological Science*, *15*(1), 20–26.

Vecera, S. P., Vogel, E. K., & Woodman, G. F. (2002). Lower region: A new cue for figure-ground assignment. *Journal of Experimental Psychology: General*, *131*(2), 194–205.

Wertheimer, M. (1923/1955). Gestalt theory. In: W.D. Ellis (Ed.), *A source book of Gestalt psychology* (pp. 1–16). London: Routhedge and Kegan Paul. (Originally published in German, 1923, London.)

Wong, E., & Weisstein, N. (1982). A new perceptual context-superiority effect: Line segments are more visible against a figure than against a ground. *Science*, *218*, 587–589.

Yantis, S., & Hillstrom, A. P. (1994). Stimulus-driven attentional capture: Evidence from equiluminant visual objects. *Journal of Experimental Psychology: Human Perception and Performance*, *20*(1), 95–107.

N. Srinivasan (Ed.)
Progress in Brain Research, Vol. 176
ISSN 0079-6123

CHAPTER 3

Long-range neural coupling through synchronization with attention

Georgia G. Gregoriou[1], Stephen J. Gotts[2], Huihui Zhou[3] and Robert Desimone[3,*]

[1]*Department of Basic Sciences, Medical School, University of Crete, Heraklion, Crete, Greece*
[2]*Laboratory of Brain and Cognition, National Institute of Mental Health (NIMH), National Institutes of Health, Bethesda, MD, USA*
[3]*McGovern Institute for Brain Research, Massachusetts Institute of Technology (MIT), Cambridge, MA, USA*

Abstract: In a crowded visual scene, we typically employ attention to select stimuli that are behaviorally relevant. Two likely cortical sources of top-down attentional feedback to cortical visual areas are the prefrontal (PFC) and posterior parietal (PPC) cortices. Recent neurophysiological studies show that areas in PFC and PPC process signals about the locus of attention earlier than in extrastriate visual areas and are therefore likely to mediate attentional selection. Moreover, attentional selection appears to be mediated in part by neural synchrony between neurons in PFC/PPC and early visual areas, with phase relationships that seem optimal for increasing the impact of the top-down inputs to the visual cortex.

Keywords: attention; frontal eye field; area V4; synchrony; top-down; lateral intraparietal area

Introduction

When exploring the world around us, our visual system is confronted with more objects than it can process at any given moment. As a result, we are only aware of a limited number of objects, typically those that are a subject of our attention. Research on the neural mechanisms of visual attention in the last two decades has provided new insights into how neural systems allow us to monitor selectively particular objects or locations while blocking out all distracting information. Attention limits visual processing to objects or locations that are relevant to behavior by selectively enhancing their representation. In electrophysiological studies this is typically seen in enhanced visual responses or increased sensitivity of individual neurons to locations or objects of interest at the expense of distracting stimuli (Luck et al., 1997; McAdams and Maunsell, 2000; Moran and Desimone, 1985; Motter, 1994; Reynolds et al., 1999; Treue and Maunsell, 1996). We originally proposed that top-down attentional feedback biased the competition between multiple stimulus representations in the cortex (Desimone and Duncan, 1995). More recent neurophysiological and modeling studies have formalized and quantified this "biased competition" idea and suggest that the competition between stimulus representations is more generally a form of contrast normalization in the cortex (Lee and

*Corresponding author.
Tel.: 617-3240141; Fax: 617-4524119;
E-mail: desimone@mit.edu

DOI: 10.1016/S0079-6123(09)17603-3

Maunsell, 2009; Reynolds et al., 1999; Reynolds and Heeger, 2009).

In addition to enhanced firing rates with attention, recent studies have found that attention can also change the relative timing of spikes in populations of neurons (Bichot et al., 2005; Fries et al., 2001; Saalmann et al., 2007; Steinmetz et al., 2000). Cells with receptive fields (RFs) at the attended location (Fries et al., 2001) as well as cells selective for the attended feature (Bichot et al., 2005) synchronize their activity in the gamma-frequency range (above 30 Hz). Given that cells have short integration times, even small increases in synchrony in a given population can result in pronounced firing rate changes in downstream neurons (Börgers and Kopell, 2008; Murthy and Fetz, 1994; Salinas and Sejnowski, 2000). Consequently, synchrony can act as another potential amplifier of behaviorally relevant signals. Indeed, recent modeling studies show how synchronized activity for attended stimuli could result in the filtering of responses to distracters (Borgers et al., 2008; Tiesinga et al., 2008; Zeitler et al., 2008).

Although both synchrony and firing rates have been shown to be modulated by attention in the visual cortex the exact mechanisms and sources of this modulation in the brain are less clear. Two likely sources of top-down feedback are the prefrontal cortex (PFC) and posterior parietal cortex (PPC) (Corbetta and Shulman, 2002; Desimone and Duncan, 1995; Gottlieb et al., 1998; Miller and Cohen, 2001; Thompson and Bichot, 2005; Thompson and Schall, 2000). Here, we review recent physiological evidence coming from simultaneous recordings in different cortical areas that support the role of PFC and PPC in enhancing and synchronizing visual cortex responses with attention. More generally, the results suggest that phase-coupled gamma-frequency oscillations play an important role in communication across brain regions.

Interactions between PFC and area V4 in attention

Object recognition in monkeys depends on the "ventral stream" visual areas, which includes the pathway from V1 through V2 and V4 to inferior temporal cortex. Cells in area V4 are selective for features such as color, orientation, and shape (Desimone and Schein, 1987; Desimone et al., 1985; Gallant et al., 1993, 1996; Pasupathy and Connor, 1999; Schein and Desimone, 1990) and they modulate their activity with attention to spatial locations as well as to specific visual features (Connor et al., 1996; Luck et al., 1997; McAdams and Maunsell, 1999, 2000; Mehta et al., 2000; Moran and Desimone, 1985; Motter, 1994; Reynolds et al., 1999; Williford and Maunsell, 2006). Moreover, recent reports have shown that attention increases neuronal synchronization in area V4 (Bichot et al., 2005; Fries et al., 2001, 2008).

PFC plays an important role in executive function, including the control of attention (Duncan, 1986; Miller and Cohen, 2001; Rossi et al., 2009; Stoet and Snyder, 2009). Lesions or deactivation of areas within the PFC impair attentional selection (Wardak et al., 2006) as well as the ability to switch attention in a flexible manner (Rossi et al., 2007) and have been reported to induce neglect in human patients (Heilman and Valenstein, 1972). One area in particular within the PFC, the frontal eye field (FEF), has been implicated in the control of spatial orienting not only via saccades (Bruce and Goldberg, 1985; Hanes and Schall, 1996; Schall, 1991) but also via covert deployment of attention (Thompson et al., 1997, 2005). FEF has direct reciprocal connections with visual cortical areas including area V4 (Barbas and Mesulam, 1981; Barone et al., 2000; Schall et al., 1995; Stanton et al., 1995; Ungerleider et al., 2008) and it is thus well suited to influence visual processing in the context of attention. Indeed, it has been shown that electrical stimulation of FEF can improve detection thresholds in an attention task and increases responses of V4 neurons to a stimulus in their RF (Moore and Armstrong, 2003; Moore and Fallah, 2001) mimicking the effects of spatial attention on behavior and neuronal responses in V4.

To test whether the FEF might be responsible for the effects of attention on neuronal responses and synchrony in V4, we recorded simultaneously from the two areas while monkeys were performing a covert attention task (Gregoriou et al.,

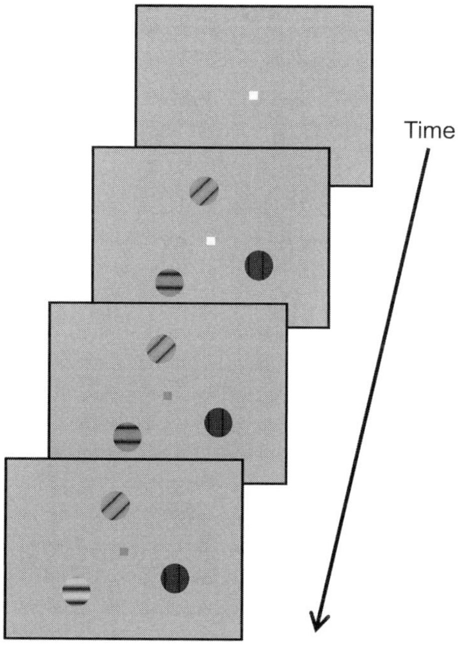

Fig. 1. Behavioral task. The monkeys had to hold a bar to initiate the trial and subsequently fixate the white fixation spot at the center of the screen. After successful fixation three sinusoidal drifting gratings (red, blue, and green) appeared on the screen, at positions distributed radially around the fixation spot at 120° intervals. The fixation spot was subsequently replaced by a small square cue that matched the color of one of the gratings indicating the color of the stimulus to be attended. The monkeys were required to shift their attention to the target stimulus while maintaining fixation of the cue and monitor the target for a color change. The animals were rewarded with a drop of juice for releasing the bar when the target changed color. On any given trial one or both of the distracter stimuli could also change color before the target but the monkeys were trained to ignore the distracters' change. If the monkeys released the bar to the distracter change, failed to maintain fixation, or did not respond to the target color change within 600 ms, the trial was aborted. (See Color Plate 3.1 in color plate section.)

2009). In the task, three colored gratings appeared in the visual field, and one of the gratings was in the joint RF of the cells in V4 and FEF (Fig. 1). A short time after the gratings appeared, a central cue instructed the monkey about which colored grating to attend (the target). The monkey was rewarded for releasing a bar when the target stimulus changed color, ignoring similar changes in the distracters.

To examine whether attention modulated neuronal responses we compared responses in trials where the target appeared inside the RF of the recorded neurons and in trials in which the target appeared outside the RF (Fig. 2). Neurons in both FEF and V4 showed enhanced responses with attention inside their RF. However, we found that the effect of attention on firing rate occurred significantly earlier in FEF compared to V4 (at 80 ms after cue onset in the FEF and at 130 ms after cue onset in V4) which is consistent with the idea that FEF is a source of feedback that modulates V4 responses with attention.

In addition to enhanced firing rates with attention, we also found enhanced synchrony in both areas in the gamma-frequency range (30–60 Hz). These results are consistent with previous reports on the effect of attention in area V4 (Fries et al., 2001, 2008) and show that neurons in FEF too show enhanced synchronization in the gamma range with attention. These findings suggest that neurons in FEF and V4 which encode the location of the behaviorally relevant stimulus synchronize their activity and could thus increase their impact on postsynaptic neurons in their target areas.

Increases in firing rate and synchrony within each area, however, do not establish a functional link between the two areas. If FEF and V4 are functionally coupled during attention then activity in the two areas should be correlated and exhibit increased phase locking with attention, as revealed by enhanced inter-area coherence. Using different measures of coherence (spike-field, field-field, and spike-spike) Gregoriou et al. (2009) indeed found that gamma-frequency coherence between V4 and FEF signals increased with attention for sites with overlapping RFs (Fig. 3). Interestingly, this effect of attention on coupled oscillations between the areas was proportionately larger than the one measured within areas. Importantly, there was no effect of attention on inter-area coherence when there was no overlap between the V4 and FEF RFs. This result suggests that the functional coupling between the two areas is spatially selective and in this particular paradigm becomes prominent only between sites with overlapping RFs. Although the design of the task allowed the animal to use both color and

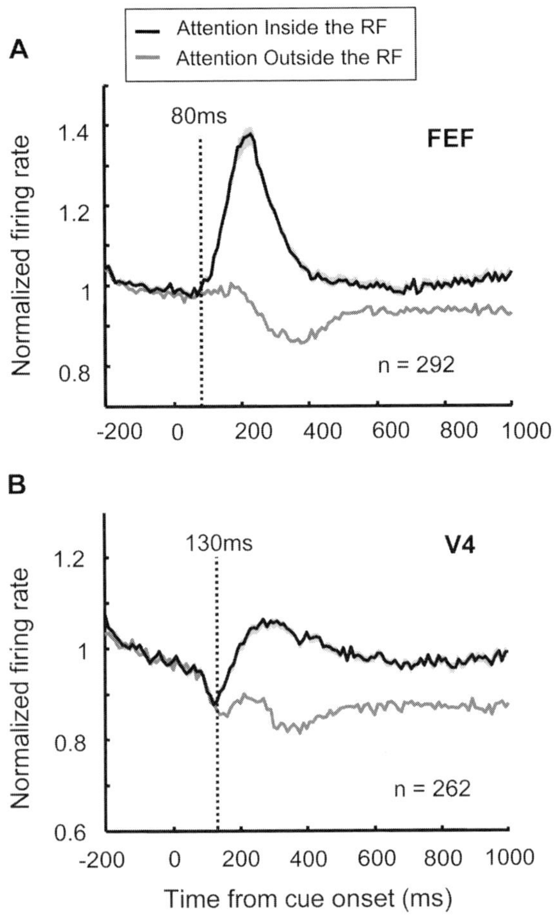

Fig. 2. Attentional effect on firing rate. Normalized firing rate of FEF neurons (A) and V4 neurons (B) averaged over the population of recorded visually responsive cells in each area. Black lines show responses when attention was directed inside the receptive field of the recorded neurons, gray lines show responses when attention was directed outside the receptive field. Shaded area over the lines indicates the standard error of the mean (±) at each time point. Dotted vertical lines show the latency of the attentional effect at the population level. Adapted from Gregoriou et al. (2009).

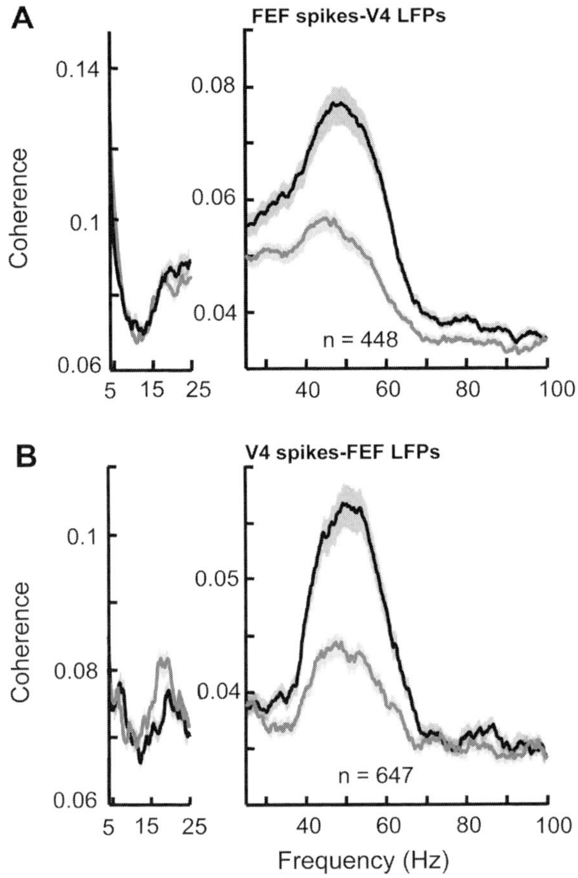

Fig. 3. Enhancement of synchronization with attention across FEF and V4. (A) Spike-field coherence between spikes from FEF and LFPs from V4 averaged across all pairs with overlapping receptive fields. (B) Spike-field coherence between spikes from V4 and LFPs from FEF averaged across all pairs with overlapping receptive fields. Tapers providing smoothing of ±10 Hz were used for spectral estimation of higher frequencies (right part of each graph, 25–100 Hz) and tapers providing smoothing of ±3 Hz were used for lower frequencies (left part of each graph, <25 Hz). Conventions as in Fig. 2. Adapted from Gregoriou et al. (2009).

location for selecting the target stimulus, the strong spatial selectivity of the attention effects underlines the importance of spatial location in target selection.

If a common oscillatory input to the two areas were responsible for causing these coupled oscillations, then gamma synchrony in FEF and V4 would be expected to have a zero phase lag. While we found that the relative phase lag between spikes and local field potentials (LFPs) within each area was close to zero for gamma frequencies (40–60 Hz), the relative phase between spikes in one area and the maximum depolarization of the gamma oscillations in the LFPs in the other area showed a shift by about half a gamma cycle (140–150°) (Fig. 4). This phase shift corresponds to a time delay of about 8–13 ms, and examination of frequency bands other than gamma for which

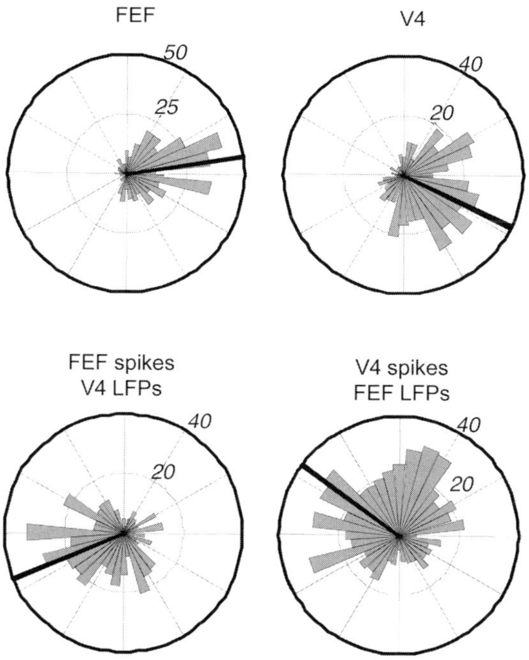

Fig. 4. Distribution of average relative phase (40–60 Hz) across the population of recorded pairs of signals, between FEF spikes and FEF LFPs, V4 spikes and V4 LFPs, FEF spikes and V4 LFPs, and V4 spikes and FEF LFPs. Black solid lines indicate the median of the distribution. Adapted from Gregoriou et al. (2009).

above-chance coherence could be measured (beta and theta frequencies) revealed the same consistent time delay. Although one cannot rule out the possibilities that the oscillatory coupling between FEF and V4 is due to a common input that has the necessary delays, or that the true delays include integer multiples of the cycle durations and are mediated by indirect pathways from FEF to V4, a direct functional coupling between the two areas with an 8–13 ms transmission delay would seem the most parsimonious explanation for all the results. Such an interpretation is also supported by previous studies that measured visual response latencies across different visual areas. Visual response latencies in V1 and V2 as well as between other areas in the ventral visual stream that are directly connected have been shown to differ by a similar amount of time ($\sim$10 ms) indicating that conduction times and synaptic delays could account for this delay (Nowak and Bullier, 1997). Taken together, the results raise the tantalizing possibility that the phase of gamma oscillations is time-shifted to allow spikes produced in one area to arrive at the time of maximum depolarization in the other area, accounting for the latency of information transfer between the two areas. This phase relationship was the same in both attention conditions indicating that it reflects more general principles of communication between the two areas under visual stimulation. However, an increase in synchrony with attention of the sort observed in our study would result in enhanced phase locking between activities in the two areas, with more spikes from one area arriving at the right time to have a larger impact on the other area and therefore bias activity for the attended stimulus.

A Granger causality analysis supported the idea that FEF was the initiator of the coupled oscillations across the two areas. Granger causality analysis provides a statistical measure of the relative strength of influences of one area upon another. It does that by essentially testing whether past values of one signal help predict future values of another signal (Geweke, 1982; Granger, 1969). In agreement with the hypothesis that FEF initiates the gamma oscillations, we found that although significant influences with attention were found in both directions (from V4 to FEF and from FEF to V4) for gamma frequencies, the attentional effects on the Granger causality values appeared significantly earlier in the FEF to V4 direction than the reverse direction (Fig. 5). However, later in the trial these effects became significantly larger for the V4 to FEF direction indicating that while the FEF to V4 (top-down) input predominates when attention is directed to the location of interest, enhanced bottom-up input from V4 may sustain activity in FEF later in the trial when attention is maintained on the target and further visual processing is required. An analysis of the relative latencies of attentional effects on firing rates and LFP gamma power in the two areas suggested that firing rate changes in FEF initiated the attentional effects on synchrony within and across areas.

The findings described above extend the results of previous studies which have established the

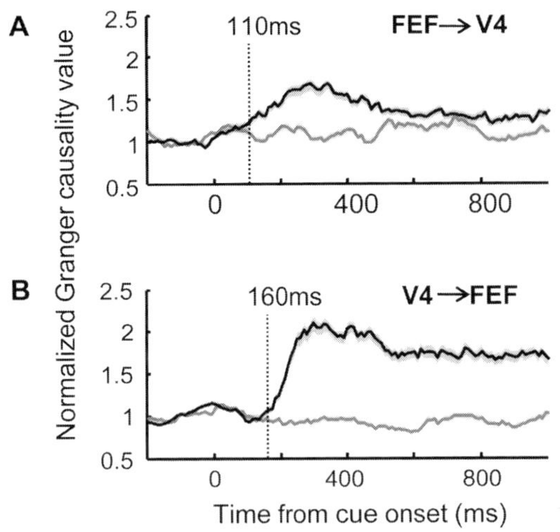

Fig. 5. Directional influences between FEF and V4. Population average of normalized Granger causality values averaged between 40 and 60 Hz across all combinations of FEF-V4 LFPs. Plots for each direction of influence, FEF→V4 (A), V4→FEF (B) are shown. Conventions as in Fig. 2. Adapted from Gregoriou et al. (2009).

role of FEF in attentional selection and have led to the proposal that the FEF holds a saliency map which encodes the behavioral significance of the stimuli (Thompson and Bichot, 2005). Naturally, other brain structures that project to V4 and that have also been implicated in attention, such as the PPC (Andersen et al., 1990; Goldberg et al., 2006; Lewis and Van Essen, 2000), are likely to contribute to the attentional effects on gamma synchrony and firing rates in V4.

Interactions between PPC and area MT in attention

Despite the compelling evidence that PFC plays an important role in attentional control, unilateral lesions of PFC do not permanently abolish the ability of monkeys to attend to visual stimuli, particularly when attention is maintained on the same stimulus across several trials (Rossi et al., 2007). This suggests that other cortical areas contribute to top-down feedback, with PPC a likely candidate. Electrophysiological studies in monkeys have reported modulation of posterior parietal neuronal responses with attention (Bisley and Goldberg, 2003; Constantinidis and Steinmetz, 2001; Gottlieb et al., 1998; Lynch et al., 1977; Robinson et al., 1978), and it has been proposed that the lateral intraparietal area (LIP) in PPC holds a saliency map that guides attentional selection, much like the FEF (Gottlieb, 2007). In agreement with this idea it has been shown that inactivation of LIP delays the discrimination of visual targets in the hemifield contralateral to the inactivated site (Wardak et al., 2004), and in humans, PPC lesions cause hemispatial neglect (Mesulam, 1981) and inability to filter out distracters (Friedman-Hill et al., 2003).

Direct evidence supporting the idea that PPC provides top-down feedback to extrastriate cortical areas was found in a study that employed simultaneous recordings in LIP and area MT (Saalmann et al., 2007), two areas which share reciprocal connections. Monkeys performed a delayed match to sample task in which both spatial and feature-based attention were manipulated. The monkeys were required to match the location and the orientation of the sample and the test stimuli. The sample and test stimuli could either appear at the same location inside the common RF (i.e., attention inside the RF) or the sample could appear outside and the test stimulus inside the RF ("attention elsewhere" condition). When both sample and test stimuli occurred at the same position, they could either have different orientations (i.e., spatial attention only) or the same orientation (i.e., both spatial and feature-based attention). Both areas showed significant increases in firing rate when attention was directed inside the RF. Attentional effects in LIP occurred earlier than in MT, consistent with the hypothesis that it is a source of feedback to MT. Moreover, in contrast to MT in which responses were mainly modulated by the spatial locus of attention, LIP responses were modulated by attention to both features and locations. Responses of LIP neurons to the test stimulus with the preferred orientation were enhanced when it matched the orientation of the sample. This is in agreement with the idea of a saliency map which integrates information about features from feature-selective areas and sends topographically organized attentional feedback to visual cortical areas (Gottlieb, 2007; Itti and Koch, 2001; Thompson and Bichot, 2005).

To test whether neural activity was synchronized across areas, Saalmann et al calculated coherence between LFPs in LIP and MT. Enhanced coherence was found between 20 and 35 Hz for the condition where attention was directed inside the common RF in both "spatial" and "spatial and feature-based" attention compared to the condition where attention was directed outside the RF. Coherence between a subset of spike trains in the two areas was also reported, confirming that spiking activities in the two areas are synchronized. Interestingly, the phase between LIP and MT spike trains indicated that LIP leads MT by 5–7 ms, which could be accounted for conduction and synaptic delays between the two areas. This time lag could ensure that signals from LIP arrive in MT at the depolarizing phase of the local oscillations maximizing the likelihood of spike generation.

Saalmann et al. also calculated the percentage of MT spikes preceded by LIP spikes in the different attention conditions. Attention appeared to cause a 10% increase in the number of MT spikes preceded by LIP spikes within 10 ms. This percentage accounted for a considerable amount of the overall increase in the firing rate of MT neurons with attention, a finding which confirms that attention does not simply lead to overall increases in the firing rate but that it has a direct effect on the relative timing of spikes causing more spikes from one area to be phase locked to activity in the other area.

Conclusion

The results from the two studies reviewed here, reveal similar general principles that govern the interaction of FEF and LIP with early visual areas in attention. Both FEF and LIP are well suited to provide top-down attentional feedback to V4 and MT, respectively, as shown by the earlier onset of attentional effects in parietal and prefrontal activities. This feedback is manifested in the oscillatory coupling of neural activity between the interconnected areas. The results from both studies showed that the coupled oscillations are shifted in time by 8–13 ms between FEF and V4 and 5–7 ms between LIP and MT, which could reflect the time necessary for spikes from one area to reach the other so that action potentials arrive in each area at its most excitable phase. This could maximize the probability of spike generation in the receiving area and could therefore amplify the impact of inputs corresponding to the attended stimulus over less coherent inputs corresponding to the unattended stimulus (Fig. 6). The difference in the time lag found in the two studies could be explained by the shorter distance between LIP and MT compared to the longer distance FEF-V4 connections and by the relative strength of connections. It should be noted however, that the frequency range within which enhanced phase locking was observed was different in the two cases. Whereas enhanced oscillatory coupling between FEF and V4 was found between 40 and 60 Hz, the same effect was observed in lower frequencies for LIP-MT coupling (20–35 Hz). It is possible that this diversity reflects differences in the tasks employed in the two studies. Early, "evoked" gamma-band activity as well as lower frequency beta-band synchrony has been associated with template matching and working memory processes (Herrmann et al., 2004; Tallon-Baudry et al., 2001), which were present in the task used by Saalmann et al. The late, "induced" gamma-band synchrony seen by Gregoriou et al. is more likely to reflect sustained attention to a stimulus (Fries et al., 2008). The degree to which these processes and their underlying mechanisms may differ remains to be elucidated in future studies.

Synchrony in different frequency bands has been suggested to mediate different attentional processes. More specifically, a study undertaken to elucidate the role of PFC and LIP in bottom-up, stimulus-driven and top-down, goal-directed attention showed enhanced coherence between LIP and PFC in frequencies 22–34 Hz in bottom-up attention, whereas with top-down attention, coherence was greater between the two areas in somewhat higher frequencies, 35–55 Hz (Buschman and Miller, 2007). The authors suggested that an extended network of areas participating in top-down processes synchronizes in lower frequencies

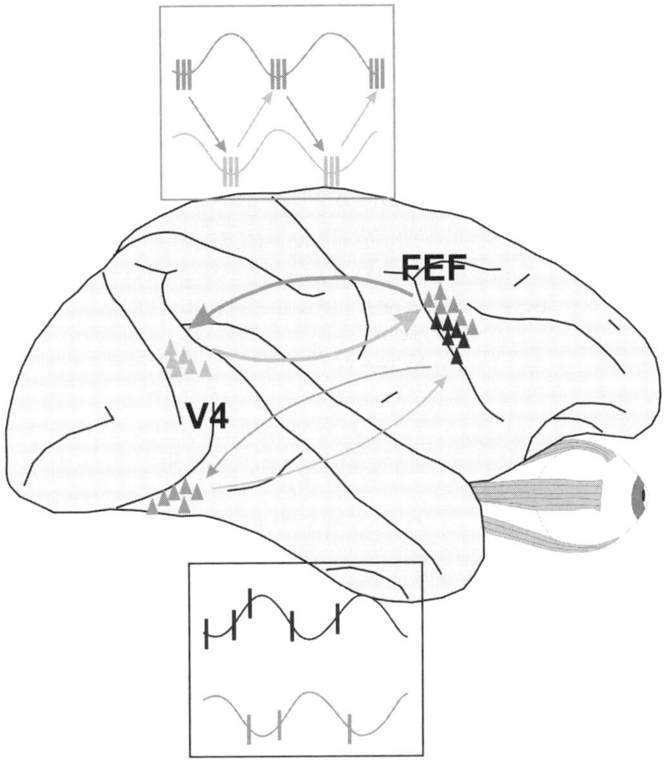

Fig. 6. Schematic illustration of inter-areal neuronal communication in attention. Dark red and light red triangles illustrate neurons in FEF and V4, respectively, encoding the attended stimulus (red book on the right), whereas dark blue and light blue triangles illustrate FEF and V4 neurons, respectively, encoding the unattended stimulus (blue book). The vertical lines in the boxes above and below the schematic brain illustrate action potentials of neurons in the four groups. Arrows indicate propagation of action potentials between areas along the projecting axons. Coherent spikes which arrive at the phase of maximal excitability increase the probability of generating spikes in the receiving area (red box). Note the phase relationship between excitability fluctuations in the two areas which facilitates neuronal communication. Less coherent spikes corresponding to the unattended stimulus (blue box) are less effective in triggering spikes in the receiving area and result in weak communication between the areas and a weaker representation of the unattended stimulus. (See Color Plate 3.6 in color plate section.)

which are more robust to conduction delays and would thus be better suited to mediate long-range or polysynaptic communication in the brain (Engel et al., 2001; Kopell et al., 2000). In contrast, synchrony in higher frequencies, in the gamma range, during bottom-up attention was suggested to reflect local computations underlying the enhancement of sensory representations (Buschman and Miller, 2007; Kopell et al., 2000). A number of studies have found long-range synchronization across distant brain areas in frequencies lower than gamma (Brovelli et al., 2004; Pesaran et al., 2008; Roelfsema et al., 1997; Sirota et al., 2008; von Stein et al., 2000) providing support to this idea. Our results (Gregoriou et al., 2009) which show synchronization of activity in gamma frequencies within each area and strong coupling between FEF and V4 in the gamma range are in agreement with the proposal that gamma synchronization can be viewed as a local phenomenon observed within an area or across areas that are monosynaptically connected (von Stein et al., 2000).

Top-down inputs (from FEF to V4) are dominant at the onset of attention to a location possibly mediating attentional selection, but the bottom-up inputs (from V4 to FEF) come to predominate later in the trial when further visual processing is required during sustained attention. Indeed, once the relevant stimulus has been selected, the brain needs to insulate its sensory representation from other inputs competing for effective visual processing. Modeling studies have shown that more coherent inputs which are oscillating in the gamma range can render less coherent inputs ineffective and can thus "lock" the representation of the attended stimulus by filtering out competing inputs (Borgers et al., 2008; Tiesinga et al., 2008; Zeitler et al., 2008).

The dynamic nature of selective interactions across brain areas in the course of attention shows that long-range oscillatory coupling between distant parts of the brain controls the activity in selective neuronal populations by setting the optimal phase difference which will facilitate neuronal communication (Fries, 2005; Womelsdorf and Fries, 2007). In a network of fixed anatomical connections such a mechanism of neuronal communication could provide the basis for the dynamic control of interactions among the subset of neuronal populations that are most relevant to the task at hand. Future studies should aim to elucidate the role of different frequencies in oscillatory coupling and their relevance to behavior.

Acknowledgments

This work was supported by NEI grants EY017292 and EY017921 to Robert Desimone. Stephen J. Gotts was supported in part by MH64445 from the National Institutes of Health (USA) and by the NIMH Intramural Research Program.

References

Andersen, R. A., Asanuma, C., Essick, G., & Siegel, R. M. (1990). Corticocortical connections of anatomically and physiologically defined subdivisions within the inferior parietal lobule. *Journal of Comparative Neurologica*, *296*, 65–113.

Barbas, H., & Mesulam, M. M. (1981). Organization of afferent input to subdivisions of area 8 in the rhesus monkey. *Journal of Comparative Neurology*, *200*, 407–431.

Barone, P., Batardiere, A., Knoblauch, K., & Kennedy, H. (2000). Laminar distribution of neurons in extrastriate areas projecting to visual areas V1 and V4 correlates with the hierarchical rank and indicates the operation of a distance rule. *Journal of Neuroscience*, *20*, 3263–3281.

Bichot, N. P., Rossi, A. F., & Desimone, R. (2005). Parallel and serial neural mechanisms for visual search in macaque area V4. *Science*, *308*, 529–534.

Bisley, J. W., & Goldberg, M. E. (2003). Neuronal activity in the lateral intraparietal area and spatial attention. *Science*, *299*, 81–86.

Borgers, C., Epstein, S., & Kopell, N. J. (2008). Gamma oscillations mediate stimulus competition and attentional selection in a cortical network model. *Proceedings of the National Academy of Sciences of the United States of America*, *105*, 18023–18028.

Börgers, C., & Kopell, N. (2008). Gamma oscillations and stimulus selection. *Neural Computation*, *20*, 383–414.

Brovelli, A., Ding, M., Ledberg, A., Chen, Y., Nakamura, R., & Bressler, S. L. (2004). Beta oscillations in a large-scale sensorimotor cortical network: Directional influences revealed by Granger causality. *Proceedings of the National Academy of Sciences of the United States of America*, *101*, 9849–9854.

Bruce, C. J., & Goldberg, M. E. (1985). Primate frontal eye fields. I. Single neurons discharging before saccades. *Journal of Neurophysiology*, *53*, 603–635.

Buschman, T. J., & Miller, E. K. (2007). Top-down versus bottom-up control of attention in the prefrontal and posterior parietal cortices. *Science*, *315*, 1860–1862.

Connor, C. E., Gallant, J. L., Preddie, D. C., & Van Essen, D. C. (1996). Responses in area V4 depend on the spatial relationship between stimulus and attention. *Journal of Neurophysiology*, *75*, 1306–1308.

Constantinidis, C., & Steinmetz, M. A. (2001). Neuronal responses in area 7a to multiple-stimulus displays: I. Neurons encode the location of the salient stimulus. *Cerebral Cortex*, *11*, 581–591.

Corbetta, M., & Shulman, G. L. (2002). Control of goal-directed and stimulus-driven attention in the brain. *Nature Reviews Neuroscience*, *3*, 201–215.

Desimone, R., & Duncan, J. (1995). Neural mechanisms of selective visual attention. *Annual Reviews of Neuroscience*, *18*, 193–222.

Desimone, R., & Schein, S. J. (1987). Visual properties of neurons in area V4 of the macaque: Sensitivity to stimulus form. *Journal of Neurophysiology*, *57*, 835–868.

Desimone, R., Schein, S. J., Moran, J., & Ungerleider, L. G. (1985). Contour, color and shape analysis beyond the striate cortex. *Vision Research*, *25*, 441–452.

Duncan, J. (1986). Disorganization of behavior after frontal-lobe damage. *Cognitive Neuropsychology*, *3*, 271–290.

Engel, A. K., Fries, P., & Singer, W. (2001). Dynamic predictions: Oscillations and synchrony in top-down processing. *Nature Reviews Neuroscience*, *2*, 704–716.

Friedman-Hill, S. R., Robertson, L. C., Desimone, R., & Ungerleider, L. G. (2003). Posterior parietal cortex and the filtering of distractors. *Proceedings of the National Academy of Sciences of the United States of America*, *100*, 4263–4268.

Fries, P. (2005). A mechanism for cognitive dynamics: Neuronal communication through neuronal coherence. *Trends in Cognitive Science*, *9*, 474–480.

Fries, P., Reynolds, J. H., Rorie, A. E., & Desimone, R. (2001). Modulation of oscillatory neuronal synchronization by selective visual attention. *Science*, *291*, 1560–1563.

Fries, P., Womelsdorf, T., Oostenveld, R., & Desimone, R. (2008). The effects of visual stimulation and selective visual attention on rhythmic neuronal synchronization in macaque area V4. *Journal of Neuroscience*, *28*, 4823–4835.

Gallant, J. L., Braun, J., & Van Essen, D. C. (1993). Selectivity for polar, hyperbolic, and Cartesian gratings in macaque visual cortex. *Science*, *259*, 100–103.

Gallant, J. L., Connor, C. E., Rakshit, S., Lewis, J. W., & Van Essen, D. C. (1996). Neural responses to polar, hyperbolic, and Cartesian gratings in area V4 of the macaque monkey. *Journal of Neurophysiology*, *76*, 2718–2739.

Geweke, J. (1982). Measurement of linear-dependence and feedback between multiple time-series. *Journal of the American Statistical Association*, *77*, 304–313.

Goldberg, M. E., Bisley, J. W., Powell, K. D., & Gottlieb, J. (2006). Saccades, salience and attention: The role of the lateral intraparietal area in visual behavior. *Progress in Brain Research*, *155*, 157–175.

Gottlieb, J. (2007). From thought to action: The parietal cortex as a bridge between perception, action, and cognition. *Neuron*, *53*, 9–16.

Gottlieb, J. P., Kusunoki, M., & Goldberg, M. E. (1998). The representation of visual salience in monkey parietal cortex. *Nature*, *391*, 481–484.

Granger, C. W. J. (1969). Investigating causal relations by econometric models and cross-spectral methods. *Econometrica*, *37*, 424–438.

Gregoriou, G. G., Gotts, S. J., Zhou, H., & Desimone, R. (2009). High frequency long range coupling between prefrontal cortex and visual cortex during attention. *Science*, *324*, 1207–1210.

Hanes, D. P., & Schall, J. D. (1996). Neural control of voluntary movement initiation. *Science*, *274*, 427–430.

Heilman, K. M., & Valenstein, E. (1972). Frontal lobe neglect in man. *Neurology*, *22*, 660–664.

Herrmann, C. S., Munk, M. H., & Engel, A. K. (2004). Cognitive functions of gamma-band activity: Memory match and utilization. *Trends in Cognitive Science*, *8*, 347–355.

Itti, L., & Koch, C. (2001). Computational modelling of visual attention. *Nature Reviews Neuroscience*, *2*, 194–203.

Kopell, N., Ermentrout, G. B., Whittington, M. A., & Traub, R. D. (2000). Gamma rhythms and beta rhythms have different synchronization properties. *Proceedings of the National Academy of Sciences of the United States of America*, *97*, 1867–1872.

Lee, J., & Maunsell, J. H. (2009). A normalization model of attentional modulation of single unit responses. *PLoS ONE*, *4*, e4651.

Lewis, J. W., & Van Essen, D. C. (2000). Corticocortical connections of visual, sensorimotor, and multimodal processing areas in the parietal lobe of the macaque monkey. *Journal of Comparative Neurology*, *428*, 112–137.

Luck, S. J., Chelazzi, L., Hillyard, S. A., & Desimone, R. (1997). Neural mechanisms of spatial selective attention in areas V1, V2, and V4 of macaque visual cortex. *Journal of Neurophysiology*, *77*, 24–42.

Lynch, J. C., Mountcastle, V. B., Talbot, W. H., & Yin, T. C. T. (1977). Parietal lobe mechanisms for directed visual attention. *Journal of Neurophysiology*, *40*, 362–389.

McAdams, C. J., & Maunsell, J. H. (1999). Effects of attention on orientation-tuning functions of single neurons in macaque cortical area V4. *Journal of Neuroscience*, *19*, 431–441.

McAdams, C. J., & Maunsell, J. H. (2000). Attention to both space and feature modulates neuronal responses in macaque area V4. *Journal of Neurophysiology*, *83*, 1751–1755.

Mehta, A. D., Ulbert, I., & Schroeder, C. E. (2000). Intermodal selective attention in monkeys. I: Distribution and timing of effects across visual areas. *Cerebral Cortex*, *10*, 343–358.

Mesulam, M. M. (1981). A cortical network for directed attention and unilateral neglect. *Annals of Neurology*, *10*, 309–325.

Miller, E. K., & Cohen, J. D. (2001). An integrative theory of prefrontal cortex function. *Annual Reviews of Neuroscience*, *24*, 167–202.

Moore, T., & Armstrong, K. M. (2003). Selective gating of visual signals by microstimulation of frontal cortex. *Nature*, *421*, 370–373.

Moore, T., & Fallah, M. (2001). Control of eye movements and spatial attention. *Proceedings of the National Academy of Sciences of the United States of America*, *98*, 1273–1276.

Moran, J., & Desimone, R. (1985). Selective attention gates visual processing in the extrastriate cortex. *Science*, *229*, 782–784.

Motter, B. C. (1994). Neural correlates of attentive selection for color or luminance in extrastriate area V4. *Journal of Neuroscience*, *14*, 2178–2189.

Murthy, V., & Fetz, E. E. (1994). Effects of input synchrony on the firing rate of a 3-conductance cortical neuron model. *Neural Computation*, *6*, 1111–1126.

Nowak, L. G., & Bullier, J. (1997). The timing of information transfer in the visual system. In K. S. Rockland, J. H. Kaas, & A. Peters (Eds.), *Cerebral Cortex* (pp. 205–241). New York: Plenum Press.

Pasupathy, A., & Connor, C. E. (1999). Responses to contour features in macaque area V4. *Journal of Neurophysiology*, *82*, 2490–2502.

Pesaran, B., Nelson, M. J., & Andersen, R. A. (2008). Free choice activates a decision circuit between frontal and parietal cortex. *Nature*, *453*, 406–409.

Reynolds, J. H., Chelazzi, L., & Desimone, R. (1999). Competitive mechanisms subserve attention in macaque areas V2 and V4. *Journal of Neuroscience*, *19*, 1736–1753.

Reynolds, J. H., & Heeger, D. J. (2009). The normalization model of attention. *Neuron*, *61*, 168–185.

Robinson, D. L., Goldberg, M. E., & Stanton, G. B. (1978). Parietal association cortex in the primate: Sensory mechanisms and behavioral modulations. *Journal of Neurophysiology*, *41*, 910–932.

Roelfsema, P. R., Engel, A. K., Konig, P., & Singer, W. (1997). Visuomotor integration is associated with zero time-lag synchronization among cortical areas. *Nature*, *385*, 157–161.

Rossi, A. F., Bichot, N. P., Desimone, R., & Ungerleider, L. G. (2007). Top down attentional deficits in macaques with lesions of lateral prefrontal cortex. *Journal of Neuroscience*, *27*, 11306–11314.

Rossi, A. F., Pessoa, L., Desimone, R., & Ungerleider, L. G. (2009). The prefrontal cortex and the executive control of attention. *Experimental Brain Research*, *192*, 489–497.

Saalmann, Y. B., Pigarev, I. N., & Vidyasagar, T. R. (2007). Neural mechanisms of visual attention: How top-down feedback highlights relevant locations. *Science*, *316*, 1612–1615.

Salinas, E., & Sejnowski, T. J. (2000). Impact of correlated synaptic input on output firing rate and variability in simple neuronal models. *Journal of Neuroscience*, *20*, 6193–6209.

Schall, J. D. (1991). Neuronal activity related to visually guided saccades in the frontal eye fields of rhesus monkeys: Comparison with supplementary eye fields. *Journal of Neurophysiology*, *66*, 559–579.

Schall, J. D., Morel, A., King, D. J., & Bullier, J. (1995). Topography of visual cortex connections with frontal eye field in macaque: Convergence and segregation of processing streams. *Journal of Neuroscience*, *15*, 4464–4487.

Schein, S. J., & Desimone, R. (1990). Spectral properties of V4 neurons in the macaque. *Journal of Neuroscience*, *10*, 3369–3389.

Sirota, A., Montgomery, S., Fujisawa, S., Isomura, Y., Zugaro, M., & Buzsaki, G. (2008). Entrainment of neocortical neurons and gamma oscillations by the hippocampal theta rhythm. *Neuron*, *60*, 683–697.

Stanton, G. B., Bruce, C. J., & Goldberg, M. E. (1995). Topography of projections to posterior cortical areas from the macaque frontal eye fields. *Journal of Comparative Neurology*, *353*, 291–305.

Steinmetz, P. N., Roy, A., Fitzgerald, P. J., Hsiao, S. S., Johnson, K. O., & Niebur, E. (2000). Attention modulates synchronized neuronal firing in primate somatosensory cortex. *Nature*, *404*, 187–190.

Stoet, G., & Snyder, L. H. (2009). Neural correlates of executive control functions in the monkey. *Trends in Cognitive Science*, *13*, 228–234.

Tallon-Baudry, C., Bertrand, O., & Fischer, C. (2001). Oscillatory synchrony between human extrastriate areas during visual short-term memory maintenance. *Journal of Neuroscience*, *21*, RC177.

Thompson, K. G., & Bichot, N. P. (2005). A visual salience map in the primate frontal eye field. *Progress in Brain Research*, *147*, 251–262.

Thompson, K. G., Bichot, N. P., & Schall, J. D. (1997). Dissociation of visual discrimination from saccade programming in macaque frontal eye field. *Journal of Neurophysiology*, *77*, 1046–1050.

Thompson, K. G., Biscoe, K. L., & Sato, T. R. (2005). Neuronal basis of covert spatial attention in the frontal eye field. *Journal of Neuroscience*, *25*, 9479–9487.

Thompson, K. G., & Schall, J. D. (2000). Antecedents and correlates of visual detection and awareness in macaque prefrontal cortex. *Vision Research*, *40*, 1523–1538.

Tiesinga, P., Fellous, J. M., & Sejnowski, T. J. (2008). Regulation of spike timing in visual cortical circuits. *Nature Reviews Neuroscience*, *9*, 97–107.

Treue, S., & Maunsell, J. H. (1996). Attentional modulation of visual motion processing in cortical areas MT and MST. *Nature*, *382*, 539–541.

Ungerleider, L. G., Galkin, T. W., Desimone, R., & Gattass, R. (2008). Cortical connections of area V4 in the macaque. *Cerebral Cortex*, *18*, 477–499.

von Stein, A., Chiang, C., & Konig, P. (2000). Top-down processing mediated by interareal synchronization. *Proceedings of the National Academy of Sciences of the United States of America*, *97*, 14748–14753.

Wardak, C., Ibos, G., Duhamel, J. R., & Olivier, E. (2006). Contribution of the monkey frontal eye field to covert visual attention. *Journal of Neuroscience*, *26*, 4228–4235.

Wardak, C., Olivier, E., & Duhamel, J. R. (2004). A deficit in covert attention after parietal cortex inactivation in the monkey. *Neuron*, *42*, 501–508.

Williford, T., & Maunsell, J. H. (2006). Effects of spatial attention on contrast response functions in macaque area V4. *Journal of Neurophysiology*, *96*, 40–54.

Womelsdorf, T., & Fries, P. (2007). The role of neuronal synchronization in selective attention. *Current Opinion in Neurobiology*, *17*, 154–160.

Zeitler, M., Fries, P., & Gielen, S. (2008). Biased competition through variations in amplitude of gamma-oscillations. *Journal of Computational Neuroscience*, *25*, 89–107.

N. Srinivasan (Ed.)
Progress in Brain Research, Vol. 176
ISSN 0079-6123

CHAPTER 4

Visual streams and shifting attention

James M. Brown*

Department of Psychology, University of Georgia, Athens, GA, USA

Abstract: Understanding the relationship between bottom-up and top-down processing in visual perception and attention is challenging. An important part of that challenge is studying the roles the parvocellular (P) and magnocellular (M) retino-geniculo-cortical pathways play in visual processing and attention. The P pathway provides the dominant initial input to the ventral stream which plays an important role in object processing and is assumed to be relatively more involved in object-based attention. The faster responding M pathway provides the dominant initial input to the dorsal stream which plays an important role in processing movement and spatial location information and is assumed to be relatively more involved in space-based attention. To gain insight into the relationship between M/dorsal and P/ventral activity and deploying visual attention, we used a covert cuing paradigm to manipulate attention while bottom-up and top-down perceptual stimulus variables created M/dorsal and P/ventral-biased conditions. One study examined the object advantage, where responses are faster for within- relative to equidistant between-object shifts of attention. Visual stream contributions to object- and spaced-based attention were revealed using psychophysically equiluminant conditions expected to reduce M/dorsal activity. Other studies investigating visual stream contributions to location-based inhibition of return (IOR) used IOR magnitude as an indicator of the ease or difficulty of deploying spatial attention. Greater IOR was found under P/ventral-biased conditions. Less IOR was found under M/dorsal-biased conditions. The results support the use of M/dorsal and P/ventral-biased conditions as a valuable strategy for studying the relationship between visual stream activity and shifting attention.

Keywords: visual pathways; dorsal/M stream; ventral/P stream; inhibition of return; shifting attention; exogenous attention

Introduction

The amount of research devoted to understanding visual attention has exploded in the past 30 years. One good example of this is the recent change in the name of the longstanding journal Perception & Psychophysics to Attention, Perception, & Psychophysics and the rationale behind it (Wolfe, 2009)! From this explosion many different ideas and theories have emerged about our visual attention abilities. The dynamics of how attention is utilized by our visual system during sensory-perceptual processing is complex and can be viewed from both bottom-up and top-down perspectives. For example, we can allocate attention in a top-down (i.e., endogenous cue) manner to locations or objects in our field of view depending on our goals, expectations, and experience (e.g., when looking for a friend in a crowd).

*Corresponding author.
Tel.: +706-542-8045; Fax: +706-542-3275;
E-mail: jmbrown@uga.edu

DOI: 10.1016/S0079-6123(09)17604-5

At the same time our attention can be drawn to different locations or objects in a bottom-up (i.e., exogenous cue) manner (e.g., when something suddenly moves or flashes). The research discussed here, with one exception, involves visual attention of the bottom-up, exogenous variety.

The motivation behind the research reviewed here is to understand the relationship between attention and bottom-up and top-down processes in visual perception by studying the roles the parvocellular (P) and magnocellular (M) retino-geniculo-cortical pathways play in visual processing and attention. The P pathway provides the dominant, initial feed-forward input to the ventral (a.k.a. "what") stream into the temporal lobe which plays a major role in object processing. The M pathway provides the dominant initial feed-forward input to the dorsal (a.k.a. "where") stream into the parietal lobe which plays a major role in spatial processing (Haxby et al., 1991; Ungerleider and Haxby, 1994; Ungerleider and Mishkin, 1982). The strategy is to selectively activate the P/ventral and M/dorsal streams using P- and M-biased stimuli and observe how shifting attention is affected. In all the experiments to be discussed (except one), a covert cuing paradigm is used where observers are instructed to refrain from moving their eyes and, in some experiments, the time between cue and target stimuli is too short to allow for eye movements. The combined results of the research I will review (1) provide convergent evidence of the importance of M/dorsal activity to shifting attention and P/ventral activity to attentive processing, (2) suggest shifting attention is more difficult with relatively greater P/ventral involvement, and (3) are consistent with models proposing fast M/dorsal feed-forward signals guide subsequent P/ventral visual processing (Bullier, 2001, 2006; Kveraga et al., 2007) and the deployment of visual attention (Vidyasagar, 1999, 2005).

Why this approach?

What motivated this research to begin with? As a new graduate student in Naomi Weisstein's lab in 1979–1980, there was a lot of excitement about ongoing research showing a close relationship between the spatial and temporal frequency response of the visual system and the perception and processing of figure and ground. It was already known the relatively slower responding P pathway plays a primary role in processing color, texture, shape, and higher spatial frequency (i.e., detailed) information and the relatively faster responding M pathway plays a greater role in processing movement, location, and lower spatial frequency (i.e., lower resolution) information (Livingstone and Hubel, 1987, 1988). A strategy used in Weisstein's lab was to manipulate sensory stimulus variables to see how figure/ground segregation and perception were effected, and conversely, to examine how the perception of a region as figure or ground influenced the sensory response to stimuli presented there (see Weisstein and Wong, 1986, 1987). For example, higher spatial frequencies bias a region to be seen as figure, in front, and lower spatial frequencies bias a region to be seen as ground, behind (Brown and Weisstein, 1988b; Klymenko and Weisstein, 1986; Klymenko et al., 1989). Conversely, sharp-edged line segments are discriminated and detected better in figure than ground while blurry lines (i.e., with high spatial frequencies removed) are detected better in ground than figure (Brown and Weisstein, 1988a; Wong and Weisstein, 1982, 1983). These findings strongly indicate an association between P/ventral processing and figure perception and between M/dorsal processing and ground perception.

One question was how attention might be related to these discoveries considering figure regions are usually what we are paying attention to and ground regions are usually unattended or ignored. Is it possible that P/ventral and M/dorsal activity is more associated with attended and unattended processing respectively? It was from this question about the relationship between these visual streams and attention the current research got its start.

What is the relationship between M/dorsal and P/ventral stream activity and shifting visual attention?

Endogenous cues

The first study we conducted addressing this question used an endogenous (i.e., symbolic) cue directing observers where to attend (Srinivasan and Brown, 2006). Our targets were sharp (P-biased) and blurred (M-biased) line segments similar to those used in the Wong and Weisstein figure/ground study (Wong and Weisstein, 1983). Targets appeared either left or right of fixation 100 ms after a cue appeared at fixation. The cue was either neutral (i.e., a plus sign) indicating the target was equally likely to appear left or right, or an arrow indicating with an 80% probability the side the target would appear. At short cue-to-target time intervals responses should be faster at cued (valid) positions and slower at uncued (invalid) positions because attention must reorient to the target after being misled by an invalid cue. In the first experiment we measured simple reaction time (RT) for detecting a target. As shown in Fig. 1a, typical cuing effects were found for both targets with the shortest RTs on validly cued trials, the longest RTs on invalidly cued trials, and with RTs in-between on neutral cue trials.

At first glance these results do not seem to support an attended-P/ventral, unattended-M/dorsal relationship. However, from a stimulus and task perspective, there was no need to attend to the spatial frequency content of the stimuli to detect them. In a second experiment requiring a discrimination response, observers had to attend to the spatial frequency content to perform the task (i.e., press one key for the sharp target, one for the blurred one). As Fig. 1b shows, responses to the P-biased sharp target again showed cuing effects reflecting the influence/allocation of attention, while M-biased blurred target responses did not. Responses to the blurred target were just as fast whether it appeared at the cued or uncued position indicating they were not influenced by attention.

There is another interpretation of the results that is also consistent with an attended-P/ventral and unattended-M/dorsal relationship. It is possible the discrimination results reflect having to process the details of the cues to know which cue was presented each trial. Such processing has likely required attending to and utilizing higher spatial frequency information. When attention was directed to the cued position it may also have been directing higher spatial frequency mechanisms to process that position at the same time.

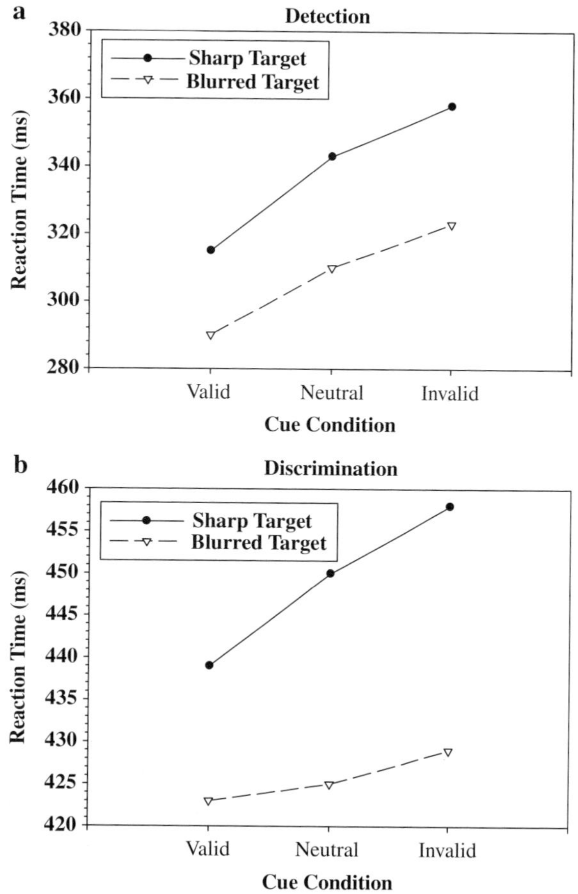

Fig. 1. (a) Detection and (b) discrimination RTs (adapted from Srinivasan and Brown, 2006).

This results in responses to the sharp target (with higher spatial frequencies present) showing a benefit for attention being directed to the cued position and a cost when it is not. Conversely, by the nature of the endogenous cuing task, attention is being directed toward higher spatial frequency mechanisms during cue processing and, therefore, not being directed toward lower spatial frequency mechanisms. With the lower spatial frequency mechanisms being the unattended mechanisms they are able to respond quickly to the blurred target wherever it appears resulting in no effect of cuing. While this alternative account is related to task demands associated with cue processing, it is still consistent with the P/attended versus M/unattended viewpoint. Some other examples of evidence for an attended-P/ventral, unattended-M/dorsal relationship comes from covert cuing studies by Yeshurun and colleagues (Yeshurun, 2004; Yeshurun and Carrasco, 1998; Yeshurun and Levy, 2003).

Exogenous cues

Starting with Posner's early studies (Posner and Cohen, 1984; Posner et al., 1985), research on allocating visual attention has demonstrated that exogenous, bottom-up cues can produce covert shifts off attention. In general, responses at short cue-to-target intervals are facilitated (Lambert and Hockey, 1991; McAuliffe and Pratt, 2005; Pratt et al., 2001) while responses at longer intervals (e.g., greater than 300 ms) are inhibited (e.g., see Klein, 2000 for a review). Whether cues are exogenous or endogenous, in covert cuing experiments using manual responses observers direct their eyes toward a fixation stimulus and refrain from moving them while cues and targets appear at different locations over time. The remaining experiments to be discussed used exogenous cues to draw attention to them before a target appeared. The first series of experiments used an inhibition of return (IOR) paradigm with long cue-to-target intervals. The last series used an object-based (OB) versus space-based (SB) attention paradigm and a short cue-to-target interval. With both paradigms M/dorsal and P/ventral-biased stimulus conditions were used to examine how visual stream activity affected shifting attention.

IOR as an indicator of shifting attention

Research on IOR has used many different methods and measures and there is a vast literature attempting to elucidate the underlying mechanisms, processes, and purposes of IOR (e.g., see Berlucchi, 2006; Klein, 2000; Lupiáñez et al., 2006). The research described here was not studying the nature of IOR per se, rather this attention phenomenon was used as a measure of shifting visual attention. All the IOR experiments to be discussed (except one) used the same cues, targets, cue-to-target timing, and general procedure. Based on the literature, IOR was expected because of the long, 1450 ms cue-to-target stimulus onset asynchrony (SOA) used. Thus, IOR magnitude was used as an indicator of the ease or difficulty of deploying spatial attention.

We examined the relationship between M/dorsal and P/ventral activity and shifting attention by manipulating bottom-up and top-down perceptual stimulus variables to create M/dorsal and P/ventral-biased conditions. Our primary bottom-up stimulus variable was target spatial frequency. Compared to the vast IOR literature, our choice of different spatial frequency Gabor patches (1, 4, and 12 cpd) as cues and targets was unique. This lower-level variable allowed us to create M-biased (1 cpd) and P-biased (4, 12 cpd) conditions based on the different sensitivities of the M and P pathways to spatial frequency (Leonova et al., 2003). Cues and targets were presented either alone (our no object, baseline condition) or in the context of 2-D or 3-D objects. Thus, the presence of objects was our higher-level, top-down stimulus variable expected to produce greater involvement of P/ventral processing.

Stimuli in all conditions were well above threshold and appeared above and below fixation. Only a simple detection response to the onset of a target was required which meant all target (e.g., spatial frequency content, orientation, and contrast) and context attributes (e.g., 2-D vs. 3-D) were irrelevant to the task. Targets appeared on 80% of the trials with responses withheld on target absent (i.e., catch) trials. A refixation stimulus was used to insure attention was drawn away from the cue before a target appeared. Other than the absence of a target on catch trials, the sequence of events in each trial was the same. A black fixation plus sign appeared indicating a trial could be started with a key press. A second after initiating a trial a cue appeared for 900 ms. Between cue offset and target onset the fixation stimulus was black for 200 ms, changed to white for 150 ms, then back to black for 200 ms. The target was visible until a response was made or 1500 ms elapsed on catch trials. There was a 750 ms blank inter-trial interval.

Spatial frequencies were tested in pairs (1+12 cpd, 1+4 cpd, 4+12 cpd) in all experiments using different groups of participants. Cue and target spatial frequency was equally likely to be the same or different from trial to trial so cue frequency was not predictive of target frequency. For example, for the 1+12 cpd pair, on trials when 1 cpd was the target, half the time the cue was 1 cpd and half the time it was 12 cpd. The same was true when 12 cpd was the target. The IOR results (in Figs. 4, 6–8) are presented as a function of target spatial frequency collapsed across cue spatial frequency.

To reduce possible influences due to proposed specializations of the left and right visual fields related to both the perceptual processing of the spatial frequency content of stimuli and attention (Christman and Niebauer, 1997; Goodale and Milner, 1992; Ivry and Robertson, 1998; Kosslyn et al., 1994; Roth and Hellige, 1998), we chose to present stimuli above and below fixation. While differences in upper versus lower visual field processing related to visual perception (Cameron et al., 2002; Carrasco et al., 2001) and attention (Carrasco et al., 2004; He et al., 1996) have been reported, the contributions of the P/ventral and M/dorsal streams to these effects is unknown. However, both Previc (1990, 1998) and Milner and Goodale (1995, 2007) have proposed upper/lower visual field biases related to the P/ventral and M/dorsal streams. Previc (1990) proposes relative biases toward the P/ventral stream in the upper visual field and the M/dorsal stream in the lower visual field as a functional difference associated with visual perception and action in near (peripersonal) and far (extrapersonal) space respectively. Milner and Goodale's (1995, 2007) distinction between vision for perception (P/ventral) versus vision for action (M/dorsal) also suggests an M/dorsal functional bias in the lower visual field that may be primarily related to visuomotor control (Danckert and Goodale, 2001). Although processing in near and far space and visuomotor control might seem unimportant to our IOR task, it is possible these visual field biases associated with the visual streams might lead to visual field differences in IOR under P/ventral and M/dorsal-biased stimulus conditions.

The first experiment was our no object, baseline condition where, other than the fixation stimulus, cues and targets appeared in a blank field (Brown and Guenther, 2009; Guenther and Brown, 2007). From the attended-P/ventral and unattended-M/dorsal perspective, we hypothesized IOR would be more associated with P/ventral processing because it is an attention phenomenon and thus IOR magnitude should be greater for P-biased, higher spatial frequency (4 and 12 cpd) targets. The faster M response to the abrupt onsets of the cue, refixation stimulus, and target should facilitate localizing where spatial attention is covertly deployed over time and thus, less IOR was predicted for the M-biased low spatial frequency (1 cpd) target. Overall IOR was greater for the higher spatial frequencies (4 and 12 cpd) when paired with 1 cpd, but there were no spatial frequency differences for the 4+12 cpd pair (see Fig. 2). The interaction of spatial frequency with visual field for the 1+12 cpd and 1+4 cpd pairs revealed the most surprising finding. Not only was IOR significantly reduced for 1 cpd in the lower visual field for the 1+12 cpd pair (14 ms), it was absent for the 1+4 cpd pair (5 ms)! This is the first time we are aware of where IOR has not been found using an exogenous cuing paradigm and a long SOA. These results support the association of greater IOR with increased P/ventral activity

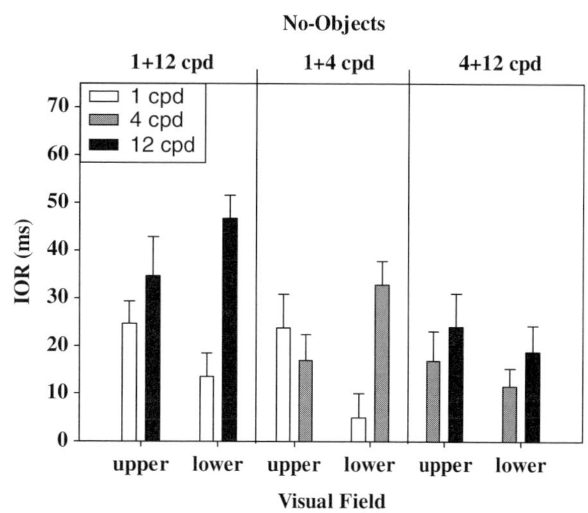

Fig. 2. IOR for no object experiments for three different target spatial frequency pairs (adapted from Brown and Guenther, 2009).

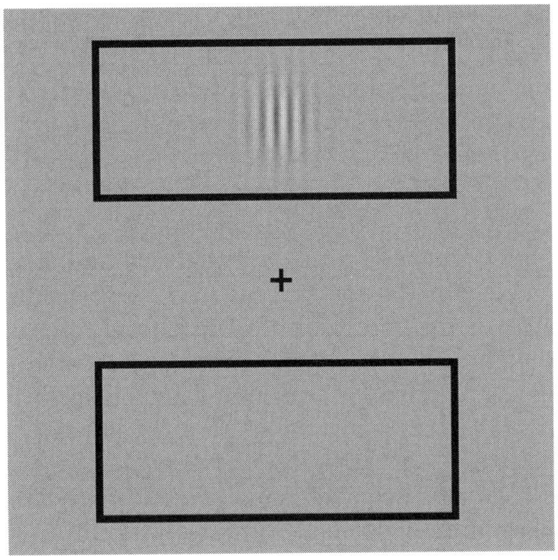

Fig. 3. Example of 2-D object display.

and less IOR with increased M/dorsal activity. The implications of these results are discussed with those of the following experiments at the end of this section.

Next we attempted to increase the P/ventral processing bias by adding 2-D and 3-D objects to the display. While increased IOR might be expected with objects compared to without them (Leek et al., 2003; McAuliffe et al., 2001), how would this higher-level perceptual variable interact with the lower-level spatial frequency differences in IOR found without objects? The same spatial frequency pairs were tested in both 2-D and 3-D conditions. In the 2-D experiment (see Fig. 3 for an example) participants ran in both no object and 2-D object conditions in a counterbalanced order. Replicating the original no object experiment, there was greater IOR for 4 and 12 cpd compared to 1 cpd without objects (see left side of Figs. 4a, b). The effect of spatial frequency for the 4+12 cpd pair was also significant (Fig. 4c) unlike the original experiment. A small but significant increase in IOR with 2-D objects was found for the 1+4 cpd (8 ms) and 4+12 cpd (11 ms) pairs only. The most obvious and important finding is the similarity in results for the no object and 2-D conditions. In particular, the visual field by target spatial frequency interaction was significant for all pairs, similar to the 1+12 cpd and

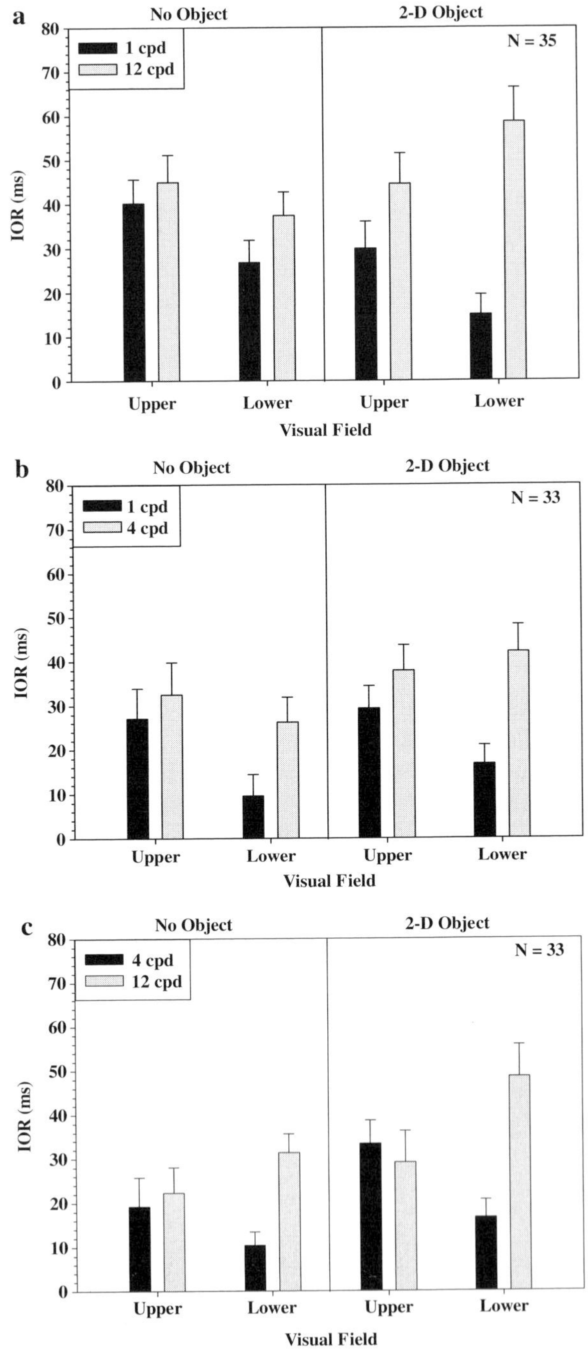

Fig. 4. IOR for no object and 2-D object conditions for target spatial frequency pairs: (a) 1+12 cpd, (b) 1+4 cpd, and (c) 4+12 cpd.

1+4 cpd pairs in the original experiment. This interaction was due to greater IOR to the higher compared to the lower spatial frequency in the lower visual field (see Figs. 2 and 4a–c). The similar trends for no object and 2-D conditions means the 2-D objects had a minimal influence on the results. These visual field influences are discussed further later in comparison with the results of the 3-D experiments covered next.

The 3-D experiments included five variations (Brown and Guenther, 2009; Guenther et al., 2009). In the first three, cues and targets appeared on the front face of 3-D cubes (see Fig. 5) where the luminance of the front face was the same as the background in the no object and 2-D experiments. For the cubes to stand out the background behind them had to be of a different luminance, so as a control for background luminance, the cubes were set against a lighter and a darker background in separate experiments. The results were identical so only the light background results are presented in Fig. 6. The 3-D objects changed the pattern of IOR in three important ways: (1) Overall IOR magnitude increased. (2) The pattern of IOR as a function of target spatial frequency changed. (3) The spatial frequency differences in IOR in the lower visual field noted earlier in the no object and 2-D conditions were eliminated. The increased magnitude and changes in the patterns of IOR are attributed to an increased P/ventral response due to the 3-D objects. The efficiency of allocating attention was hindered by the interaction of this

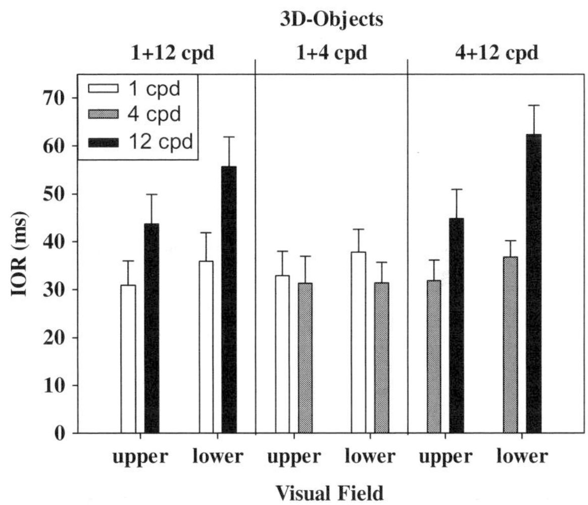

Fig. 6. IOR for 3-D object condition for three different target spatial frequency pairs (adapted from Brown and Guenther, 2009).

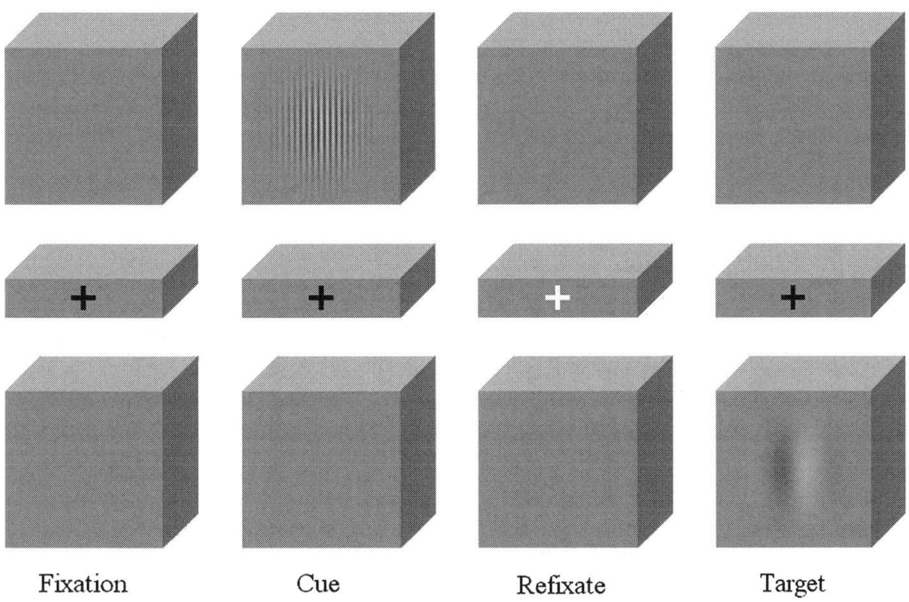

Fig. 5. Example of 3-D object condition (light background) (adapted from Brown and Guenther, 2009).

higher-level variable with the lower-level variable of spatial frequency. Shifting attention back to cued objects was slowed causing an increase in IOR magnitude.

The last two 3-D experiment variations tested alternative interpretations that were consistent with an increase in P/ventral activity but were not related to the perceived three-dimensionality of the objects. One alternative account was that the edges of the 3-D objects introduced many high spatial frequencies into the display which could have produced increased P/ventral activity. While a similar argument might be made for the 2-D objects, the 3-D objects had oblique orientations, different luminance regions, and most importantly, produced different results. A direct test was made by blurring the 3-D objects so they still appeared 3-D, but spatial frequencies above 3 cpd were removed. As Fig. 7 shows, the results were unchanged ruling out spurious high spatial frequencies as the cause of increased P/ventral activity.

The fifth experiment investigated the possibility that the results with 3-D objects were due to cues and targets being perceived and processed as texture on the objects. If so, this could have caused increased P/ventral activity given the important role the P/ventral stream plays in texture processing (Livingstone and Hubel, 1987, 1988). To test this idea, the 3-D objects were positioned to the left side of the display while cues and targets appeared in blank space above and below fixation, just like the no object condition. With the cues and targets spatially separated from the objects the results should replicate the results of the no object condition if the texture account is correct. If however, the perception of the 3-D objects is what is causing increased P/ventral activity, then the results should replicate the previous 3-D conditions. As Fig. 8 shows, the results replicated the previous 3-D conditions ruling out the texture account.

In a final IOR experiment, more traditional cues and targets and a more traditional placeholder paradigm was used. The cue was a small (0.4°) white square 600 ms in duration. The target was a slightly larger (0.6°) white square and the cue-to-target SOA was 800 ms. The primary sensory manipulation was the temporal nature of the stimuli with cue onset/offset and target onset either abrupt or ramped (Guenther, 2008; Guenther and Brown, 2009). The abrupt condition was our M-biased condition because it should create a stronger M response compared to the ramped condition due to the transient nature of the stimuli (Breitmeyer and Julesz, 1975; Breitmeyer, 1984; Tolhurst, 1975). By default the

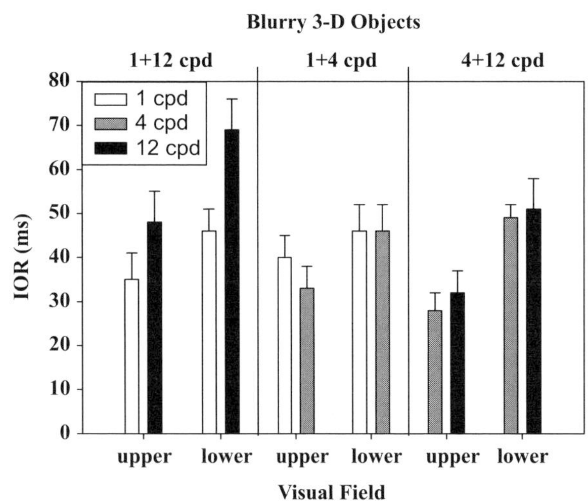

Fig. 7. IOR for 3-D blurry object condition for three different target spatial frequency pairs.

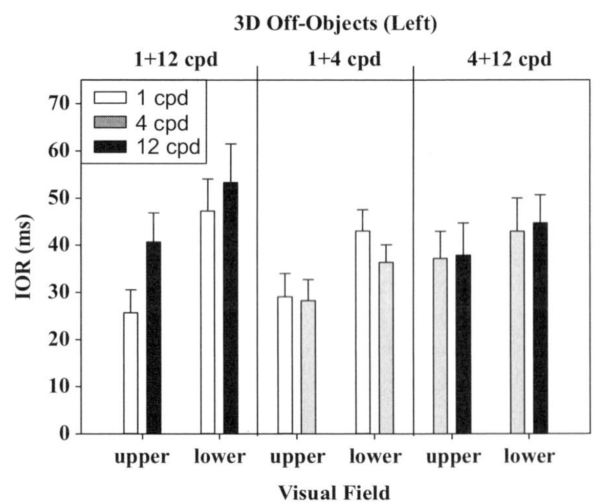

Fig. 8. IOR for 3-D off-object condition where objects appeared on the left side of display and cues and targets appeared above and below fixation in the center of the display.

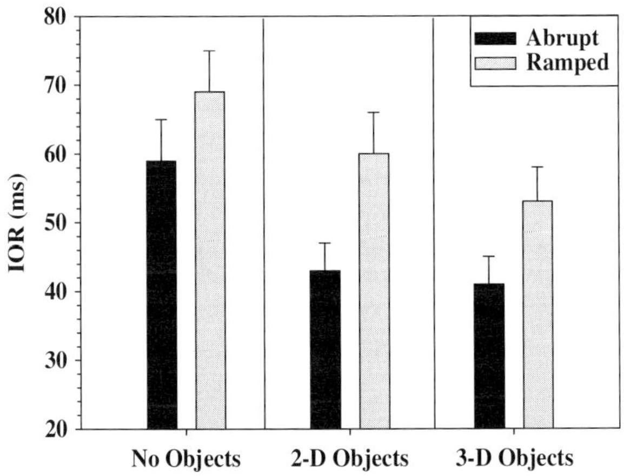

Fig. 9. IOR from abrupt versus ramped onset experiment using traditional cues and targets under no object, 2-D, and 3-D object conditions (adapted from Guenther and Brown, 2009).

ramped condition was our P/ventral-biased condition because in actuality these conditions might be better described as strong versus weak M-biased. In the ramped condition cue and target luminance increased to peak over the first 100 ms and cue luminance decreased to background luminance over the last 100 ms. Based on our previous results, the P/ventral-biased ramped condition was predicted to produce greater IOR than the abrupt. Again objects (2-D and 3-D) were used as a higher-level perceptual variable to bias processing toward the P/ventral stream and compared to a no object condition. If P/ventral activity increased with objects then changes in the magnitude and/or pattern of IOR should occur compared to the no object condition. Overall the ramping manipulation did produce a significant increase in IOR, but it did not produce greater IOR when combined with the 2-D and 3-D contexts (see Fig. 9). The lack of an effect of 2-D objects is consistent with our previous study with spatial frequency targets. The lack of an influence of 3-D objects suggests our previous 3-D context effects may be somehow related to the spatial frequency specific targets used. Despite the lack of context effects, the results do provide convergent evidence of greater P/ventral activity being associated with greater IOR. It also supports the tactic of varying lower-level sensory and higher-level perceptual variables to probe the interaction of attention and P/ventral and M/dorsal stream activity.

The purpose of these IOR studies was to explore the relationship between the M/dorsal and P/ventral visual streams and shifting attention. Our approach uses both bottom-up and top-down stimulus variables to create M/dorsal and P/ventral-biased conditions and measures IOR magnitude as an indicator of the ease or difficulty of shifting attention. The overall results indicate shifting attention from one location to another is more difficult with increased P/ventral activity, at least within this paradigm (including the abrupt/ramped experiment). In the no object conditions spatial frequency differences was the only stimulus variable to bias processing toward M/dorsal versus P/ventral streams. The results showed a consistent pattern of IOR with smaller spatial frequency differences in the upper visual field and larger differences in the lower visual field (except for 4+12 cpd in the original experiment). Why would there be less IOR for the lower versus higher spatial frequency in the lower, but not the upper visual field? Why did 2-D objects have no effect on this pattern of IOR, while 3-D objects eliminated it?

Any theoretical account will need to consider upper versus lower visual field differences in sensory, perceptual, and attentional processing. While there is evidence of upper and lower visual field differences in visual (Cameron et al., 2002; Carrasco et al., 2001) and attentional (Carrasco et al., 2004; He et al., 1996) processing, it is not clear the exact roles the P/ventral and M/dorsal streams play in these differences. Previc's (1990) proposal that the P/ventral and M/dorsal streams play greater roles in upper and lower visual field processing respectively combined with Milner and Goodale's (Danckert and Goodale, 2001; Milner, 1995; Milner and Goodale, 2007) proposed bias toward the M/dorsal stream in the lower visual field may provide a framework for understanding the visual field differences in IOR found. From his perspective visual field differences are related to the task demands associated with perceptual processing in near (i.e., lower visual field) and far space (i.e., upper visual field) emphasizing the "dorsal and ventral pathways differ more in their processing strategies in different regions of visual space than in the particular types of information they process" (p. 521). Thus, both stimulus and task variables influencing M/dorsal and P/ventral activity could lead to upper versus lower visual field differences in visual and attentional processing.

In relating Previc's (1990) perspective to the IOR paradigm and spatial frequency specific targets used here, the no object condition would primarily be a SB attention task. An important function of SB attention is orienting to new or threatening events in our environment, while the mechanisms underlying IOR increase the efficiency of allocating attention by inhibiting us from returning to recently attended locations (Klein, 2000; Klein and MacInnes, 1999). How might processing strategy differences associated with the spatial frequency content of stimuli interact with allocating SB attention in near and far space? From a survival perspective, visual attention may have evolved to process events occurring in near space with a higher priority, a greater urgency, compared to those in far space. For example, if a predator moves or emerges from some tall grass nearby, it would be important for survival to be able to quickly reorient attention there even if we had just looked at or attended to that position. Such events would be most associated with lower spatial frequency information, making it advantageous not to be inhibited to respond to lower spatial frequency events in near space. By comparison, the minimal or reduced threat posed by lower spatial frequency information in far space and by higher spatial frequency information in both near and far space would make it efficient to inhibit returning to it. This perspective provides an account of the pattern of IOR found to the different spatial frequencies in the upper and lower visual fields in the no object condition. Little or no IOR was found to be 1 cpd in the lower visual field while substantial IOR was found to be 4 and 12 cpd in the lower visual field. In the upper visual field there was little difference in IOR between 1, 4, and 12 cpd. This same account is consistent with the results of the 2-D object conditions. The question is why did 3-D objects change the results the way they did?

We propose that the changes in IOR with 3-D objects are due to increased P/ventral and OB attention involvement due to the presence of the objects (Brown and Guenther, 2009; Guenther and Brown, 2007; Guenther et al., 2009). With P/ventral activity playing an important role in object processing it can also be expected to play an important role in OB attention. While P/ventral activity can be influenced by lower-level stimulus variables like spatial frequency and their temporal characteristics, OB attention may require more object-like stimuli for it to have an influence. There are clearly some discrepancies between our use of the term object and the effects we have found. For example, stimuli like our 2-D objects have been used as objects in attention research (see next section below). However, in our IOR experiments using non-traditional spatial frequency targets, 2-D and 3-D objects had different effects suggesting interactions of the perceived three-dimensionality of the objects and the bottom-up stimulus variable of spatial frequency. With these caveats in mind how might we account for the 3-D object effects?

First, the effect of 3-D objects on overall IOR magnitude is consistent with our hypothesis of greater IOR with increased P/ventral activity. Second, by combining Previc's (1990) idea of P/ventral and M/dorsal processing strategy differences in near and far space with proposed contributions of P/ventral and M/dorsal streams to spatial attention (Vidyasagar, 1999) and object recognition (Bar, 2003; Kveraga et al., 2007) we may provide a reasonable account for the changes in the pattern of IOR as a function of spatial frequency and visual field with 3-D objects.

Bar and colleagues propose that global object shape is rapidly transmitted via low spatial frequencies by the M/dorsal stream to prefrontal cortex. From there, initial candidate object activity is fed back to the temporal lobe where it is integrated with incoming detailed information via the P/ventral stream to facilitate recognition (Bar, 2003, 2004; Bar et al., 2006; Kveraga et al., 2007). Although recognition was not required in our experiments, the constant presence of the 3-D objects might be expected to create such a cycle of low spatial frequency, M/dorsal mediated activity associated with them being fed back to the P/ventral stream. We propose this fast M/dorsal mediated activity is also used by the P/ventral stream to quickly "tag" objects, marking their presence and position in the field of view (for related ideas see Fazl et al., 2009; Watson and Humphreys, 1997).[1] Tagging results from an interaction of M/dorsal and P/ventral processing and plays an important role in allocating and shifting attention to objects. It might also be speculated to play a role in allocating OB attention to objects in the field of view.

Now consider how tagging objects might increase IOR to higher spatial frequencies like 4 and 12 cpd as was found. Once tagged, returning attention to objects to obtain further high spatial frequency information would normally be superfluous. So, the increased IOR to high spatial frequencies with objects we found is a reflection of increased processing efficiency. As noted above, the minimal amount of IOR to 1 cpd in the lower visual field without objects can be attributed to the strong survival value associated with being able to quickly return attention to a previously inspected location when a low spatial frequency event occurs in near space. However, in line with Previc's notion of differences in processing strategies, once tagged, the urgency to respond to a low spatial frequency stimulus appearing on or in an object in near space is eliminated so IOR increased to 1 cpd in the lower visual field to levels comparable to the higher frequencies. Thus tagging is an outcome of an interaction of M/dorsal and P/ventral processing allowing more efficient deployment of attention.

There is at least one potential problem with this account because of the results from the 3-D experiment where the objects were set off to the left of the display. Targets appeared in blank space in that experiment just like the no object conditions, yet the results replicated the other 3-D conditions. This was interpreted as evidence of the 3-D objects increasing overall P/ventral activity and, therefore, overall IOR. This interpretation is still viable. However, if we use a strict interpretation of the proposed tagging function such that it specifically involves objects themselves, then there should have been no influence on IOR in nearby space, but one was found. Further tests of the tagging account of the 3-D effects are currently underway.

How are SB and OB attention related to the M/dorsal and P/ventral streams?

While every object simultaneously occupies a location in space (or sequence of locations if moving), psychophysical (Leek et al., 2003; Tipper et al., 1994), functional imaging (Corbetta et al., 2000; Muller and Kleinschmidt, 2003; Serences et al., 2004), and neurological patient research (Corbetta et al., 2000; Egly et al., 1994a, b; Muller and Kleinschmidt, 2003; Serences et al., 2004) indicates two dynamically interactive attention systems associated with SB and OB processing. OB attention is assumed to involve greater P/ventral stream activity because of its relatively greater role in object processing (Haxby et al., 1991; Ungerleider and Haxby, 1994; Ungerleider and Mishkin, 1982), while SB attention is assumed to involve greater M/dorsal stream activity because of its relatively greater role in spatial processing (Haxby et al., 1991; Ungerleider and Haxby, 1994; Ungerleider and Mishkin, 1982). This distinction is consistent with research on P/ventral and M/dorsal activity and attention (e.g., Cheng et al., 2004; Di Russo and Spinelli, 1999; Marois et al., 2000; Yeshurun, 2004) as well as

[1]Two points should be noted about our tagging function and Fazl et al.'s (2009) attentional shroud. First, their shroud is mediated via P/ventral pathway activity while our tag is proposed to be mediated via the M/dorsal pathway. This might suggest their model fits our results better from the perspective of increased P/ventral activity leading to increased IOR, but it is not clear at this time how their proposal would account for the visual field differences. Second, although in some instances they used 2-D like objects to illustrate the build up and decay of attentional shrouds, their model's emphasis (related to it being P/ventral mediated) is how attention operates on surfaces. Despite the previous sentence, this may be why we found such striking differences in IOR between our 2-D and 3-D objects because the 3-D objects did have surfaces in different orientations, whereas the 2-D objects could be considered a 2-D frame with a hole in the middle (i.e., not a surface). Future studies are clearly needed.

research indicating M/dorsal stream activity guides or facilitates spatial attention (Vidyasagar, 1999), visual processing (Bullier, 2001), and object recognition (Kveraga et al., 2007).

In a recent study we assessed the contribution of M/dorsal activity to SB and OB attention by using an exogenous cuing paradigm with equiluminant and non-equiluminant stimuli (Boyd et al., 2007; Brown et al., 2009). M/dorsal pathway activity was expected to be reduced with equiluminant stimuli because of its poor sensitivity to wavelength (Livingstone and Hubel, 1987). While the perception of figure-ground (Koffka, 1935; Livingstone and Hubel, 1987), depth (Brown and Koch, 2000; Livingstone and Hubel, 1987), motion (Cavanagh et al., 1987), illusory contours (Brigner and Gallagher, 1974; Brussell et al., 1977; Frisby and Clatworthy, 1975; Gregory, 1977), and visual phantoms (Brown, 2000) are impaired at equiluminance, this was the first study to examine how reduced M/dorsal activity would influence the allocation of SB and OB attention.

Pairs of vertically and horizontally oriented rectangles were used as objects. Eight different object/background color combinations were tested in separate experiments. White/black, white/gray, white/red, and white/green were non-equiluminant conditions. Green/red and red/green combinations were tested twice, once set physically equiluminant and again when set psychophysically equiluminant using a minimal flicker technique (for details see Brown et al., 2009). In each trial a brief cue drew

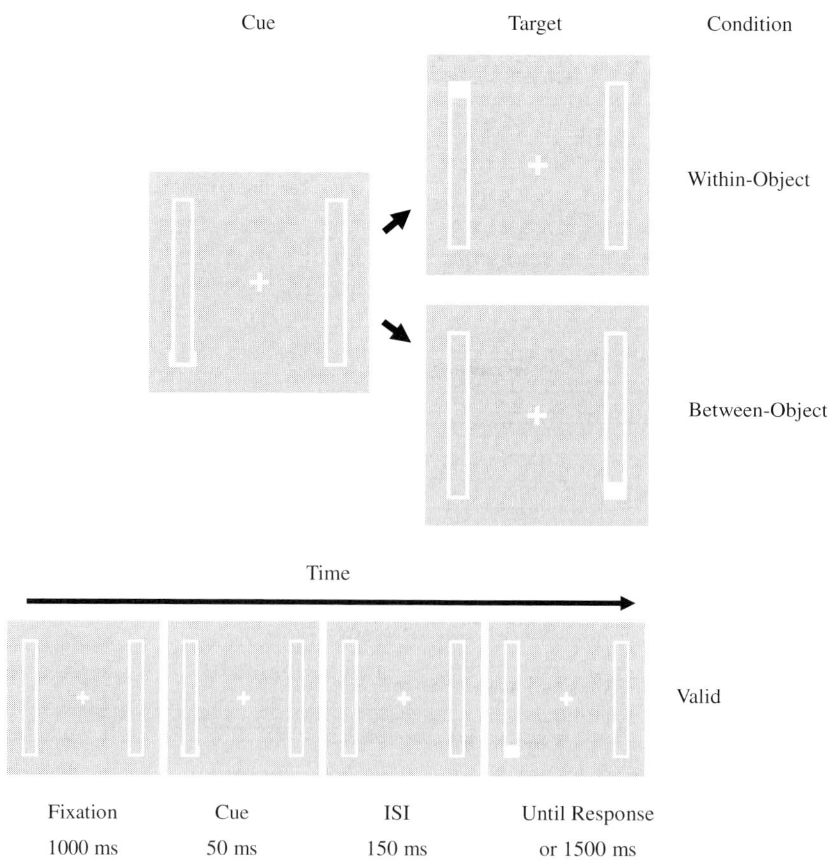

Fig. 10. An example illustrating cuing conditions and timing parameters for object- and space-based attention experiments (adapted from Brown et al., 2009).

attention to the end of one of the objects. A target appeared following the cue on 80% of the trials. On 75% of the target-present trials the cue appeared at the cued location. On the remaining target-present trials the target appeared equally often at the other end of the cued object (invalid within-object) or at the adjacent location in the nearby object (invalid between-object) (see Fig. 10). RTs from validly

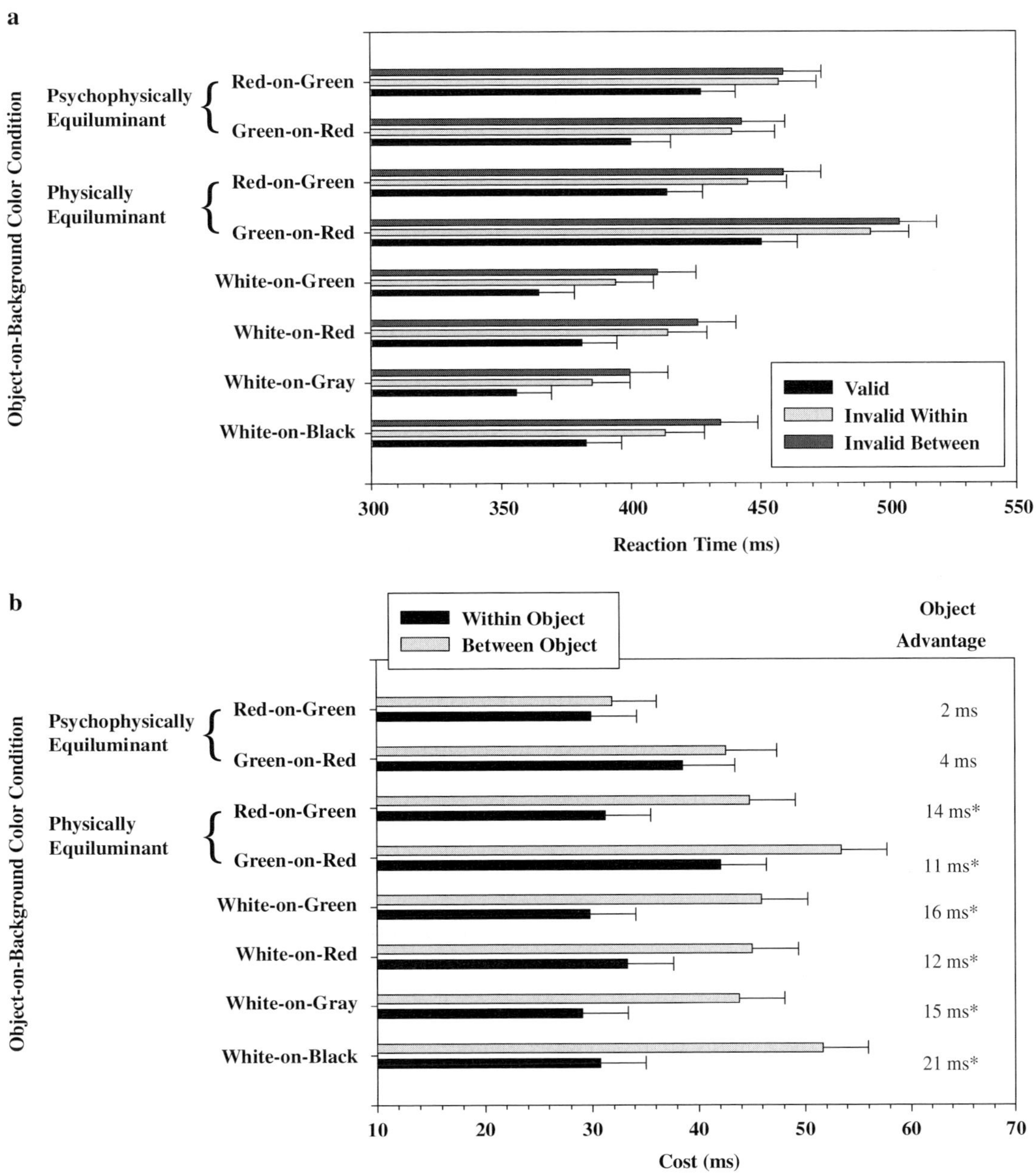

Fig. 11. (a) Reaction times and (b) costs for various object-on-background color conditions tested in separate experiments (* = significant object advantage) (adapted from Brown et al., 2009).

cued trials were subtracted as a baseline from invalid within- and between-object RTs to calculate a cost for shifting attention within versus between objects respectively. The cost for within-object shifts is nearly universally found to be less than for between-object shifts. This is commonly referred to as an object advantage because, even though the spatial distance between cue and target is identical for both conditions, within-object shifts are faster.

The perspective of the present study was to consider the attention operations involved with shifting both OB and SB attention, including disengaging from the cue, shifting, and engaging the target (Posner, 1980). Brown and Denney's (2007) recent evidence that the object advantage may be due to disengaging OB attention to shift to another object is of particular relevance. From their perspective, within- and between-object shifts would involve shift and engage operations for both OB and SB attention. Thus, the disengage operation is the primary operation differentiating within- versus between-object shifts. SB attention must always disengage whether shifting within or between objects. OB attention need only disengage during between-object shifts because during a within-object shift it remains within the cued object. Equiluminance (both physical and psychophysical) had its expected sensory effect creating longer RTs for equiluminant compared to non-equiluminant conditions (Breitmeyer, 1984; Burr and Corsale, 2001) (see Fig. 11a). Despite this sensory influence on RTs, all conditions showed a validity effect with RTs shorter on valid compared to invalid trials. The key question was how equiluminance would influence costs for within- and between-object shifts (i.e., the object advantage).

All non-equiluminant conditions plus the two physically equiluminant conditions produced a significant object advantage. Thus, while physically and psychophysically equiluminant conditions produced a similar sensory effect, they produced different effects on attention. From this result we can also infer that longer RTs do not automatically mean less of an object advantage. Of most importance, for the first time ever that we are aware of, the two psychophysically equiluminant conditions eliminated the object advantage (see Fig. 11b). At present, other theoretical perspectives on the object advantage including spreading-attention (Abrams and Law, 2000; Avrahami, 1999; Brown et al., 2006), biased-competition (Vecera, 1994, 2000; Vecera and Behrmann, 2001; Vecera and Flevaris, 2005), and prioritization (Shomstein and Yantis, 2002, 2004) cannot account for equiluminance eliminating the object advantage. Our account can, by combining SB and OB attention engage, disengage, and shift operations with M/dorsal and P/ventral activity being relatively more involved with SB and OB attention respectively. Remember, longer between-object RTs are due to OB attention disengaging from the cued object to shift to the target (Brown and Denney, 2007). Logically then, OB attention disengaging must create the object advantage because SB attention is disengaging during both between- and within-object shifts. The OB attention disengage operation and thus between-object shifts, should not be influenced much at equiluminance because of the P/ventral streams greater sensitivity to wavelength information. The M/dorsal mediated SB attention disengage operation should be influenced the most at equiluminance which means within-object shifts should be interfered with the most. By interfering with within-object shifts the costs for within- and between-object shifts became more similar which eliminated the object advantage.

Conclusions

Faced with the enormous challenges posed by trying to understand and explain visual perception and attention, researchers have used many different tasks and approaches. The approach described here is attempting to unravel the complex interplay between the M/dorsal and P/ventral streams and shifting attention. Using standard attention tasks and stimulus conditions biased toward processing by one stream or the other, we are beginning to make in-roads into understanding their contributions.

References

Abrams, R. A., & Law, M. B. (2000). Object-based visual attention with endogenous orienting. *Perception & Psychophysics, 62*, 818–833.

Avrahami, J. (1999). Objects of attention, objects of perception. *Perception & Psychophysics, 61*, 1604–1612.

Bar, M. (2003). A cortical mechanism for triggering top-down facilitation in visual object recognition. *Journal of Cognitive Neuroscience, 15*, 600–609.

Bar, M. (2004). Visual objects in context. *Nature Reviews Neuroscience, 5*, 617–629.

Bar, M., Kassam, K. S., Ghuman, A. S., Boshyan, J., Schmidt, A. M., Dale, A. M., et al. (2006). Top-down facilitation of visual recognition. *Proceedings of the National Academy of Sciences, 103*, 449–454.

Berlucchi, G. (2006). Inhibition of return: A phenomenon in search of a mechanism and a better name. *Cognitive Neuropsychology, 23*, 1065–1074.

Boyd, M. C., Guenther, B. A., & Brown, J. M. (2007). Investigating the role of the magnocellular pathway in object- and location-based attention. *Journal of Vision, 7*, 1078.

Breitmeyer, B., & Julesz, B. (1975). The role of on and off transients in determining the psychophysical spatial frequency response. *Vision Research, 15*, 411–415.

Breitmeyer, B. G. (1984). *Visual masking: An integrative approach*. Oxford: Clarendon.

Brigner, W. L., & Gallagher, M. B. (1974). Subjective contour: Apparent depth or simultaneous brightness contrast? *Perceptual and Motor Skills, 38*, 1047–1053.

Brown, J. M. (2000). Fundus pigmentation and equiluminant moving phantoms. *Perceptual and Motor Skills, 90*, 963–973.

Brown, J. M., Breitmeyer, B. G., Leighty, K. A., & Denney, H. I. (2006). The path of visual attention. *Acta Psychologica, 121*, 199–209.

Brown, J. M., & Denney, H. I. (2007). Shifting attention into and out of objects: Evaluating the processes underlying the object advantage. *Perception & Psychophysics, 69*, 606–618.

Brown, J. M., & Guenther, B. A. (2009). Magnocellular and parvocellular pathway influences on location-based inhibition-of-return. *Attention, Perception, & Psychophysics* (submitted).

Brown, J. M., Guenther, B. A., Narang, S., & Siddiqui, A. (2009). Eliminating an object advantage. *Attention, Perception, & Psychophysics* (submitted).

Brown, J. M., & Koch, C. (2000). Influences of occlusion, color, and luminance on the perception of fragmented pictures. *Perceptual and Motor Skills, 90*, 1033–1044.

Brown, J. M., & Weisstein, N. (1988a). A phantom context effect: Visual phantoms enhance target visibility. *Perception & Psychophysics, 43*, 53–56.

Brown, J. M., & Weisstein, N. (1988b). A spatial frequency effect on perceived depth. *Perception & Psychophysics, 44*, 157–166.

Brussell, E. M., Stober, S. R., & Bodinger, D. M. (1977). Sensory information and subjective contour. *The American Journal of Psychology, 90*, 145–156.

Bullier, J. (2001). Integrated model of visual processing. *Brain Research Reviews, 36*, 96–107.

Bullier, J. (2006). What is fed back? In J. L. van Hemmen & T. J. Sejnowski (Eds.), *23 problems in systems neuroscience* (pp. 103–132). New York: Oxford University Press.

Burr, D. C., & Corsale, B. (2001). Dependency of reaction times to motion onset on luminance and chromatic contrast. *Vision Research, 41*, 1039–1048.

Cameron, E. L., Tai, J. C., & Carrasco, M. (2002). Covert attention affects the psychometric function of contrast sensitivity. *Vision Research, 42*, 949–967.

Carrasco, M., Giordano, A. M., & McElree, B. (2004). Temporal performance fields: Visual and attentional factors. *Vision Research, 44*, 1351–1365.

Carrasco, M., Talgar, C. P., & Cameron, E. L. (2001). Characterizing visual performance fields: Effects of transient covert attention, spatial frequency, eccentricity, task and set size. *Spatial Vision, 15*, 61–75.

Cavanagh, P., MacLeod, D. I. A., & Anstis, S. M. (1987). Equiluminance: Spatial and temporal factors and the contribution of blue-sensitive cones. *Journal of the Optical Society of America A, 4*, 1428–1438.

Cheng, A., Eysel, U. T., & Vidyasagar, T. R. (2004). The role of the magnocellular pathway in serial deployment of visual attention. *European Journal of Neuroscience, 20*, 2188–2192.

Christman, S. D., & Niebauer, C. L. (1997). The relation between left-right and upper-lower visual field asymmetries. In: S. D. Christman (Ed.), *Cerebral asymmetries in sensory and perceptual processing* (pp. 263–296). Amsterdam: Elsevier.

Corbetta, M., Kincade, J. M., Ollinger, J. M., McAvoy, M. P., & Shulman, G. L. (2000). Voluntary orienting is dissociated from target detection in human posterior parietal cortex. *Nature Neuroscience, 3*, 292–297.

Danckert, J., & Goodale, M. A. (2001). Superior performance for visually guided pointing in the lower visual field. *Experimental Brain Research, 137*, 303–308.

Di Russo, F., & Spinelli, D. (1999). Spatial attention has different effects on the magno- and parvocellular pathways. *NeuroReport, 10*, 2755.

Egly, R., Driver, J., & Rafal, R. D. (1994a). Shifting visual attention between objects and locations: Evidence from normal and parietal lesion subjects. *Journal of Experimental Psychology: General, 123*, 161–177.

Egly, R., Rafal, R., Driver, J., & Starrveveld, Y. (1994b). Covert orienting in the split brain reveals hemispheric specialization for object-based attention. *Psychological Science, 5*, 380–382.

Fazl, A., Grossberg, S., & Mingolla, E. (2009). View-invariant object category learning, recognition, and search: How spatial and object attention are coordinated using surface-based attentional shrouds. *Cognitive Psychology, 58*, 1–48.

Frisby, J. P., & Clatworthy, J. L. (1975). Illusory contours: Curious cases of simultaneous brightness contrast. *Perception, 4*, 349–357.

Goodale, M. A., & Milner, A. D. (1992). Separate visual pathways for perception and action. *Trends in Neurosciences*, *15*, 20–25.

Gregory, R. L. (1977). Vision with isoluminant colour contrast: 1. A projection technique and observations. *Perception*, *6*, 113–119.

Guenther, B. A. (2008). Influences of abrupt vs. ramped stimulus presentation on location-based inhibition of return. Unpublished master's thesis, University of Georgia, Athens, Georgia, USA.

Guenther, B. A., & Brown, J. M. (2007). Exploring parvocellular and magnocellular pathway contributions to location-based inhibition of return. *Journal of Vision*, *7*, 541.

Guenther, B. A., & Brown, J. M. (2009). Influences of abrupt vs. ramped stimulus presentation on location-based inhibition of return. *Attention, Perception, & Psychophysics* (submitted).

Guenther, B. A., Narang, S., Siddiqui, A., & Brown, J. M. (2009). Exploring the causes of object effects on location based inhibition of return when using spatial frequency specific cues and targets. Naples, FL: Vision Sciences Society.

Haxby, J. V., Grady, C. L., Horwitz, B., Ungerleider, L. G., Mishkin, M., Carson, R. E., et al. (1991). Dissociation of object and spatial visual processing pathways in human extrastriate cortex. *Proceedings of the National Academy of Sciences*, *88*, 1621–1625.

He, S., Cavanagh, P., & Intriligator, J. (1996). Attentional resolution and the locus of visual awareness. *Nature*, *383*, 334–337.

Ivry, R. B., & Robertson, L. C. (1998). *The two sides to perception*. Cambridge, MA: The MIT Press.

Klein, R. M. (2000). Inhibition of return. *Trends in Cognitive Sciences*, *4*, 138–146.

Klein, R. M., & MacInnes, W. J. (1999). Inhibition of return is a foraging facilitator in visual search. *Psychological Science*, *10*, 346–352.

Klymenko, V., & Weisstein, N. (1986). Spatial frequency differences can determine figure-ground organization. *Journal of Experimental Psychology Human Perception and Performance*, *12*, 324–330.

Klymenko, V., Weisstein, N., Topolski, R., & Hsieh, C. H. (1989). Spatial and temporal frequency in figure-ground organization. *Perception & Psychophysics*, *45*, 395–403.

Koffka, K. (1935). *Principles of Gestalt psychology*. New York: Harcourt, Brace and World.

Kosslyn, S. M., Anderson, A. K., Hillger, L. A., & Hamilton, S. E. (1994). Hemispheric differences in sizes of receptive fields or attentional biases? *Neuropsychology*, *8*, 139–147.

Kveraga, K., Boshyan, J., & Bar, M. (2007). Magnocellular projections as the trigger of top-down facilitation in recognition. *Journal of Neuroscience*, *27*, 13232.

Lambert, A., & Hockey, R. (1991). Peripheral visual changes and spatial attention. *Acta Psychologica*, *76*, 149–163.

Leek, E. C., Reppa, I., & Tipper, S. P. (2003). Inhibition of return for objects and locations in static displays. *Perception & Psychophysics*, *65*, 388–395.

Leonova, A., Pokorny, J., & Smith, V. C. (2003). Spatial frequency processing in inferred PC- and MC-pathways. *Vision Research*, *43*, 2133–2139.

Livingstone, M. S., & Hubel, D. H. (1987). Psychophysical evidence for separate channels for the perception of form, color, movement, and depth. *Journal of Neuroscience*, *7*, 3416–3468.

Livingstone, M. S., & Hubel, D. H. (1988). Segregation of form, color, movement, and depth: Anatomy, physiology, and perception. *Science*, *240*, 740–749.

Lupiáñez, J., Klein, R. M., & Bartolomeo, P. (2006). Inhibition of return: Twenty years after. *Cognitive Neuropsychology*, *23*, 1003–1014.

Marois, R., Leung, H. C., & Gore, J. C. (2000). A stimulus-driven approach to object identity and location processing in the human brain. *Neuron*, *25*, 717–728.

McAuliffe, J., & Pratt, J. (2005). The role of temporal and spatial factors in the covert orienting of visual attention tasks. *Psychological Research*, *69*, 285–291.

McAuliffe, J., Pratt, J., & O'Donnell, C. (2001). Examining location-based and object-based components of inhibition of return in static displays. *Perception & Psychophysics*, *63*, 1072–1082.

Milner, A. D. (1995). Cerebral correlates of visual awareness. *Neuropsychologia*, *33*, 1117–1130.

Milner, A. D., & Goodale, M. A. (1995). *The visual brain in action*. Oxford: Oxford University Press.

Milner, A. D., & Goodale, M. A. (2007). Two visual systems re-viewed. *Neuropsychologia*, *46*, 774–785.

Muller, N. G., & Kleinschmidt, A. (2003). Dynamic interaction of object-and space-based attention in retinotopic visual areas. *Journal of Neuroscience*, *23*, 9812–9816.

Posner, M. I. (1980). Orienting of attention. *The Quarterly Journal of Experimental Psychology*, *32*, 3–25.

Posner, M. I., & Cohen, Y. (1984). Components of visual orienting. *Attention and performance X*, 531–556.

Posner, M. I., Rafal, R. D., Choate, L. S., & Vaughan, J. (1985). Inhibition of return: Neural basis and function. *Cognitive Neuropsychology*, *2*, 211–228.

Pratt, J., Hillis, J., & Gold, J. M. (2001). The effect of the physical characteristics of cues and targets on facilitation and inhibition. *Psychonomic Bulletin and Review*, *8*, 489–495.

Previc, F. H. (1990). Functional specialization in the lower and upper visual fields in humans: Its ecological origins and neurophysiological implications. *Behavioral and Brain Sciences*, *13*, 519–575.

Previc, F. H. (1998). The neuropsychology of 3-D space. *Psychological Bulletin*, *124*, 123–164.

Roth, E. C., & Hellige, J. B. (1998). Spatial processing and hemispheric asymmetry: Contributions of the transient/magnocellular visual system. *Journal of Cognitive Neuroscience*, *10*, 472–484.

Serences, J. T., Schwarzbach, J., Courtney, S. M., Golay, X., & Yantis, S. (2004). Control of object-based attention in human cortex. *Cerebral Cortex*, *14*, 1346–1357.

Shomstein, S., & Yantis, S. (2002). Object-based attention: Sensory modulation or priority setting. *Perception & Psychophysics*, *64*, 41–51.

Shomstein, S., & Yantis, S. (2004). Configural and contextual prioritization in object-based attention. *Psychonomic Bulletin and Review*, *11*, 247–253.

Srinivasan, N., & Brown, J. M. (2006). Effects of endogenous spatial attention on the detection and discrimination of spatial frequencies. *Perception*, *35*, 193–200.

Tipper, S. P., Weaver, B., Jerreat, L. M., & Burak, A. L. (1994). Object-based and environment-based inhibition of return of visual attention. *Journal of Experimental Psychology: Human Perception and Performance*, *20*, 478–499.

Tolhurst, D. J. (1975). Sustained and transient channels in human vision. *Vision Research*, *15*, 1151–1155.

Ungerleider, L. G., & Haxby, J. V. (1994). "What" and "where" in the human brain. *Current Opinion in Neurobiology*, *4*, 157–165.

Ungerleider, L. G., & Mishkin, M. (1982). Two cortical visual streams. In D. J. Ingle, M. A. Goodale, & R. J. W. Mansfield (Eds.), *Analysis of Visual Behavior* (pp. 549–586). Cambridge, MA: MIT Press.

Vecera, S. P. (1994). Grouped locations and object-based attention: Comment on Egly, Driver, and Rafal (1994). *Journal of Experimental Psychology: General*, *123*, 316–320.

Vecera, S. P. (2000). Toward a biased competition account of object-based segregation and attention. *Brain and Mind*, *1*, 353–384.

Vecera, S. P., & Behrmann, M. (2001). Attention and unit formation: A biased competition account of object-based attention. In T. F. Shipley & P. J. Kellman (Eds.), *From fragments to objects: Segregation and grouping in vision* (pp. 145–180). Amsterdam: North-Holland.

Vecera, S. P., & Flevaris, A. V. (2005). Attentional control parameters following parietal-lobe damage: Evidence from normal subjects. *Neuropsychologia*, *43*, 1189–1203.

Vidyasagar, T. R. (1999). A neuronal model of attentional spotlight: Parietal guiding the temporal. *Brain Research Brain Research Reviews*, *30*, 66–76.

Vidyasagar, T. R. (2005). Attentional gating in primary visual cortex: A physiological basis for dyslexia. *Perception*, *34*, 903–911.

Watson, D. G., & Humphreys, G. W. (1997). Visual marking: Prioritizing selection for new objects by top-down attentional inhibition of old objects. *Psychological Review*, *104*, 90–122.

Weisstein, N., & Wong, E. (1986). Figure-ground organization and the spatial and temporal responses of the visual system. In E. C. Schwab & H. C. Nusbaum (Eds.), *Pattern recognition by humans and machines: Visual perception* (pp. 31–64). Orlando: Academic Press.

Weisstein, N., & Wong, E. (1987). Figure-ground organization affects the early visual processing. In M. A. Arbib & A. R. E. Hansen (Eds.), *Vision, brain, and cooperative computation* (pp. 209–230). Cambridge, MA: MIT Press.

Wolfe, J. M. (2009). A new beginning. *Attention, Perception, & Psychophysics*, *71*, 1.

Wong, E., & Weisstein, N. (1982). A new perceptual context-superiority effect: Line segments are more visible against a figure than against a ground. *Science*, *218*, 587.

Wong, E., & Weisstein, N. (1983). Sharp targets are detected better against a figure, and blurred targets are detected better against a background. *Journal of Experimental Psychology: Human Perception and Performance*, *9*, 194–201.

Yeshurun, Y. (2004). Isoluminant stimuli and red background attenuate the effects of transient spatial attention on temporal resolution. *Vision Research*, *44*, 1375–1387.

Yeshurun, Y., & Carrasco, M. (1998). Attention improves or impairs visual performance by enhancing spatial resolution. *Nature*, *396*, 72–75.

Yeshurun, Y., & Levy, L. (2003). Transient spatial attention degrades temporal resolution. *Psychological Science*, *14*, 225.

N. Srinivasan (Ed.)
Progress in Brain Research, Vol. 176
ISSN 0079-6123

CHAPTER 5

Covert attention effects on spatial resolution

Marisa Carrasco[1,*] and Yaffa Yeshurun[2]

[1]Department of Psychology & Center for Neural Science, New York University, New York, NY, USA
[2]Department of Psychology & Institute of Information Processing and Decision Making, University of Haifa, Haifa, Israel

Abstract: First, we review the characteristics of endogenous (sustained) and exogenous (transient) spatial covert attention. Then we examine the effects of these two types of attention on spatial resolution in a variety of tasks, such as acuity, visual search, and texture segmentation. Both types of covert attention enhance resolution; directing attention to a given location allows us to better resolve the fine details of the visual scene at that location. With exogenous attention, but not with endogenous attention, this is the case even when enhanced spatial resolution hampers performance. The enhanced resolution at the attended location comes about at the expense of lower resolution at the unattended locations.

Keywords: covert attention; exogenous attention; transient attention; endogenous attention; sustained attention; texture segmentation; spatial resolution; visual search; acuity

Each time we open our eyes we are confronted with an overwhelming amount of information. Despite this fact, we have the clear impression of understanding what we see. This requires separating the wheat from the chaff, selecting relevant information out of the irrelevant noise. Attention is what turns looking into seeing, allowing us to select a certain location or aspect of the visual scene and to prioritize its processing. Such selection is necessary because the limits on our capacity to absorb visual information are severe. They may be imposed by the fact that there is a fixed amount of overall energy consumption available to the brain, and by the high-energy cost of the neuronal activity involved in cortical computation. Attention is crucial in optimizing the use of the system's limited resources, by enhancing the representation of objects appearing at the relevant locations or composed of relevant features while diminishing the representation of objects appearing at the less relevant locations, or composed of less relevant aspects of our visual environment.

The processing of sensory input is facilitated by knowledge and assumptions about the world, by the behavioral state of the organism, and by the (sudden) appearance of possibly relevant information in the environment. For example, spotting a friend in a crowd is much easier if you know two types of information: where to look and what to look for. Indeed, numerous studies have shown that directing attention to a spatial location or to distinguishing features of a target can enhance its discriminability and the neural response it evokes.

*Corresponding author.
Tel.: +1-212-998-8328; Fax: +1-212-995-4349;
E-mail: marisa.carrasco@nyu.edu

DOI: 10.1016/S0079-6123(09)17605-7

Spatial covert attention

Attention can be allocated by moving one's eyes toward a location, or by attending to an area in the periphery without actually directing one's gaze toward it. This peripheral deployment of attention, known as covert attention, aids us in monitoring the environment, and can inform subsequent eye movements. Cognitive, psychophysical, electrophysiological, and neuroimaging studies provide evidence for the existence of covert attention in both humans and nonhuman primates. Humans deploy covert attention routinely in many everyday situations, such as searching for objects, driving, crossing the street, playing sports, and dancing, as well as in social situations such as when deception about intentions is desired, in competitive activities like sports, or when moving the eyes would provide a cue to intentions that the individual wishes to conceal.

Covert attention improves perceptual performance — accuracy and speed — on many detection, discrimination, and localization tasks. Moreover, covert attention affects performance and appearance of objects in several tasks mediated by dimensions of early vision, such as contrast sensitivity (reviewed in Carrasco, 2006; Reynolds and Chelazzi, 2004), spatial resolution, and acuity.

In this chapter we review a series of psychophysical studies showing that when spatial attention is directed to a given location, performance improves in visual search, texture segmentation, and acuity tasks, which are limited by spatial resolution. For instance, when attending to a location observers can resolve information that is unresolvable without attending to that location, and can discriminate finer details than they can without directing attention to the cued location. The finding that attention improves spatial resolution has inspired neuronal models that implement the role of visual attention in object recognition (Deco and Zihl, 2001), and has been captured in computational models proposing that interactions among visual filters result in both increased gain and sharpened tuning (Lee et al., 1999).

Spatial attention: endogenous and exogenous

A growing body of behavioral evidence demonstrates that there are two covert attention systems that deal with facilitation and selection of information: "endogenous" and "exogenous". The former is a voluntary system that corresponds to our ability to willfully monitor information at a given location; the latter is an involuntary system that corresponds to an automatic orienting response to a location where sudden stimulation has occurred. Endogenous attention is also known as "sustained" attention and exogenous attention is also known as "transient" attention. These terms refer to the temporal nature of each type of attention: whereas observers seem to be able to sustain the voluntary deployment of attention to a given location for as long as needed to perform the task, the involuntary deployment of attention is transient, meaning it rises and decays quickly (Muller and Rabbitt, 1989; Nakayama and Mackeben, 1989). The different temporal characteristics and degrees of automaticity of these systems suggest that they may have evolved for different purposes and at different times — the transient, exogenous system may be phylogenetically older.

To investigate covert attention, it is necessary to keep both the task and the stimuli constant across conditions while manipulating attention. Psychophysical studies have shown that we can differentially engage endogenous and exogenous attention by using different spatial cues. In the endogenous condition, a central cue — typically an arrow at the center of the visual field — points to the most likely location of the subsequent target. In the exogenous condition, a brief peripheral cue is typically presented next to one of the target locations. A central cue directs attention in a goal- or conceptually driven fashion in about 300 ms and engages endogenous, sustained attention. Because about 200–250 ms are needed for goal-directed saccades to occur (Mayfrank et al., 1987), the stimulus onset asynchrony (SOA) for the sustained cue may allow observers to make an eye movement toward the cued location. Thus, to verify that the outcome of this manipulation is due to covert attention one has to ensure that eye movements do not take place. In our studies, we used an infrared camera to monitor the observers' eyes, ensuring that central fixation is maintained throughout each trial. A peripheral cue presented in a location near the relevant location draws attention in a stimulus-driven, automatic manner in about 100 ms and engages exogenous attention in a transient manner, even when the cue is uninformative with regard to the target location or identity.

Covert attention affects spatial resolution

The "resolution hypothesis" states that attention can enhance spatial resolution. The following sets of studies have provided evidence for this hypothesis. In these studies we have employed peripheral or central cues to manipulate either exogenous or endogenous attention in a variety of tasks, such as acuity, visual search, and texture segmentation, which are mediated by spatial resolution. Figure 1 includes an example of experimental trials with central or peripheral cues, to manipulate sustained or transient attention respectively, and a texture segmentation task.

Acuity tasks

Acuity tasks are designed to measure the observer's ability to resolve fine details. Performance in

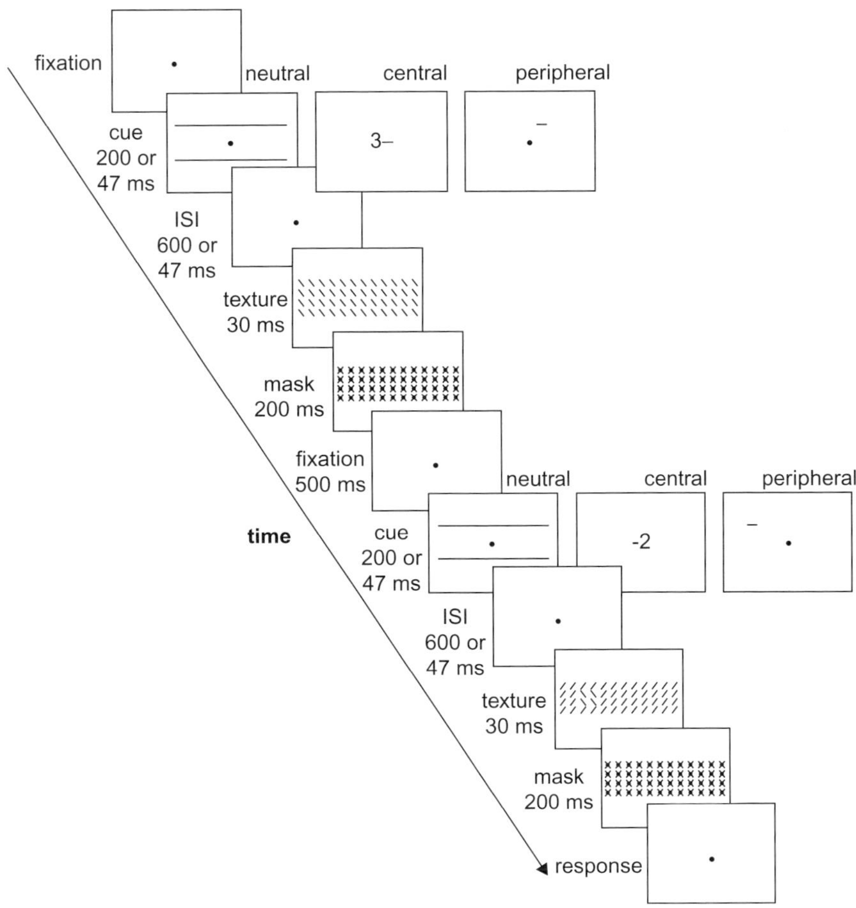

Fig. 1. Schema of the frame sequence in a typical trial with a central (sustained attention) or peripheral (transient attention) cue in a 2IFC texture segmentation task. The participants had to indicate which of the two intervals included a texture target whose orientation was orthogonal to that of the texture background. In this example the target is present in the second interval. The peripheral cue is a small horizontal bar appearing above the target location, and the central cue is composed of a digit indicating the eccentricity at which the target may appear and a line indicating the hemifield in which the target may appear. Adapted from Yeshurun et al. (2008).

some of these tasks, like the detection of a small gap in a Landolt-square, is limited by the retinal mosaic, while in other tasks, like identification of offset direction with Vernier targets, it is limited by cortical processes (e.g., Levi et al., 1985; Olzak and Thomas, 1986). By combining such tasks with attentional cueing we were able to demonstrate that directing transient attention to the target location improves performance in both acuity and hyperacuity tasks even when a suprathreshold target is presented without distracters. Specifically, we investigated whether covert attention can enhance spatial resolution via signal enhancement in a visual acuity task. We used a suprathreshold target (Landolt-square), which appeared at one of four possible eccentricities along the vertical or horizontal meridian and asked observers to indicate which side of the Landolt-square had a gap (Yeshurun and Carrasco, 1999). When a peripheral cue indicates the location of the upcoming target, observers' performance improves in terms of both speed and accuracy; they are able to detect a smaller gap appearing on a Landolt-square. Similarly, directing attention to the location of a Vernier target allowed observer to identify smaller horizontal offsets (Fig. 2; Yeshurun and Carrasco, 1999).

The same pattern of results is found whether or not a mask follows a target; that is, when all sources of added external noise-distracters, global masks, and local masks- have been eliminated from the display (Fig. 3; Carrasco et al., 2002). The decrement in performance with eccentricity is more pronounced along the vertical than horizontal meridian. The magnitude of the cueing effect increased with eccentricity but the magnitude of this effect was similar at different isoeccentric locations (Carrasco et al., 2002; Yeshurun and Carrasco, 1999). The finding that this effect becomes more pronounced as target eccentricity increases is consistent with the idea that attention enhances spatial resolution.

It is worth noting that the magnitude of the attentional effect is similar when comparing performance at the cued location with a central-neutral cue (a small circle at the center of the display) or with a distributed-neutral cue (four copies of the peripheral cue, simultaneously presented at the centers of each of the four quadrants). This finding rules out the possibility that the results are due to the fact that the central-neutral cue reduces the extent of the attentional spread.

It has long been postulated that attention helps manage limited resources and that the benefit exerted at the attended location is often accompanied by a cost at the unattended location(s). Indeed, this trade-off in processing is present with simple displays and in tasks mediated by early vision. For instance, both exogenous (Pestilli and Carrasco, 2005; Pestilli et al., 2007) and endogenous (Ling and Carrasco, 2006a) attention enhance contrast sensitivity at the attended location at the expense of decreasing sensitivity at the unattended location.

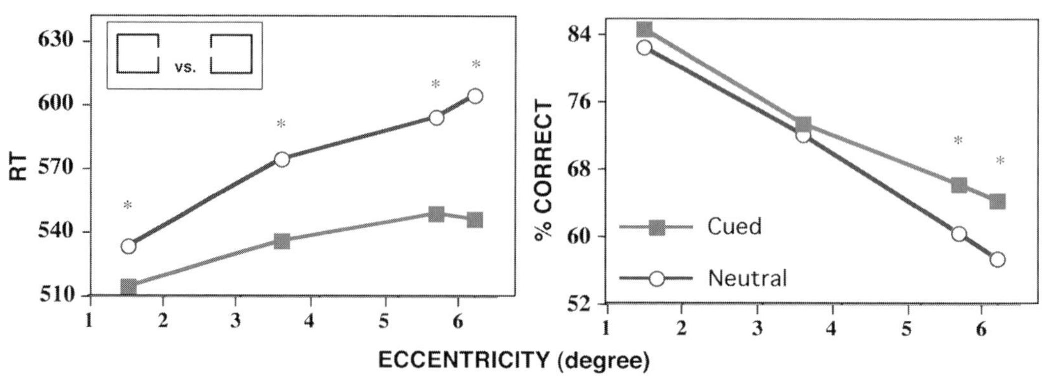

Fig. 2. RT (left panel) and accuracy (right panel) for detection of a gap in a Landolt-square (inset). Adapted from Yeshurun and Carrasco (1999).

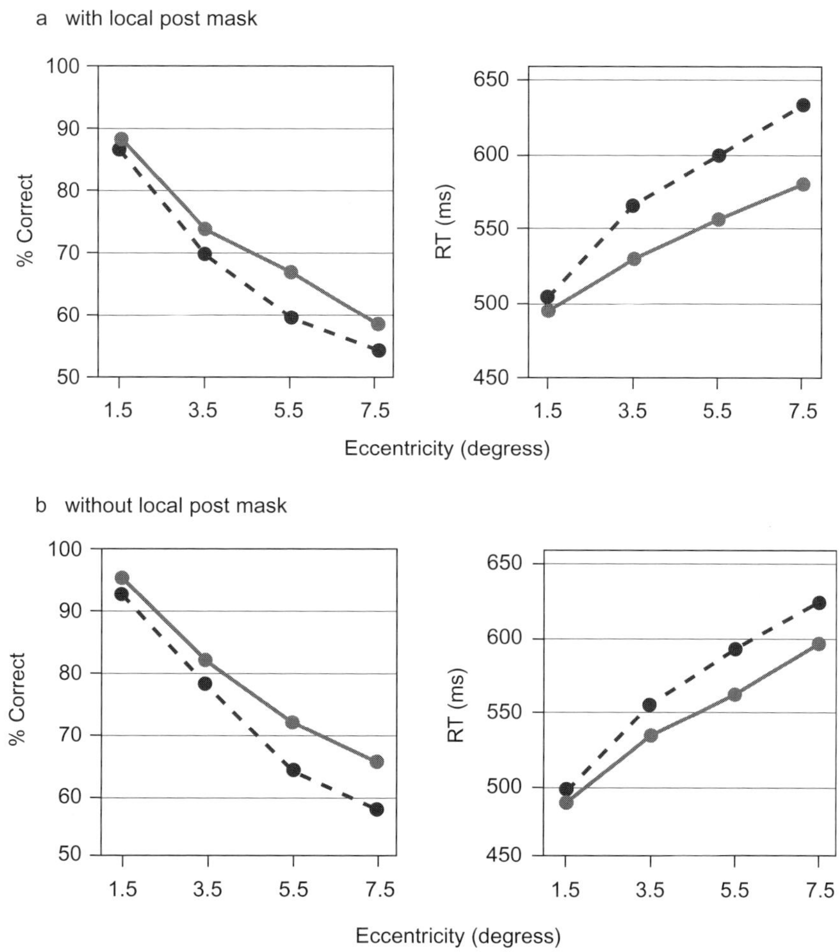

Fig. 3. Accuracy and RT for detection of a gap in a Landolt-square as a function of eccentricity: (a) with a local mask following the Landolt-square and (b) without a local mask. Continuous gray line indicates cued condition and the dashed black line indicates neutral condition. Adapted from Carrasco et al. (2002).

Once we established that covertly attending to a stimulus location increases spatial acuity (Carrasco et al., 2002; Yeshurun and Carrasco, 1999), we investigated whether increased spatial acuity is coupled with a decreased acuity at unattended locations (Montagna et al., 2009). We measured the effects of exogenous (transient, involuntary) and endogenous (sustained, voluntary) attention on observers' acuity thresholds for a Landolt gap resolution task at both attended and unattended locations, and compared the pattern of their trade-offs by maintaining task and stimuli identical while selectively engaging either type of attention. The fact that the attentional effect was evaluated against a neutral baseline condition for each type of attention allowed us to establish whether it represented a benefit, a cost, or both.

Spatial covert attention was manipulated via cues preceding stimulus presentation (Fig. 4). On each trial, a pre-cue either indicated a specific stimulus location (cued trials) or indicated both stimulus locations (neutral trials). Different types of cues selectively engaged either exogenous (peripheral uninformative cue) or endogenous (central informative cue) attention. Observers reported the location of a gap (top or bottom side) in the target Landolt-square indicated by a response cue following stimuli offset. The two

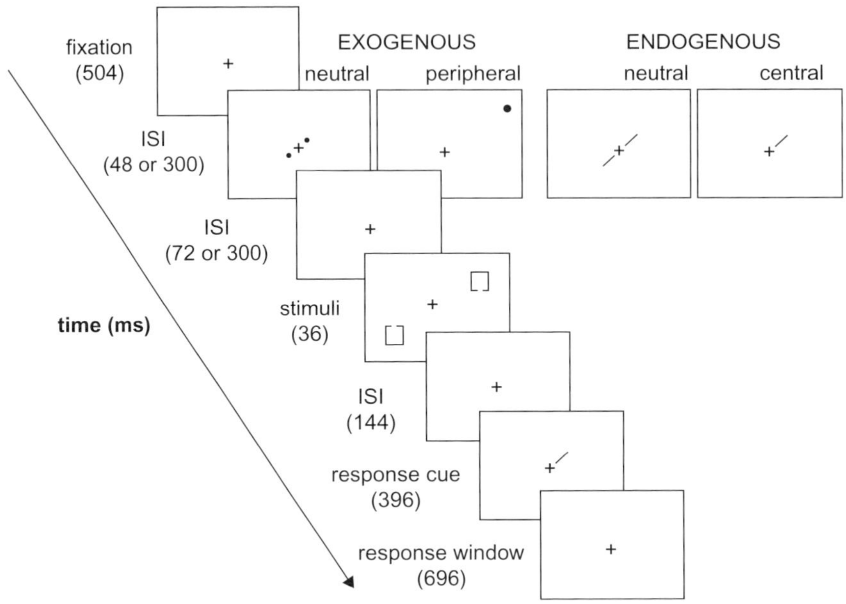

Fig. 4. Trial sequence. The trial sequence was identical for the exogenous and endogenous attention conditions except for the spatiotemporal characteristics of the peripheral and central cues. Adapted from Montagna et al. (2009).

attentional conditions, exogenous and endogenous, were blocked per session and each had its corresponding neutral cue baseline condition to quantify the magnitude of the attentional effects.

Gap-size thresholds (75% localization accuracy) were measured for each attention condition (exogenous and endogenous) and each cueing condition (cued, neutral, and uncued). For exogenous attention, observers were informed that the peripheral cue was uninformative, that is, it was not predictive of target location or gap side. For endogenous attention, observers were informed that the cue would indicate the target location on 70% of the central-cue trials, and were instructed to allocate their voluntary attention to the cued location. For both exogenous and endogenous attention, acuity thresholds were lower in the cued and higher in the uncued condition compared to the neutral baseline condition (Fig. 5). Both types of attention increased acuity at the attended and decreased it at unattended locations relative to a neutral baseline condition. The fact that acuity trade-offs emerge for very simple, non-cluttered displays, in which only two stimuli are competing for processing challenges the idea that perceptual processes are of unlimited capacity (e.g., Palmer et al., 2000), or that attentional selection is required only once the perceptual load exceeds the capacity limit of the system (e.g., Lavie, 1995). On the contrary, it suggests that trade-offs are a mandatory and basic characteristic of attentional allocation and that such a mechanism has a general effect across different stimulus and task conditions.

Visual search

In a visual search task, observers are typically required to detect the presence of a predefined target appearing among other nonrelevant items; for instance, a red vertical line appearing among red tilted lines in a feature search, or a red vertical line appearing among red tilted and blue vertical lines (e.g., Treisman, 1985). It was previously demonstrated that performance in visual search tasks, for both features and conjunctions, deteriorates as the target is presented at farther peripheral locations (Carrasco et al., 1995). This reduction in performance is attributed to the poorer spatial resolution at the periphery (e.g.,

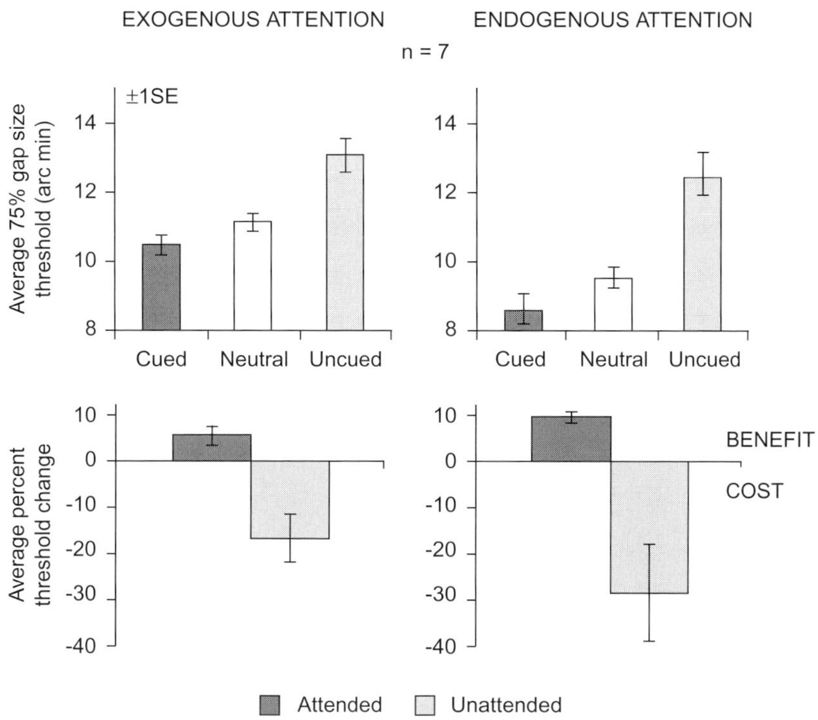

Fig. 5. Average gap-size thresholds (75% localization accuracy) for both exogenous (upper-left panel) and endogenous (upper-right panel) attention for the cued, neutral, and uncued conditions. The lower panels depict the average percent change in acuity thresholds at cued and uncued locations as compared to the neutral condition for exogenous (left) and endogenous (right) attention. Values below zero indicate a cost in acuity, whereas values above zero indicate a benefit. Error bars show ±1 SE. Adapted from Montagna et al. (2009).

Carrasco et al., 1995, 1998; Carrasco and Frieder, 1997). We have found that when observers direct their attention to the target location prior to the onset of the search display, the performance deterioration with target eccentricity is significantly reduced for both features and conjunctions (Carrasco and Yeshurun, 1998; Fig. 6). The ability of the peripheral cue to reduce this performance decrement supports the resolution hypothesis because it implies that attention can reduce resolution differences between the fovea and the periphery.

Texture segmentation

We performed a crucial test of the resolution hypothesis by exploring the effects of transient attention on a task in which performance is diminished by heightened resolution (Yeshurun and Carrasco, 1998). If attention indeed enhanced resolution, performance at the attended location should be impaired rather than improved. The task is a basic texture segmentation task that involves the detection of a texture target embedded in the background of an orthogonal orientation (Fig. 7). Observers' performance in this task does not peak when the target is presented at foveal locations, where resolution is highest. Instead, performance peaks at mid-peripheral locations, and drops as the target appears at more central or farther peripheral locations (e.g., Gurnsey et al., 1996; Joffe and Scialfa, 1995; Kehrer, 1989). Moreover, when the scale of the texture is manipulated, performance peaks at different eccentricities. Enlarging the scale of the texture shifts the peak of performance to farther locations, whereas decreasing this scale shifts the peak of performance toward the center

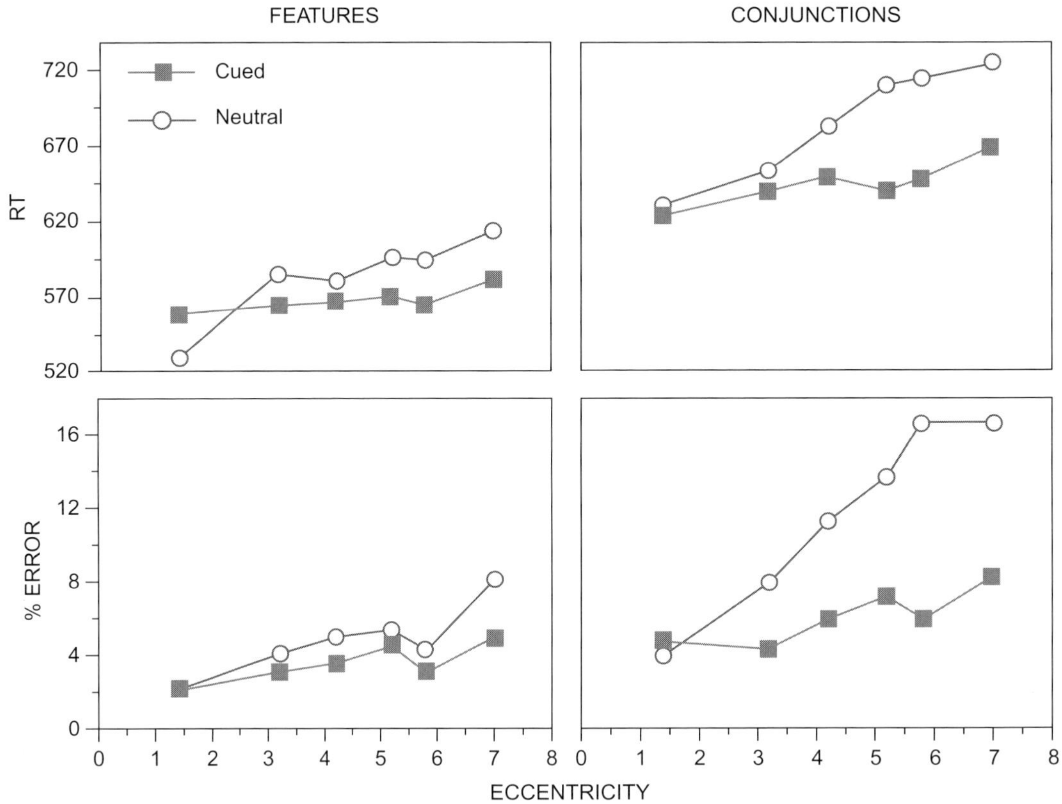

Fig. 6. RT and error rate for feature search (left panel — a search for a red vertical line appearing among red tilted lines) and conjunction search (right panel — a search for a red vertical line appearing among red tilted and blue vertical lines). Adapted from Carrasco and Yeshurun (1998).

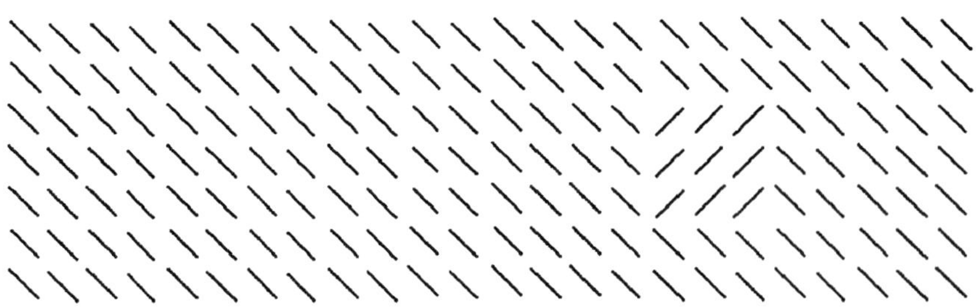

Fig. 7. Example of the texture stimuli used in Yeshurun and Carrasco (1998).

(Gurnsey et al., 1996; Joffe and Scialfa, 1995; Kehrer, 1989).

The finding that in this texture segmentation task performance drops at central locations — central performance drop (CPD) — is attributed to a mismatch between the average size of spatial filters at the fovea and the scale of the texture (Gurnsey et al., 1996; Kehrer, 1997). There is ample evidence that we process visual stimuli by means of parallel spatial filters. These are low-level analyzers that are tuned to a specific band of spatial frequency and orientation (e.g., De Valois and De Valois, 1988; Graham, 1989; Phillips and Wilson, 1984). It has been suggested that the size of these filters at the fovea may be too small for the scale of the texture, as if spatial resolution at the fovea is too high for the task. At more peripheral regions, the filters' average size increases gradually, and is presumably optimal around the peak of performance. At farther locations, the filters are too big and their low resolution limits performance. Consequently, the finding that performance with a larger texture scale peaks at farther eccentricities may reflect the fact that the processing of this enlarged texture requires larger filters that are more abundant at farther eccentricities, and vice versa (Gurnsey et al., 1996; Kehrer, 1997).

We hypothesized that if attention indeed enhances spatial resolution, attending to the target location should enhance performance at the periphery, where the resolution is too low, but should impair performance at the fovea, where the resolution is already too high for the task. Moreover, if attention enhances resolution by effectively decreasing the average size of filters at the attended location (e.g., Moran and Desimone, 1985; Reynolds and Desimone, 1999), then for a larger texture scale, attention should impair performance for a wider range of eccentricities; for a smaller texture scale, attention should impair performance in a narrower range of eccentricities. This is due to the fact that with a larger texture scale the mismatch between the texture scale and the size of the filters would extend farther toward the periphery and vice versa (Yeshurun and Carrasco, 1998). To test these predictions we combined peripheral cues with this texture segmentation task. On the cued trials a peripheral cue indicated the target location prior to its appearance, allowing observers to focus their attention, in advance, on the target location without having time to move their eyes to the location. On the neutral trials a pair of lines, appearing above and below the display, indicated that the target was equally likely to appear at any location. The texture target appeared at any of 17 possible eccentricities, and the scale of the texture was manipulated by viewing the display from three different distances – 228, 57, or 28 cm (see neutral and peripheral conditions in Fig. 1).

For all three viewing distances the pattern of the results conformed to the resolution hypothesis (Fig. 8). Accuracy was higher for the cued than the neutral trials at the more peripheral locations but was lower at central locations. Hence, attending to the target location improved performance at peripheral locations, where the resolution was too low for the scale of the texture, but impaired performance in central locations, where the resolution was already too high. Moreover, as predicted, with a larger texture scale (middle panel), performance was impaired in a larger range of eccentricities (0–5°), compared to the medium texture scale (0–1°, left panel). Similarly, with a smaller texture scale (right panel), performance was impaired at a smaller range of eccentricities (0–0.66°). This study demonstrated that (a) attention helps performance that is limited by resolution that is too low, but hinders performance that is limited by resolution that is too high; (b) the range of eccentricities in which attention hinders performance depends on the scale of the texture and the average size of the filters at a given eccentricity. Although no other existing model of attention could predict an attentional impairment, this impairment is predicted by the resolution hypothesis (Yeshurun and Carrasco, 1998).

We obtain the same pattern of results when we present the texture along the vertical rather than the horizontal meridians. Interestingly, when the texture was presented along the vertical meridian performance peaked at farther eccentricities in the lower than in the upper vertical meridian, indicating that resolution was higher in the lower half. Furthermore, the peripheral cue affected performance along the vertical meridian uniformly, indicating that the degree of enhanced resolution brought about by transient attention was constant along the vertical meridian (Talgar and Carrasco, 2002). Consistent with findings in contrast sensitivity (Cameron et al., 2002; Carrasco et al., 2001), performance on texture segmentation indicates that the vertical meridian asymmetry for spatial resolution is

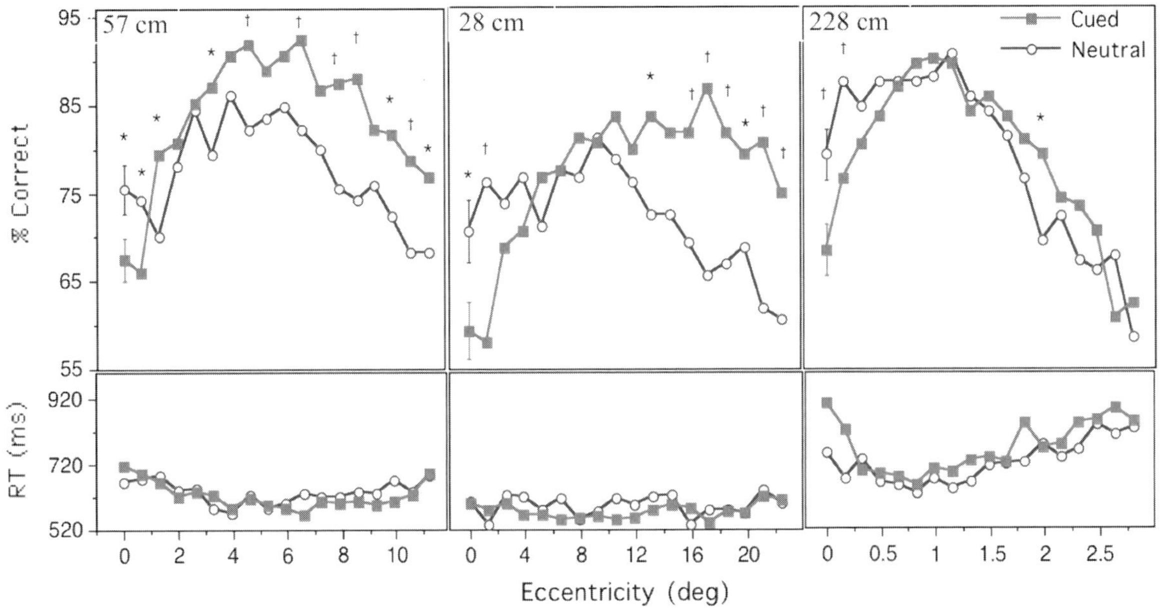

Fig. 8. Observers' performance as a function of target eccentricity and cueing condition for the three viewing distances. Because viewing distance varied, the eccentricity values (abscissa) differ in the three panels. Adapted from Yeshurun and Carrasco (1998).

determined by visual, not attentional, constraints. These findings shed light on the nature of the attentional mechanism by lending strong support to the hypothesis that attention enhances the spatial resolution at the attended location, possibly by reducing the average size of the corresponding filters.

We conducted another study to investigate the level of visual processing at which these attentional effects take place (Yeshurun and Carrasco, 2000). At the level of the visual cortex, texture segmentation theoretically involves passage of visual input through two layers of spatial linear filters, separated by a point-wise nonlinearity. The first-order linear filters are assumed to perform a more local analysis of spatial frequency and orientation, and are thought to correspond to simple cortical cells in area V1. The second-order linear filters are considered to be of a larger scale and assumed to perform a more global analysis on the output of the first-order filters plus the intermediate nonlinearity (e.g., Bergen and Landy, 1991; Fogel and Sagi, 1989; Graham et al., 1992; Malik and Perona, 1990; Sutter et al., 1989, 1995). To assess the level of processing at which attention affects spatial resolution we used textures of a different nature (Yeshurun and Carrasco, 2000). These textures were composed of narrow-band stimuli, ensuring that only filters of a specific scale were activated (Fig. 9; Graham et al., 1992).

By manipulating the spatial-frequency content of the texture we were able to replicate our previous findings (Yeshurun and Carrasco, 1998), demonstrating that these effects are robust and can generalize to textures of a very different nature. More importantly, we could differentially stimulate first or second-order filters of various scales. We found that the pattern of the attentional effects on texture segmentation depended only on the second-order frequency of the texture. As can be seen in Fig. 10, the attentional effect was the same regardless of the first-order content: for both the low-frequency (top-left panel) and the high-frequency (top-right) conditions, a significant interaction emerged; accuracy was higher for cued trials than neutral trials at more peripheral eccentricities, but accuracy was lower at central locations (0–2°). In contrast, the attentional effect differed when the second-order content was

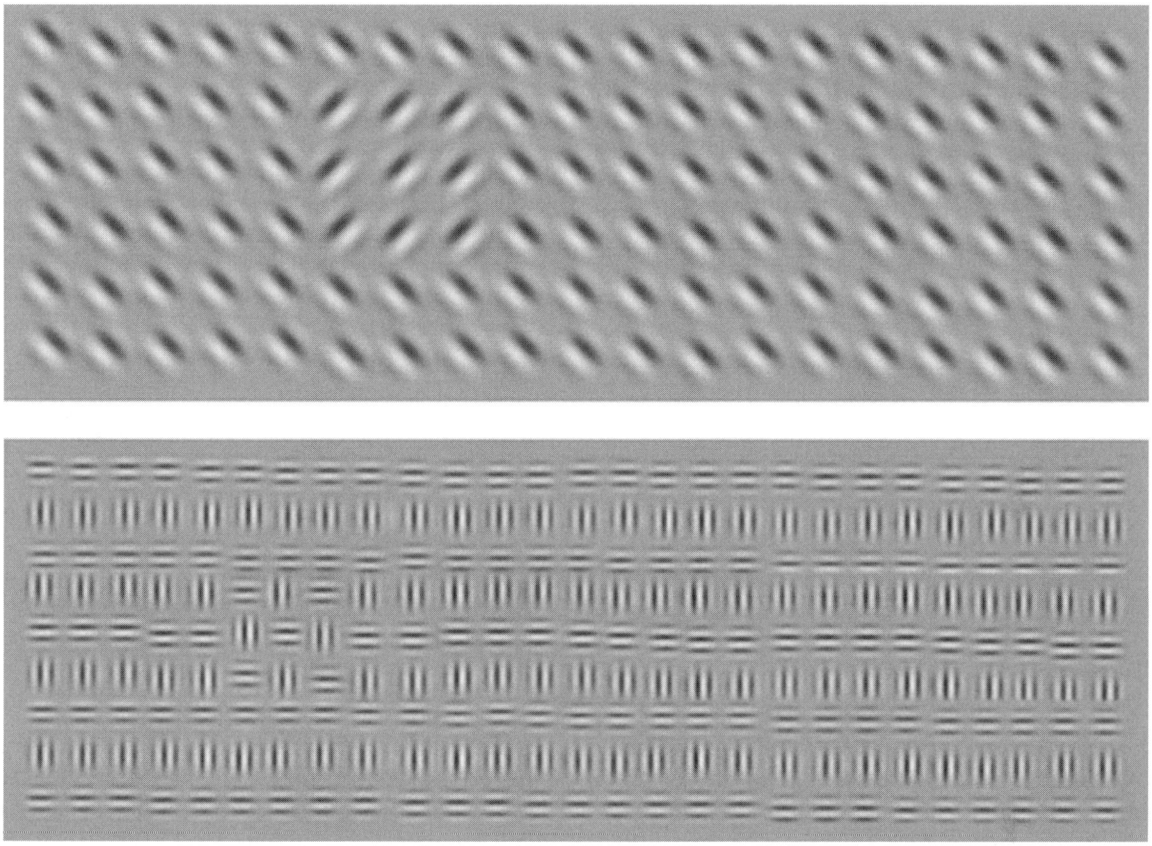

Fig. 9. An example of the first-order (top) and second-order (bottom) textures used in Yeshurun and Carrasco (2000).

varied: attention impaired performance in a greater range of eccentricities for the low-frequency (bottom-left) than the high-frequency (bottom-right) conditions (0–7.76° vs. 0–3.33°), and an attentional benefit emerged only for the high-frequency condition. This suggests that attention operates at the second stage of filtering, possibly by reducing the size of the second-order filters, resulting in enhanced spatial resolution. This finding indicates that attention can modulate processing as early as at the primary visual cortex. Thus, these attentional effects suggest a link between task performance (behavior) and physiological studies demonstrating attentional modulation of activity in area V1, either by means of single cell recording (Ito and Gilbert, 1999; Motter, 1993) or by fMRI (Brefczynski and DeYoe, 1999; Gandhi et al., 1999; Kastner and Ungerleider, 2000; Martinez et al., 1999).

To test directly whether covert attention enhances spatial resolution by increasing sensitivity to high spatial frequencies, we employed a cueing procedure in conjunction with selective adaptation (Carrasco et al., 2006). The selective adaptation procedure is used to assess the spatiotemporal properties of the visual system. It has long been demonstrated that prolonged exposure to one type of stimulus reduces sensitivity to those stimulus parameters and other similar stimuli, thus allowing for the selective adaptation for a particular variable or set of variables, such as spatial frequency and orientation (Blakemore and Campbell, 1969; Graham, 1989; Movshon and Lennie, 1979; Saul and Cynader, 1989). While keeping the stimulus content identical, we manipulated the availability of spatial-frequency information by reducing observers' sensitivity to a range of frequencies.

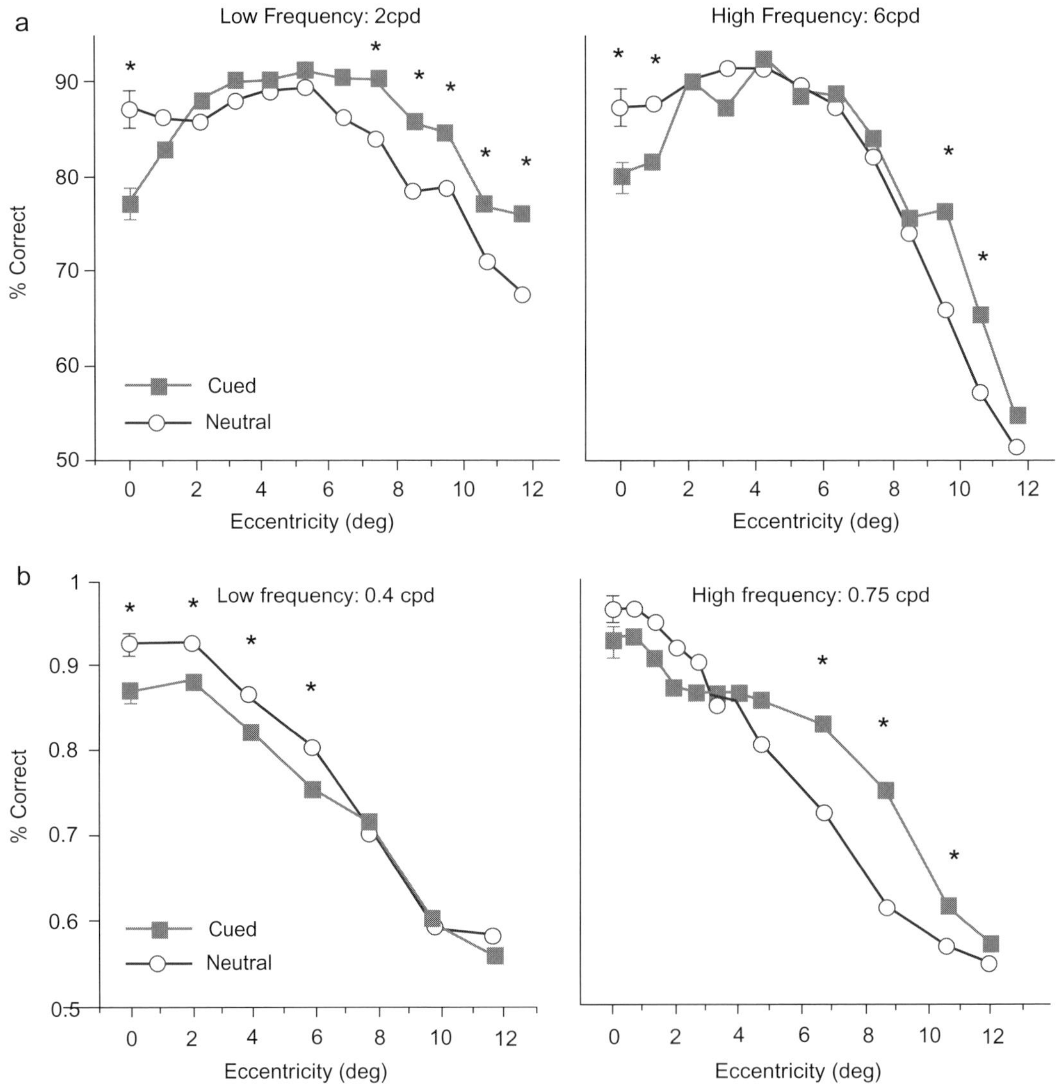

Fig. 10. Performance with first-order (a) and second-order (b) textures of low (left) or high (right) frequency as a function of cueing condition and target eccentricity. Adapted from Yeshurun and Carrasco (2000).

At central locations when high-frequency non-optimal filters participate in the normalization process the weakened response of the optimal filters would result in the CPD. Thus by adapting to high spatial frequencies, the nonoptimal filters would be removed from the normalization process and the CPD would be diminished. Furthermore, were the central attentional impairment (Talgar and Carrasco, 2002; Yeshurun and Carrasco, 1998, 2000) due to an increased sensitivity to high frequencies and a reduced sensitivity to lower frequencies, adapting to high spatial frequencies should eliminate the attentional impairment at central locations and diminish the benefit in the peripheral locations. If the contribution of the nonoptimal high frequencies is diminished in the normalization process, cueing the target location could no longer inhibit the optimal filters for the scale of the texture and performance would not be impaired, that is, no central attentional impairment would emerge.

Observers performed a 2-AFC discrimination task after selectively adapting to 0-cpd (baseline), 1-cpd (low spatial frequency), or 8-cpd (high spatial frequency). The results indicate that the CPD was present in the baseline and the low-spatial-frequency neutral conditions but was eliminated in the high-spatial-frequency neutral condition (Fig. 11). Furthermore, the central attentional impairment present in the baseline and low-frequency exogenous cueing conditions was eliminated in the high-frequency exogenous cueing condition. In other words, we found that by adapting to low spatial frequencies, performance in this texture segmentation task does not change. However, by adapting to high spatial frequencies, the CPD is diminished and the central attentional impairment is eliminated. These results indicate that the CPD is primarily due to the dominance of high-spatial-frequency responses, and that transient covert attention enhances spatial resolution by increasing sensitivity to higher spatial frequencies.

In another study we examined the adaptability of transient attention regarding spatial resolution. In particular, we investigated whether the scale of the information that attracts attention (the size of the attentional cue) can modulate the effects of transient attention on the spatial resolution at the attended location (Yeshurun and Carrasco, 2008). Various studies have manipulated the size of the attended region by employing cues of different sizes or dual tasks (e.g., Goto et al., 2001; Greenwood and Parasuraman, 2004; Hock et al., 1998; Müller et al., 2003). These studies have found that the larger the attended region, the lower the resolution. Although these studies manipulated sustained attention, they suggest that transient attention may also be able to modulate its effect on spatial resolution as a function of the cue size, so that the larger the cue the lower the resolution.

To test this hypothesis, we used a texture segmentation task that was similar to the one employed in our previous studies (e.g., Yeshurun and Carrasco, 1998; Fig. 7), and systematically manipulated the size of the attentional cue (Fig. 12). If the gradual increase in the size of the attentional cue leads to a gradual resolution decrement, then performance at central locations should gradually improve and at peripheral locations should gradually deteriorate as the cue size increases. Moreover, as cue size increases the eccentricity at which performance peaks should gradually shift to nearer eccentricities reflecting the gradual decrease in resolution, with the performance peak of the largest cue being at the nearest eccentricity (as it designates the largest area — the whole display). Alternatively, if transient attention does not alter its operation based on the size of the attentional cue, its effect on spatial resolution should not change in a gradual fashion with changes in cue size.

The findings consistently replicated the attentional enhancement of spatial resolution reported previously with a small cue (Carrasco et al., 2006;

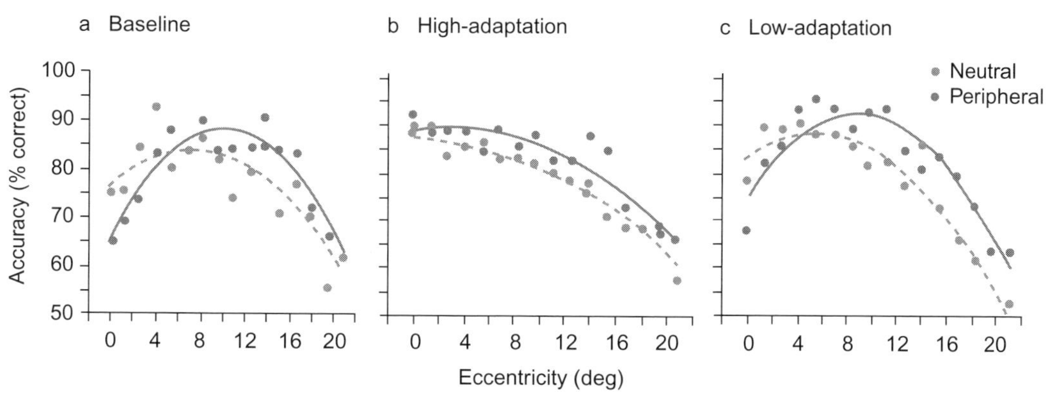

Fig. 11. Observers' performance as a function of cue type and target eccentricity. (a) Baseline, (b) high-spatial-frequency adaptation grating, and (c) low-spatial-frequency adaptation grating. Adapted from Carrasco et al. (2006).

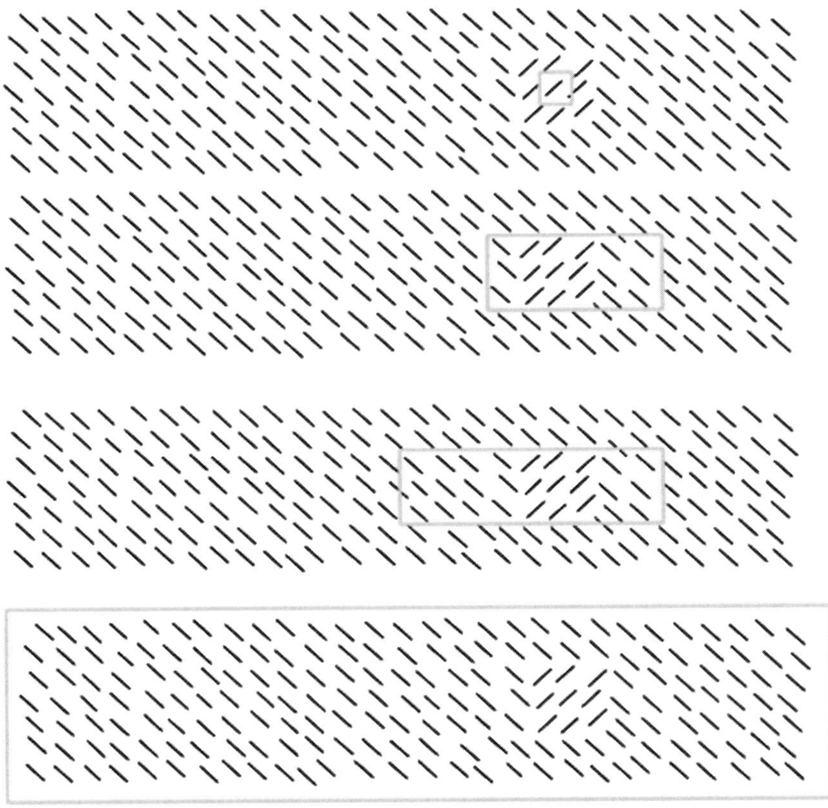

Fig. 12. An example of cues of different sizes and the textures used in Yeshurun and Carrasco (2008). The largest cue (bottom) was similar to the neutral cue employed previously (e.g., Yeshurun and Carrasco, 1998), and since it carried no information regarding the target location this cue served as the baseline to which performance with smaller cues was compared.

Talgar and Carrasco, 2002; Yeshurun and Carrasco, 1998), but there was no evidence of gradual resolution decrement with large cues. Specifically, a differential effect was found for the different cue sizes, but it mainly reflects an attentional effect for the small cue sizes and no effect for larger cues (Fig. 13). There was no gradual change in performance with increasing cue size. These findings indicate that in this texture segmentation task, transient attention exerts its effects on spatial resolution only when it is directed to a small region by a small cue. There is no evidence that transient attention can flexibly lower resolution when it is attracted to a broader spatial region by large cues.

The texture segmentation studies described thus far employed a peripheral cue to measure the effects of transient attention. Transient attention increases spatial resolution even when it is detrimental to the task at hand. Improved resolution due to transient attention is advantageous because most everyday tasks — such as reading, searching for small objects, or identifying fine details — benefit from heightened resolution. Thus, an attentional mechanism that increases spatial resolution by default can be very effective. However, in certain situations resolution enhancement is not beneficial. For example, when a more global assessment of a scene is required (e.g., viewing an impressionist painting) enhancing resolution is not optimal. Likewise, a high-resolution analysis of the scene will not provide optimal results when navigating through the world under poor atmospheric conditions (e.g., fog or haze). We wondered how sustained attention, given its top-down nature, would affect performance in a

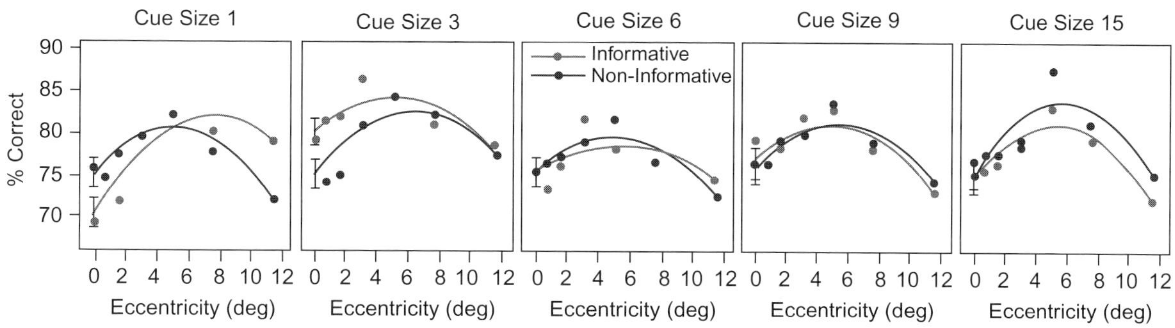

Fig. 13. Observers' performance as a function of cue size and target eccentricity. "Informative" refers to the trials in which the cue carried some information regarding the target location (the larger the cue the less precise this information is). "Noninformative" refers to the trials in which the cue carried no information regarding the target location (the largest cue). The number of cue size indicates the number of texture columns encompassed by the cue frame. Adapted from Yeshurun and Carrasco (2008).

texture segmentation task in which enhanced spatial resolution is detrimental to performance.

In a recent study (Yeshurun et al., 2008) we employed a central cue to test whether sustained attention can also affect performance in a texture segmentation task, and whether this effect will be similar to that found with peripheral cues. In some of the experiments of this study the texture segmentation task was the same as the one employed with transient attention in previous studies (Talgar and Carrasco, 2002; Yeshurun and Carrasco, 1998, 2008; Fig. 1). In other experiments the texture was modified from a homogeneous to a heterogeneous background to preclude the need for a post-mask and thus ensure that performance is limited only by spatial factors (Fig. 14). The average orientation of line elements in the texture display was $\pm 45^\circ$ from vertical, the actual orientation of each line element was chosen at random from a uniform distribution of orientations. As the range of sampled orientations around the mean increases, the target patch becomes harder to detect. The resulting texture stimuli were very similar to the ones used by Potechin and Gurnsey (2003). With these texture stimuli we used a Yes–No detection task rather than the 2IFC task employed before. The central cue was composed of a digit indicating the eccentricity at which the target may appear and a line indicating the quadrant in which the target may appear.

The pattern of results was very similar for both types of texture stimuli and tasks: sustained attention, like transient attention, can affect texture segmentation. However, in contrast to transient attention, the effects of sustained attention did not vary as a function of eccentricity (Fig. 15). Directing sustained attention to the target location improved performance at all eccentricities (unless performance was at chance level). There was no attentional impairment at central locations. These findings indicate that the attentional benefit that emerged in both experiments is robust and can be generalized to different textures and tasks.

In this study we also evaluated the contribution of location uncertainty at the decisional level to the effect of sustained attention. We compared the effect of the central pre-cues with the effect of post-cues, which indicate the target location *after* the offset of the texture display. Spatial post-cues, like post-masks, are considered to effectively reduce location uncertainty (e.g., Carrasco et al., 2000; Carrasco and Yeshurun, 1998; Kinchla et al., 1995; Luck et al., 1994, 1996; Lu and Dosher, 2004; Smith, 2000). Both pre- and post-cues reduce location uncertainty, as both allow the observer to assign lower weights to information extracted from the non-cued locations; however, only the pre-cues allow for a change in the quality of the texture representation due to the advanced allocation of attention to the location of the upcoming target. Thus, any additional benefit yielded by pre-cues compared to post-cues could be ascribed to an attentional modulation of the

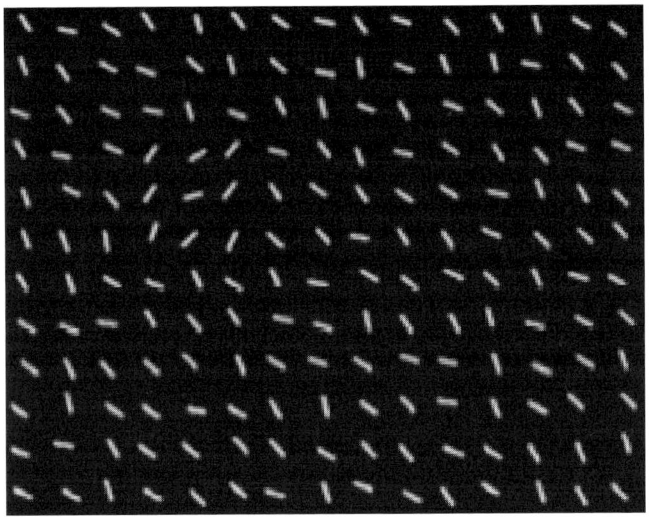

Fig. 14. An example of the heterogeneous textures used in Yeshurun et al. (2008).

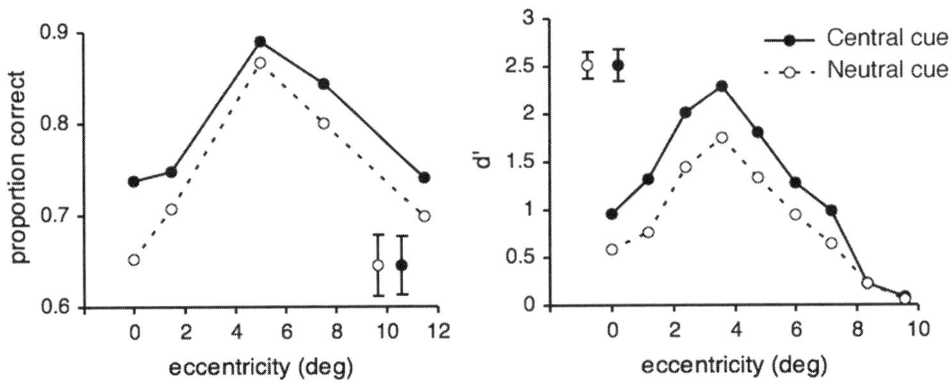

Fig. 15. Observers' performance as a function of cue condition and target eccentricity, for texture stimuli with homogeneous (left panel; see Fig. 7) or heterogeneous (right panel; see Fig. 14) background. Adapted from Yeshurun et al. (2008).

quality of the texture representation rather than to the mere reduction of location uncertainty at the decisional stage. The results showed that performance with the central pre-cue, which triggers sustained attention, was significantly higher than performance with its neutral condition, whereas performance for the central post-cue was only marginally higher than its neutral condition. Moreover, the central pre-cue elicited a significantly better performance than the central post-cue. These results indicate that the benefit of the central pre-cue went well beyond the mere effect of location uncertainty at the decisional stage — it improved the quality of the texture representation.

Discussion

The various studies we described thus far were designed to test the effects of transient and sustained attention on performance by employing peripheral and central pre-cues, respectively. The studies of transient attention clearly demonstrate that transient attention can affect performance in various basic tasks like acuity and texture segmentation. Directing transient attention to the target location reduced performance differences between the center and the periphery in visual search tasks (Carrasco and Yeshurun, 1998), improved performance in tasks that were limited by acuity or hyperacuity (Carrasco et al., 2001; Montagna et al., 2009; Yeshurun and Carrasco, 1999), and improved or impaired texture segmentation depending on the combination of the eccentricity of the texture target and the scale of the texture (Carrasco et al., 2006; Talgar and Carrasco, 2002; Yeshurun and Carrasco, 1998, 2000, 2008).

It is important to note that the effects of transient attention on acuity measures could not be accounted for by many of the prominent hypotheses regarding the attentional mechanism like shifts in the decisional criterion, location uncertainty reduction, or reduction of external noise (e.g., Dosher and Lu, 2000; Eckstein et al., 2002; Kinchla et al., 1995; Lu and Dosher, 2004; Shiu and Pashler, 1994) for the following reasons: because the peripheral cue did not convey information regarding the correct response and only indicated the target location (Carrasco et al., 2002; Yeshurun and Carrasco, 1999), or conveyed no information regarding either the correct response or the target location (Montagna et al., 2009), it did not associate a higher probability with one of the responses and observers could not rely on its presence to reach a decision. Moreover, the target was presented alone, without other items to introduce external noise, and it was a suprathreshold target that could not be confused with the blank at the other locations (Yeshurun and Carrasco, 1999). Additionally, we found similar results with and without a local post-mask (Carrasco et al., 2002). In contrast to these attentional mechanisms, the improved performance in acuity tasks could be accounted for by the resolution hypothesis suggesting that transient attention enhances the spatial resolution at the attended location.

The alternative mechanisms of attention mentioned above also fail to account for the effects of transient attention on texture segmentation, namely the attentional impairment of performance at central locations (Carrasco et al., 2006; Talgar and Carrasco, 2002; Yeshurun and Carrasco, 1998, 2000, 2008), because all alternative hypotheses would predict a benefit on performance throughout all eccentricities. Only the resolution hypothesis predicts the attentional impairment of performance at central locations, and therefore, the findings of the texture segmentation studies lend strong support to the resolution hypothesis.

The resolution hypothesis is in line with other psychophysical studies suggesting that attention allows a fine-scale analysis. For instance, Morgan et al. (1998) measured orientation thresholds in a visual search task. They presented a Gabor patch in one of two possible orientations, with or without distracters, and found that when distracters were present, spatially cueing target location reduced orientation thresholds to the level found when the target was presented alone. The authors suggested that focusing attention on the target location reduced thresholds through the operation of a smaller scaled "stimulus analyzer" (Morgan et al, 1998, p. 368). Likewise, when Tsal and Shalev (1996) studied the effects of cueing attention on the perceived length of short lines, they found that a briefly presented line is judged to be shorter when its location was known in advance. They suggested that the attended line was perceived as shorter because the processing of an attended stimulus is mediated by smaller "attentional receptive fields" (Tsal and Shalev, 1996, p. 242).

The resolution hypothesis is also consistent with a comparative study that evaluated the effects of spatial covert attention on Landolt acuity as a function of different SOAs for human and nonhuman primates (Golla et al., 2004). The findings for both species demonstrate a consistent enhanced acuity when the target location was pre-cued as compared to a no-cue condition (i.e., when there was no temporal or spatial indication for both trial onset and target location). As was the case in the psychophysical studies with humans described above (Carrasco et al., 2002; Montagna et al., 2009; Yeshurun and Carrasco, 1999), the attentional effect increased with eccentricity in human and nonhuman primates.

There may be several ways in which this attentional enhancement of spatial resolution is accomplished. First, attention may, in effect, reduce the size of receptive fields at the attended area. This hypothesis is consistent with neurophysiological studies on endogenous attention, demonstrating that a neuron's response to its preferred stimulus is greatly reduced when the preferred stimulus is not attended, and an attended, non-preferred stimulus is also presented within the neuron's receptive field. These findings suggest that attention contracts the cell's receptive field around the attended stimulus (e.g., Anton-Erxleben et al., 2009; Moran and Desimone, 1985; Reynolds and Desimone, 1999; Womelsdorf et al., 2006).

Alternatively, attention may enhance resolution by increasing the sensitivity of the smallest receptive fields at the attended area (Balz and Hock, 1997), which in turn may inhibit the sensitivity of the larger receptive fields at the same area. At central locations, when high-frequency nonoptimal filters participate in the normalization process, the weakened response of the optimal filters results in the CPD. Indeed, adapting to high spatial frequencies resulted in a diminished CPD probably due to the fact that the nonoptimal filters were removed from the normalization process. Furthermore, adapting to high spatial frequencies also eliminated the attentional impairment at central locations. Because the contribution of the nonoptimal high frequencies was diminished in the normalization process, cueing the target location could no longer inhibit the optimal filters and performance could not be impaired, that is, there was no central attentional impairment. These results support the hypothesis that the CPD is primarily due to the dominance of high-spatial-frequency responses, and that covert attention enhances spatial resolution by increasing sensitivity to higher spatial frequencies (Carrasco et al., 2006).

Like transient attention, sustained attention affects performance in basic visual tasks mediated by spatial resolution tasks (Montagna et al., 2009; Yeshurun et al., 2008). Unlike transient attention, directing sustained attention to the target location via central pre-cues improved texture segmentation at both central and peripheral locations. This finding could not be accounted for by uncertainty reduction because when we compared performance with central pre- and post-cues we found that performance with the pre-cue was significantly better than performance with the post-cue. The effects of sustained attention on texture segmentation could be accounted for by an attentional mechanism that is capable of either enhancement or decrement of spatial resolution to optimize performance. According to this view, sustained attention optimized performance at all eccentricities via resolution enhancement at the periphery where performance is limited by a resolution that is too low, and via resolution decrement at central locations where performance is limited by a resolution that is too high. This view of sustained attention portrays a highly adaptive mechanism that can adjust its operation on a trial-by-trial basis. Note, however, that the eccentricity-independent effects of sustained attention could also be attributed to an attentional mechanism that affects texture segmentation by improving the signal to noise ratio at all eccentricities through means other than resolution modification, like reduction of external noise at early levels of processing (e.g., Dosher and Lu, 2000; Lu and Dosher, 2004), possibly via distracter suppression (e.g., Shiu and Pashler, 1994).

The finding that sustained attention affects texture segmentation in a different manner than transient attention is consistent with studies demonstrating differential effects for sustained and transient attention. For instance, Briand and Klein (1987) and Briand (1998) found that with peripheral cues, but not with central cues, the effects of attention were larger for a conjunction search than for a feature search. Another study that tested the effects of sustained and transient attention under low-noise versus high-noise conditions reported that sustained attention could affect performance only under high-noise conditions, but not under low-noise conditions (e.g., Dosher and Lu, 2000). Transient attention, however, could operate under both low-noise and high-noise conditions (Lu and Dosher, 1998,

2000). A more recent study has shown that both sustained and transient attention increase contrast sensitivity, even in low-noise conditions, but whereas the former is mediated by a contrast-gain mechanism, the latter seems to be mediated by both contrast-gain and response-gain mechanisms (Ling and Carrasco, 2006b). Moreover, a population-coding model that estimates attentional effects on population contrast response given psychophysical data indicates that whereas sustained attention changes population contrast response via contrast gain, transient attention changes population contrast response via response gain (Pestilli et al., 2009).

Some studies dealing with the effects of attention on temporal aspects of processing also show differential effects for sustained and transient attention. For instance, involuntary allocation of attention (via peripheral noninformative cues) impairs temporal order judgment, whereas voluntary allocation of attention (via central informative cues) improves it (Hein et al., 2006). Furthermore, a recent study employing a speed-accuracy trade-off procedure, which enables conjoint measures of discriminability and temporal dynamics, showed that with central cues, the attentional benefits increased with cue validity while costs remained relatively constant. However, with peripheral cues, the benefits and the costs were comparable across the range of cue validities (Giordano et al., 2009).

Finally, in line with the idea of limited resources, we have demonstrated an attentional trade-off for spatial resolution: our ability to resolve small details in a stimulus increases at the attended location, while decreasing elsewhere for both exogenous and endogenous attention (Montagna et al., 2009). This trade-off was measured for spatial acuity thresholds and was found even in impoverished, non-cluttered displays in which only two stimuli (one target and one distracter) appear at known locations to compete for processing resources. This finding suggests that the cost in acuity at unattended locations may be a mandatory consequence of the attentional allocation of resources to the attended location. Together with the effects of covert attention on contrast sensitivity (Ling and Carrasco, 2006a; Pestilli and Carrasco, 2005; Pestilli et al., 2007), this study suggests that visual processing trade-offs are a general mechanism of attentional allocation, whose perceptual consequences affect several basic visual dimensions, and it supports the idea that spatial covert attention helps regulate the expenditure of cortical computation.

Conclusions

Attentional facilitation in visual tasks reflects a combination of mechanisms such as signal enhancement, noise exclusion, and decisional factors. In this chapter we described a set of studies on sustained and transient covert attention that support one of these mechanisms — signal enhancement via enhanced resolution. These studies employ different tasks, like gap detection, visual search, and texture segmentation, and different stimuli, like squares, Vernier stimuli, textures composed of many line segments or Gabor patches. Yet all of them suggest the same conclusion — directing attention to the target location allows us to better resolve the fine details of the visual scene.

References

Anton-Erxleben, K., Stephan, V. M., & Treue, S. (2009). Attention reshapes center-surround receptive field structure in macaque cortical area MT. *Cerebral Cortex*, in print (doi:10.1093/cercor/bhp002).

Balz, G. W., & Hock, H. S. (1997). The effect of attentional spread on spatial resolution. *Vision Research*, *37*, 1499–1510.

Bergen, J. R., & Landy, M. S. (1991). Computational modeling of visual texture segregation. In M. S. Landy & J. A. Movshon (Eds.), *Computational models of visual processing* (pp. 253–271). Cambridge, MA: MIT Press.

Blakemore, C. B., & Campbell, F. W. (1969). On the existence of neurons in the human visual system selectively sensitive to the orientation and size of retinal images. *American Journal of Physiology*, *203*, 237–260.

Brefczynski, J. A., & DeYoe, E. A. (1999). A physiological correlate of the 'spotlight' of visual attention. *Nature Neuroscience*, *2*, 370–374.

Briand, K. A. (1998). Feature integration and spatial attention: More evidence of a dissociation between endogenous and exogenous orienting. *Journal of Experimental Psychology: Human Perception and Performance*, *24*, 1243–1256.

Briand, K. A., & Klein, R. M. (1987). Is Posner's "beam" the same as Treisman's "glue"? On the relation between visual orienting and feature integration theory. *Journal of Experimental Psychology: Human Perception and Performance*, *13*, 228–241.

Cameron, E. L., Tai, J. C., & Carrasco, M. (2002). Covert attention affects the psychometric function of contrast sensitivity. *Vision Research*, *42*, 949–967.

Carrasco, M. (2006). Covert attention increases contrast sensitivity: Psychophysical, neurophysiological, and neuroimaging studies. In S. Martinez-Conde, S. L. Macknik, L. M. Martinez, J. M. Alonso, & P. U. Tse (Eds.), *Visual perception. Part I. Fundamentals of vision: Low and mid-level processes in percetion – Progress in Brain Research* (pp. 33–70). Amsterdam: Elsevier.

Carrasco, M., Evert, D. L., Chang, I., & Katz, S. M. (1995). The eccentricity effect: Target eccentricity affects performance on conjunction searches. *Perception & Psychophysics*, *57*, 1241–1261.

Carrasco, M., & Frieder, K. S. (1997). Cortical magnification neutralizes the eccentricity effect in visual search. *Vision Research*, *37*, 63–82.

Carrasco, M., Loula, F., & Ho, Y.-X. (2006). How attention enhances spatial resolution: Evidence from selective adaptation to spatial frequency. *Perception & Psychophysics*, *68*, 1004–1012.

Carrasco, M., McLean, T. L., Katz, S. M., & Frieder, K. S. (1998). Feature asymmetries in visual search: Effects of display duration, target eccentricity, orientation and spatial frequency. *Vision Research*, *38*, 347–374.

Carrasco, M., Penpeci-Talgar, C., & Eckstein, M. (2000). Spatial attention increases contrast sensitivity across the CSF: Support for signal enhancement. *Vision Research*, *40*, 1203–1215.

Carrasco, M., Talgar, C. P., & Cameron, E. L. (2001). Characterizing visual performance fields: Effects of transient covert attention, spatial frequency, eccentricity, task and set size. *Spatial Vision*, *15*, 61–75.

Carrasco, M., Williams, P. E., & Yeshurun, Y. (2002). Covert attention increases spatial resolution with or without masks: Support for signal enhancement. *Journal of Vision*, *2*, 467–479.

Carrasco, M., & Yeshurun, Y. (1998). The contribution of covert attention to the set-size and eccentricity effects in visual search. *Journal of Experimental Psychology: Human Perception and Performance*, *24*, 673–692.

Deco, G., & Zihl, J. (2001). A neurodynamical model of visual attention: Feedback enhancement of spatial resolution in a hierarchical system. *Journal of Computational Neuroscience*, *10*, 231–253.

De Valois, R. L., & De Valois, K. K. (1988). *Spatial vision*. New York: Oxford University Press.

Dosher, B. A., & Lu, L. (2000). Mechanisms of perceptual attention in precuing of location. *Vision Research*, *40*(10–12), 1269–1292.

Eckstein, M. P., Shimozaki, S. S., & Abbey, C. K. (2002). The footprints of visual attention in the Posner cueing paradigm revealed by classification images. *Journal of Vision*, *2*, 25–45.

Fogel, I., & Sagi, D. (1989). Gabor filters as texture discriminator. *Biological Cyber*, *61*, 103–113.

Gandhi, S. P., Heeger, D. J., & Boynton, G. M. (1999). Spatial attention affects brain activity in human primary visual cortex. *Proceedings of the National Academy of Sciences of the United States of America*, *96*, 3314–3319.

Giordano, A. M., McElree, B., & Carrasco, M. (2009). On the automaticity and flexibility of covert attention: A speed-accuracy trade-off analysis. *Journal of Vision*, *9*(3), 30, 1–10.

Golla, H., Ignashchenkova, A., Haarmeier, T., & Thier, P. (2004). Improvement of visual acuity by spatial cueing: A comparative study in human and non-human primates. *Vision Research*, *44*(13), 1589–1600.

Goto, M., Toriu, T., & Tanahashib, J. (2001). Effect of size of attended area on contrast sensitivity function. *Vision Research*, *41*, 1483–1487.

Graham, N. (1989). *Visual pattern analyzers*. New York: Oxford University Press.

Graham, N., Beck, J., & Sutter, A. (1992). Nonlinear processes in spatial-frequency channel models of perceived texture segregation: Effects of sign and amount of contrast. *Vision Research*, *32*, 719–743.

Greenwood, P. M., & Parasuraman, R. (2004). The scaling of spatial attention in visual search and its modification in healthy aging. *Perception & Psychophysics*, *66*(1), 3–22.

Gurnsey, R., Pearson, P., & Day, D. (1996). Texture segmentation along the horizontal meridian: nonmonotonic changes in performance with eccentricity. *Journal of Experimental Psychology: Human Perception and Performance*, *22*, 738–757.

Hein, E., Rolke, B., & Ulrich, R. (2006). Visual attention and temporal discrimination: Differential effects of automatic and voluntary cueing. *Vision Cognition*, *13*(1), 20–50.

Hock, H. S., Balz, G. W., & Smollon, W. (1998). Attentional control of spatial scale: Effects on self-organized motion patterns. *Vision Research*, *38*, 3743–3758.

Joffe, K. M., & Scialfa, C. T. (1995). Texture segmentation as a function of eccentricity, spatial frequency and target size. *Spatial Vision*, *9*, 325–342.

Kastner, S., & Ungerleider, L. G. (2000). Mechanisms of visual attention in the human cortex. *Annual Review of Neuroscience*, *23*, 315–341.

Kehrer, L. (1989). Central performance drop on perceptual segregation tasks. *Spatial Vision*, *4*, 45–62.

Kehrer, L. (1997). The central performance drop in texture segmentation: A simulation based on a spatial filter model. *Biological Cyber*, *77*, 297–305.

Kinchla, R. A., Chen, Z., & Evert, D. L. (1995). Pre-cue effects in visual search: Data or resource limited? *Perception & Psychophysics*, *57*(4), 441–450.

Ito, M., & Gilbert, C. D. (1999). Attention modulates contextual influences in the primary visual cortex of alert monkeys. *Neuron*, *22*, 593–604.

Lavie, N. (1995). Perceptual load as a necessary condition for selective attention. *Journal of Experimental Psychology: Human Perception and Performance*, *21*, 451–468.

Lee, D. K., Itti, L., Koch, C., & Braun, J. (1999). Attention activates winner-take-all competition among visual filters. *Nature Neuroscience*, *2*, 375–381.

Levi, D. M., Klein, S. A., & Aitsebaomo, A. P. (1985). Vernier acuity, crowding and cortical magnification. *Vision Research*, *25*(7), 963–977.

Ling, S., & Carrasco, M. (2006a). When sustained attention impairs perception. *Nature Neuroscience*, *9*, 1243–1245.

Ling, S., & Carrasco, M. (2006b). Sustained and transient covert attention enhance the signal via different contrast response functions. *Vision Research*, *46*, 1210–1220.

Lu, Z.-L., & Dosher, B. A. (1998). External noise distinguishes attention mechanisms. *Vision Research*, *38*(9), 1183–1198.

Lu, Z.-L., & Dosher, B. A. (2000). Spatial attention: Different mechanisms for central and peripheral temporal precues? *Journal of Experimental Psychology: Human Perception and Performance*, *26*, 1534–1548.

Lu, Z.-L., & Dosher, B. A. (2004). Spatial attention excludes external noise without changing the spatial frequency tuning of the perceptual template. *Journal of Vision*, *4*(10), 10, 955–966.

Luck, S. J., Hillyard, S. A., Mouloua, M., & Hawkins, H. L. (1996). Mechanisms of visual-spatial attention: Resource allocation or uncertainty reduction? *Journal of Experimental Psychology: Human Perception and Performance*, *22*, 725–737.

Luck, S. J., Hillyard, S. A., Mouloua, M., Woldorff, M. G., Clark, V. P., & Hawkins, H. L. (1994). Effects of spatial cuing on luminance detectability: Psychophysical and electrophysiological evidence for early selection. *Journal of Experimental Psychology: Human Perception and Performance*, *20*, 887–904.

Malik, J., & Perona, P. (1990). Preattentive texture discrimination with early vision mechanisms. *Journal of the Optical Society of America A*, *7*, 923–932.

Martinez, A., Anllo-Vento, L., Sereno, M. I., Frank, L. R., Buxton, R. B., Dubowitz, D. J., et al. (1999). Involvement of striate and extrastriate visual cortical areas in spatial attention. *Nature Neuroscience*, *2*(4), 364–369.

Mayfrank, L., Kimmig, H., & Fischer, B. (1987). In J. K. O'Regan & A. Levy-Schoen (Eds.), *Eye movements: From physiology to cognition* (pp. 37–45). New York: North-Holland.

Montagna, B., Pestilli, F., & Carrasco, M. (2009). Attention trades off spatial acuity. *Vision Research*, *49*, 735–745.

Moran, J., & Desimone, R. (1985). Selective attention gates visual processing in the extrastriate cortex. *Science*, *229*, 782–784.

Morgan, M. J., Ward, R. M., & Castet, E. (1998). Visual search for a tilted target: Tests of spatial uncertainty models. *Quarterly Journal of Experimental Psychology*, *51A*, 347–370.

Motter, B. M. (1993). Focal attention produces spatially selective processing in visual cortical areas V1, V2, and V4 in the presence of competing stimuli. *Journal of Neurophysiology*, *70*, 909–919.

Movshon, J. A., & Lennie, P. (1979). Pattern-selective adaptation in visual cortical neurones. *Nature*, *278*, 850–852.

Muller, H. J., & Rabbitt, P. M. (1989). Reflexive and voluntary orienting of visual attention: Time course of activation and resistance to interruption. *Journal of Experimental Psychology: Human Perception and Performance*, *15*, 315–330.

Müller, N. G., Bartelt, O. A., Donner, T. H., Villringer, A., & Brandt, S. A. (2003). A physiological correlate of the "zoom lens" of visual attention. *The Journal of Neuroscience*, *23*(9), 3561–3565.

Nakayama, K., & Mackeben, M. (1989). Sustained and transient components of focal visual attention. *Vision Research*, *29*, 1631–1647.

Olzak, L. A., & Thomas, J. P. (1986). Seeing spatial patterns. In K. R. Boff, L. Kaufman, & J. P. Thomas (Eds.), *Handbook of perception and human performance* (Vol. 1, pp. 1–65). New York: Wiley.

Palmer, J., Verghese, P., & Pavel, M. (2000). The psychophysics of visual search. *Vision Research*, *40*, 1227–1268.

Pestilli, F., & Carrasco, M. (2005). Attention enhances contrast sensitivity at cued and impairs it at uncued locations. *Vision Research*, *45*, 1867–1875.

Pestilli, F., Ling, S., & Carrasco, M. (2009). A population-coding model of attention's influence on contrast response: Estimating neural effects from psychophysical data. *Vision Research*, *49*, 1144–1153.

Pestilli, F., Viera, G., & Carrasco, M. (2007). How do attention and adaptation affect contrast sensitivity? *Journal of Vision*, *7*(7), 1–12.

Phillips, G. C., & Wilson, H. R. (1984). Orientation bandwidths of spatial mechanisms measured by masking. *Journal of the Optical Society of America A*, *1*, 226–232.

Potechin, C., & Gurnsey, R. (2003). Backward masking is not required to elicit the central performance drop. *Spatial Vision*, *16*, 393–406.

Reynolds, J. H., & Chelazzi, L. (2004). Attentional modulation of visual processing. *Annual Review of Neuroscience*, *27*, 611–647.

Reynolds, J. H., & Desimone, R. (1999). The role of neural mechanisms of attention in solving the binding problem. *Neuron*, *24*, 19–29.

Saul, A. B., & Cynader, M. S. (1989). Adaptation in single units in visual cortex: The tuning of aftereffects in the spatial domain. *Visual Neuroscience*, *2*, 593–607.

Shiu, L.-P., & Pashler, H. (1994). Negligible effect of spatial precuing on identification of single digits. *Journal of Experimental Psychology: Human Perception and Performance*, *20*, 1037–1054.

Smith, P. L. (2000). Attention and luminance detection: Effects of cues, masks, and pedestals. *Journal of Experimental Psychology: Human Perception and Performance*, *26*, 1401–1420.

Sutter, A., Beck, J., & Graham, N. (1989). Contrast and spatial variables in texture segregation: Testing a simple spatial-frequency channels model. *Perception & Psychophysics*, *46*, 312–332.

Sutter, A., Sperling, G., & Chubb, C. (1995). Measuring the spatial frequency selectivity of second-order texture mechanisms. *Vision Research*, *35*, 915–924.

Talgar, C. P., & Carrasco, M. (2002). Vertical meridian asymmetry in spatial resolution: Visual and attentional factors. *Psychonomic Bulletin and Review*, *9*, 714–722.

Treisman, A. (1985). Preattentive processing in vision. *Computer Vision, Graphics, and Image Processing*, *31*, 156–177.

Tsal, Y., & Shalev, L. (1996). Inattention magnifies perceived length: The attentional receptive field hypothesis. *Journal of Experimental Psychology: Human Perception and Performance*, *22*, 233–243.

Womelsdorf, T., Anton-Erxleben, K., Pieper, F., & Treue, S. (2006). Dynamic shifts of visual receptive fields in cortical area MT by spatial attention. *Nature Neuroscience*, *9*, 1156–1160.

Yeshurun, Y., & Carrasco, M. (1998). Attention improves or impairs visual perception by enhancing spatial resolution. *Nature*, *396*, 72–75.

Yeshurun, Y., & Carrasco, M. (1999). Spatial attention improves performance in spatial resolution tasks. *Vision Research*, *39*, 293–306.

Yeshurun, Y., & Carrasco, M. (2000). The locus of attentional effects in texture segmentation. *Nature Neuroscience*, *3*, 622–627.

Yeshurun, Y., & Carrasco, M. (2008). The effects of transient attention on spatial resolution and the size of the attentional cue. *Perception & Psychophysics*, *70*(1), 104–113.

Yeshurun, Y., Montagna, B., & Carrasco, M. (2008). On the flexibility of sustained attention and its effects on a texture segmentation task. *Vision Research*, *48*(1), 80–95.

N. Srinivasan (Ed.)
Progress in Brain Research, Vol. 176
ISSN 0079-6123

CHAPTER 6

Focused and distributed attention

Narayanan Srinivasan*, Priyanka Srivastava, Monika Lohani and Shruti Baijal

Centre of Behavioural and Cognitive Sciences, University of Allahabad, Allahabad, India

Abstract: Recent studies on attention have emphasized distinctions between focused and distributed attention. Distributed attention has been shown to play a key role in obtaining statistical information or processing global aspects of a scene. In addition to differences in information processing, focused and distributed attention differ in terms of thc way they interact with emotions. We review findings that indicate close relationship between focused attention and sad emotions as well as distributed attention and happy emotions. Given the potentially close relationship between attention and consciousness, these two types of attention may differ in terms of processes leading to awareness. We review different positions on the relationship between attention and consciousness and arguments for the existence of opposition between attention and awareness that have been made based on findings with color afterimages. We discuss our studies on attention and afterimages indicating the close linkage between different types of attention and awareness as indicated by differences in the strength of afterimages based on the type of attention deployed.

Keywords: focused attention; distributed attention; emotions; awareness; afterimages

Focused attention

The process of selecting information from the visual field for identification and awareness has been visualized in terms of a spotlight (Posner, 1980) or zoom lens (Eriksen and Yeh, 1985). Selective attention is theorized in terms of the stage at which the selection occurs. Early selection theories (Broadbent, 1958) argue that selection occurs at an early stage in perceptual processing and directing attention to a particular location or object typically enhances information processing at that location or for that object. Late selection theories argue that selection occurs after identification of stimuli to choose appropriate actions or responses (Deutsch and Deutsch, 1963). Intermediate views on the stage at which selection occurs have also been proposed (Treisman, 1960). In general, selective attention focused toward a location or object or an action results in better performance.

Studies based on visual search (Treisman and Gelade, 1980) have led to a two-stage model consisting of a preattentive stage and an attentive stage. Preattentive processing can be defined as quick and basic feature analysis of the visual field, on which attention can subsequently operate. These basic featural computations are combined or bound together through focused attention enabling object identification (Treisman and Gelade, 1980). An alternate way to think of attention would be in terms of the load theory of attention. Focusing on a task at hand can prevent

*Corresponding author.
Tel./Fax: +915322460738;
E-mail: nsrini@cbcs.ac.in, ammuns@yahoo.com

DOI: 10.1016/S0079-6123(09)17606-9

task-irrelevant stimuli from reaching awareness (early selection) when the processing of task-relevant stimuli involves a high level of perceptual load that consumes all available capacity. In contrast, when processing of the task-relevant stimuli places lower demands (low load) on the perceptual system, spare capacity or processing resources leads to the perception of irrelevant stimuli as proposed by late selection theories (Lavie, 1995; Lavie et al., 2004). In contrast to focused attention, attention could be distributed over visual space to enable processing of multiple stimuli. We discuss the concept of distributed attention in the next section. We also discuss the role of focused and distributed attention in terms of emotional information processing as well as awareness.

Distributed attention

The concept of distributed attention has been proposed to explain aspects of information processing that cannot be accounted by focused attention. Treisman (2006) has discussed the significance of two types of distinct attentional allocations that lead to differences in processing, with focused attention enabling detailed analysis of specific features and objects and distributed attention facilitating global registration of scene properties. Ariely (2001) showed that visual system represents statistical properties when sets of similar objects are presented. He showed that mean size of discs of various sizes could be perceived more accurately compared to their individual sizes. Moreover, it was later shown that mean judgment was more compatible with tasks that require distributing attention globally compared to a task that requires focusing attention to individual items in the display (Chong and Treisman, 2005b). Even the variation of inherent properties of the distribution of sizes also did not affect mean judgments (Chong and Treisman, 2003). These findings points to separate mechanisms underlying distributed attention system.

It is possible that the distributed attention mechanisms are recruited when focused attention fails to benefit perception. For example, it was shown that even with the poor identification of individual items such as orientation signals in crowded displays, the visual system accurately estimates the average tilt (Parkes et al., 2001). The extraction of the statistical properties appears to be a robust process and it applies to many stimulus dimensions, including orientation (Dakin and Watt, 1997; Parkes et al., 2001), motion speed (Atchley and Andersen, 1995; Watamaniuk and Duchon, 1992), and motion direction (Williams and Sekuler, 1984). Moreover, the mean size can be computed almost as efficiently as the size of a single item (Chong and Treisman, 2003). Mean judgment accuracy also remains good under difficult perceptual conditions, such as brief set exposure duration, or the insertion of a delay between two sets that need to be compared based on mean judgments. In addition, increasing the numerosity and density of the elements of the multiple item display did not impair the performance on the mean judgment task (Ariely, 2001; Chong and Treisman, 2005a). There is also evidence that extraction of statistical properties is not an automatic process and can be modulated by features of the previously attended item (de Fockert and Marchant, 2008).

An alternative account for the lack of set size effects on computing statistical information is the subsampling strategy which has been used as an alternate explanation for the findings of mean judgment of size (Myczek and Simons, 2008). A number of simulations were performed where subsets were selected at random and on average those subsets had a mean size similar to that of the entire set. As a result, simulations based on subset-averaging of one or two items were very similar to the performance of participants who were instructed to average the entire set. However, the strategy of subset-averaging may not explain all the findings on distributed attention based tasks of mean computation. For example, in dual task conditions, the task of mean judgment benefited from tasks requiring distributed or global attention (pop-out search) compared to focused attention task (conjunction search) (Chong and Treisman, 2005b). This emphasizes parallel processing mechanisms for distributed attention contrary to serial processing mechanisms for focused attention. The parallel accounts for distributed attention were also confirmed when an advantage in mean judgment was observed with successive presentation compared to simultaneous presentation of sets (Chong and Treisman, 2005b). In addition to statistical information, distributed attention might be linked to happy emotions and the link between differences in focused and distributed attention in the context of emotions is discussed in the next section.

Scope of attention and emotions

Given the profound social significance, emotions play a significant role in modulating cognitive processes including attention and perception. Studies investigating the emotion–attention interaction, with dot probe (Mogg et al., 1997), visual search (Vuilleumier et al., 2001), and stroop task (MacKay et al., 2004) show emotional stimuli capture and direct attention more readily than neutral stimuli. Imaging studies have also shown amplified response for emotional stimuli compared to neutral stimuli (Stormark et al., 1995). The effects of emotions on cognitive processes like attention and memory are emotion specific (Bradley et al., 2000; Eastwood et al., 2001, 2003; Frischen et al., 2008; Gupta and Srinivasan, 2009; Ohman et al., 2001; Srinivasan and Gupta, submitted; Srinivasan and Hanif, in press; Vuilleumier, 2001).

Several studies have shown that emotional expressions capture attention and interfere with the ongoing task even though they are not relevant to the current task (Vuilleumier et al., 2001; White, 1996). Negative emotional expressions have shown more interference than positive emotional expression indicating more effective attention capture by negative emotional expressions (Yantis, 1996). Studies using visual search task (Eastwood et al., 2001; also see Williams et al., 2005) have shown that sad faces were detected faster than happy faces among neutral faces. In another study using visual search, participants required to count features embedded in negative, positive, and neutral schematic faces, took longer time with negative faces compared to positive or neutral faces (Eastwood et al., 2003). These findings indicate that faces with sad expression may capture attention faster and also holds attention for a longer period of time.

In addition to attention capture, emotions also interact with the scope of attention. Control of attention has been shown to be influenced by the current affective state of the observer (Hasher et al., 2007; Oaksford et al., 1996). It has been long hypothesized that arousal during negative states is associated with a constriction of attentional focus (Derryberry and Reed, 1998). The narrowing of attention is sometimes referred as "weapon focus" where attention is focused at the expense of encoding peripheral details (Christianson and Loftus, 1990). However, studies on positive emotion show that positive emotional stimuli broaden the scope of attentional processes according to the broaden-and-build theory (Fredrickson, 2004; Fredrickson and Branigan, 2005; Wadlinger and Issacowitz, 2006). Broaden-and-build theory proposes that a primary function of positive emotion is to broaden people's thought-action repertoires (Fredrickson, 2001, 2003), increasing their flexibility and enhancing their global scope. Effect of positive affect on creative and more generative mindset shows greater cognitive flexibility across diverse situations (Estrada et al., 1994, 1997), intuitive judgments (Bolte et al., 2003), decision making (Isen, 2001), and creative problem solving (Isen et al., 1987, 1985).

Evidence of broadening of attention comes from a study by Fredrickson (2003) in which a particular emotion was induced by showing participants small evocative film clips. For example, emotion of joy was elicited by showing a herd of playful penguins waddling and sliding on the ice, sadness was elicited with scenes of death and funerals, serenity was elicited with clips of peaceful nature scene, and neutral scenes were used to elicit no emotion. Using global–local visual processing tasks, they measured whether participants saw the big picture or focused on the smaller details. The participants' task was to judge which of the two comparison figures is more similar to a standard figure. One comparison figure resembled the standard in global configuration and the other in local, detailed elements. They found that people who experienced positive emotions (as assessed by self-report or electromyographic signals from the face) tend to choose the global configuration, suggesting a broadened pattern of thinking.

Similarly, another study (Fredrickson and Branigan, 2005) measured the scope of attention and thought-action repertoires as a function of positive emotion and showed that relative to neutral and negative emotions, positive emotion broadens the scope of attention and thought-action repertoires by showing global bias. In their study, temporary states of emotion (amusement, contentment, neutrality, anger, and anxiety) were induced by showing movies followed by the identification of the hierarchical visual stimuli. Participants showed biased selection to global shape when it was followed by a positive emotional state compared to negative and neutral state. Thought-action repertoires were evaluated by the open-ended twenty statements test, which showed that people experiencing positive emotions have more numerous thought-action urges than people experiencing negative emotions.

Rowe et al. (2007) investigated the role of positive emotion in broadening the scope of visual attentional filter and reducing the selectivity. They found that positive emotion results in a fundamental change in the breadth of attentional allocation to both external and internal conceptual space. In their study, they measured the effect of positive emotion on two different cognitive domains: semantic search (remote associate task) and visual selective attention (Erickson flanker task). In remote associate task, participants were asked to override typically semantic associations to find semantically distant or remote associations, whereas in Erickson flanker task, participants were presented a target with flanking distractors and task was to selectively attend the central target while ignoring the distractors. In the conceptual domain, relative to the neutral and sad mood, positive affect was associated with increased capacity to generate remote associates for the familiar words (Isen, 2001). In the visuospatial domain, positive affect impaired the visual selective attention by increasing processing of spatially adjacent flanking distractors, suggesting an increase in the scope of visuospatial attention.

Similarly using a flanker task, Fenske and Eastwood (2003) found flanker effect for happy faces but not sad faces indicating that sad faces lead to narrowing of attention and potentially happy faces might lead to broadening of attention. Srinivasan and Gupta (submitted) have investigated the scope of attention on emotional information by manipulating perceptual load. The participants were shown emotional stimuli (happy, sad, and neutral faces) in the background (distractor) with a letter string consisting of six letters at the center. Participants were required to report the color of the string in the low-load condition and a specific target letter in the high-load condition. The experiment with different load conditions was immediately followed by a surprise recognition test for the distractor faces. The results showed better recognition memory for sad faces compared to happy faces during more focused attention in the high-load condition. In addition, happy faces were recognized better compared to sad faces in the case of distributed attention in the low-load as well as high-load conditions. These results indicate that sad and happy faces interact differently with attention. Sad faces are associated with focused attention while happy faces are associated with distributed attention.

Another study by Srivastava and Srinivasan (2008) has investigated the role of emotional stimuli in shifts of visual attention between objects. In their study, participants were presented with happy and sad stimuli using attentional dwell time paradigm. Attentional dwell time paradigm is method of displaying two targets in sequence at different location with variable temporal separations between two targets. Two experiments were conducted by manipulating emotional faces as T1 or T2 in separate experiments. In the first experiment, emotional stimuli (T1) were followed by the neutral target (T2). The result showed less impairment for neutral T2 performance when it was preceded by the happy faces compared to sad faces. This could be due to lesser attentional resources or broadening of attention associated with happy faces. To investigate whether happy stimuli demand less attentional resources, second experiment was conducted using emotional stimuli as T2 preceded by neutral T1. Result showed better identification of the happy faces than sad faces, indicating less attentional demand for happy stimuli compared to sad face.

In agreement with the differences in the scope of attention, emotion identification has been shown to be associated with differences in processing of hierarchical information (Srinivasan and Hanif, in press). In this study, participants were shown hierarchical letters followed by emotional faces. The task was to identify the emotion present in a face as soon as possible followed by the reporting of the preceding target that occurred either at the global or local level. Happy faces preceded by global target identification were faster than local target identification. Once again these results indicate close relationship between perceptual processing strategies associated with differences in the scope of attention and emotions. In addition to emotion identification, differences in scope of attention have been linked to approach and avoidance behavior (Förster et al., 2006). Approach behavior has been associated with global processing due to the broadening of the scope of attention and avoidance behavior is associated with local processing due to the narrowing of the scope of attention (Förster et al., 2006). Differences in processing global–local processing have also been reciprocally linked differences in regulatory focus with promotion focus linked to global processing and prevention focus linked to local processing (Förster and Higgins, 2005).

These finding support the theories that argue for emotion–attention interactions and more specifically show that reciprocal links between emotions and the scope of attention. Better performance for positive information in presence of global stimuli compared to local stimuli or vice versa supports the theory of positive emotion (Fredrickson, 2003). It also indicates that broadening of attention requires less attention (Srivastava and Srinivasan, 2008) therefore interfere less with the subsequent target processing. In addition to differences related to the nature of information processing as well as emotions, focused and distributed attention might be linked differences in awareness. In the next section, we discuss the relationship between attention and awareness in the context of different types of attention.

Types of attention and awareness

The role of attention in awareness is a central question in the cognitive sciences (James, 1890). One of the earliest discoveries reflecting this idea comes from observations that when people were asked to attend to two events at the same time, they typically became conscious of only one event at any given moment in time (Broadbent, 1958; Cherry, 1953). Findings from many different paradigms have led to views arguing for a strong relationship between attention and consciousness (Mack and Rock, 1998; Rensink et al., 1997). It has been suggested that attention may be necessary for consciousness. It is now widely accepted that the understanding of consciousness rests upon appreciation of the brain networks that subserve attention (Posner, 1994). Given the close relationship between attention and consciousness, a model of cortical-thalamic network implicated in the studies of visual attention was proposed for the study of consciousness (Crick, 1994).

Studies that have provided compelling evidence for the close link between attention and awareness have used the paradigm of inattentional blindness (Mack and Rock, 1998). In their experiments, observers were briefly presented with a cross and were asked to judge, out of the vertical or horizontal components (that differed slightly in length), which of the two was longer. In a critical trial, an irrelevant stimulus was flashed in one of the quadrants formed by the cross. After the trial, observers were asked to perform a recognition task to test whether they could identify the unexpected target. With their attention focused on the discrimination task, a large number of observers failed to notice the target stimulus. Around 25% of participants said that they did not notice the unexpected stimulus that appeared parafoveally while the cross was presented at fixation. Interestingly around 75% of the participants reported not perceiving the target stimulus that appeared at fixation with the cross presented parafoveally. Observers failed to report the irrelevant stimulus when they were not aware that such a stimulus might appear, although the unidentified stimulus would have been visible under normal conditions. Mack and Rock (1998) argued that in the absence of attention, the irrelevant stimuli never rose to the level of conscious perception. We may not consciously perceive objects that we have not attended.

The lack of attention leading to inattentional blindness is also used to explain the failure of change detection in several change blindness (CB) experiments (Grimes, 1996; Rensink, 2002; Rensink et al., 1997). Grimes (1996) tracked observers' eye movements while they viewed scenes for 10 s, in a change detection experiment. Scenes were altered during eye movements, and a single object was changed either in size, color, or location or they could disappear. Observers failed to detect these changes because the changed object was not attended and thus not consciously perceived. CB is the phenomena where we fail to perceive large changes, in our surroundings as well as in experimental conditions. Change could be in existence, properties, semantic identity, and spatial layout. Attention is required to perceive change and in the absence of localized transient motion signals (that may attract or grab attention) attention is directed by high level of interest (Rensink et al., 1997). Only when attention is focused on an object, a change in the object is usually perceived. The contents of visual short-term memory are simply over written with succeeding stimuli without focused attention (Rensink, 2002).

However, inattentional blindness fails to explain convincingly the results of Simons and Levin (1997) or Rensink et al. (1997) experiments in which stimuli is presented for a very long time. In their CB experiments, observers may have attended to the object and yet not detected changes to them. CB studies do show that more information is available than what is reported. For example, it has been shown that performance on a localization task was above chance level even in undetected trials (Fernandez-Duque and Thornton, 2000). In addition, response times are longer in failed change detection trials in which change actually occurred (Williams and Simons, 2000). Change detection has been shown for changes in the background (Driver et al., 2001). More interesting are claims of mindsight in which observers claimed to sense the change before they were aware of the change suggesting that sensing could be a different form of awareness (Rensink, 2004).

A slightly different perspective on the close relationship between attention and consciousness is provided by the studies in which load was manipulated and awareness of stimuli were evaluated (Cartwright-Finch and Lavie, 2006; Lavie, 2006). One was an inattentional blindness task in which the primary task was easy (low load) or difficult (high load). They found that inattentional blindness was more in the high-load condition compared to the low-load condition (Lavie, 2006). They also performed a change detection study in which the primary task (low load or high load) was presented at fixation and change between two scenes had to be detected at peripheral locations. Once again, change detection was better in the low-load condition compared to the high-load condition indicating that focused attention is necessary and plays a critical role in awareness (Lavie, 2006). In addition to better performance, a recent study has shown that attention can alter phenomenal appearance (Carrasco et al., 2004). They showed that the contrast of an attended (using an exogenous cue) grating was higher than the contrast of the unattended grating indicating once again the critical role of focused attention in awareness.

While acknowledging the close relationship between attention and consciousness, a large number of recent studies have convincingly argued that attention is different from consciousness (LaBerge, 1995; Baars, 1997; Hardcastle, 1997; Naccache et al., 2002; Crick and Koch, 2003; Lamme, 2003; Woodman and Luck, 2003; Kentridge et al., 2004). According to Lamme (2003), consciousness operates prior to attention. Attentional selection operates on conscious stimuli leading to verbal report or store for later conscious, typically verbal access. Unconscious stimuli are outside the control of attention.

According to Dehaene et al. (2006), consciousness and top-down attention can be thought of in terms of a 2×2 matrix in which one of the dimensions is bottom-up stimulus strength (weak or sufficiently strong) and the other is top-down attention (absent or present). They identified four classes of processing: subliminal-unattended, subliminal-attended, preconscious, and conscious. These different types of processes are subserved by different neural networks. Conscious processing refers to the case in which stimulus strength is high and top-down attention is present. This class is characterized by reportability, intense activation, and long-range interaction across cortical areas. They also argue that the subliminal (unattended) is characterized by absence of priming, is typically not affected by top-down attention, and can be characterized as essentially feed-forward processes in the brain. Unlike the subliminal (unattended) processes, the processes in the other subliminal class are supposed to show stronger activation and short-term priming. Both the subliminal types of processes are not associated with reportability. The preconscious, mainly sensorimotor in nature, display priming effects and are also not reportable in the absence of top-down attention. They also argue that global synchronization is characteristic of conscious processes and local synchronization is characteristic of preconscious processes (Dehaene et al., 2006).

In a similar vein, Koch and Tsuchiya (2007) have proposed a fourfold classification scheme in which attention and consciousness are different. Certain processes are analyzed in terms of whether top-down attention is necessary or not and whether they give rise to consciousness, resulting in a 2×2 matrix of possibilities. Some processes like early rapid vision do not need attention and may not give rise to consciousness. This will also cover a significant amount of unconscious information processing. Some processes may need attention and will give rise to consciousness. Some processes like priming and thoughts may require attention and may not give rise to consciousness. It is quite possible that some processes benefit from attentional processing without the involvement of consciousness. The most interesting possibility is the case of processes for which attention is not required but gives rise to consciousness.

Argument for the potentially opposite effects of attention and awareness have been made based on findings from studies in which the lack of attention resulted in better performance (Kanai and Verstraten, 2006; Li et al., 2002). In an experiment using stimuli in which the direction of motion was ambiguous, priming effect was reduced when attention was distracted using a task in between the presentation of the prime and the ambiguous motion stimulus (Kanai and Verstraten, 2006). The role of attention on the ability to identify meaningful categories has been investigated with a dual task paradigm involving a difficult visual search task in which observers had to search for an odd element in an array of five randomly rotated Ls or Ts as well as a scene/object categorization task (Li et al., 2002). Participants performed better with categorization of objects present in natural scenes like animal versus non-animals and vehicle versus non-vehicles and such quick categorizations involving meaningful stimuli has been argued to occur with almost no attention (Li et al., 2002).

Although several accounts have described the relationship between attention and consciousness in terms of attended versus unattended as well as conscious versus unconscious, it is important to consider the effects of different types of attention and consciousness. One way in which consciousness has been characterized is in terms of primary consciousness and access consciousness (Block, 2005). Primary consciousness refers to the phenomenal aspects of experience, i.e., qualia. Access consciousness refers to the functional aspects of consciousness, which is related to cognitive processes like executive attention, planning, and voluntary control that facilitates its subjective nature and reportability.

Wyart and Tallon-Baudry (2008) recorded magnetoencephalographic signals while human subjects performed a task in which faint gratings were presented at an attended or unattended location (on some trials no stimulus was presented). After each trial, participants indicated which of two orientations they thought matched the previously presented grating and whether they had seen the grating. Trials were classified as aware (grating was detected and orientation was identified correctly) or as unaware (grating was not detected and orientation was identified at chance level). Spatial attention increased the likelihood of conscious report: more gratings were consciously seen at the attended location ($\sim 50\%$) than at the unattended location ($\sim 40\%$). Attention also shortened reaction times on the orientation discrimination task for consciously seen gratings, but not for unseen gratings. Additionally, the gamma band power changes reflected in separate frequency and time ranges represented attention and awareness-related activity, which was found to be independent of each other although both correlated with conscious report. The awareness-related gamma power changes represented phenomenal awareness that represents raw neural representation of perceptual information (van Gaal and Fahrenfort, 2008).

Melloni et al. (2007) investigated the neural correlates of access awareness, which relates to the ability to report about the phenomenal representations and found that conscious report is selectively correlated with increased phase coupling in gamma band activity across occipital, parietal, and frontal areas rather than power changes. These results indicate that different forms of consciousness may be associated with different types of attention and different neural mechanisms (van Gaal and Fahrenfort, 2008).

Typically, the notion of "attention" used in many of the studies exploring the relationship between attention and awareness is focused attention. Given that different types of attention provide different kinds of information (Ariely, 2001; Chong and Treisman, 2003, 2005a; Treisman, 2006), they may also result in differences in awareness. The phenomenal awareness associated with the reports of participants who have claimed to see more than they could verbally report in iconic memory experiments might be linked with distributed attention and access consciousness might be more closely linked to focused attention. While change detection in an object might depend on focused attention, the feeling associated with sensing the change (without accompanying the detection of change itself) in change detection might depend on processes associated with distributed attention (Rensink, 2004).

Color afterimages and attention

One important methodology that has been used to study awareness is through adaptation and afterimages (Kirschfeld, 1999). An afterimage is complementary to the original pattern in both brightness and color and such afterimages are thus called negative afterimages (Suzuki and Grabowecky, 2003). For example, an afterimage occurs after the adaptation to a particular stimulus (color) for a prolonged period of time, e.g., prolonged looking at a red square produces a green afterimage. Awareness of afterimage as measured by the strength of the afterimage gets affected by manipulation of focused attention during adaptation (Suzuki and Grabowecky, 2003). Focused attention to the adapting stimulus reduces the strength of the afterimage. More specifically, strength of afterimages has been shown to be modulated by spatial spread of attention and level at which the stimulus structure is being processed (Baijal and Srinivasan, submitted).

Suzuki and Grabowecky (2003) showed two overlapped triangles to the participants for 7–10 s during adaptation period. Both the triangles were afterimage inducers. The task was to selectively attend to one of the superimposed triangles (on the basis of color or motion). The results indicated that the attended triangle produced weaker afterimage. The effect was further confirmed by the demonstration of delayed onset of afterimage when the afterimage inducer was attended (when participants reported change in color of the inducer) compared to unattended (when participants performed a digit counting task away from adaptor). Even when attention was manipulated during the formation of afterimages rather than during adaptation, focused attention produced deleterious effects on the strength of afterimage (Lou, 2001). In their study, participants were asked to attend to either of the afterimages and attended afterimage was found to disappear from the awareness faster than unattended afterimage. These findings have been used to argue that attention may have opposing effects on awareness (Koch and Tsuchiya, 2007). However, it is yet not clear how different manipulations of attention affect awareness.

One way to view attention is in terms of processing load (Cartwright-Finch and Lavie, 2006). Theeuwes et al. (2004) argued that low processing load ensues broadening of the attentional window. If low load leads to increase in the scope of attention, then manipulation of processing load also provides a way to investigate the effect of attention on awareness. To observe the effect of processing load on the inducing stimulus and afterimage formation, attention was manipulated in a study with color afterimages using a central task with differing attentional demands/resources during the adaptation period. In the afterimage formation period, attention was manipulated by instructing the participants to attend a particular afterimage to see the effect of voluntary attention on the strength of afterimages.

The stimuli consisted of two triangles of opposite orientations superimposed on each other, forming a star against the black background (Fig. 1). One of the component triangles was green and the other was orange. The triangles were presented along with a constant stream of letters in the centre for 30 s. Load was manipulated using a 0-back task (low load) and 2-back task (high load). The participant had to count the number of occurrences of a given target letter in 0-back task. In the 2-back task, participants had to count the number of times a current letter was the same as the one before the previous letter. This was followed by afterimage formation period where a gray screen with fixation mark was displayed. Blue and pink colored afterimages were formed for orange and green triangles respectively. Participants were instructed to attend to the blue afterimage on half of the trials and pink on the other half trials. Since the onset was immediate after the removal of the adapting stimulus (given the long adaptation periods used), it was not measured. Participants were instructed to press the assigned key as soon as one of the afterimages (attended or unattended) disappeared and press the corresponding key on the reappearance of any of the disappeared triangles. The sequence of frames in a particular trial is shown in Fig. 2.

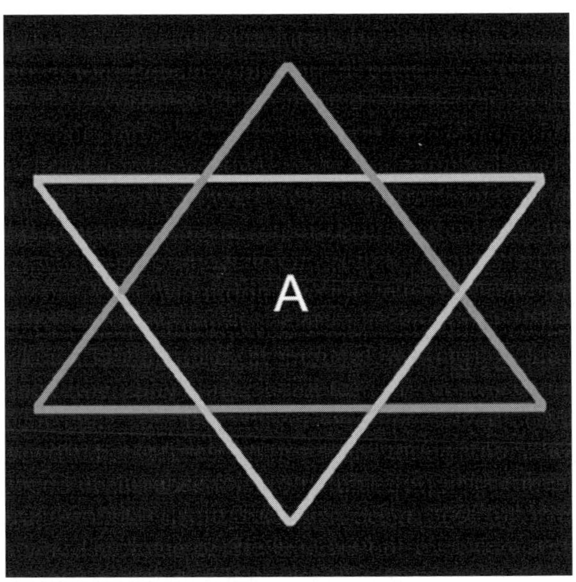

Fig. 1. Afterimage inducer display.

In all the trials, the attended afterimage disappeared first consistent with the findings from Lou (1999). The durations of the afterimage was measured and the results are shown in Fig. 3. There was a significant effect of processing load, during the adaptation stage, on afterimage durations with longer durations for the 2-back condition compared to the 0-back condition (see Fig. 3). The result adds to the previous findings (Suzuki and Grabowecky, 2003) that attention weakens the afterimage and delayed the appearance of the afterimage. It is not simply the lack or presence of attention that affects afterimage formation but processing load also determines the duration of afterimages. Low processing load associated with broadening of attention (Theeuwes et al., 2004) may have resulted in better distractor (the inducers) processing leading to the lesser afterimage durations compared with the high processing load condition. The results indicate that attention and working memory play a critical role in the formation and duration of afterimages.

An explicit distinction was made between different types of attention based on their effects on perceptual awareness (Baijal and Srinivasan, submitted). The participants performed a central task with small, large, local, or global letters and a blue square as adapting stimulus for 20 s. Once the

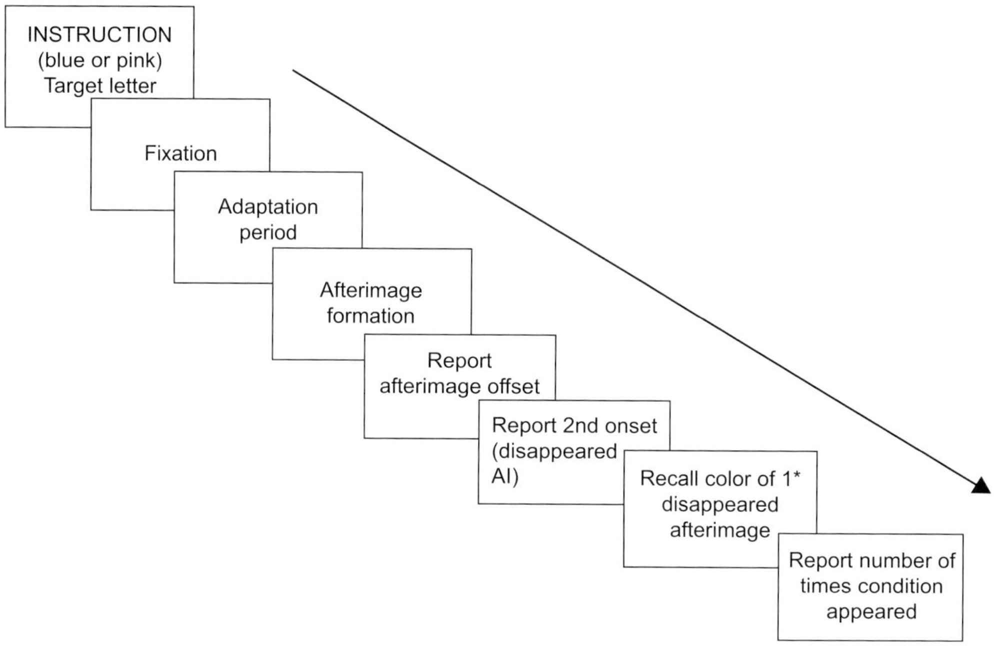

Fig. 2. Sequence of events in a given trial.

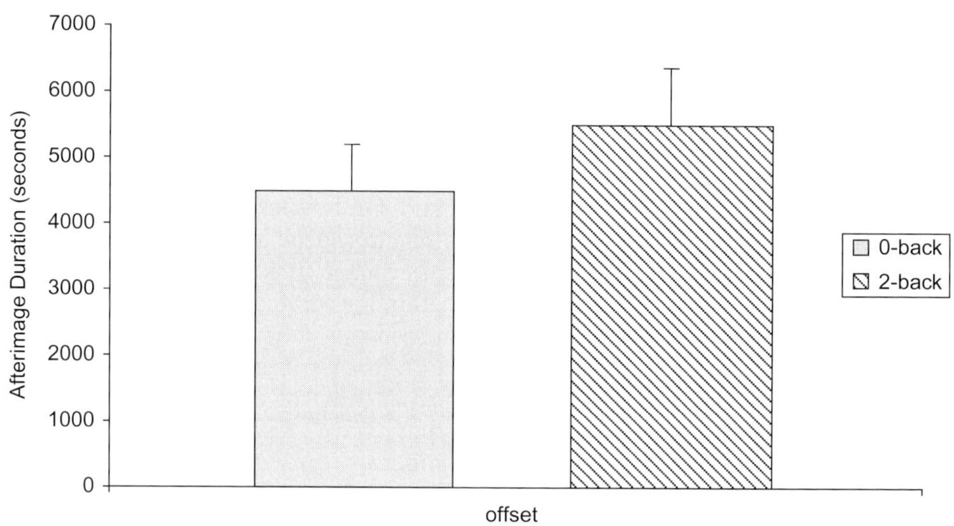

Fig. 3. Afterimage duration as a function of the load.

inducing stimuli was removed resulting in the color afterimage, the participants indicated the onset and offset of the afterimage. It was observed that the increase in spatial spread of attention (modulated by the central task) results in decrease of afterimage duration. However, in terms of levels of processing, global processing produced larger afterimage durations with stimuli controlled for spatial extent. The results suggest that focused or distributed attention produce different effects on awareness, possibly through their differential interactions with polarity-dependent and independent processes involved in the formation of color afterimages.

Attention has been shown to have contrasting effects on color afterimages (Lou, 1999, 2001; Suzuki and Grabowecky, 2003; Tsuchiya and Koch, 2005) and the aftereffects of motion, tilt, and depth (Chaudhuri, 1990; Rose et al., 2003; Spivey and Spirn, 2000). With color afterimages, attention reduces the strength of the afterimages. However, with motion aftereffects, focused attention increases the strength of the aftereffects. A possible explanation of these contrasting effects has been proposed by Suzuki and Grabowecky (2003), according to which attention may affect polarity-dependent and polarity-independent processes differently, thereby leading to different effects on adaptation. Polarity-independent processes in the visual system that play a critical role in contrast adaptation have been postulated to underlie the effect of attention on negative color afterimages (Suzuki and Grabowecky, 2003). The effect of attention on the motion, tilt, and depth aftereffects in this view may depend on polarity-dependent processes.

Yet another possible explanation of the effect of attention on color afterimages is provided with a model based on two different systems, a boundary contour system (BCS) and a feature contour system (FCS) (Wede and Francis, 2007a, b). According to this model, more attention on adapting stimuli generates stronger aftereffects in the orientation-dependent and polarity-independent BCS, resulting in the delayed and weaker color afterimages produced in the polarity-dependent FCS that underlies the formation of color afterimages. In the context of this model, distributed attention may weaken boundaries thereby resulting in weaker aftereffects in BCS. This would enable stronger color afterimages in the FCS. There is some evidence that global processing is more dependent on low spatial frequency processing (Badcock et al., 1990; Shulman and Wilson, 1987). This would further result in stronger afterimages based on aftereffects in the FCS (Georgeson and Turner, 1985; Wede and Francis, 2007a). While the mechanisms proposed above to explain the effects of differences in attention on color afterimages are tentative, the results of the study with afterimages and other paradigms do clearly show that differences in attention do matter for awareness.

Conclusions

Focused and distributed attention mechanisms differ in terms of the nature of information processing. In addition, they also differ in terms of emotional processing with close links between sad and focused attention as well as happy and distributed attention. Given the natural links between attention and awareness, it is important to consider different types of attention for understanding the relationship between attention and awareness. The results show that not only focused and distributed attention differ in terms of differences in information processing but also may result in differences in awareness. Further investigations, particularly those in which brain activity is simultaneously monitored while different forms of attentional mechanisms are recruited to generate attention-dependent awareness, are needed to understand the relationship between attention and awareness.

References

Ariely, D. (2001). Seeing sets: Representation by statistical properties. *Psychological Science*, *12*, 157–162.

Atchley, P., & Andersen, G. (1995). Discrimination of speed distributions: Sensitivity to statistical properties. *Vision Research*, *35*, 3131–3144.

Baars, B. J. (1997). *In the theater of consciousness: The workspace of the mind*. Oxford, England: Oxford University Press.

Badcock, J. C., Whitworth, F. A., Badcock, D. R., & Lovegrove, W. J. (1990). Low frequency filtering and the processing of local-global stimuli. *Perception*, *19*, 617–629.

Baijal, S. & Srinivasan, N. (submitted). Types of attention matter for awareness: A study with color afterimages.

Block, N. (2005). Two neural correlates of consciousness. *Trends in Cognitive Sciences*, *9*, 46–52.

Bolte, A., Goschke, T., & Kuhl, J. (2003). Emotion and intuition: effects of positive and negative mood on implicit judgments of semantic coherence. *Psychological Science*, *14*, 416–421.

Bradley, B. P., Mogg, K., & Miller, N. H. (2000). Covert and overt orienting of attention to emotional faces in anxiety. *Cognition and Emotion*, *14*, 789–808.

Broadbent, D. E. (1958). *Perception and Communication*. London: Pergamon Press.

Carrasco, M., Ling, S., & Read, S. (2004). Attention alters appearance. *Nature Neuroscience*, *7*, 308–313.

Cartwright-Finch, U., & Lavie, N. (2006). The role of perceptual load in inattentional blindness. *Cognition*, *102*, 321–340.

Chaudhuri, A. (1990). Modulation of the motion aftereffect by selective attention. *Nature*, *344*, 60–62.

Cherry, E. C. (1953). Some experiments on the recognition of speech, with one and with two ears. *Journal of the Acoustic Society of America*, *25*, 975–979.

Chong, S. C., & Treisman, A. (2003). Representation of statistical properties. *Vision Research*, *43*, 393–404.

Chong, S. C., & Treisman, A. (2005a). Statistical processing: Computing the average size in perceptual groups. *Vision Research*, *45*, 891–900.

Chong, S. C., & Treisman, A. (2005b). Attentional spread in the statistical processing of visual displays. *Perception & Psychophysics*, *67*, 1–13.

Christianson, S. A., & Loftus, E. (1990). Some characteristics of people's traumatic memories. *Bulletin of the Psychonomic Society*, *28*, 195–198.

Crick, F. (1994). *The astonishing hypothesis*. New York: Scribner's.

Crick, F., & Koch, C. (2003). A framework for consciousness. *Nature Neuroscience*, *6*, 119–126.

Dakin, S., & Watt, R. J. (1997). The computation of orientation statistics from visual texture. *Vision Research*, *37*, 3181–3192.

de Fockert, J. W., & Marchant, A. P. (2008). Attention modulates set representation by statistical properties. *Perception & Psychophysics*, *70*(5), 789–794.

Dehaene, S., Changeux, J., Naccache, L., Sackur, J., & Sergent, C. (2006). Conscious, preconscious, and subliminal processing: A testable taxonomy. *Trends in Cognitive Sciences*, *10*, 204–211.

Derryberry, D., & Reed, M. A. (1998). Anxiety and attentional focusing: Trait, state and hemispheric influences. *Personality and Individual Differences*, *25*, 745–761.

Deutsch, J. A., & Deutsch, D. (1963). Attention: Some theoretical considerations. *Psychological Review*, *70*, 80–90.

Driver, J., Davis, G., Russell, C., Turatto, M., & Freeman, E. D. (2001). Segmentation, attention and phenomenal visual objects. *Cognition*, *80*, 61–95.

Eastwood, J. D., Smilek, D., & Merikle, P. M. (2001). Differential attention guidance by unattended faces expressing positive and negative emotion. *Perception & Psychophysics*, *63*, 1000–1013.

Eastwood, J. D., Smilek, D., & Merikle, P. M. (2003). Negative facial expression captures attention and disrupts performance. *Perception & Psychophysics*, *65*, 352–358.

Eriksen, C. W., & Yeh, Y. (1985). Allocation of attention in visual field. *Journal of Experimental Psychology: Human Perception and Performance*, *11*, 583–597.

Estrada, C. A., Isen, A. M., & Young, M. J. (1994). Positive affect influences creative problem solving and reported source of practice satisfaction in physicians. *Motivation and Emotion*, *18*, 285–299.

Estrada, C. A., Isen, A. M., & Young, M. J. (1997). Positive affect facilitates integration of information and decreases anchoring in reasoning among physicians. *Organizational Behavior and Human Decision Processes*, *72*, 117–135.

Fenske, M. J., & Eastwood, J. D. (2003). Modulation of focused attention by faces expressing emotion: Evidence from flanker tasks. *Emotion*, *3*, 327–343.

Fernandez-Duque, D., & Thornton, I. M. (2000). Change detection without awareness: Do explicit reports underestimate the representation of change in visual system? *Visual Cognition*, *7*, 323–344.

Förster, J., Friedman, R. S., Özelsel, A., & Denzler, M. (2006). Enactment of approach and avoidance behavior influences the scope of perceptual and conceptual attention. *Journal of Experimental Social Psychology*, *42*, 133–146.

Förster, J., & Higgins, E. T. (2005). How global versus local perception fits regulatory focus. *Psychological Science*, *16*, 631–636.

Fredrickson, B. L. (2001). The role of positive emotions in positive psychology: The broaden-and-build theory of positive emotions. *American Psychology*, *56*(3), 218–226.

Fredrickson, B. L. (2003). The value of positive emotions. *American Scientist*, *91*, 330–335.

Fredrickson, B. L. (2004). The broaden and build theory of positive emotion. *Philosophical Transactions: Biological Sciences (The Royal Society of London)*, *359*, 1367–1377.

Fredrickson, B. L., & Branigan, C. (2005). Positive emotions broaden the scope of attention and thought-action repertoires. *Cognition and Emotion*, *19*, 313–332.

Frischen, A., Eastwood, J. D., & Smilek, D. (2008). Visual search for faces with emotional expressions. *Psychological Bulletin*, *134*(5), 662–676.

Georgeson, M. A., & Turner, R. S. (1985). Afterimages of sinusoidal, square-wave and compound gratings. *Vision Research*, *25*, 1709–1720.

Grimes, J. (1996). On the failure to detect changes in scenes across saccades. In K. Akins (Ed.), *Vancouver studies in cognitive science. Vol. 5. Perception* (pp. 89–110). New York: Oxford University Press.

Gupta, R., & Srinivasan, N. (2009). Emotions help memory for faces: Role of whole and parts. *Cognition and Emotion*, *23*, 807–816.

Hardcastle, V. G. (1997). Attention versus consciousness: A distinction with a difference. *Cognitive Studies: Bulletin of the Japanese Cognitive Science Society*, *4*, 56–66.

Hasher, L., Lustig, C., & Zacks, R. T. (2007). Inhibitory mechanisms and the control of attention. In A. Conway, C. Jarrold, M. Kane, A. Miyake, & J. Towse (Eds.), *Variation in working memory* (pp. 227–249). New York: Oxford University Press.

Isen, A. M. (2001). An influence of positive affect on decision making in complex situations: Theoretical issues with

practical implications. *Journal of Consumer Psychology, 11*(2), 75–85.
Isen, A. M., Daubman, K. A., & Nowicki, G. P. (1987). Positive affect facilitates creative problem solving. *Journal of Personality and Social Psychology, 52*, 1122–1131.
Isen, A. M., Johnson, M. M. S., Mertz, E., & Robinson, G. F. (1985). The influence of positive affect on the unusualness of word associations. *Journal of Personality and Social Psychology, 48*, 1413–1426.
James, W. (1890). *The principles of psychology* (Vol. 1). New York: Holt. (Reprinted in 1950 by Dover Press, New York).
Kanai, R., & Verstraten, F. A. (2006). Attentional modulation of perceptual stabilization. *Proceedings of the Royal Society of London. B: Biological sciences, 273*, 1217–1222.
Kentridge, R. W., Heywood, C. A., & Weiskrantz, L. (2004). Spatial attention speeds discrimination without awareness in blindsight. *Neuropsychologia, 42*, 831–835.
Kirschfeld, K. (1999). Afterimages: A tool for defining the neural correlate of visual consciousness. *Consciousness and Cognition, 8*, 462–483.
Koch, C., & Tsuchiya, N. (2007). Attention and consciousness: Two distinct brain processes. *Trends in Cognitive Sciences, 11*, 16–22.
LaBerge, D. (1995). *Attentional processing*. Cambridge, MA: Harvard University Press.
Lamme, V. A. F. (2003). Why visual awareness and attention are different? *Trends in Cognitive Sciences, 7*, 12–18.
Lavie, N. (1995). Perceptual load as a necessary condition for selective attention. *Journal of Experimental Psychology: Human Perception and Performance, 21*, 451–468.
Lavie, N. (2006). The role of perceptual load in visual awareness. *Brain Research, 1080*, 91–100.
Lavie, N., Hirst, A., De Fockert, J. W., & Viding, E. (2004). Load theory of selective attention and cognitive control. *Journal of Experimental Psychology: General, 133*, 339–354.
Li, F. F., van Rullen, R., Koch, C., & Perona, P. (2002). Rapid natural scene categorization in the near absence of attention. *Proceedings of the National Academy of Sciences, 99*, 9596–9601.
Lou, L. (1999). Selective peripheral fading: Evidence for inhibitory sensory effect of attention. *Perception, 28*, 519–526.
Lou, L. (2001). Effects of voluntary attention on structured afterimages. *Perception, 30*, 1439–1448.
Mack, A., & Rock, I. (1998). *Inattentional blindness*. Cambridge, MA: MIT Press.
Mackay, D. G., Shafto, M., Taylor, J. K., Marian, D. E., Abrams, l., & Dyer, J. R. (2004). Relations between emotion, memory, and attention: Evidence from taboo Stroop, lexical decision, and immediate memory tasks. *Memory & Cognition, 32*, 474–488.
Melloni, L., Molina, C., Pena, M., Torres, D., Singer, W., & Rodriguez, E. (2007). Synchronization of neural activity across cortical areas correlates with conscious perception. *Journal of Neuroscience, 27*, 2858–2865.
Mogg, K., Bradley, B. P., de Bono, J., & Painter, M. (1997). Time course of attentional bias for threat information in nonclinical anxiety. *Behaviour Research and Therapy, 35*, 297–303.
Myczek, K., & Simons, D. J. (2008). Better than average: Alternatives to statistical summary representations for rapid judgments of average size. *Perception & Psychophysics, 70*, 772–788.
Naccache, L., Blandin, E., & Dehaene, S. (2002). Unconscious masked priming depends on temporal attention. *Psychological Science, 13*, 416–424.
Oaksford, M., Morris, F., Grainger, B., & Williams, J. M. G. (1996). Mood, reasoning, and central executive processes. *Journal of Experimental Psychology: Learning, Memory, and Cognition, 22*, 477–493.
Ohman, A., Lundqvist, D., & Esteves, F. (2001). The face in the crowd revisited: a threat advantage with schematic stimuli. *Journal of Perception of Social Psychology, 80*, 381–396.
Parkes, L., Lund, J., Angelucci, A., Solomon, J., & Morgan, M. (2001). Compulsory averaging of crowded orientation signals in human vision. *Nature Neuroscience, 4*, 739–744.
Posner, M. I. (1980). Orienting of attention. *Quarterly Journal of Experimental Psychology, 32*, 3–25.
Posner, M. I. (1994). Attention: The mechanisms of consciousness. *Proceedings of the National Academy of Sciences, 91*, 7398–7403.
Rensink, R. A. (2002). Change detection. *Annual Review of Psychology, 53*, 245–277.
Rensink, R. A. (2004). Visual sensing without seeing. *Psychological Science, 15*, 27–32.
Rensink, R. A., O'Regan, J. K., & Clark, J. J. (1997). To see or not to see: The need for attention to perceive changes in scenes. *Psychological Science, 8*, 368–373.
Rose, D., Bradshaw, M. F., & Hibbard, P. B. (2003). Attention affects the stereoscopic depth aftereffect. *Perception, 32*, 635–640.
Rowe, G., Hirsh, J. B., & Anderson, A. K. (2007). Positive affect increases the breadth of attentional selection. *Proceedings of the National Academy of Sciences, 104*, 383–388.
Shulman, G. L., & Wilson, J. (1987). Spatial frequency and selective attention to local and global information. *Perception, 16*, 89–101.
Simons, D. J., & Levin, D. T. (1997). Change blindness. *Trends in Cognitive Sciences, 1*, 261–267.
Spivey, M. J., & Spirn, M. J. (2000). Selective visual attention modulates the direct tilt aftereffect. *Perception & Psychophysics, 62*, 1525–1533.
Srinivasan, N., & Gupta, R. (submitted). Time course of visual attention for emotional faces.
Srinivasan, N., & Hanif, A. (in press). Global-happy and local-sad: Perceptual processing affects emotion identification. *Cognition and Emotion.*
Srivastava, P., & Srinivasan, N. (2008). Emotional information modulates the temporal dynamics of visual attention. *Perception, 37*, . ECVP Abstract, S11
Stormark, K. M., Nordby, H., & Hugdahl, K. (1995). Attentional shifts to emotionally charged cues: Behavioural and ERP data. *Cognition and Emotion, 9*, 507–523.
Suzuki, S., & Grabowecky, M. (2003). Attention during adaptation weakens negative afterimages. *Journal of*

Experimental Psychology: Human Perception and Performance, 29(4), 793–807.

Theeuwes, J., Kramer, A. F., & Belopolsky, A. V. (2004). Attentional set interacts with perceptual load in visual search. *Psychonomic Bulletin and Review, 11*, 697–702.

Treisman, A. (1960). Contextual cues in selective listening. *Quarterly Journal of Experiment Psychology, 12*, 242–248.

Treisman, A. (2006). How deployment of attention determines what we see. *Visual Cognition, 14*, 411–443.

Treisman, A., & Gelade, G. (1980). A feature-integration theory of attention. *Cognitive Psychology, 12*, 97–136.

Tsuchiya, N., & Koch, C. (2005). Continuous flash suppression reduces negative afterimages. *Nature Neuroscience, 8*, 1096–1101.

van Gaal, S., & Fahrenfort, J. J. (2008). The relationship between visual awareness, attention and report. *Journal of Neuroscience, 28*(21), 5401–5402.

Vuilleumier, P., Armony, J. L., Driver, J., & Dolan, R. J. (2001). Effects of attention and emotion on face processing in the human brain: An event-related fMRI study. *Neuron, 30*, 829–841.

Wadlinger, H. A., & Issacowitz, D. M. (2006). Positive mood broadens visual attention to positive stimuli. *Motivation and Emotion, 30*, 89–101.

Watamaniuk, S. N. J., & Duchon, A. (1992). The human visual system averages speed information. *Vision Research, 32*, 931–942.

Wede, J., & Francis, G. (2007a). Attentional effects on afterimages: Theory and data. *Vision Research, 47*, 2249–2258.

Wede, J., & Francis, G. (2007b). Cortical dynamics of negative afterimages: Spatial properties of the inducer [Abstract]. *Journal of Vision, 7*(9), 277.

White, M. (1996). Anger recognition is independent of spatial attention. *New Zealand Journal of Psychology, 25*, 30–35.

Williams, D. W., & Sekuler, R. (1984). Coherent global motion percepts from stochastic local motions. *Vision Research, 24*, 55–62.

Williams, M. A., Moss, S. A., Bradshaw, J. L., & Mattingley, J. B. (2005). Look at me, I'm smiling: Visual search for threatening and non threatening facial expressions. *Visual Cognition, 12*, 29–50.

Williams, P., & Simons, D. J. (2000). Detecting changes in novel 3D objects: Effects of change magnitude, spatiotemporal continuity, and stimulus familiarity. *Visual Cognition, 7*, 297–322.

Woodman, G. F., & Luck, S. J. (2003). Dissociations among attention, perception, and awareness during object-substitution masking. *Psychological Science, 14*, 605–611.

Wyart, V., & Tallon-Baudry, C. (2008). Neural dissociation between visual awareness and spatial attention. *Journal of Neuroscience, 28*, 2667–2679.

Yantis, S. (1996). Attentional capture in vision. In A. F. Kramer, M. G. H. Coles, & G. D. Logan (Eds.), *Converging operations in the study of visual selective attention*. Washington, DC: American Psychological Association.

N. Srinivasan (Ed.)
Progress in Brain Research, Vol. 176
ISSN 0079-6123

CHAPTER 7

The functional architecture of divided visual attention

Kimron Shapiro*

Wolfson Centre for Clinical and Cognitive Neuroscience, School of Psychology, Bangor University, Bangor, Gwynedd, UK

Abstract: When we identify a visual object such as a word or letter our ability to detect a second object is impaired if it appears within 500 ms of the first. This outcome has been named the 'attentional blink' (AB) and has been the topic of numerous research reports since 1992 when the first AB paper was published. During the first decade of research on this topic, the focus has been on 'behavioural' approaches to understanding the AB phenomenon, with manipulations made on stimulus parameters (e.g. type and spatial distribution), nature of the stimuli (uni-modal or cross-modal) and importantly the role of masking. More recently, researchers have begun to focus on neurophysiological underpinnings of the AB studying patients with focal lesions and using approaches such as ERP, TMS, fMRI and MEG. My chapter presents the results of a number of such neurophysiological techniques, suggesting that localisation, in combination with activation and synchronisation methods have begun to unravel a dynamic temporo-parietal frontal network of structures involved in the AB.

Keywords: attention; attentional blink; event-related potential; functional imaging; magnetoencephalography

Introduction

Performing two tasks in close temporal proximity results in deficits in the second task. Such deficits have been studied under the heading of the psychological refractory period (PRP) where targets from the same or different modalities are presented, without masks, with a speeded response required on both. Whereas reaction time (RT) to the first target remains unchanged as a function of the delay between the two tasks (stimulus onset asynchrony or SOA), RT to the second task becomes exponentially larger the closer in time the two targets are presented.

A similar outcome occurs when two targets are required to be detected or identified as part of a rapid serial visual presentation (RSVP) with both masked by the preceding and succeeding stream items (see Fig. 1). When the two targets occur in close temporal proximity, i.e. separated by less than approximately 500 ms, identification or detection of the second is adversely affected after the first has been correctly identified or detected (see Fig. 2). This phenomenon has been named the attentional blink (AB; Raymond et al., 1992) and has been studied extensively since its inception (cf. Shapiro, 1994). Various accounts have

*Corresponding author.
Tel.: +44 (0)1248 383626; Fax: +44 (0)1248 382599;
E-mail: k.shapiro@bangor.ac.uk

DOI: 10.1016/S0079-6123(09)17607-0

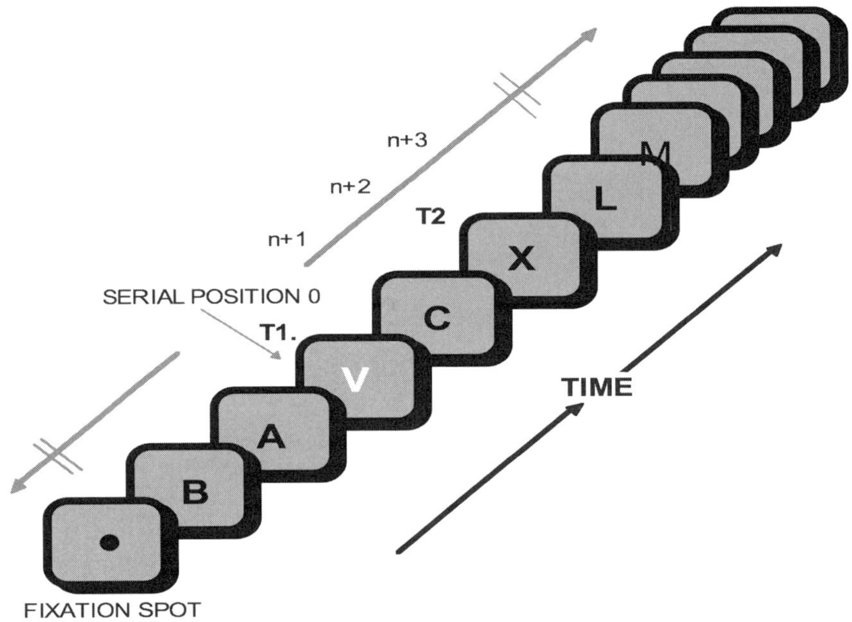

Fig. 1. Schematic representation of the rapid serial visual presentation method used to study the attentional blink. T1 appears as the only white letter; T2 is the letter 'X' which is presented on 50% of trials in one of the eight serial positions following T1.

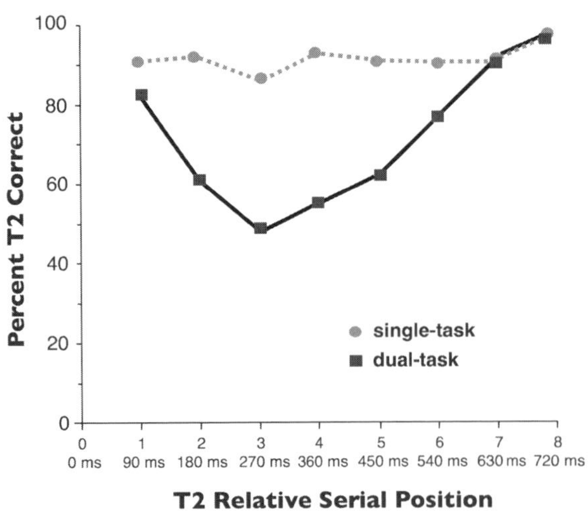

Fig. 2. Typical findings characterising the results of an attentional blink experiment. Percent T2 correct responses are plotted on the *Y* axis, whereas the relative serial position and SOA values between T1 and T2 are plotted on the *X* axis. Results from the single-target (control) condition are plotted with circles and the results of the dual-target (experimental) condition are plotted with squares.

been advanced to explain the AB (Bowman and Wyble, 2007; Chun and Potter, 1995; Olivers and Nieuwenhuis, 2005; Shapiro et al., 1994), with all generally advancing the notion that with short SOAs the second target is unable to be processed into a durable form of storage and/or report as a result of the attentional demands of the first.

The purpose of the present chapter is not to review the extensive literature on the AB, including the various accounts referred to above, but instead to answer a circumscribed set of questions with recent evidence drawn from the published AB literature as well as from studies in various stages of preparation prior to publication. These questions are drawn from multiple approaches used to study the AB including behavioural, electrophysiological and neuroimaging. The questions addressed by the following chapter are as follows.

1. What can be concluded from empirical evidence that the AB can be attenuated or even abolished under certain circumstances?

2. How does evidence from both behavioural and neurophysiological AB experiments support the 'over-investment' hypothesis, which argues that the AB occurs due to the investment of too much attention to the first target task?
3. How do recent electrophysiological results inform us about the relationship between processing the first target and the ensuing effect on the second and how does long-range synchronisation between the two targets correlate with the AB outcome and define an AB 'network'?
4. What do functional imaging results tell us about brain regions involved in the AB and how do they relate to conscious experience?

When no AB occurs

In the face of the large number of empirical studies showing the AB to be a robust outcome, a handful of studies are very informative in their finding of an absence of an AB. Such studies help to define the boundary conditions of what produces an AB and in turn constrain theories attempting to explain the phenomenon. In one such study Drew and Shapiro (2006) found evidence that the mask's effectiveness on both T1 and T2 could be reduced by a manipulation known to produce a conceptual 'blindness' for the second occurrence of a repeated stimulus. Repetition blindness (RB; Kanwisher, 1987), as the phenomenon is known, can be obtained with letters or even words that form sentences. Participants in such experiments often fail to report the second occurrence of a repeated letter or word, even when doing so in the case of a word affects the grammaticality of the sentence. The 'token individuation' account of RB argues that while the physical attributes of the second occurrence of the target are perceived (i.e. 'typed') the representation (or 'token') for the second occurrence fails to be manifest, as a token already exists from the first occurrence of the target. The logic of the experiment by Drew and Shapiro was that if RB could be found to reduce the effectiveness of the T2 mask, as masking typically is a prerequisite for obtaining an AB, this would demonstrate that masking in the AB is occurring at a conceptual rather than perceptual level, given that the perceptual requirements of masking were met.

As shown in Fig. 3, the item occurring in the RSVP stream before T1 and after T1 were the same in the experimental condition, whereas they were different in the control condition. The results revealed an ~10% attenuation of the AB at Lag 3 consistent with the above claim that the AB is operating at a level in the information processing stream further 'upstream' than perception. This result is consistent with the 'interference' hypothesis suggested by Shapiro et al. (1994) which holds that the AB occurs after the target has been processed into visual short-term memory (VSTM) but is unable to be successfully retrieved due to competition from other items in VSTM, i.e. the first target and its mask. However, the data is also consistent with the hypothesis proposed by Chun and Potter (1995), which argues that the AB is due to a difficulty consolidating the first target into VSTM, in turn placing a greater demand on T2 for the same consolidation. According to this account, anything that facilitates consolidation is beneficial: in particular, removing the mask from T1. Nevertheless, the attenuation of the AB arising from this manipulation sheds light on the nature of the 'masking' that occurs in this paradigm relative to the more traditional role played by masks in other paradigms.

In another demonstration of a modification to the basic AB paradigm, which revealed an attenuation of the blink, Martin et al. (submitted) altered two of the parameters in the canonical AB paradigm. Although there have been many AB experiments, varying parameters such as speed of the RSVP presentation (between 6 and 20 items per second) and the nature of the specific stimuli (i.e. letters, digits), all that the present author is aware of kept whatever parameters were chosen constant throughout the experiment. Martin et al., on the other hand, varied presentation speed and stimulus size — separately — within and across trials. In one temporal manipulation, although the SOA between each target and its respective mask was held constant at the canonical value of 100 ms,

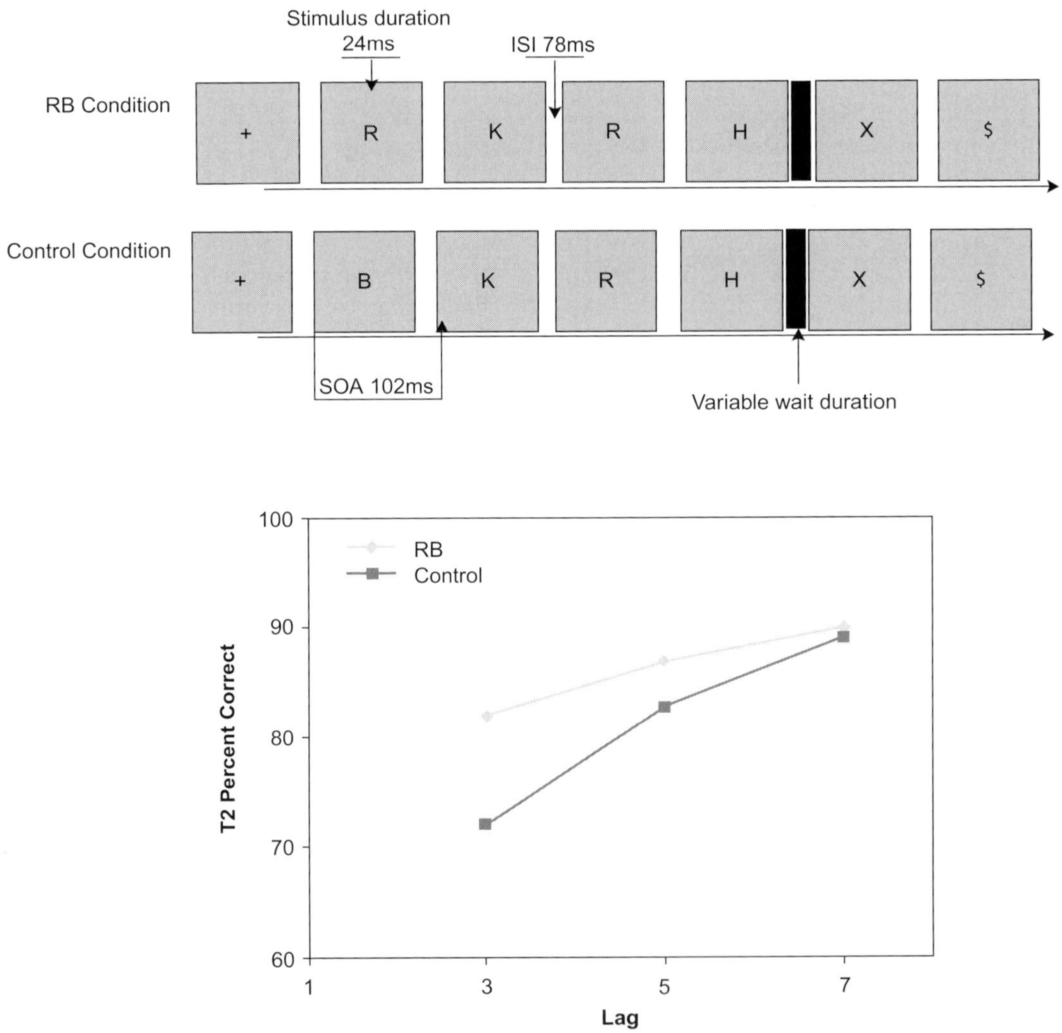

Fig. 3. Top panel shows schematically the stimulus sequence for the RB (top) and Control (bottom) conditions. The bottom panel shows the results for the same two conditions plotted as a function of lag (*X* axis) and T2 percent correct (*Y* axis).

the SOA for other parts of the RSVP stream was varied around a mean of 85 ms, with a range of from 17 to 153. The three conditions generated by varying different parts of the stream were either the items before T1, the items between T1 and T2 or both. In a second spatial manipulation, we varied the size of the stimuli, maintaining the canonical size (18 pt. font) of the two targets and their masks, but varied the size of the other RSVP items between 14 and 24. As with the temporal manipulation, three conditions saw the font size varied before T1, between T1 and T2 or both. To recap, in both temporal and spatial manipulations, both targets and their masks were identical and held at canonical AB parameters, but in the former the SOA was varied creating what we refer to as a temporal discontinuity and in the latter the font size was varied creating what we refer to as a spatial discontinuity. Although the location within the RSVP stream of the occurrence of the discontinuity had a similar effect on AB magnitude relative to the canonical blink, the results of the temporal and spatial manipulations were in dramatically opposite directions (see

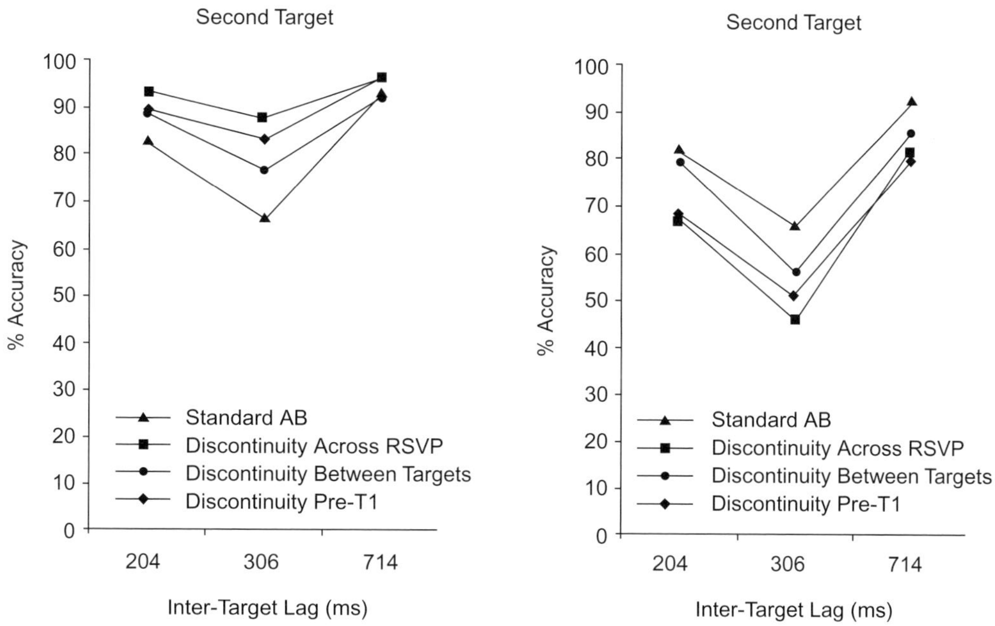

Fig. 4. Left panel shows the results for the temporal discontinuity condition plotting T2 accuracy on the *Y* axis as a function of Lag on the *X* axis. The right panel shows the results for the spatial discontinuity condition plotted the same way. Triangles represent the standard AB condition; squares represent discontinuity across the RSVP stream; circles represent discontinuity between T1 and T2; and diamonds represent discontinuity pre-T1.

Fig. 4). Whereas temporal discontinuity attenuated the AB, spatial discontinuity exacerbated it. The location manipulation resulted in the least effect occurring when the discontinuity occurred between T1 and T2, with the second largest effect occurring before T1 and the largest effect occurring when the discontinuity occurred in both. We interpret this outcome in the following way. In both manipulations, the discontinuity likely provides an alerting signal that draws attention to the temporal or spatial nature of the discontinuity, respectively. However, whereas the spatial discontinuity engages spatial attention that conflicts with the spatial judgement required of the targets (letter identity), temporal discontinuity leaves only the alerting trace, which facilitates performance due to the boost in attention. We are currently using an MEG approach to evaluate a different account, where we entertain the idea that each discontinuity introduces a different kind of neural 'noise' to the brain, with each having a different outcome. Stochastic noise, as it is referred, has been shown to have beneficial effects on target detection by seemingly paradoxically increasing the signal-to-noise ratio.

The over-investment hypothesis

In a final demonstration where the AB is attenuated, my colleagues and I examined the 'over-investment' hypothesis (Olivers and Nieuwenhuis, 2005). Olivers and his colleagues have proposed that the AB occurs because too much attention is allocated to the RSVP stream in the AB paradigm, leaving too little attention for T2. In a dramatic demonstration to support this counterintuitive claim, Olivers showed that non-contingent background music (i.e. distracting stimulus) was able to significantly attenuate the AB, reasoning that attention to the music prevented over-investment prior to T2. Arend et al. (2006) sought to evaluate this claim using a more controlled background distractor as well as keeping the distracting task within the same (visual) modality. For one condition Arend et al.

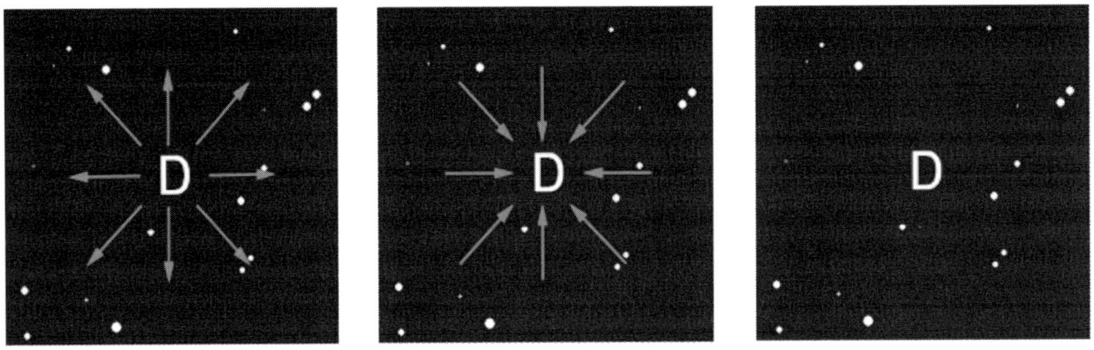

Fig. 5. The left panel shows a schematic of the 'outward' experimental condition; the middle panel a schematic of the 'inward' condition; and the right panel the control (static) condition.

created a moving background distractor field of dots that emanated from behind a standard AB task occurring at fixation. The background field moved to the screen's edge then disappeared ('motion outward' condition), and occurred after the fixation point was removed and before the RSVP stream commenced. Participants did not have to respond in any way to the background task. In a second ('motion inward') condition the identical field of moving dots emanated from the screen's edge and moved toward the screen's centre where the RSVP stream was presented. This condition was established to enable us to examine the influence of the direction of motion. A final ('static control') condition employed the same number of dots but they remained static on the screen to control for the presence versus absence of motion. We found a significant attenuation of the blink at various Lags 2 and 3 in the 'motion outward' condition, relative to the 'static control' condition, with slightly more of a blink occurring in the 'motion inward' condition (Fig. 5; Lag 2), which then showed similar behaviour to the 'motion outward' condition by the next lag (Lag 3). We view this as a replication and extension of Olivers and Nieuwenhuis (2005), where we are able to show that the same outcome occurs even when the distractor (i.e. visual field) is presented in the same modality.

My colleagues and I (Vogels et al., submitted) subsequently went on to examine the neural substrate of the over-investment hypothesis using both ERP and fMRI approaches. Using fMRI, we created a WM task that required the allocation of attention prior to the AB task. Participants were required to remember either two digits (low-load condition) or four digits (high-load condition) that had to be stored until the end of the trial and matched to a test digit shown prior to the instruction to recall T1 and T2 from the AB task (Fig. 6, Panel A). We anticipated that the high-load condition would yield a greater BOLD signal and might paradoxically reveal less of an AB than the low-load condition. Instead we found that trials on which the blink did not occur (no-AB trials) were associated with a higher BOLD signal, regardless of whether in the low- or high-load condition (Fig. 7, Panel B), than was revealed on trials in both load conditions when an AB did occur. The BOLD increase was witnessed in a variety of areas as shown in Fig. 7 (Panel A), some of which (e.g. MFG and OTPJ) have been shown to be active during the AB task. This result is consistent with the prediction from the over-investment hypothesis that anything distracting attention from the T1 task will benefit detection in the AB task. To complete the picture of the other results, there was a greater degree of blink in the high- as compared to the low-load condition but only at the short lag, i.e. in the middle of the AB (Fig. 6, Panel B). Finally, the high-load condition revealed worse performance on the WM task than did the low-load condition and did so at both short and long lags (Fig. 6, Panel C).

Using event-related potentials (ERPs), my colleagues and I (Martin et al., in submitted)

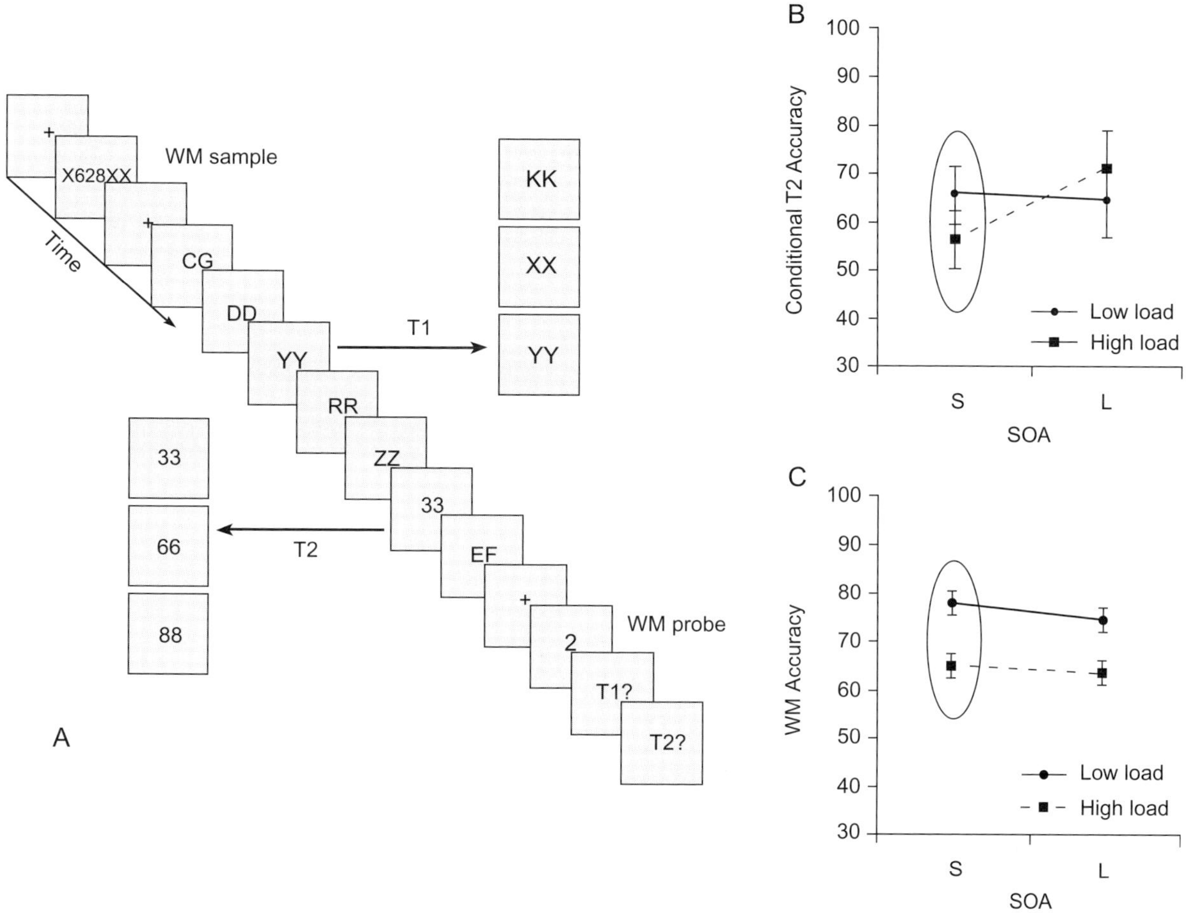

Fig. 6. (A) Schematic representation of trial structure. (B, C) Behavioural performance on measures of interest. Oval encircles performance to which analyses will be confined, namely within the blink-sensitive interval (short SOA). (B) Mean identification accuracy for the second target, conditional on correct first target identification (T2/T1). (C) Mean accuracy for WM probe. Bars denote standard error of the mean.

evaluated the over-investment hypothesis from a different approach. We reasoned that the contingent negative variation (CNV) component, which is known to index preparedness to respond to a target in response to a ready signal, should show increased negativity during an interval prior to the onset of T1. Furthermore, this should only be evident in a condition where the AB is attenuated by a manipulation designed to prevent resources from being committed to the RSVP stream, as specified by the over-investment hypothesis, and only on trials when T2 cannot be reported, i.e. an AB occurred. To effect such a manipulation we turned to the procedure used by Arend et al. (2006) where outward peripheral motion attenuated the AB. The CNV elicited by preparation for T1 in this condition was compared to no-motion control condition (Arend et al.), which produced a normal AB. The CNV was measured during a 1000-ms interval prior to T1 onset in a between-subjects design (see Fig. 8, Panel A). The results as shown in Fig. 8 (Panel B) reveal the opposite in so far as we observed an increased negativity — and only in the motion condition — but on trials when no AB occurred. We take this as evidence against the over-investment hypothesis as 'no-AB' trials should have shown a diminished CNV to T1. The difference

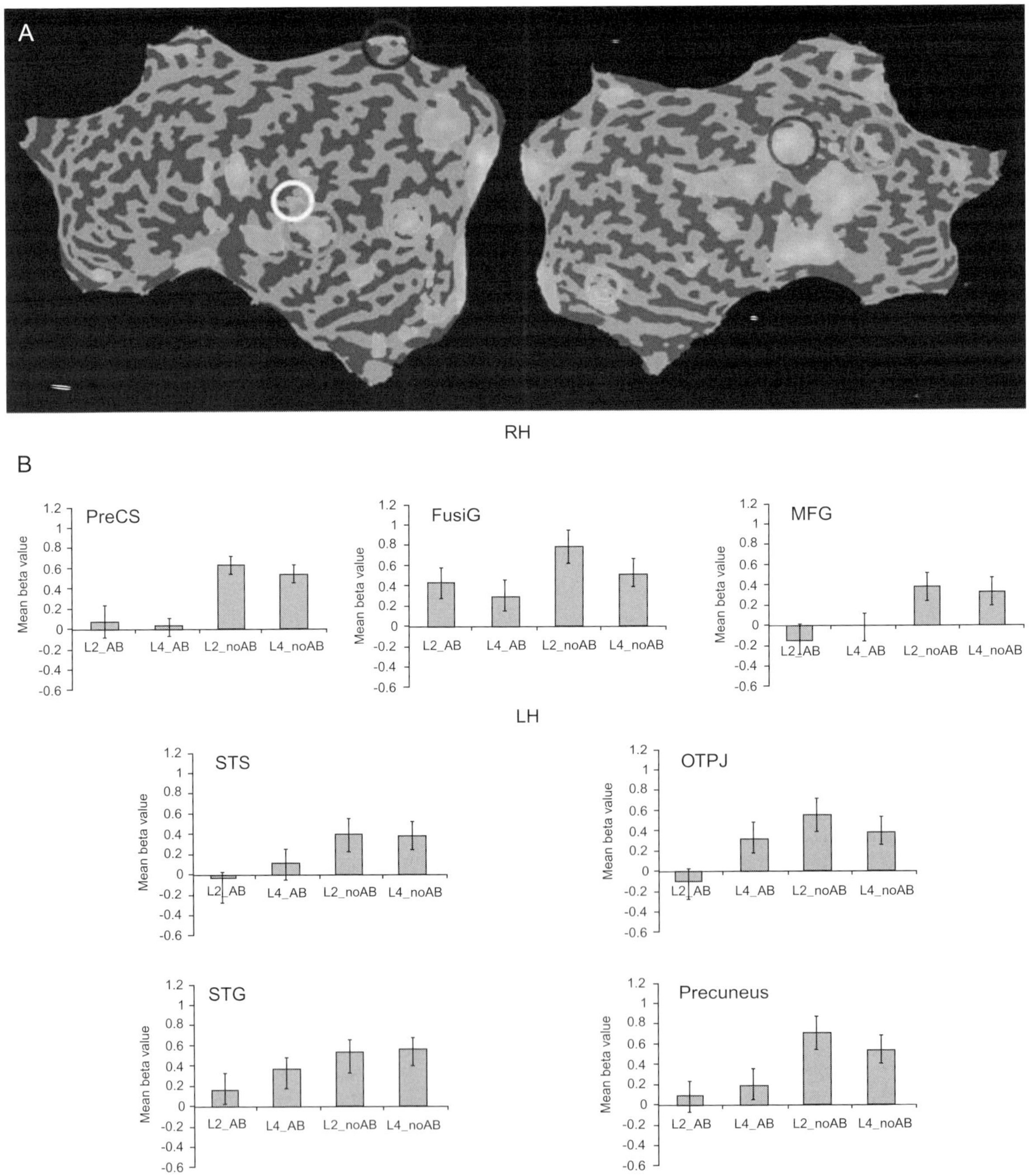

Fig. 7. Cortex-based group analysis of the experiment. (A) Contrast for the main effect of blink. Contrast maps showing averaged blink and no-blink trial evoked activity during encoding using a contrast threshold value of $P<0.01$, uncorrected. To protect against false positives, a cluster filter correction was implemented. Group-averaged random effect activation maps are superimposed on a flattened MNI template brain. On the flattened template, light and dark grey regions indicate gyri and sulci, respectively. Colour indicates t-value: $t(16)>2.95$ to >8 (red to yellow), positive activity. Marked clusters represent ROIs described in (B). (B) Mean parameter estimates of the peak voxel of selected ROIs for the attentional blink (AB) and no-attentional blink (no-AB) conditions of the main effect of blink during the encoding phase. LH, left hemisphere; RH, right hemisphere; PreCS, precentral sulcus (purple); FusiG, fusiform gyrus (cyan); MFG, middle frontal gyrus (red); STS, superior temporal sulcus (pink); OTPJ, occipitotemporoparietal junction (green); STG, superior temporal gyrus (yellow); Precuneus (dark blue). Error bars ± SE. (See Color Plate 7.7 in color plate section.)

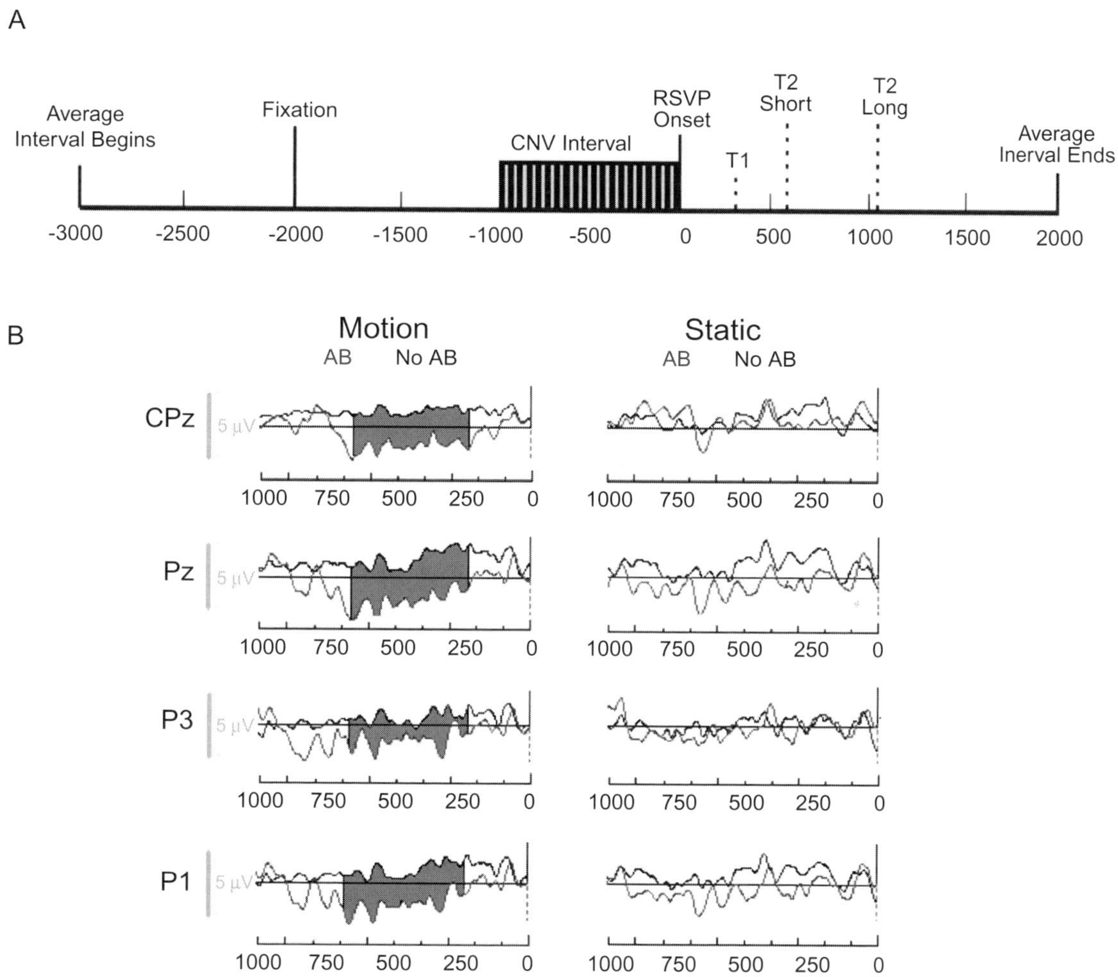

Fig. 8. Top panel: Critical temporal markers for averaged ERP waveforms. As shown the CNV was measured between fixation and RSVP onset. Bottom panel: Respective ERP waveforms for dual-target trials. Shown are AB versus no-AB trials for the motion and static conditions. Vertical bars mark the temporal interval analysed to reflect CNV amplitude. Red colouring indicates a statistically significant difference between AB and No-AB trials as revealed by post-hoc comparisons. (See Color Plate 7.8 in color plate section.)

between the fMRI and ERP approaches can be only indirectly compared as the two approaches also used different experimental manipulations designed to attenuate the AB.

Neural synchronisation and the AB

Using another approach, magnetoencephalography (MEG), my colleagues and I (Shapiro et al., 2006) decided to examine the neural basis of the general resource model underlying the AB, i.e. attentional resources to T1 preclude the availability of resources to T2. In order to minimise any extraneous noise being introduced into the MEG measurement, we designed a variant of the standard AB paradigm where participants looked for any occurrence of two pre-specified targets; the letters 'X' and 'O'. As shown in Panel A of Fig. 9, there were five trial types that could occur: two 'single-target' conditions where either an X or an O appeared, one condition where neither targets would appear, and two 'dual-target' conditions where both targets would appear,

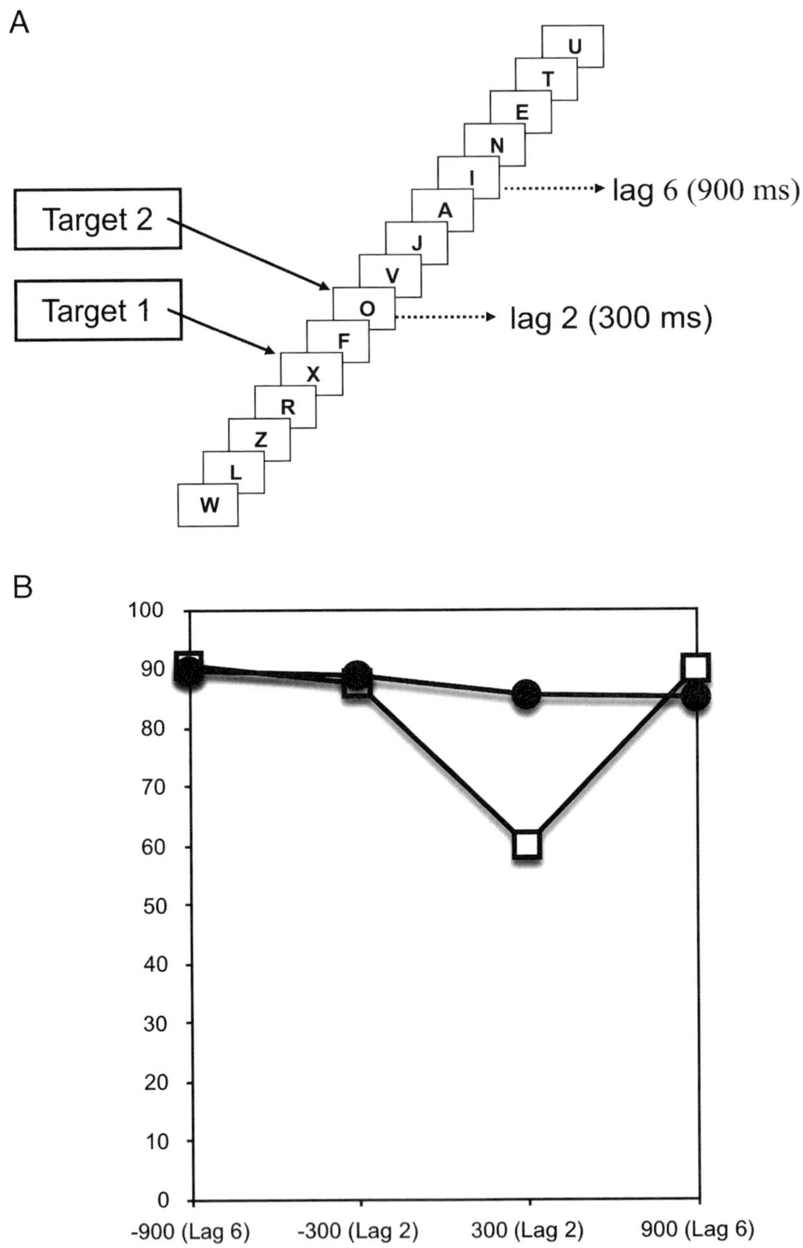

Fig. 9. Top panel: Schematic of stimulus stream showing Target 1 and 2, with Target 2 shown at the short lag (2; 300 ms) and long lag (3; 900 ms). Bottom panel: Single-target and dual-target responses of 10 participants. In dual-target conditions, negative and positive lags refer to performance for T1 and T2, respectively. In the single-target conditions, negative and positive lags refer to the lag of the single target as a function of where the other target would have occurred had it been presented.

separated either by a short lag (300 ms; the middle of the AB interval) or by a long lag (900 ms; outside the AB interval). The dependent variable was the amplitude of strongest signal emanating from any area of the brain with the goal being to characterise the modulation of the amplitude as a function of the trial type as described above. The behavioural results, as show in Fig. 9 (Panel B)

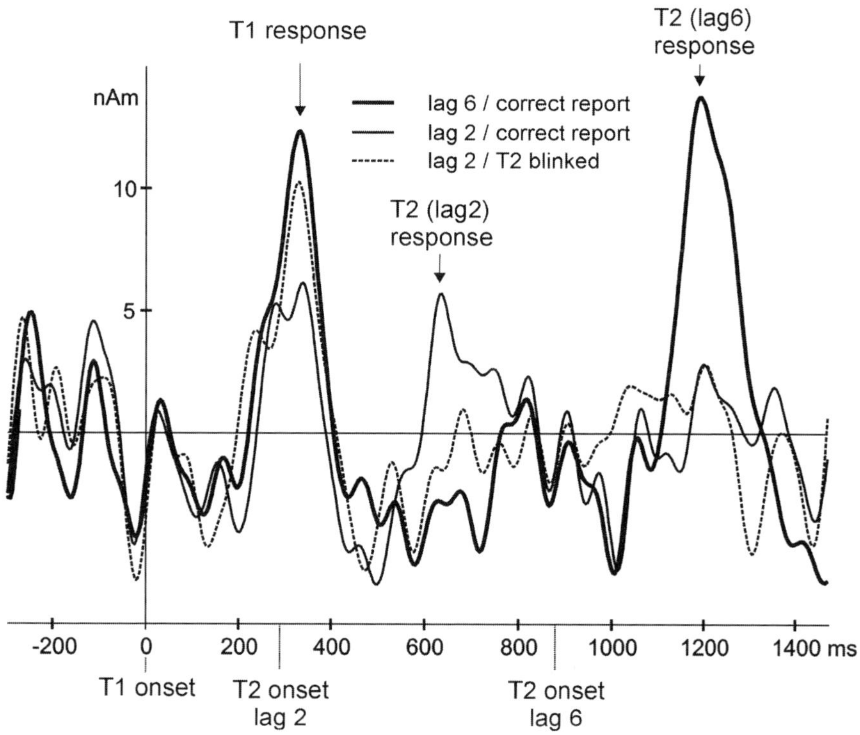

Fig. 10. Target-related activation on (1) lag 6 trials in which both targets can be reported (bold); (2) lag 2 trials in which both targets can be reported (non-bold) and (3) lag 2 trials in whichT2 is not reported, i.e. a 'blink' occurs (dashed). Waveforms represent sources with strongest target-related responses averaged across all participants and target letters.

reveal an AB occurred as exemplified by the difference between the (positive) SOA of 300 ms (short lag) versus 900 ms (long lag).[1] Turning to the electrophysiological results (Fig. 10), one of the most significant findings lending support to a general resource model can be seen when trials are post-categorised into T2 correct (no AB) and T2 incorrect (AB). T2 amplitude at the long lag is similar in amplitude to T1, as would be expected given that at the long lag T2 performance — where attention is presumed available — is generally as good as T1. On the other hand, at the short lag where attention is presumed by all theoretical accounts of the AB to be less available, performance is worse for T2 correct trials relative to the long-lag trials even though the behavioural outcome is the same as that at the long lag, i.e. participants are correct in their T2 judgement. Interestingly, also at the short lag, T2 incorrect trials reveal a further reduction in amplitude, suggesting a further effect of the lack of attention. Support for a general resource model of the AB was provided as we discovered T1 amplitude varied as a function of whether T2 was correct or incorrect: T1 amplitude was less for the former and greater for the latter. We followed this up with a correlation between T1 amplitude and the magnitude of the AB[2] and discovered that there was a significant correlation ($r = 0.74$), suggesting that the more attention put to T1 the less was left for T2. We note here that the fMRI and MEG data are consistent with the over-investment hypothesis but the CNV data are not. It is possible the latter approach is not measuring

[1]Negative SOAs refer to T1 performance at the corresponding short and long lags.

[2]AB magnitude was calculated as the area under the curve denoting the AB function.

'attention' to the RSVP task but more research must be conducted to verify this.

We performed another analysis on the data from the previous experiment, which resulted in our uncovering a correlation between changes in synchronisation/de-synchronisation and the occurrence of the AB (Gross et al., 2004). Long-range synchronisation has been suggested (Varela et al., 2001) as a mechanism by which communication among non-adjacent brain areas may be accomplished in a rapid and dynamic manner. Synchronisation is defined as phase-locked oscillatory activity between two or more cortical areas. The purpose of the analysis described here was to determine if long-range synchronisation is a potential mechanism able to account for when the AB occurs. In order to measure synchronisation, we first performed a time–frequency analysis and determined that at approximately 400 ms post-target occurrence, at 15 Hz, i.e. in the beta range, there existed a significant increase in power when a target was identified correctly (see Fig. 11). This enhancement distinguishes target from non-target processing and was used to localise the brain areas involved in target processing in the next step and as shown in Fig. 12. These (bilateral) areas were determined to be the posterior parietal, temporal, occipital and frontal lobes, as well as the (right) anterior cingulate gyrus. All the areas identified have been shown to be involved in many tasks requiring attention and specifically in the AB task (Shapiro et al., 2003). In the third step, we used the neural areas identified in Step 2 to characterise two 'networks' based on two distinct trial types. First, a 'distractor-related' network based on trials when no targets were presented and second, a 'target-related' network, when

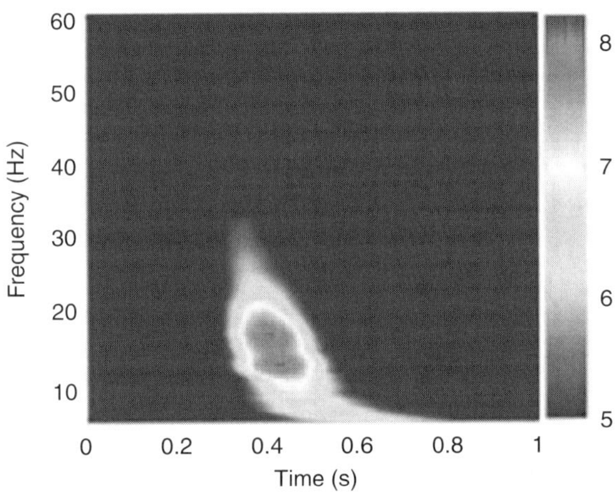

Fig. 11. TFR for the distractor condition subtracted from the target condition. Time 0 marks the onset of the target. The TFR represents the average across subjects and channels and is displayed in units of standard deviation of the baseline (thresholded at a value of 5). TFRs have been normalised for each frequency before averaging. (See Color Plate 7.11 in color plate section.)

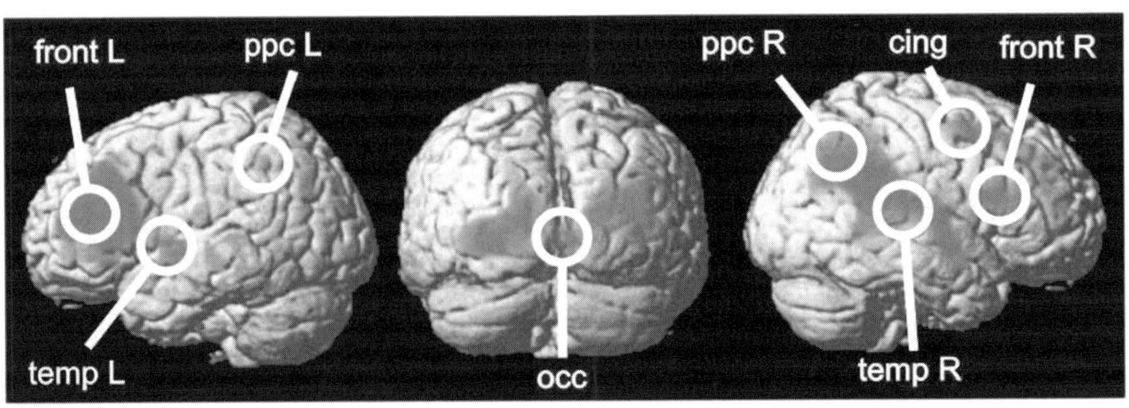

Fig. 12. Localisation of the time–frequency target component displayed in Fig. 1. Functional maps of oscillatory power in the beta band were computed for each subject. The functional maps were spatially normalised by using SPM99, and a permutation analysis with SnPM99 was performed. Only areas with a significance of $P<0.01$ (corrected) are shown. The maximum of each ROI is marked and labeled and was used for further computations. A single occipital ROI was used.

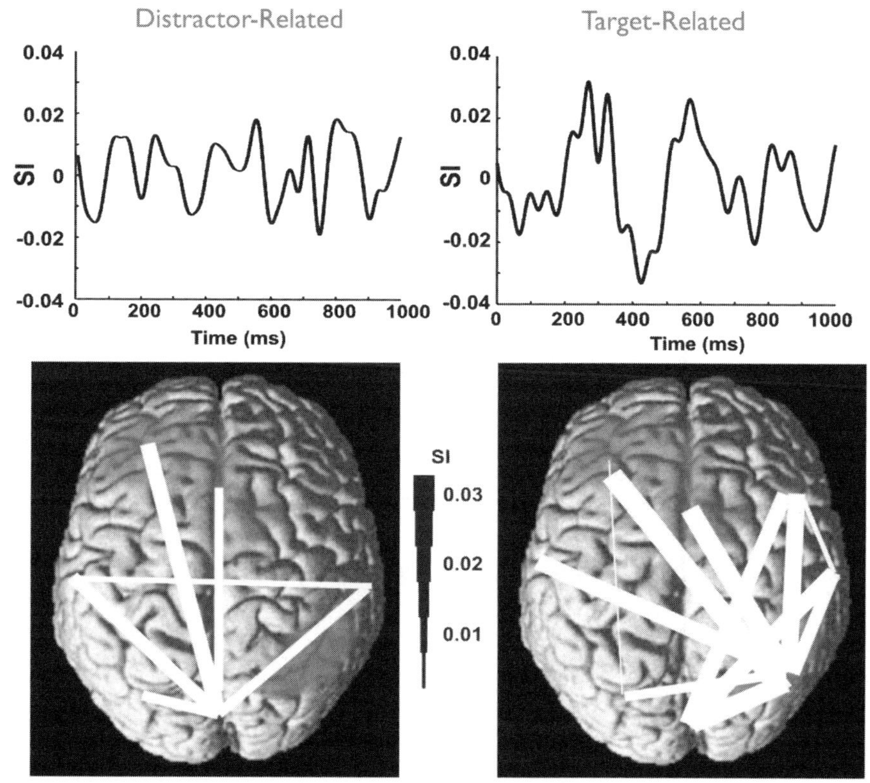

Fig. 13. Classification of stimulus- and target-related connections. Top panel: SI for one subject for a typical stimulus-related (left, occipital to posterior parietal left) and a typical target-related (right, frontal left to posterior parietal right) connection. SI was computed based on sensor groups that are most sensitive to a given region. Bottom panel: The stimulus-related (left) and target-related (right) networks are shown with linewidth coding for the strength of synchronisation at 260 ms. (See Color Plate 7.13 in color plate section.)

two targets were presented (see Fig. 13). The distractor-related network is largely centred on visual cortex, linking it primarily with (left) temporal and (left) frontal areas. In stark contrast, the target-related network is centred on (right) posterior parietal cortex and links this area primarily with (left) temporal, (left) frontal and (right) anterior cingulate cortices.

To characterise the dynamic nature of the target-related network, we then post-categorised all trials into four trial types: distractor, single-target, dual-target when no AB occurred and dual-target when an AB did occur and examined the modulation of long-range synchronisation at 15 Hz among the elements involved in this network. As shown in Fig. 14, whereas the distractor trials revealed no significant modulation of synchronisation, all other trials on which there was a first target (T1) showed increased synchronisation to this target relative to the distractor (only) baseline. Examining synchronisation to the second target (T2), we found increased synchronisation on trials when T2 could be reported, i.e. no AB occurred, relative to trials when an AB did occur. The latter revealed increased synchronisation over the baseline (distractor) trial type. Perhaps the most interesting — and unexpected — finding was that the masks on both T1 and T2 revealed significant de-synchronisation on no-AB trials relative to AB trials with the latter showing more de-synchronisation than either single-target or distractor trial types. Based on many published reports, e.g. Raymond et al. (1992), Seiffert and Di Lollo (1997), demonstrating the importance of masking on both T1 and T2 to the production of the AB, we interpret this outcome to suggest that

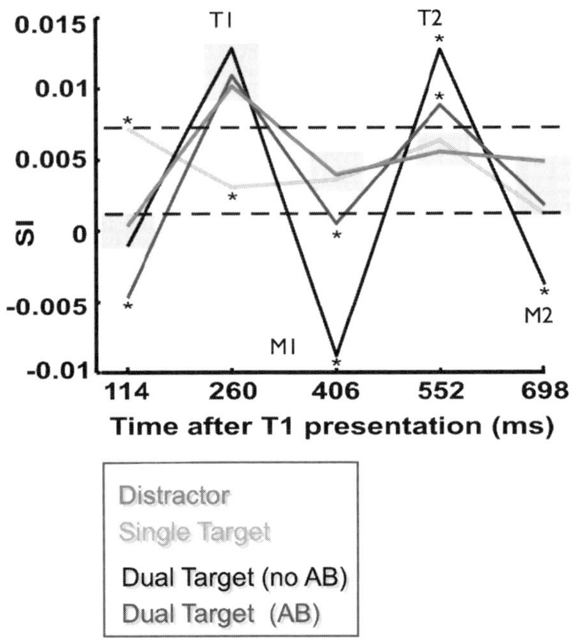

Fig. 14. SI for the components of five successive stimuli. The x axis specifies time after presentation of the first target. Each point represents the mean SI in a 60-ms window centered at 260 ms after the respective stimulus. Values at 260 ms quantify the network synchronisation to the first target, and values at 114 ms represent the network synchronisation corresponding to the distractor preceding the first target. Conditions are colour-coded (black, no-AB; red, AB; blue, target; green, distractor). The dashed lines mark the extent of SI in trials containing only distractors. Points marked with an asterisk are significantly different from their neighbours at the same position ($P < 0.05$, Kruskal–Wallis test), whereas points within the same shaded area are not significantly different. Negative values arise from the filtering of the SI time courses. (See Color Plate 7.14 in color plate section.)

the trials on which T2 could be detected, i.e. no AB occurred, were due to the uncoupling (i.e. desynchronisation) of each target to its respective mask thus preventing the mask from overwriting the target.

In a final attempt to understand the wider implications of the synchronisation modulation described above we performed a final analysis. The rationale behind this analysis was that there must be a mechanism by which top-down and bottom-up processing can interact and we reasoned that synchronisation is a suitable candidate for such a mechanism. In the particular case of the task demands in our experiment, participants would need to engage a top-down mechanism, such as the definition of the targets for which they were searching, with a bottom-up mechanism that encoded the perceptual information from each target candidate, i.e. letter in the stimulus stream. We reasoned that, if synchronisation was the mechanism by which these two opposite but complementary forms of information processing could interact, then we should expect to see changes in the degree of synchronisation to a given target as a function of the expectation of when that target was anticipated to occur. We were able to assess the ‘anticipation’ by virtue of the fact that the first target was scheduled to occur (randomly) at position 4, 5 or 6 following the start of the RSVP stream. Thus we reasoned that if the target did not occur at position 4 there would be an increased expectation that it would appear at position 5 and, if not, then a further increase at position 6. Accordingly, we expected the degree of synchronisation to rise with each successive possible target position. As is shown in the middle top graph of Fig. 15 our prediction was confirmed, revealing a pattern of increasing synchronisation as the location of actual position of the target's occurrence increased (Gross et al., 2006). Nakatani

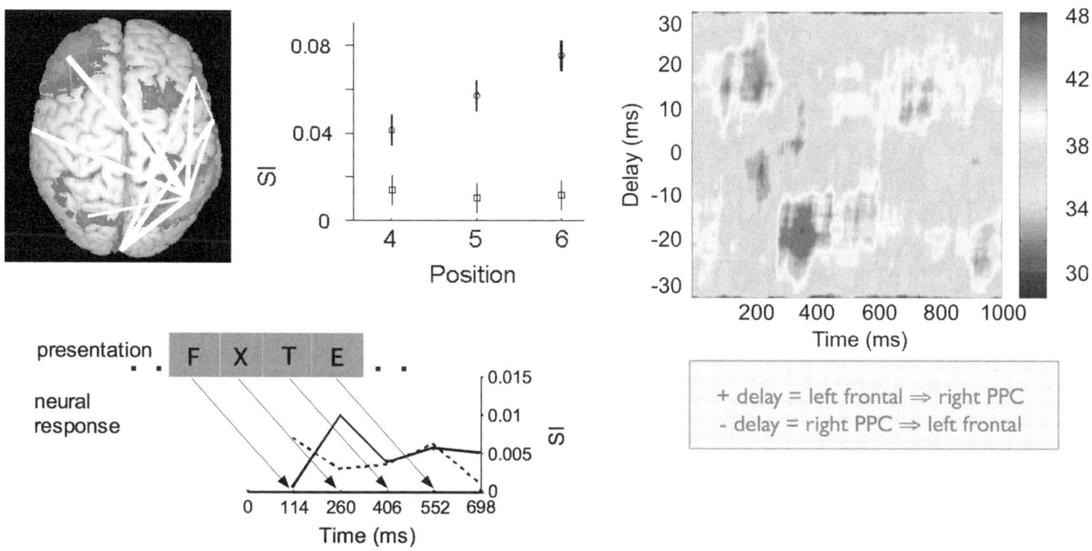

Fig. 15. Top left panel: Connections in the target-related network. Functional maps of oscillatory power in the beta-band were computed for each subject. The functional maps were spatially normalised using SPM99 and a permutation analysis using SnPM99 was performed. Only areas with a significance below $P<0.05$ (corrected) are shown. Lines mark connections for which the phase synchronisation is significantly modulated by target presentation. The displayed connections form the target-related network. Top middle panel: Modulation of phase synchronisation (SI) by targets at different positions in the presentation stream. The mean of 11 points surrounding the maximum (at about 260 ms) and the minimum (at about 114 ms) was computed for all subjects and connections of the target-related network for targets (circles) and distractors (boxes). Lines extending from the mean indicate the standard error. The modulation (difference of synchronisation and desynchronisation) increases with the position only for target trials. Top right panel: Y axis shows delay between left prefrontal and right PPC activation with positive delays indicating left frontal preceding right PPC and negative delays the reverse relative to time since target onset shown on X axis. Bottom panel: Modulation of phase synchronisation (SI) by targets as compared with distractors. The solid line shows the SI in trials where a target occurs, whereas the dashed line shows the SI in trials with only distractors. The X axis specifies time relative to target onset. Each point represents the mean SI in a 60-ms-long window centered at 260 ms after the respective stimulus. For illustration, part of a possible letter sequence (with target X) is shown in the upper part of the panel. Synchronisation values at 260 ms quantify the network synchronisation to the target. At 114 ms a reduced synchronisation is evident. This may represent the network response to a distractor that is followed by a target. The panel illustrates that target X is already partly processed at 114 ms and obviously affects the processing of the distractor by reducing the synchronisation. (See Color Plate 7.15 in color plate section.)

et al. (2005) found a similar outcome using EEG but in the gamma frequency. We were also able to examine the timing of activity in each of the two loci involved in the fronto-temporo-parietal network to compute a relative delay in when the activity occurred in each area. As shown in the upper rightmost part of Fig. 15 we determined that at 200 ms post-target there is a de-synchronisation with the phase difference suggesting the flow of information from left frontal to right posterior parietal. As frontal areas are known to be involved in working memory, decoupling of left frontal and right posterior parietal areas might prevent the distractor from entering later stages of processing, which might cause interference with target processing. Following this 200 ms de-synchronisation — at 300 ms post-target — there is an increase in synchronisation between these same two areas but with a reverse phase indicating the flow of information from right posterior parietal to frontal. The increased synchronisation at approximately 300 ms with an opposite direction of information flow in turn might represent the passing of the target identity information (obtained in parietal cortex) to further processing and/or storage in frontal areas. In summary, we believe the results of this experiment demonstrate the important role played by synchronisation in co-ordinating higher cognitive activities such as attention and perception.

Although various experiments as described above have suggested the AB can be attenuated (or increased) by various manipulations, it is important to ask the question: Is there any processing of an unreported target during the AB interval? To answer this question, Luck et al. (1996) (see also Vogel et al., 1998) employed an ERP approach using the N400 waveform as an index of processing to a target that cannot be reported, i.e. produced an AB. The N400 component occurs in response to a violation of an expected outcome, given a particular context. For example, a small N400 is seen to the final word of the sentence, 'The man wore blue trousers and a green bucket', whereas a small N400 would be seen to the final word of the corresponding sentence, 'The man wore blue trousers and a green shirt'. To adapt this approach to the AB paradigm we set a 'context' before each RSVP stream (experimental trial), as shown in Fig. 16, by

Stimulus Type	Time (ms)	Related Trial	Unrelated Trial
Context Word	1000	**RAZOR**	**WHEEL**
Blank	1000		
Distractor	83	PNVCSZP	KDSWPVZ
Distractor	83	GRSDPKN	VNMCPKL
Distractor	83	BVCPLMS	FEPMCNV
Distractor	83	DSPWTFR	VPMTDZM
Distractor	83	RLDJHGK	HJDLGFP
Distractor	83	SPLDJMF	DFPLJKH
T1	83	**3333333**	**4444444**
Distractor	83	WDPTBNF	GHJDMVT
Distractor	83	SCDPVBF	HDVCBNM
T2	83	XSHAVEX	XJEWELX
Distractor	83	FDLNLKB	NMCVPHJ
Distractor	83	DLJJCNW	DCVPBJM
Distractor	83	WPSCDSN	PCNBVLK
Distractor	83	DPWVCPB	NPMTVDK
Distractor	83	CBNDPNJ	BRTFPMF
Distractor	83	RTPMVBC	JLSDCDK
Distractor	83	TWSCLMN	LKSDVCP
Distractor	83	LJVBCMH	DKKHNVP
Distractor	83	RMVCPKL	WKLDMZP
Distractor	83	DPNMNVZ	CPNHVGB
Blank	1000		
Response Cue	2000	?	?
Blank	2000		

Fig. 16. Example stimulus sequences for trials on which the second target was either related or unrelated to the context word. Note that the probe word was drawn in red and all other items were drawn in blue. Also note that the probe word was flanked by Xs, when necessary, to create a total of 7-characters in the string.

displaying a word for 1000 ms. The context of this word could then be either congruent with T2, producing a small N400, or incongruent producing a large N400. The logic is that to produce an N400, T2 would have to be identified to a semantic, i.e. meaning, level of awareness. To cancel all extraneous 'noise' we then subtracted congruent T2 words from incongruent T2 words to produce a difference score. We produced this N400 difference score for both single-target trials (report T2 only) and dual-target trials (report T1 then T2) using only three lag positions; two on either side of the AB and one in the middle of the AB interval. The T1 task required participants to detect whether the only string of digits was 'even or odd'. As can be seen in the left panel of Fig. 17, the behavioural data revealed a characteristic AB with the dual-trial condition revealing reduced T2 identification at Lag 3 relative to the single-target control. Strikingly, the electrophysiological data (right panel of Fig. 17) shows no reduced N400 at any lags suggesting that T2 was processed to a semantic level of awareness in spite of its failure to be reported. The implications of this finding are that conscious awareness requires an additional step beyond semantic awareness and that stimuli experiencing an AB fail to reach this stage of processing.

Functional imaging and the AB

In a final series of studies to be described in the present chapter, Shapiro and his colleagues investigated the neurophysiological substrate of the AB using fMRI. A prior study reported by Marois et al. (2004) found a plausible brain-behaviour correspondence between the behavioural fate of 'place' (scene) targets presented during the AB interval and activity in a particular part of the brain — the parahippocampal 'place area (PPC) — sensitive to such targets. Marois et al. revealed that (T2) targets that could not be identified correctly, i.e. revealed an AB, showed less BOLD activation in the PPC on trials when the AB occurred than on trials when the AB did not occur. The departure point for the investigation by Shapiro et al. (2007) was the fact

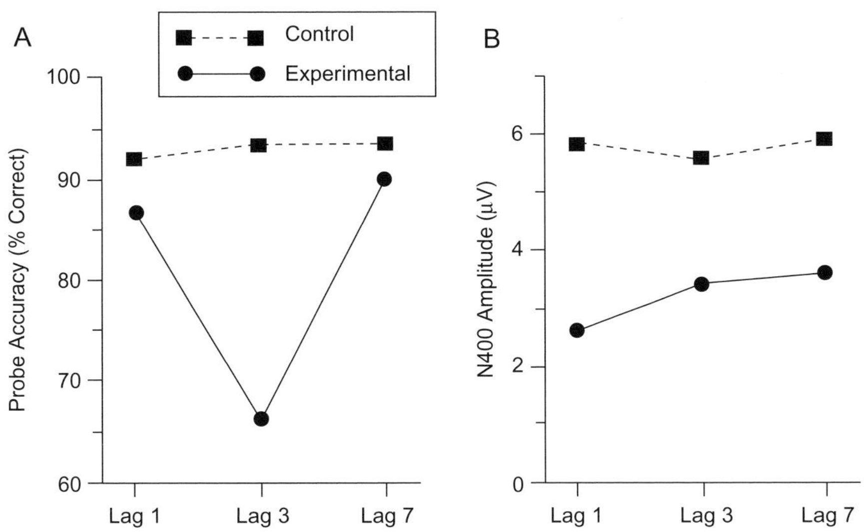

Fig. 17. (A) Probe discrimination accuracy as a function of lag for the experimental and control conditions. These values reflect only the trials on which the first target was correctly discriminated (first-target accuracy was 96% correct overall, with no effect of lag). (B) Mean N400 amplitude as a function of lag for probe words in the experimental and control conditions, measured from the unrelated–related difference waves and averaged across electrode sites. N400 amplitude was computed as the mean amplitude between 300 and 500 ms post-stimulus, relative to a 200-ms pre-stimulus baseline, at the F3, Fz, F4, C3, Cz, C4, P3, Pz and P4 electrode sites.

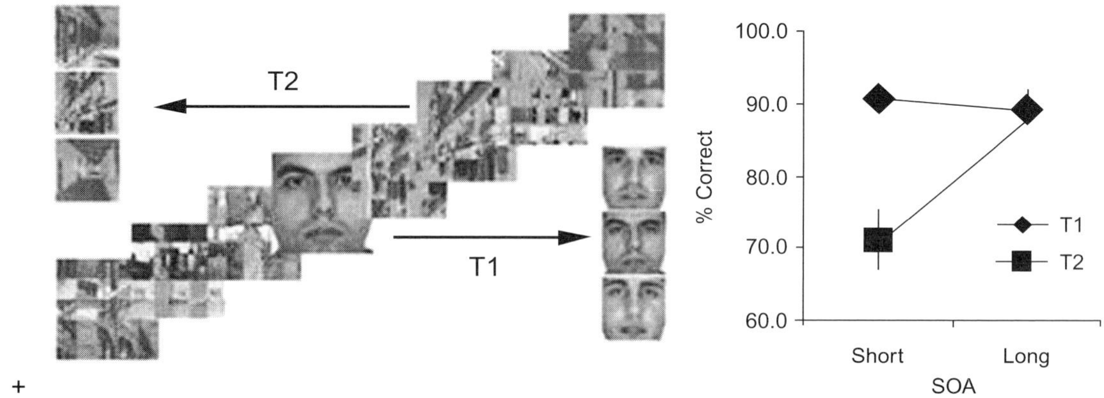

Fig. 18. Left panel: Schematic representation of a typical trial. Right panel: T1 and conditional T2 accuracy (*y* axis) as a function of SOA (*x* axis).

that the T1–T2 interval (~450 ms) used by Marois et al. was near the recovery point of the interval typically revealing an AB in the canonical AB paradigm (~100–500 ms). As described in his report, Marois and his colleagues chose this interval to enable participants to perform the T2 task at a reasonable level of performance, given the difficulty of the target task. Using a similar paradigm to Marois et al., as shown in Fig. 18 (left), the T1 task was to select which of three possible 'faces' was presented, whereas the T2 task was to choose which of three possible 'scenes' was presented. An AB was revealed as is shown in Fig. 18 (right), when T2 was able to be reported less accurately at the short versus long SOA, relative to T1, which could be reported accurately at both SOAs. Using a region of interest (ROI) approach, Shapiro et al. measured activity in a variety of neural areas (see Fig. 19) but the finding of primary interest was that participants showed an increased BOLD response on trials in the PPA when an AB occurred (Fig. 19; Panel B). This stands in stark contrast to the result by Marois et al. and Kranczioch et al. (2005). In an attempt to understand the disparate results by investigators who used highly similar paradigms, Johnston et al. (2007) set up a replication of the experiment by Shapiro et al. (2007) but in addition manipulated the contrast of the second target to create a difficult T2 task (low contrast) or easy target task (high contrast). Johnston et al. were able to replicate the result of Shapiro et al. in revealing more BOLD activity in PPA on AB trials as compared to no-AB trials in the high contrast condition during the AB interval (Fig. 20; Panels B, C). However, in the low contrast condition and outside the AB interval, more BOLD activity was revealed on no-AB trials (relative to AB trials), replicating the results of Marois et al. (2004). To summarise the results of Johnston et al., these investigators were able to conclude that perceptual difficulty leads to more activity in that part of the brain responsible for processing the particular (T2) stimulus used on trials when the target is perceived than when it is not perceived but under conditions of full attention, i.e. outside the AB interval. On the other hand, when attention is in short supply (i.e. during the AB interval), the brain has to work harder to perceive targets that are not fully perceived than those that are.

To summarise, the present chapter has reviewed recent and past work on the AB phenomenon to address the following four questions.

1. What can be concluded from empirical evidence that the AB can be attenuated or even abolished under certain circumstances?
2. How does evidence from both behavioural and neurophysiological AB experiments support the 'over-investment' hypothesis, which argues that the AB occurs due to the investment of too much attention to the first target task?

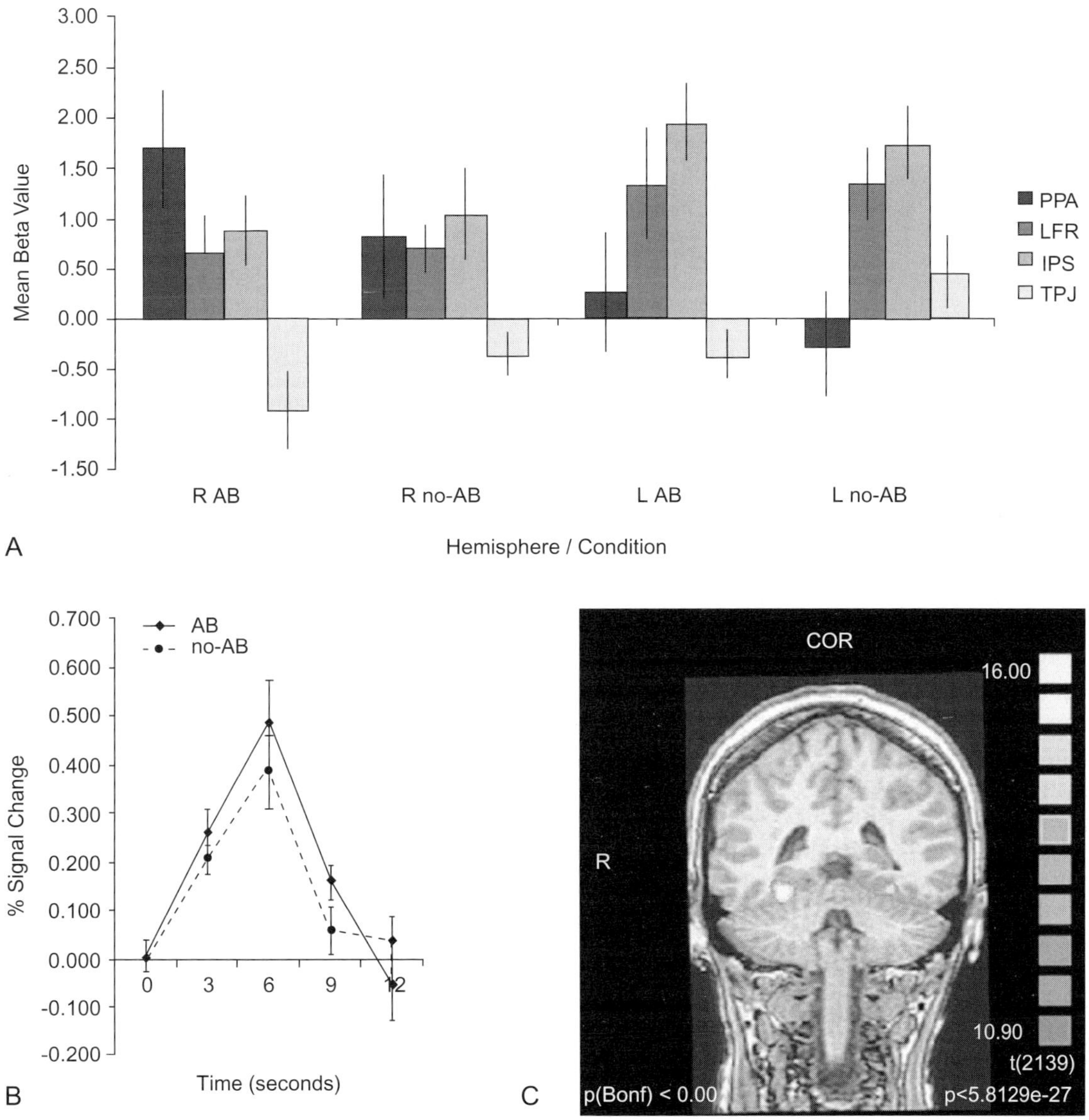

Fig. 19. (A) Mean parameter estimates for the short SOA, attentional blink and no-attentional blink conditions, separated into right and left hemisphere for each region of interest. Error bars represent the standard error of the mean. (B) Percent signal change versus time (post-stimulus onset) for the short SOA, attentional blink and no-attentional blink conditions extracted from the PPA ROI. Error bars represent the standard error of the mean. (C) A representative coronal slice of one participant showing activations in PPA as a result of the localiser scans. The functional imaging result is shown superimposed on an anatomical MRI data set. (See Color Plate 7.19 in color plate section.)

3. How do recent electrophysiological results inform us about the relationship between processing the first target and the ensuing effect on the second and how does long-range synchronisation between the two targets correlate with the AB outcome and define an AB 'network'?
4. What do functional imaging results tell us about brain regions involved in the AB and how do they relate to conscious experience?

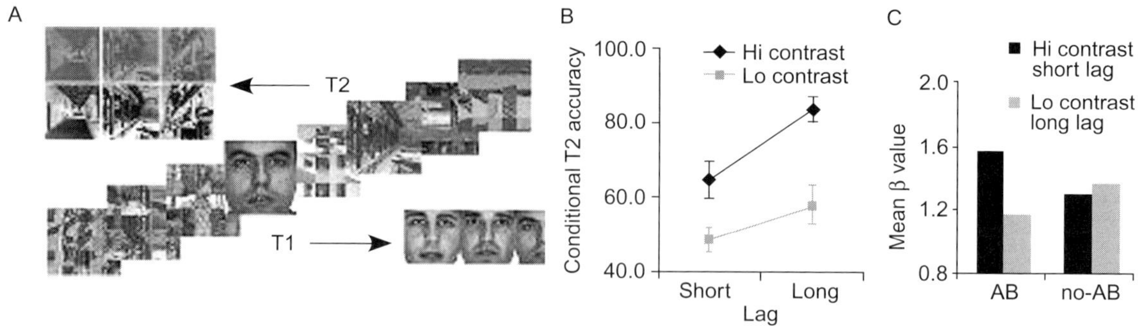

Fig. 20. (A) Schematic representation of the trial structure showing low-contrast and high-contrast (T2) targets. (B) T2 mean identification accuracy conditional on correct first target identification. (C) Mean b estimates obtained from each participant's parahippocampal place area for the short-lag high-contrast and long-lag low-contrast, AB and no-AB trials.

In addressing these questions it is hoped the reader has a better understanding of the parameters producing (or not) the robust finding referred to as an AB, both new as well as old theoretical accounts of the AB, and perhaps most importantly how investigations of this phenomenon provide a useful approach to understanding the boundary conditions of conscious experience. The present chapter was not intended to be an encompassing treatment of the many theories purporting to account for the AB; but instead, a 'primer' of recent empirical results using AB methodology. Before concluding, the reader is referred to a recent theoretical account of the AB (Dehaene et al., 2003) which provides a framework in which to place the AB that encompasses the notion of conscious experience. Dehaene and his colleagues suggest that conscious perception occurs when visual stimuli enter into a 'global workspace' where multiple areas of the brain are linked together. Within this framework the AB occurs when the first target is able to ignite this global workspace but the second target is unable to do so until processing of the first target has been completed.

References

Arend, I., Johnston, S. J., & Shapiro, K. L. (2006). Task-irrelevant visual motion and flicker attenuate the attentional blink. *Psychonomic Bulletin and Review*, *13*(4), 600–607.

Bowman, H., & Wyble, B. (2007). The simultaneous type, serial token model of temporal attention and working memory. *Psychological Review*, *114*(1), 38–70.

Chun, M. M., & Potter, M. C. (1995). A two-stage model for multiple target detection in rapid serial visual presentation. *Journal of Experimental Psychology: Human Perception and Performance*, *21*(1), 109–127.

Dehaene, S., Sergent, C., & Changeux, J. P. (2003). A neuronal network model linking subjective reports and objective physiological data during conscious perception. *Proceedings of the National Academy of Sciences of the United States of America*, *100*(14), 8520–8525.

Drew, T., & Shapiro, K. L. (2006). Representational masking and the attentional blink. *Perception & Psychophysics*, *13*(4), 513–528.

Gross, J., Schmitz, F., Schnitzler, I., Kessler, K., Shapiro, K., Hommel, B., et al. (2004). Modulation of long-range neural synchrony reflects temporal limitations of visual attention in humans. *Proceedings of the National Academy of Sciences of the United States of America*, *101*(35), 13050–13055.

Gross, J., Schmitz, F., Schnitzler, I., Kessler, K., Shapiro, K., Hommel, B., et al. (2006). Anticipatory control of long-range phase synchronization. *Europian Journal of Neuroscience*, *24*(7), 2057–2060.

Johnston, S. J., Shapiro, K. L., Vogels, W., & Roberts, N. J. (2007). Imaging the attentional blink: Perceptual versus attentional limitations. *Neuroreport*, *18*(14), 1475–1478.

Kanwisher, N. G. (1987). Repetition blindness: Type recognition without token individuation. *Cognition*, *27*(2), 117–143.

Kranczioch, C., Debener, S., Schwarzbach, J., Goebel, R., & Engel, A. K. (2005). Neural correlates of conscious perception in the attentional blink. *NeuroImage*, *24*(3), 704–714.

Luck, S. J., Vogel, E. K., & Shapiro, K. L. (1996). Word meanings can be accessed but not reported during the attentional blink. *Nature*, *383*(6601), 616–618.

Marois, R., Yi, D. J., & Chun, M. M. (2004). The neural fate of consciously perceived and missed events in the aftentional blink. *Neuron*, *41*(3), 465–472.

Martin, E. W., Enns, J. T., & Shapiro, K. L. (submitted). Attentional capture and the attentional blink: A dissociation of spatial and temporal discontinuity.

Martin, E. W., Klein, C., Roberts, M., Johnston, S. J., Arend, I., & Shapiro, K. L. (submitted). Task irrelevant activity facilitates attentional investment: An evaluation of the overinvestment hypothesis.

Nakatani, C., Ito, J., Nikolaev, A. R., Gong, P., & van Leeuwen, C. (2005). Phase synchronization analysis of EEG during attentional blink. *Journal of Cognitive Neuroscience*, *17*, 1969–1979.

Olivers, C. N., & Nieuwenhuis, S. (2005). The beneficial effect of concurrent task-irrelevant mental activity on temporal attention. *Psychological Science*, *16*(4), 265–269.

Raymond, J. E., Shapiro, K. L., & Arnell, K. M. (1992). Temporary suppression of visual processing in an RSVP task: An attentional blink? *Journal of Experimental Psychology: Human Perception and Performance*, *18*(3), 849–860.

Seiffert, A. E., & Di Lollo, V. (1997). Low-level masking in the attentional blink. *Journal of Experimental Psychology: Human Perception and Performance*, *23*(4), 1061–1073.

Shapiro, K. L. (1994). The attentional blink: The brain's "eyeblink". *Current Directions in Psychological Science*, *3*(3), 86–89.

Shapiro, K. L., Hillstrom, A. P., & Husain, M. (2003). Control of visuotemporal attention by inferior parietal and superior temporal cortex. *Current Biology*, *12*(15), 1320–1325.

Shapiro, K. L., Johnston, S. J., Vogels, W., Zaman, A., & Roberts, N. (2007). Increased functional magnetic resonance imaging activity during nonconscious perception in the attentional blink. *Neuroreport*, *18*(4), 341–345.

Shapiro, K. L., Raymond, J. E., & Arnell, K. M. (1994). Attention to visual pattern information produces the attentional blink in rapid serial visual presentation. *Journal of Experimental Psychology: Human Perception and Performance*, *20*(2), 357–371.

Shapiro, K. L., Schmitz, F., Martens, S., Hommel, B., & Schnitzler, A. (2006). Resource sharing in the attentional blink. *Neuroreport*, *17*(2), 163–166.

Varela, F., Lachaux, J. P., Rodriguez, E., & Martinerie, J. (2001). The brainweb: Phase synchronization and large-scale integration. *Nature Reviews Neuroscience*, *2*(4), 229–239.

Vogel, E. K., Luck, S. J., & Shapiro, K. L. (1998). Electrophysiological evidence for a postperceptual locus of suppression during the attentional blink. *Journal of Experimental Psychology: Human Perception and Performance*, *24*(6), 1656–1674.

Vogels, W., Johnston, S. J., Linden, D. E., & Shapiro, K. L. (submitted). The neural substrate of the effect of working memory on the attentional blink.

N. Srinivasan (Ed.)
Progress in Brain Research, Vol. 176
ISSN 0079-6123

CHAPTER 8

Practice begets the second target: task repetition and the attentional blink effect

Chie Nakatani[1,*], Shruti Baijal[2] and Cees van Leeuwen[1]

[1]*Laboratory for Perceptual Dynamics, RIKEN Brain Science Institute, Wako-shi, Saitama, Japan*
[2]*Centre for Behavioural and Cognitive Sciences, University of Allahabad, Allahabad, India*

Abstract: Even with unimpaired vision, observers sometimes fail to see things right before their open eyes. A typical example is the *attentional blink* effect, a period in which observers are unable to detect a target item in a sequence of stimuli, for as long as the previous one occupies their mind. Having considered a range of mechanisms proposed to explain attentional blink effect, we arrive at our preferred explanation, which ascribes the effect to a contextually motivated imbalance in the allocation of attentional resources between earlier and later target information. We interpret in this perspective our data on how the attentional blink effect changes as a result of practice.

Keywords: selective attention; visual masking; working memory; RSVP; event-related potentials

"Cannot see with open eyes (Miedomo miezu)" is a Japanese idiom which refers to a paradoxical mental state, at least for the visually unimpaired, and often appears in conjunction with the conditional clause, "if the mind is not in the right place (kokoro kokoni arazareba)". Together they are used when a person is incapable of recognizing the obvious. In fact, we sometimes literally miss a salient event before our wide open eyes. Often our "inability to see" is a matter of being distracted. We may be daydreaming, or absent-minded, as the Japanese idiom suggests. Or, perhaps our mind is focused, but focused on something else. Yet, even when we are firmly focused on an upcoming target, we observers may fail, in certain conditions, to report an event that is conspicuously visible or, technically speaking, well above the threshold of detection.

Cognitive psychologists have been reporting a wide range of such blindness phenomena. For example, when two images are presented in alternation, one a slightly modified version of the other, observers often have difficulty in finding the difference between them, even if that is a very salient, macabre one such as switching heads between two people in the foreground of the image (Change blindness; see Rensink, 2000, 2002). Another example is that when the same object is presented twice, even when each presentation is very distinct from the other, observers often report, nevertheless, that the object was presented only once (Repetition blindness; see Kanwisher, 1987, 1991). These phenomena may truly be considered eye-openers, at least for those who believe that observers'

*Corresponding author.
Tel.: +81 48 462 1111; Fax: +81 48 467 7098;
E-mail: cnakatani@brain.riken.jp

DOI: 10.1016/S0079-6123(09)17608-2

eye-witness testimonies are typically reliable sources of evidence.

The blindness phenomena, which admittedly are a rather heterogeneous class, are not mere showcase items but offer valuable research paradigms for the study of human perception. One of these that has received considerable interest in the last decades is the *attentional blink* (AB) (Broadbent and Broadbent, 1987; Raymond et al., 1992). In a typical AB experiment, a sequence of visual stimuli (e.g., alphanumeric symbols) is presented at a rapid rate, that is, with stimulus onset asynchronies (SOA) of the order of 100 ms. This way of presenting stimuli is called rapid serial visual presentation (RSVP). Two targets, T1 and T2, are embedded in the RSVP sequence; the number of intervening nontargets is varied. When, for example, T2 is presented after T1 with two nontarget presentations in-between, this is called a Lag 3 condition. The SOA between both targets (T1–T2 SOA) for a Lag 3 condition in 100 ms RSVP would be 300 ms (see Fig. 1).

Observers are then asked to report both T1 and T2. Accuracy of T1 report is only slightly worse in this case than in a corresponding task where only T1 is reported, but still is far above chance level. For T2 on the other hand, accuracy alone or conditional to correct report of T1 (T2|T1) varies, depending on the SOA between T1 and T2. When the SOA is 200–500 ms, T2 is likely to be missed, compared to conditions with SOA >500 ms. The reduction in T2 report within Lag 2–5 conditions is what the term "AB effect" refers to (see Fig. 2).

A corollary phenomenon to the AB effect is called Lag-1 sparing; a T2 presented at Lag 1, that is, immediately after T1, is detected with still high accuracy in spite of the short SOA between T1 and T2 (Potter et al., 1998). AB and Lag-1 sparing are extremely robust phenomena; they have been observed numerous times in several varieties by a large number of research groups (e.g., Chun and Potter, 1995; Di Lollo et al., 2005; Hommel et al., 2006; Jolicoeur, 1998; Vogel and Luck, 2002).

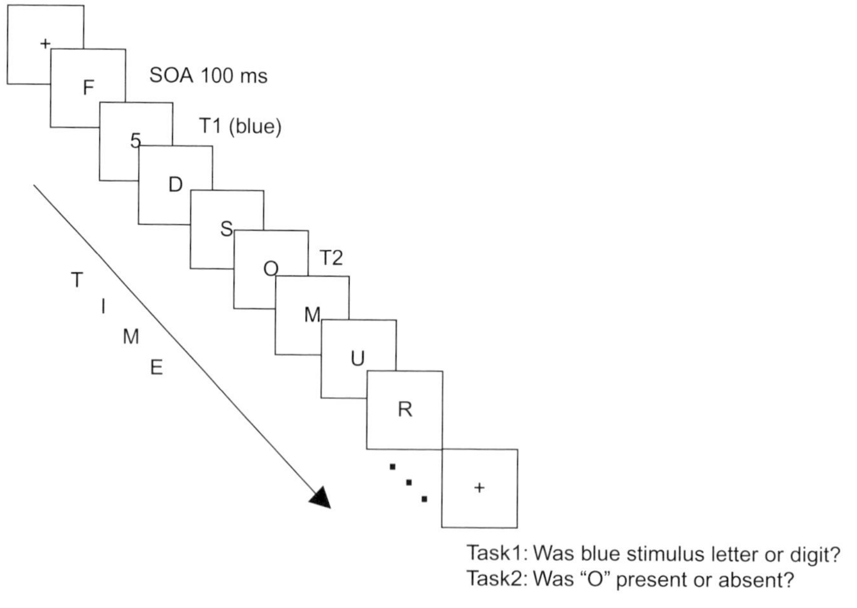

Fig. 1. The course of events in an AB task (an example taken from Nakatani et al., 2005). Twenty stimuli, 19 white and 1 blue, were presented after a fixation cross with a SOA of 100 ms. The category of the blue stimulus (letter or digit) and presence/absence of letter "O" were to be reported. The blue stimulus (T1) preceded "O" (T2). The SOA between T1 and T2 was varied between 100 and 700 ms. The control condition to the dual task condition was T2 single task, in which a blue stimulus was present, however, no task was required for it.

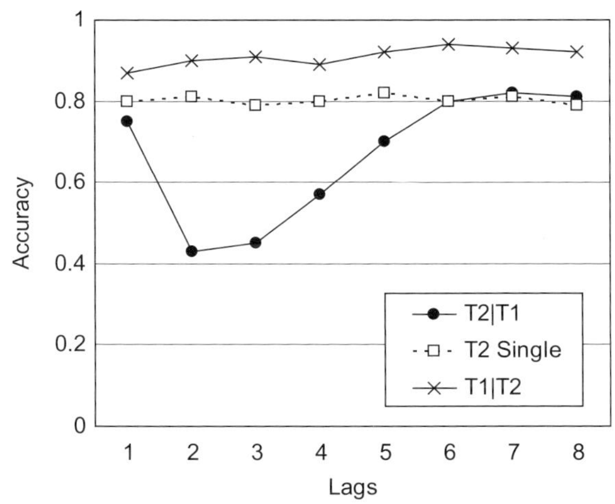

Fig. 2. Hypothetical results of an AB task. T2 accuracy conditional to correct T1 report (T2|T1) is reduced in Lags 2-5 conditions, in which the SOA between T1 and T2 is in the range of 200–500 ms. In the Lag 1 condition, T2|T1 report is spared (Lag-1 sparing). In a task where T2 report is required without T1 report (T2 single), the T2 accuracy does not show AB. T1 accuracy does not depend on T2 report (T1|T2).

In this chapter, we first provide an overview of possible factors that may contribute to the AB effect at various levels of processing. We will consider the likelihood of sensory, perceptual, and post-perceptual level contributions to the effect in light of recent empirical findings. In the second half of the chapter, we discuss whether the AB effect can be reduced by practice, and if so, what does this tell us about these possible explanations. The effects of practice will be analyzed through the study of event-related potentials (ERPs).

What causes the attentional blink effect?

Whereas AB experiments generally involve two targets, a far larger number of items that occur in the RSVP stream are nontargets, or "distractors". Presentation of distractor patterns in the temporal vicinity of targets calls to mind the visual pattern masking phenomenon. To compare the timing of both phenomena, in masking studies accuracy of target report falls down to chance level when the SOA between a target and a patterned mask is around 40 ms. For *forward* masking, in which a patterned mask precedes the target, the chance-level SOA is 0–40 ms. For *backward* masking, where a patterned mask follows the target, chance-level SOA is 0–33 ms (Breitmeyer and Ogmen, 2006). Both these time intervals are quite different from the SOAs in a typical AB task, which are of the order of 100 ms. In fact, a single target embedded in a distractor sequence with 100 ms SOA RSVP yields near-perfect performance (see, e.g., the control condition in Raymond et al, 1992). Thus, the AB effect cannot plausibly be understood as a case of masking of T2 by a distractor.

At the same time, it is known that the AB effect is avoided when the next item after T1 (T1+1) is omitted (Seiffert and Di Lollo, 1997), while omission of one item prior to T1 (T1−1) does not affect AB (Breitmeyer et al., 1999). We may consider the distractors' effect to be an indirect one, such that their presence together with T1 impairs the processing of T2, which then results in the AB effect. We might thus consider the AB as a result of spurious processing of the T1+1 distractor.

Processing of the T1+1'th item is clearly in evidence, once we realize that if the T1+1 item is the second target (T2), this leads to the Lag-1 sparing phenomenon. Whenever the Lag-1 sparing occurs, the type of information processed about the target, obviously, must be sufficient to distinguish it from a distractor. But, since observers do not know in advance whether the T+1 item is a target or not, distractors will be processed to the same level. Thus, when the second target is spared, the criterion that distinguishes it from a distractor can inform us about the level of processing in distractors. Lag-1 sparing was observed in tasks that require semantic distinctions, for instance digit targets versus letter distractors (e.g., Akyurek and Hommel, 2005). We may, therefore, conclude that T1+1 items are processed at least up to such a level.

The indirect effect of distractors may, therefore, occur at the level of working memory (Baddeley, 1986). The capacity of working memory is notoriously small; according to recent estimates it is restricted to four items, on average (Luck and Vogel, 1997). If distractor information is allowed to enter, this could cause it to overflow. Some studies had suggested that working memory is highly loaded in AB conditions (Akyurek et al.,

2007a). Recent findings, however, present counterevidence (Arnell et al., 2008; Colzato et al., 2007; Martens and Johnson, 2009). Martens and Johnson (2009), for example, reported that measures of neither working memory nor short-term memory capacity, nor general intelligence correlate with the size of the AB effect in individuals. Therefore, none of these factors are likely to be involved in the AB effect.

To reconcile these conflicting findings, Arnell et al. (2008) pointed to the need to distinguish these factors and an "executive component" of working memory. For this they calculated a test score, from which general intelligence and memory storage capacity factors were eliminated statistically by means of partial correlation techniques. The residual correlation was negative with the size of the AB effect. This result suggests that the AB effect does not depend on static storage capacity in short term or working memory but rather on its operational characteristics. The T+1 item, therefore, does not distract T2 processing merely by loading working memory with irrelevant information; rather, the presence of the distractor seems to interfere with working memory access or retrieval procedures.

Let us first consider retrieval, about which we can be brief. In the typical AB task, T1 is always reported before T2. Thus, while the first report occurs, interference may hit retrieval of the second target. Sergent and Dehaene (2004) asked observers to report T2 prior to T1. However, the AB effect was still obtained in this condition. The result demonstrates that working memory retrieval is not a major contributor to the AB effect.

Let us now consider the processes that regulate admission to working memory. What kind of information is admitted? Given that T1 is almost always reported, its probability of being admitted must be high. When the T1+1 item is the second target, as Lag-1 sparing indicates, information is also likely to be admitted. Visser et al. (1999) reviewed more than 100 papers on Lag-1 sparing and found that the Lag-1 sparing disappeared when T1 and T2 are presented in different locations or differ on more than one "dimension", such as task (e.g., T1 identification and T2 detection) or category (e.g., digit or alphabet). We may therefore think of the interface as a filter that is sensitive to spatial location and item features which determine target (e.g., category and color).

This view has led to a "filter reconfiguration" account of the AB effect. Di Lollo et al. (2005) proposed that the filter is reconfigured when mismatch between T1 and T1+1 items is sufficiently large, for example, T1 is a green digit, while T1+1 distractor is red alphabet. Filter reconfiguration delays admission to working memory, increasing the probability of T2 information to be lost due to decay or overwriting.

Nieuwenstein and colleagues questioned the filter reconfiguration account (Nieuwenstein, 2006; Nieuwenstein et al., 2005). In their study, targets were two arbitrary selected digits and distractors were alphabets or digits. Targets were presented either in green or in red, while distractors could be green, red, blue, or light gray. Thus, according to Di Lollo et al. (2005), the AB effect should be enhanced when color is different between T1 and T1+1 (e.g., green T1 and red T1+1). This result failed to occur. Nieuwenstein (2006), therefore, concluded that admission was still possible after color mismatch of T1 and T1+1. Instead, the author found high incidence of correct T2 report when one of the distractors between T1 and T2 had a target color (red or green), rather than any of the other possible colors. This is a positive indication that a filter reconfiguration triggered by an earlier feature mismatch was unlikely to have occurred.

We may, therefore, raise the possibility of a more sophisticated admission mechanism, which is sensitive to context of information itself (e.g., this "green" was previously presented as a target property). Chun (1997a, b) proposed an admission mechanism that takes into account such "episodic distinctiveness of visual input". In this model, information selected depending on item type and episodic distinctiveness is processed further in working memory, resulting into a reportable format called "object token". When episodic distinctiveness is lacking, for example, between a T1+1 distractor and a subsequent T2, the distractor could be reported instead of T2. Isaak et al. (1999) reported that T1+1 or T2+1 distractor items were wrongly reported as T2, so called "intrusion errors", in about 30% of all T2 error trials. These results clearly show the role of episodic/contextual item properties in working memory admission.

Intrusion errors may contribute to the AB effect. However, these alone are not enough to explain the AB effect because about 70% of the errors show no evidence of intrusion. This consideration leads us to presume that over and above external distractors, the internal state (or context) plays an important role in the AB effect as well. Observers may therefore have some degree of control over the AB effect. Olivers and Nieuwenhuis (2005, 2006) proposed that the AB effect results from overcommitment of resources to the task (over-investment account). For example, in Experiment 3 of their 2006 study, participants performed two sessions of the AB task. One group was given an explicit instruction to concentrate *less* on the AB task after the first session; no such instruction was given to the other. In the first, but not in the second group, AB was reduced between the two sessions. The authors explained this result by pointing out that the first instruction reduced over commitment, resulting in the availability of extra resources for T2. These studies, therefore, showed that the AB effect is under strategic control, which could be manipulated by instruction.

The overinvestment account could be criticized for its failure to specify how reduction of over-allocation is achieved. Varieties of experimental manipulations have been aimed at reducing over-allocation, such as insertion of a task irrelevant to AB targets, verbal instruction not to pay attention to stimuli, or interruption by music (Olivers and Nieuwenhuis, 2005, 2006). Yet, these do not easily converge to a common mechanism for reallocation (although at least one proposal has been made; see Chapter 7 by Shapiro in this volume). But then, the reduction of over-allocation might not rely on a single mechanism; reallocation of resources could occur through multiple subsystems. Dehaene et al. (2003) proposed a neurally plausible model of how observers arrive about judgments on what they see, which allows versatile resource reallocation among several subsystems. The model assumes two orthogonal routes of activation dispersion in the brain. One is a cortico-cortical route that roughly corresponds to a feedforward flow of visual information processing and the other is realized through a panoply of recurrent cortico-thalamic loops, contributing to the quick propagation of activity to the whole brain. The two routes combined embody a brain-wide network, termed global workspace, through which target information in local systems is accessed by other systems, including ones involved in reporting a target. In support of the model, Sergent et al. (2005) showed that target presentation evoked global brain activation during an AB task (for details, see Del Cul et al., 2007). Whereas T1 invariably evoked global activation, it was observed for T2 only when this target was reported. The authors concluded that global activation is critical for the T2 report.

In our view, this view is an interesting proposal, several aspects of which however need further elaboration. Is the global activity evoked by the target, or should it rather arise spontaneously, for instance in anticipation of the target? Nakatani et al. (2005) showed that large-scale synchronization of oscillatory brain activity *prior to* target presentation was needed for T2 detection in an AB task. Another issue is whether the activation is global or is confined to a sparse, distributed network of interconnected regions. Using magnetoencephalography (MEG) source localization, Gross et al. (2004) located several hubs of activation during an AB task, and showed that they form a sparse, distributed functional network. Despite these quibbles, there are useful elements in this theory that allow us to connect this result with Olivers and Nieuwenhuis (2006). The capacity to involve a widely distributed system points to a possible way to reallocate resources through dynamically adding connectivity in a manner determined by the internal state of the system itself. This opens the door to understanding effects of active anticipation, strategic control, and, generally, context on the AB.

How does practice affect to the attentional blink?

Opening the door to a contextual account of the AB implies, at the same time, closing the door on

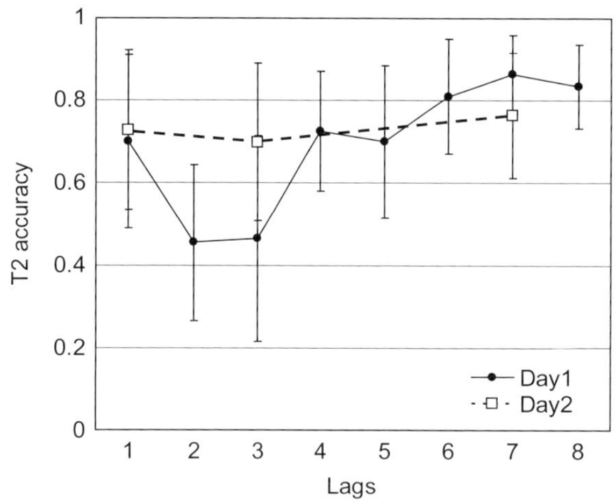

Fig. 3. Attentional blink is attenuated with practice. In the first session (solid line), participants showed the AB effect. The same participants, however, lost the effect when they were retested on another day (broken line).

efforts to find a single underlying mechanism that will explain the AB from invariable constraints on processing. In following section of this chapter, we discuss on practice effects in the AB phenomenon to understand dynamics of visual report system. Practice, by definition, modifies internal states. Thus, we should first investigate whether, indeed, the AB effect changes with practice. In one of our studies, participants repeated an AB task over two sessions about a week apart. In the first session, we obtained a robust AB effect, in combination with Lag-1 sparing. In the second session, the same participants repeated the same task with the same task instruction. Here, the AB effect had already disappeared (see Fig. 3).

We may interpret the practice effect as a form of resource reallocation. Practice leads to automatization, which generally reduces the resources required in processing. In principle, reduction of resource could happen to T1 and T2 processes simultaneously. Such a reduction may lead to reallocation of resource. In earlier discussion of this chapter, we suggested the hypothesis that internal, distinctive context to T2 is critical for its admission to working memory. Practice may reallocate extra resources to T2, which will provide it with the distinctive context needed for its admission to working memory.

To test this hypothesis, it is desirable to have a measure that distinguishes resource allocation to different stages of T2 processing. Event-related brain potentials, typically, fulfill this requirement. As its name suggest, the ERP method assumes that external or internal events, such as stimulus input and access to semantic contents, produce detectable components in electrocortical activity. To extract such event-related activity from nonevent-related ones, participants are subject to the event and their electroencephalogram (EEG) is recorded repeatedly. The multiple EEG records are aligned from the onset (or another specific time point) of the event and averaged. The averaging is to cancel out nonevent-related activities, which by definition are not time-locked to the event. The average ERPs show voltage fluctuation in real-time, and thus provide information on the real-time brain processes relevant to the event.

ERP measures as well as event-related components in MEG have been used extensively to study the AB phenomenon (Akyurek et al., 2007b; Giesbrecht et al., 2007; Kessler et al., 2005; Koivisto and Revonsuo, 2008; Kranczioch et al., 2003, 2007; Martens et al., 2006a, b; McArthur et al., 1999; Nakatani et al., submitted; Rolke et al., 2001; Sessa et al., 2007; Sergent et al., 2005; Shapiro et al., 2006; Slagter et al., 2007; Vogel et al., 1998; Vogel and Luck, 2002). These studies all share the same, simple assumption: ERP amplitude and latency reflect, respectively, resource allocated to and time spent on processing the visual stimuli. In an AB task, T1, T2, and even distractors evoke a sequence of activities. Usually the distractor components are removed (e.g., by subtracting ERPs of distractor-only conditions from ERPs in target-with-distractor conditions,) to isolate target-evoked components. Several target-evoked components appear within the 50–1000 ms episode from target onset. These are usually denoted by their polarity (*Positive* or *Negative*) and order, such as P1, N1, P2, N2, and P3. We briefly review the relevant findings on the relationship between the ERP components and T1 and T2 processing, prior to discussing ERPs in the practice effect.

T1-evoked ERPs

In general, T1 report is accurate and constant across AB parameters like T1–T2 lags (see Fig. 2). We might, therefore, infer that T1 processes are completed regardless of the fate of T2. Thus, the T1-evoked ERPs would correlate little, if not at all, with the AB effect. The first T1-evoked ERP is *P1*, which appears within 80–150 ms from T1 onset in occipital regions. Subsequent to P1, *N1* appears in the latency of 150–200 ms also in occipital regions. Amplitude and latency of P1 and N1 evoked by T1 did not correlate with accuracy of T2 report (Vogel and Luck, 2002) or T2 visibility (Sergent et al., 2005). The polarity, latency, and the scalp distribution of the T1-evoked P1 and N1 are consistent to early visual ERPs that are known to be sensitive to spatial attention allocation (Mangun, 1995; Hillyard et al., 1998). The results, therefore, suggest that spatial attention to T1 processes earlier than 200 ms is allocated similarly in AB and non-AB trials.

The next peak evoked by T1 is *P2*, which appears around 220 ms in occipito-parietal regions. Amplitude and latency of the component did not correlate with percentage of correct T2 report (Nakatani et al., submitted). Amplitude of posterior P2 of this latency, in general, is related to reentrant feedback from higher to lower visual areas (Kotsoni et al., 2007). Therefore, it may not be surprising that the amplitude, presumably reentrant feedback of T1 information, was constant regardless of the percentage of correct T2 report.

Following P2, *N2* appears around 250–300 ms in occipito-parietal regions. Sergent et al. (2005) reported that the N2 evoked by T1 did not correlate with subjective rating of T2 visibility. Also, its latency did not correlate with percentage of correct T2 report (Nakatani et al., submitted). The occipito-parietal (i.e., posterior) N2 is often related, among other things, to spatial attention for detection of a target among distractors, for example, in visual search (Folstein and Van Petten, 2008; Woodman and Luck, 1999, 2003). Thus, the results suggest that the allocation of spatial attention to T1 processing around 250 ms also did not correlate with the AB effect.

The next component, *P3*, appears around 300–600 ms from T1 onset. This component is centered in parietal regions, but widely distributed. Many studies have therefore begun to distinguish two subcomponents of the P3, based on scalp distribution and latency: *P3a*, a front-central component with a latency of 300–450 ms, and *P3b*, a parieto-temporal component with the latency of 400–800 ms (e.g., Sergent et al., 2005; Slagter et al., 2007). Sergent et al. (2005) reported that both P3a and P3b did not correlate with T2 visibility. Martens et al. (2006a) also found no correlation between P3 amplitude and the size of the AB effect. On the contrary, a positive correlation was reported using a M300 component, which is a P3 equivalent in MEG. Shapiro et al. (2006), by contrast, reported that individuals with a larger M300 showed a greater AB effect. The P3 family is generally understood to reflect post-sensory target processing such that: P3 reflects target detection at large; P3a reflects attention-driven working memory operation; and P3b reflects task relevant context updating and further memory operation (Soltani and Knight, 2000; Polich, 2007). As the contrasting reports on amplitude illustrate, it is inconclusive whether T1-evoked P3-related processes correlate with the AB effect.

Besides amplitude, Martens et al. (2006b) compared the peak latency of T1-evoked P3 between a group of observers who did not show AB (non-blinkers) and ones who did (blinkers). In their study, the peak latency was shorter in non-blinkers than in blinkers. The authors concluded that non-blinkers are able to consolidate T1 information at a faster rate than blinkers. These P3 results suggest that T1 processes in working memory or later would correlate with the AB effect.

T2-evoked ERPs

In contrast with T1-related processes, which appear generally unaffected by AB conditions, we may expect that T2-related processes are disturbed. The apparent question is, therefore, at what point this occurs. ERP studies showed that amplitude and latency of P1, N1, and P2 did not correlate with the AB effects (Vogel et al., 1998; Vogel and Luck, 2002; Sergent et al., 2005;

Nakatani et al., submitted). These results suggest that T2 processes up to 220 ms do not differ between AB and non-AB trials.

On the other hand, the amplitude of T2-evoked posterior N2 positively correlated with T2 visibility (Sergent et al., 2005) and sensitivity to T2, while its latency did not (Nakatani et al., submitted). The result suggests that spatial-attention-based T2 selection process relate closely to T2 visibility (Kiss et al., 2008; Woodman and Luck, 1999, 2003). Spatial attention may, indeed, be relevant for T2 report. For instance, as mentioned, Lag-1 sparing does not occur when Lag 1 T2 is presented in a location different from T1 (Breitmeyer et al., 1999; Seiffert and Di Lollo, 1997). Thus, we may assume that T2 report is hampered by limited availability of spatial attention around 250 ms after its onset.

Some other studies did not report such a correlation in N2, but in P3 instead. In early studies, it was reported that the stronger the AB effect, the smaller P3 amplitude (McArthur et al., 1999; Vogel and Luck, 2002). Later studies reported, however, that P3, P3a, and P3b did not change their amplitude in proportion to AB magnitude. Rather, the components appeared only when T2 was visible, that is, the component correlated with "seen" versus "unseen" states of mind in the observer (Kranczioch et al., 2003; Sergent et al., 2005). These results suggest that T2 is likely to be detected when working memory operations for T2, such as consolidation and context updating, were successful (Polich, 2007). In our terminology, this means that T2 admission is necessary but not crucial.

As for latency, Sessa et al. (2007) examined T2-evoked P3 latency in conditions without distractor items after T2, in order to observe the T2-evoked P3 component without contamination from ERPs elicited by the distractors. The latency of P3 in an AB condition was longer than that in non-AB condition. The authors concluded that increased latency implied that completion of T2 consolidation was delayed in the AB condition.

These results suggest that divergence between AB and non-AB trials could occur at around 250 ms from T2 onset, presumably, during the spatial-attention-based process of selecting T2 information for admission to working memory. The results of T2-evoked ERPs lead us to two possible scenarios to the AB effect. One case in which T2 does not receive sufficient spatial attention for an admission to working memory, and the other in which T2 is not properly consolidated for a report in working memory.

Having discussed the ERP findings, let us now return to the questions regarding the practice effect. Whereas T1 accuracy largely remains unchanged, practice improves T2 report. We, then, raised the question if the practice effect is simply due to automatization or the result of allocating extra resource to T2 processing. In the former case, we would predict that practice reduces in amplitude both T1-evoked P3, and T2-evoked N2 and P3. In the latter case, however, we would expect an *increase* in amplitude of the T2-evoked peak activity (N2 and P3). We may then ask a further question, whether the extra resource is allocated to T2 admission and/or to the T2 consolidation processes. If T2 admission process gets the extra resource, we would expect an increase in T2-evoked N2. On the other hand, if T2 consolidation process receives it, we should observe an increase in T2-evoked P3.

Studies reporting practice effects in the AB have been remarkably sparse. There is only one study, to our knowledge, in which ERPs were investigated in relation to practice effects in an AB paradigm (Slagter et al., 2007). Their study, however, compared AB task performance before and after 3 months of *meditation* training. Experienced yoga practitioners and novices took part in their experiment. After the meditation practice, the AB effect was eliminated in the yoga practitioners, and correspondingly, T1-evoked P3b decreased without a change in T2-evoked components. The AB effect was merely reduced in the novice practitioner group, and no ERP reduction whatsoever took place. The observed pattern of ERP results is inconclusive, but seems to be in accordance with automatization of both T1 and T2, kept in the balance with transfer of resources to T2, so that a null effect results. Nevertheless, the authors concluded that improvement in performance was due to mental training. Methodological issues, however, make it difficult to draw any conclusion from this study on the processes affected by practice. The long-term meditation training used by Slagter et al. (2007) was confounded with task repetition. Moreover, the interval between the first and the second session of their AB session is very long, as numerous factors could change within a 3-month period.

Resolving these methodological problems, Nakatani et al. (submitted) investigated the practice effect. Participants completed two sessions of an AB task, which were held on 2 days, separated by 1 day to 2 weeks. Between the sessions, the AB was reduced — sensitivity to T2 increased. T1-evoked P3a showed a decreasing tendency between sessions but its amplitude did not correlate with T2 sensitivity. For the components evoked by the second target, amplitude of N2 increased, with T2 sensitivity, while that of P3a or P3b did not (see Fig. 4). Nakatani et al. (submitted) found that the T2-evoked N2 amplitude was largest 1 day after the first session, but gradually decreased and came back to baseline after around 4 days. These results show that practice leads not just to automatization, but also to increases in spatial attention allocation, used in the process of admission of T2 information to working memory.

The assumption that spatial attention processes are involved in working memory admission is theoretically motivated. For instance, the COrollary Discharge of Attention Movement (CODAM) model of consciousness and attention assumes an attention control generator in the parietal region, which modulates activation of primary and associative visual cortices (Fragopanagos et al., 2005; Taylor, 2003). The CODAM model assumes that N2 reflects the control signal and amplified activation in the visual cortices. Similarly, the simultaneous type, serial token (ST^2) model assumes a blaster unit, which gives a short, but intensive attentional enhancement to post-T1 stimuli presented at the same spatial location as T1 (Bowman and Wyble, 2007; Bowman et al., 2008).

The effect of practice illustrates the context specificity of resource allocation. Flexibility is intrinsic to a contextual account. Whereas simple practice without feedback made extra resource available to the T2 admission, practice with feedback might change the flow of resources to

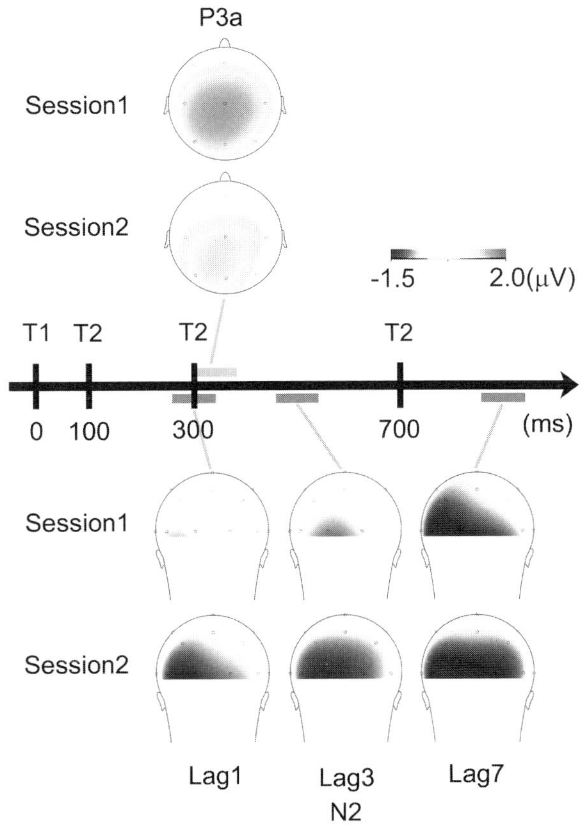

Fig. 4. Practice effects on T1-evoked P3a and T2-evoked N2. Horizontal axis shows time from T1 onset. Vertical tick marks indicate onsets of T1 and T2 in Lag 1, 3, and 7 conditions. (Above) Head figures (bird's eye view, nose pointing upward) show the scalp distributions of T1-evoked P3a in Sessions 1 and 2, averaged over a 100-ms period marked as a short horizontal bar above the horizontal axis. (Below) Head figures (rear view) show the scalp distributions of T2-evoked N2 in Sessions 1 and 2, averaged over the 100-ms period marked as short horizontal bars below the horizontal axes.

the benefit of other processing components (Nakatani and Van Leeuwen, in preparation). The results so far confirm the invaluable advice on how to overcome a moment of blindness: rest, try again, and detach the mind from what already is clearly in evidence.

List of abbreviations

AB	attentional blink
CODAM	COrollary Discharge of Attention Movement model

EEG	electroencephalography
ERPs	event-related potentials
M300	P3 equivalent MEG component
MEG	magnetoencephalography
N1	the first negative ERP component
N2	the second negative ERP component
P1	the first positive ERP component
P2	the second positive ERP component
P3	the third positive ERP component
P3a and P3b	ERP subcomponents of P3
RSVP	rapid serial visual presentation
SOA	stimulus onset asynchronies
ST^2	simultaneous type, serial token model
T1	the first target
T1+1	the next item after T1
T1−1	one item prior to T1
T1−T2 SOA	SOA between both targets
T2	the second targets
T2|T1	T2 accuracy conditional to correct report of T1

References

Akyurek, E. G., & Hommel, B. (2005). Short-term memory and the attentional blink: Capacity versus content. *Memory & Cognition*, *33*, 654–663.

Akyurek, E. G., Hommel, B., & Jolicoeur, P. (2007a). Direct evidence for a role of working memory in the attentional blink. *Memory & Cognition*, *35*, 621–627.

Akyurek, E. G., Riddell, P. M., Toffanin, P., & Hommel, B. (2007b). Adaptive control of event integration: Evidence from event-related potentials. *Psychophysiology*, *44*, 383–391.

Arnell, K. M., Stokes, K. A., Maclean, M. H., & Gicante, C. (2008). Executive control processes of working memory predict attentional blink magnitude over and above storage capacity. *Psychological Research*. DOI 10.1007/s00426-008-0200-4.

Baddeley, A. (1986). *Working memory*. Clarendon Press: Oxford.

Bowman, H., & Wyble, B. (2007). The simultaneous type, serial token model of temporal attention and working memory. *Psychological Review*, *114*, 38–70.

Bowman, H., Wyble, B., Chennu, S., & Craston, P. (2008). A reciprocal relationship between bottom-up trace strength and the attentional blink bottleneck: Relating the LC-NE and ST(2) models. *Brain Research*, *1202*, 25–42.

Breitmeyer, B. G., Ehrenstein, A., Pritchard, K., Hiscock, M., & Crisan, J. (1999). The roles of location specificity and masking mechanisms in the attentional blink. *Perception & Psychophysics*, *61*, 798–809.

Breitmeyer, B. G., & Ogmen, H. (2006). *Visual masking: Time slices through conscious and unconscious vision*. Oxford: Oxford University Press.

Broadbent, D. E., & Broadbent, M. P. (1987). From detection to identification: Response to multiple targets in rapid serial visual presentation. *Perception & Psychophysics*, *42*, 105–113.

Chun, M. M. (1997a). Temporal binding errors are redistributed by the attentional blink. *Perception & Psychophysics*, *59*, 1191–1199.

Chun, M. M. (1997b). Types and tokens in visual processing: A double dissociation between the attentional blink and repetition blindness. *Journal of Experimental Psychology: Human Perception & Performance*, *23*, 738–755.

Chun, M. M., & Potter, M. C. (1995). A two-stage model for multiple target detection in rapid serial visual presentation. *Journal of Experimental Psychology: Human Perception & Performance*, *21*, 109–127.

Colzato, L. S., Spape, M. M. A., Pannebakker, M. M., & Hommel, B. (2007). Working memory and the attentional blink: Blink size is predicted by individual differences in operation span. *Psychonomic Bulletin and Review*, *14*, 1051–1057.

Dehaene, S., Sergent, C., & Changeux, J. P. (2003). A neuronal network model linking subjective reports and objective physiological data during conscious perception. *Proceedings of the National Academy of Sciences of the United States of America*, *100*, 8520–8525.

Del Cul, A., Baillet, S., & dehaene, S. (2007). Brain dynamics underlying the nonlinear threshold for access to consciousness. *PLoS Biology*, *5*, e260.

Di Lollo, V., Kawahara, J., Shahab Ghorashi, S. M., & Enns, J. T. (2005). The attentional blink: Resource depletion or temporary loss of control?. *Psychological Research*, *69*, 191–200.

Folstein, J. R., & Van Petten, C. (2008). Influence of cognitive control and mismatch on the N2 component of the ERP: A review. *Psychophysiology*, *45*, 152–170.

Fragopanagos, N., Kockelkoren, S., & Taylor, J. G. (2005). A neurodynamic model of the attentional blink. *Brain Research: Cognitive Brain Research*, *24*, 568–586.

Giesbrecht, B., Sy, J. L., & Elliott, J. C. (2007). Electrophysiological evidence for both perceptual and postperceptual selection during the attentional blink. *Journal of Cognitive Neuroscience*, *19*, 2005–2018.

Gross, J., Schmitz, F., Schnitzler, I., Kessler, K., Shapiro, K., Hommel, B., et al. (2004). Modulation of long-range neural synchrony reflects temporal limitations of visual attention in humans. *Proceedings of the National Academy of Sciences of the United States of America*, *101*, 13050–13055.

Hillyard, S. A., Teder-Salejarvi, W. A., & Munte, T. F. (1998). Temporal dynamics of early perceptual processing. *Current Opinion in Neurobiology*, *8*, 202–210.

Hommel, B., Kessler, K., Schmitz, F., Gross, J., Akyurek, E., Shapiro, K., et al. (2006). How the brain blinks: Towards a neurocognitive model of the attentional blink. *Psychological Research*, *70*, 425–435.

Isaak, M. I., Shapiro, K. L., & Martin, J. (1999). The attentional blink reflects retrieval competition among multiple rapid serial visual presentation items: Tests of an interference model. *Journal of Experimental Psychology: Human Perception & Performance*, *25*, 1774–1792.

Jolicoeur, P. (1998). Modulation of the attentional blink by on-line response selection: Evidence from speeded and unspeeded task1 decisions. *Memory & Cognition*, *26*, 1014–1032.

Kanwisher, N. (1987). Repetition blindness: Type recognition without token individuation. *Cognition*, *27*, 117–143.

Kanwisher, N. (1991). Repetition blindness and illusory conjunctions: Errors in binding visual types with visual tokens. *Journal of Experimental Psychology: Human Perception and Performance*, *17*, 404–421.

Kessler, K., Schmitz, F., Gross, J., Hommel, B., Shapiro, K., & Schnitzler, A. (2005). Cortical mechanisms of attention in time: Neural correlates of the Lag-1-sparing phenomenon. *European Journal of Neuroscience*, *21*, 2563–2574.

Kiss, M., Van Velzen, J., & Eimer, M. (2008). The N2pc component and its links to attention shifts and spatially selective visual processing. *Psychophysiology*, *45*, 240–249.

Koivisto, M., & Revonsuo, A. (2008). Comparison of event-related potentials in attentional blink and repetition blindness. *Brain Research*, *1189*, 115–126.

Kotsoni, E., Csibra, G., Mareschal, D., & Johnson, M. H. (2007). Electrophysiological correlates of common-onset visual masking. *Neuropsychologia*, *45*, 2285–2293.

Kranczioch, C., Debener, S., & Engel, A. K. (2003). Event-related potential correlates of the attentional blink phenomenon. *Brain Research: Cognitive Brain Research*, *17*, 177–187.

Kranczioch, C., Debener, S., Maye, A., & Engel, A. K. (2007). Temporal dynamics of access to consciousness in the attentional blink. *Neuroimage*, *37*, 947–955.

Luck, S. J., & Vogel, E. K. (1997). The capacity of visual working memory for features and conjunctions. *Nature*, *390*, 279–281.

Mangun, G. R. (1995). Neuronal mechanism of visual selective attention. *Psychophysiology*, *32*, 4–18.

Martens, S., Elmallah, K., London, R., & Johnson, A. (2006a). Cuing and stimulus probability effects on the P3 and the AB. *Acta Psychologica*, *123*, 204–218.

Martens, S., & Johnson, A. (2009). Working memory capacity, intelligence, and the magnitude of the attentional blink revisited. *Experimental Brain Research*, *192*, 43–52.

Martens, S., Munneke, J., Smid, H., & Johnson, A. (2006b). Quick minds don't blink: electrophysiological correlates of individual differences in attentional selection. *Journal of Cognitive Neuroscience*, *18*, 1423–1438.

Mcarthur, G., Budd, T., & Michie, P. (1999). The attentional blink and P300. *Neuroreport*, *10*, 3691–3695.

Nakatani, C., Baijal, S., & Van Leeuwen, C. (submitted). Curbing the attentional blink: Practice keeps the mind's eye open.

Nakatani, C., Ito, J., Nikolaev, A. R., Gong, P., & Van Leeuwen, C. (2005). Phase synchronization analysis of EEG during attentional blink. *Journal of Cognitive Neuroscience*, *17*, 1969–1979.

Nakatani, C., & Van Leeuwen, C. (in preparation). Feedback and the atttentional blink effect: An ERP study.

Nieuwenstein, M. R. (2006). Top-down controlled, delayed selection in the attentional blink. *Journal of Experimental Psychology: Human Perception & Performance*, *32*, 973–985.

Nieuwenstein, M. R., Chun, M. M., Van Der Lubbe, R. H., & Hooge, I. T. (2005). Delayed attentional engagement in the attentional blink. *Journal of Experimental Psychology: Human Perception & Performance*, *31*, 1463–1475.

Olivers, C. N., & Nieuwenhuis, S. (2005). The beneficial effect of concurrent task-irrelevant mental activity on temporal attention. *Psychological Science*, *16*, 265–269.

Olivers, C. N., & Nieuwenhuis, S. (2006). The beneficial effects of additional task load, positive affect, and instruction on the attentional blink. *Journal of Experimental Psychology: Human Perception & Performance*, *32*, 364–379.

Polich, J. (2007). Updating P300: An integrative theory of P3a and P3b. *Clinical Neurophysiology*, *118*, 2128–2148.

Potter, M. C., Chun, M. M., Banks, B. S., & Muckenhoupt, M. (1998). Two attentional deficits in serial target search: the visual attentional blink and an amodal task-switch deficit. *Journal of Experimental Psychology: Learning, Memory & Cognition*, *24*, 979–992.

Raymond, J. E., Shapiro, K. L., & Arnell, K. M. (1992). Temporary suppression of visual processing in an RSVP task: an attentional blink? *Journal of Experimental Psychology: Human Perception & Performance*, *18*, 849–860.

Rensink, R. A. (2000). Seeing, sensing, and scrutinizing. *Vision Research*, *40*, 1469–1487.

Rensink, R. A. (2002). Change detection. *Annual Review of Psychology*, *53*, 245–277.

Rolke, B., Heil, M., Streb, J., & Hennighausen, E. (2001). Missed prime words within the attentional blink evoke an N400 semantic priming effect. *Psychophysiology*, *38*, 165–174.

Seiffert, A. E., & Di Lollo, V. (1997). Low-level masking in the attentional blink. *Journal of Experimental Psychology: Human Perception and Performance*, *23*, 1061–1073.

Sergent, C., Baillet, S., & Dehaene, S. (2005). Timing of the brain events underlying access to consciousness during the attentional blink. *Nature Neuroscience*, *8*, 1391–1400.

Sergent, C., & Dehaene, S. (2004). Is consciousness a gradual phenomenon? Evidence for an all-or-none bifurcation during the attentional blink. *Psychological Science*, *15*, 720–728.

Sessa, P., Luria, R., Verleger, R., & Dell'acqua, R. (2007). P3 latency shifts in the attentional blink: Further evidence for second target processing postponement. *Brain Research*, *1137*, 131–139.

Shapiro, K., Schmitz, F., Martens, S., Hommel, B., & Schnitzler, A. (2006). Resource sharing in the attentional blink. *Neuroreport*, *17*, 163–166.

Slagter, H. A., Lutz, A., Greischar, L. L., Francis, A. D., Nieuwenhuis, S., Davis, J. M., et al. (2007). Mental training affects distribution of limited brain resources. *PLoS Biology*, *5*, e138.

Soltani, M., & Knight, R. T. (2000). Neural origins of P300. *Critical Reviews in Neurobiology*, *14*, 199–224.

Taylor, J. G. (2003). Paying attention to consciousness. *Progress in Neurobiology*, *71*, 305–335.

Visser, T. A. W., Bischof, W. F., & Di Lollo, V. (1999). Attentional switching in spatial and nonspatial domains: Evidence from the attentional blink. *Psychological Bulletin*, *125*, 458–469.

Vogel, E. K., & Luck, S. J. (2002). Delayed working memory consolidation during the attentional blink. *Psychonomic Bulletin & Review*, *9*, 739–743.

Vogel, E. K., Luck, S. J., & Shapiro, K. L. (1998). Electrophysiological evidence for a postperceptual locus of suppression during the attentional blink. *Journal of Experimental Psychology: Human Perception & Performance*, *24*, 1656–1674.

Woodman, G. F., & Luck, S. J. (1999). Electrophysiological measurement of rapid shifts of attention during visual search. *Nature*, *400*, 867–869.

Woodman, G. F., & Luck, S. J. (2003). Serial deployment of attention during visual search. *Journal of Experimental Psychology: Human Perception & Performance*, *29*, 121–138.

N. Srinivasan (Ed.)
Progress in Brain Research, Vol. 176
ISSN 0079-6123

CHAPTER 9

Using biologically plausible neural models to specify the functional and neural mechanisms of visual search

Glyn W. Humphreys*, Harriet A. Allen and Eirini Mavritsaki

Behavioural Brain Sciences, School of Psychology, University of Birmingham, Birmingham, UK

Abstract: We review research from our laboratory that attempts to pull apart the functional and neural mechanisms of visual search using converging, inter-disciplinary evidence from experimental studies with normal participants, neuropsychological studies with brain lesioned patients, functional brain imaging and computational modelling. The work suggests that search is determined by excitatory mechanisms that support the selection of target stimuli, and inhibitory mechanisms that suppress irrelevant distractors. These mechanisms operate through separable though overlapping neural circuits which can be functionally decomposed by imposing model-based analyses on brain imaging data. The chapter highlights the need for inter-disciplinary research for understanding complex cognitive processes at several levels.

Keywords: visual search; visual attention; computational modelling; functional brain imaging

Introduction

The visual world presents us with a complex and dynamically changing environment where it is important to be able to select efficiently stimuli that match our current behavioural goals. Attempts to measure the efficiency of visual selection have frequently used visual search to examine which factors facilitate and which impair selection processes. Across the past 40 years, numerous studies of search have been conducted which have demonstrated that, when targets differ from distractors in terms of their basic features (their colour, shape, size and so forth), then search is relatively efficient. In such cases the time taken to find the target increases by less than 10 ms/item as the number of distractors increases. In contrast, when targets and distractors share features, then search is much less efficient, with target detection times increasing often by 30 ms or more for each distractor present (see Wolfe, 1994). These contrasting patterns of search have often been characterised in terms of a two-stage account of visual selection (e.g. Neisser, 1967). According to this two-stage account, there is a first pre-attentive stage of visual processing which operates in parallel across the visual field and codes simple visual features. Targets that differ from distractors in their coding at this stage (having different features) can be detected in a spatially parallel manner. Targets that share features with distractors will activate overlapping representations at the pre-attentive stage and require further processing

*Corresponding author.
Tel.: (0044) (0)121 414 4930; Fax: (0044) (0)121 414 4897;
E-mail: g.w.humphreys@bham.ac.uk

DOI: 10.1016/S0079-6123(09)17609-4

before they can be detected. This further processing is carried out at the second, attentive stage where there is serial scrutiny of each item — often this is conceived in terms of a serial window of attention being shifted from item to item. Due to this serial scrutiny, search rates increase to at least a minimal time required to make serial shifts of attention, and search time can be linearly related to the number of distractors present. Two-stage theories of this type remain highly influential. For example, Treisman's Feature Integration Theory (FIT; Treisman, 1998; Treisman and Gelade, 1980) maintains a distinction between feature-based pre-attentive stages and an attentional stage required to conjoin features. Wolfe's Guided Search Theory (GST) also proposes a first stage where simple features are coded and a second stage in which the items signalled as being most different from their neighbours are serially selected (Wolfe, 1994).

Other accounts, however, maintain that, rather than there being a strict dichotomy between pre-attentive and attentive stages of vision, there is a continuum of search efficiencies determined by different relations between targets and distractors. In their 'Attentional Engagement Theory' (AET), Duncan and Humphreys (1989, 1992) proposed that search efficiency was determined by the similarity between the target and the distractors (as above), and also by the similarity between distractors. High target-distractor similarity hinders search efficiency. High distractor similarity, on the other hand, can facilitate search because it enables distractors to be grouped and segmented from the target. Thus, even when targets and distractors share features search can be efficient if the distractors are homogeneous and can be grouped and rejected together (see also Humphreys et al., 1989). Duncan and Humphreys (1989, 1992) proposed that distractors were rejected together by a process of 'spreading suppression' when they grouped separately from targets. Quite similar ideas have subsequently been incorporated into traditional two-stage accounts in order to explain effects of distractor homogeneity. For example, FIT assumes that a process of distractor inhibition can be recruited which enables distractors to be rejected en masse through suppression of their common feature(s) (Treisman and Sato, 1990). In GST, distractor suppression operates through a process of lateral inhibition, so that distractors with common features will tend to suppress one another.

In addition to the process of rejecting distractors, accounts such as GST assume that search is guided by top-down activation of target features. This gives known targets a competitive edge in search tasks, enabling them rather than distractors to be selected. In terms of AET, targets are given a competitive advantage due to their having a template held in working memory which has a higher resting activation value than any template for distractors (see also Bundesen, 1990, for a similar idea expressed in terms of the target having a higher 'pertinence value'). There is clear behavioural evidence that having foreknowledge of the target makes a large difference to search, even determining whether stimuli 'pop out' or not. For example, large targets can be detected efficiently in the presence of small and medium-sized distractors, but medium targets are detected inefficiently amongst large and small distractors. Rather than this solely reflecting a difference in coding at the first pre-attentive stage, Hodsoll and Humphreys (2001) showed that efficient search depended on foreknowledge of the target. Having a template for the target enabled the feature differences to be used to guide search efficiently.

Modelling search

These ideas of search being guided by top-down activation from a target template, and also by distractor suppression, can be incorporated into more formal accounts of search including both mathematical (Bundesen, 1990) and computational models (e.g. Heinke and Humphreys, 2003; Humphreys and Müller, 1993; Mavritsaki et al., 2006). One value of such models is that, by demonstrating whether a proposed architecture can generate plausible search results, they provide an existence proof that the mechanisms of search could operate in the manner proposed. For example, since linear search functions can be generated by processes operating in a spatially parallel manner, the models demonstrate that serial processing operations are not *necessarily* required. The models also provide ways of analysing how complex processes interact to generate 'whole-system' behaviour, something that is otherwise difficult to specify. Furthermore, if the models can incorporate processes that mimic real neuronal firing, then further physiological constraints can be added to the constraints of having to capture a body of psychological data, to give a multi-level account of human performance. This is the approach we have tried to follow when implementing the sSoTS (spiking Search over Time and Space) model (Mavritsaki et al., 2006), shown in Fig. 1.

sSoTS incorporates processes proposed by nearly all the major psychological accounts of search. Within the model visual input is coded by activating topographic maps representing simple visual features. Within each map the units interact through lateral inhibition, enhancing activation for a stimulus that differs from its local neighbours. This activation is transmitted to a 'master map' that sums activity for a given location within each feature map and then feeds back this activation to 'sharpen' the competition for selection at the feature map level. In addition, top-down activation is transmitted to the feature maps, both to increase the activation for target features and to decrease activity for known distractor features. This top-down activation can give the target a competitive edge enabling it to be selected ahead of the distractors, with target selection determined by setting a threshold for units within the master map of locations. Search efficiency in the model is determined by the overlap in features shared by targets and distractors and by whether the distractors have common or different features. Search operates in a spatially parallel manner across all of the items present, with efficiency decreasing linearly as targets and distractors share features. Activation profiles for the model are shown in Fig. 2 for two 'standard' search cases: A — where the target is defined by a difference in a single feature (SF) relative to distractors (e.g. target = blue H (*italic*) amongst blue As), and B — target = a conjunction of features, each of which is shared with a distractor (target = blue H and distractors blue As and green Hs). Activation in units corresponding to the location of the target rise more rapidly in the single feature relative to the conjunction case. Figure 3a shows the RTs for units at the target location, based on the time to reach a critical threshold point (signalling a target detection response) in the single feature and conjunction (CJ) conditions as a function of the number of distractors present. Search is more efficient (increasing less with the display size) in the single feature than the conjunction case. These data capture the difference in search efficiency between single feature and conjunction search (see Fig. 3b). For illustration, data are also shown for a preview condition which uses the same items as conjunction search but presents one set of distractors for 800 ms prior to the onset of the second set of distractors, plus the target when present (see below for further discussion).

Models such as sSoTS cannot only integrate different psychological proposals but they can also generate predictions about how processing may operate at a neural level, given that the model is based on the operation of biologically plausible processing units (simulate spiking neurons). For example, the feature maps may plausibly be located within areas of early visual cortex which respond to simple visual features. However, the master map is more likely to be located within more anterior regions of cortex where there is evidence for neurons to be coded to the locations but not necessarily the features of visual stimuli (Courtney et al., 1994). If units within this location map are damaged, sSoTS predicts that search efficiency will deteriorate, particularly for targets that share features with distractors. This occurs because the model is less able to 'sharpen' any competition for selection, particularly for units on the affected ('contralesional') side. The net result is that targets on that side become difficult to detect, particularly when they share features with distractors. This is illustrated in the predictions for both reaction times to detect the target shown in Fig. 4a, and in the predicted error rates (target misses), shown in Fig. 4b.

The units within the model also operate using time parameters mimicking those of real neurons. For example, after firing, units build up a calcium parameter which reduces the future likelihood of firing for a period — units enter a refractory state.

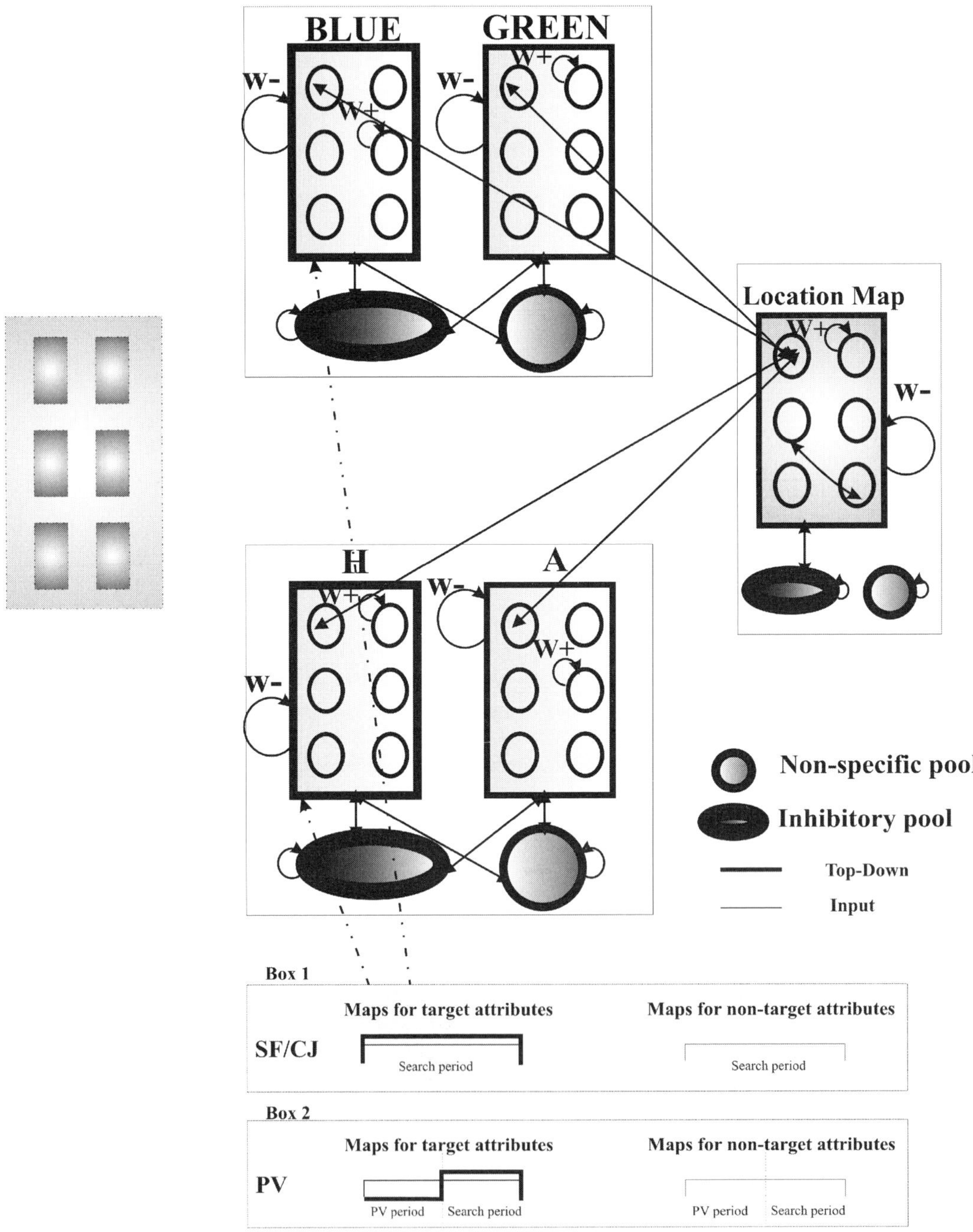

Fig. 1. The architecture of the sSoTS model. Input into the model is fed into the feature maps and from there into the location units. Units within maps, and at the same location across maps, operate in a mutually inhibitory way through the pools of inhibitory units. Activity in the location maps is fed back to the earlier maps, to bias competition for selection in favour of features that differ from their neighbours. In addition to such bottom-up biases, both excitatory and inhibitory activity can be set in a top-down manner, to facilitate target selection. Top-down excitation and inhibition help to bias search to favour the target.

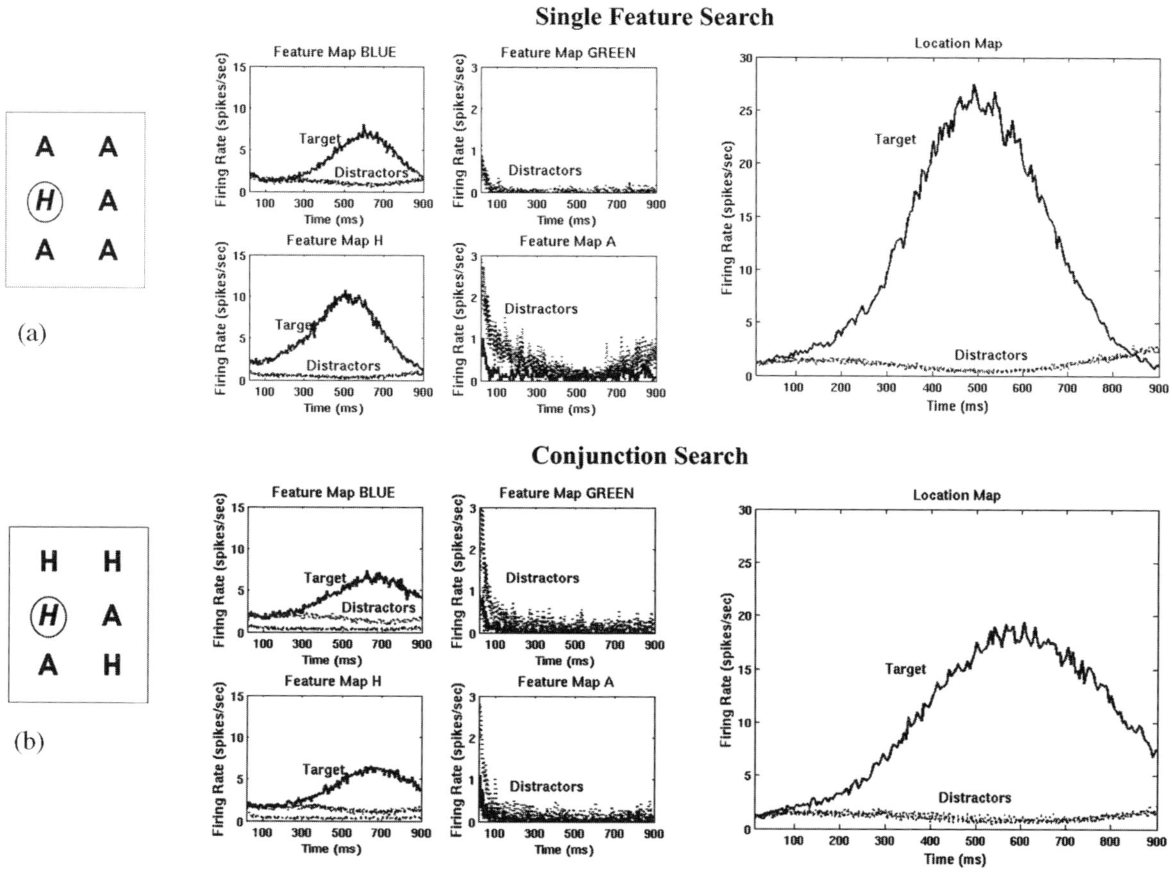

Fig. 2. (a) Activity in sSoTS plotted for four feature maps (BLUE, GREEN, H and A) and the Location map, when the target was a blue H (*italic*) and the distractors blue As. (b) Activity in the same maps for a conjunction target (blue H target, blue A and green H distractors). Activation in the Location map rises less rapidly and reaches a lower peak.

The emergent dynamics of activity lead to clear predictions being made about what might happen when the presentation of distractors is staggered over time. If there is presentation of one set of distractors prior to the other items, then activation for the initial distractors may be in a refractory state when the new items appear. Targets should be detected efficiently if they are presented at the time when distractor units are refractory, even if the distractors share their features with targets. This is illustrated in Fig. 5a.

Visual search over time

Predictions about the dynamics of visual search have been tested in studies using the 'preview' search procedure. In this procedure the presentation of distractors is staggered over time, with one set of distractors appearing before the others and the target. This staggered presentation can facilitate target selection, as illustrated in Fig. 3b, where search is shown to be as efficient in the preview condition as in the single feature condition when only the new set of items is presented. sSoTS makes a matching prediction (Figs. 3a, b). Interestingly, and again like the model, there is a distinct time course to this effect — the first set of distractors needs to be presented up to 400 ms before the other items for search to benefit (Watson and Humphreys, 1997; Watson et al., 2003). This is a striking result because the different sets of stimuli can be temporarily segmented over much shorter time intervals than this — one

(a) Simulations

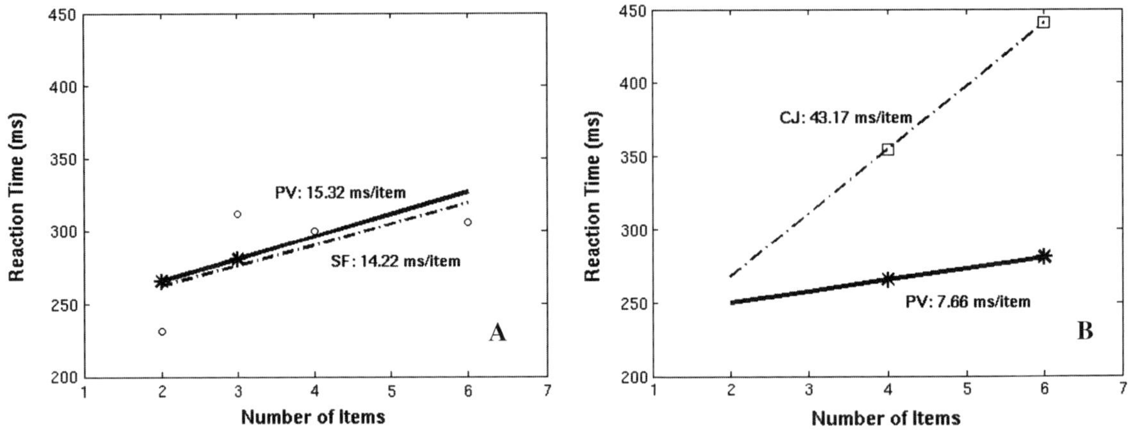

(b) Comparable human data (from Watson, Olivers & Humphreys, 2003)

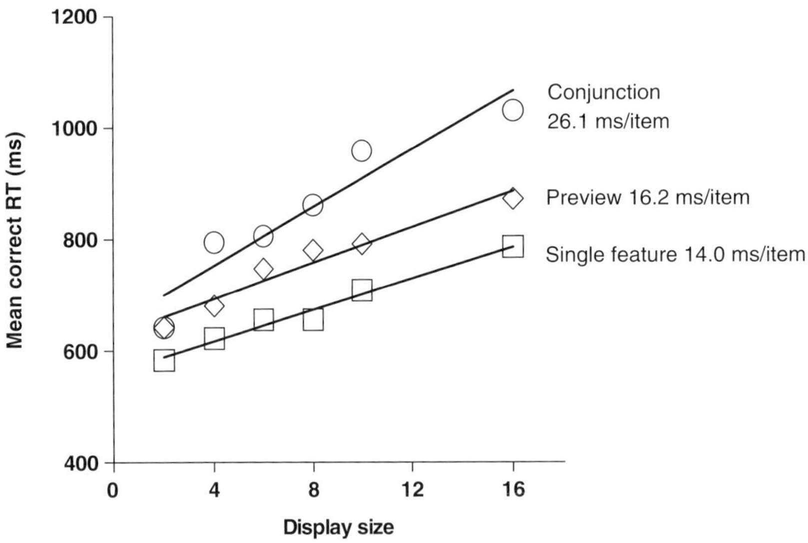

Fig. 3. (a) Simulations of single feature (SF), conjunction (CJ) and preview search (PV) in sSoTS. Note the steep slope on conjunction search even though the model operates in a spatially parallel manner. (b) Comparable human data (adapted from Watson et al., 2003). In human preview search, slopes for the preview condition very often match those in the single feature baseline (equivalent to when only the new search items are presented) and both are faster than the conjunction condition.

can see that the old and new displays differ in time, but, with a short interval, it remains difficult not to be affected by the old items. This indicates that temporal segmentation alone is not sufficient to explain performance. However, the time course does match that expressed by sSoTS (see Figs. 5a, b) — an emergent property of sSoTS's biologically plausible assumptions about the time course of the refractory state for neurons.

Other data from preview search also match other aspects of sSoTS. As we have noted, sSoTS incorporates the proposals made by psychological models that search is contingent on top-down

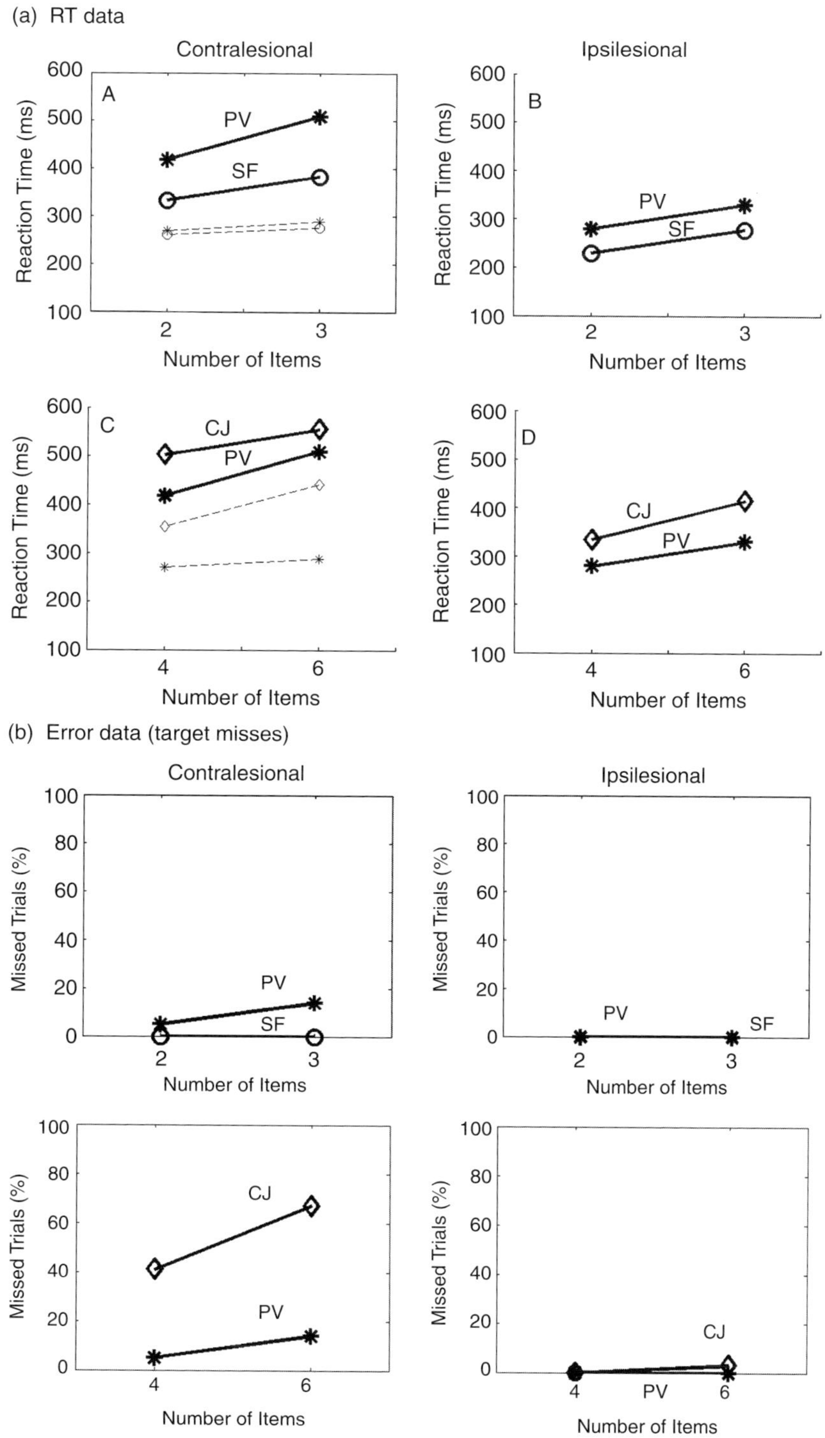

Fig. 4. Simulations of the effects of lesioning units on one side of the location map in sSoTS. (a) RT data; (b) error data (target misses). The data are shown for targets falling on the contralesional or ipsilesional side of space (side affected by lesion; side unaffected by lesion). In the top figures, the data are shown for single feature (SF) and preview search (PV) according to the number of items in the final display of the preview condition (either two or three items). In the bottom figures the data are shown for the conjunction and preview conditions, plotted against the number of items in the final display (four or six, in both preview and conjunction search). In each case the dotted lines show the results when the model was unlesioned. The results indicate that lesioning disrupts search in the conjunction and preview conditions most, for contralesional targets.

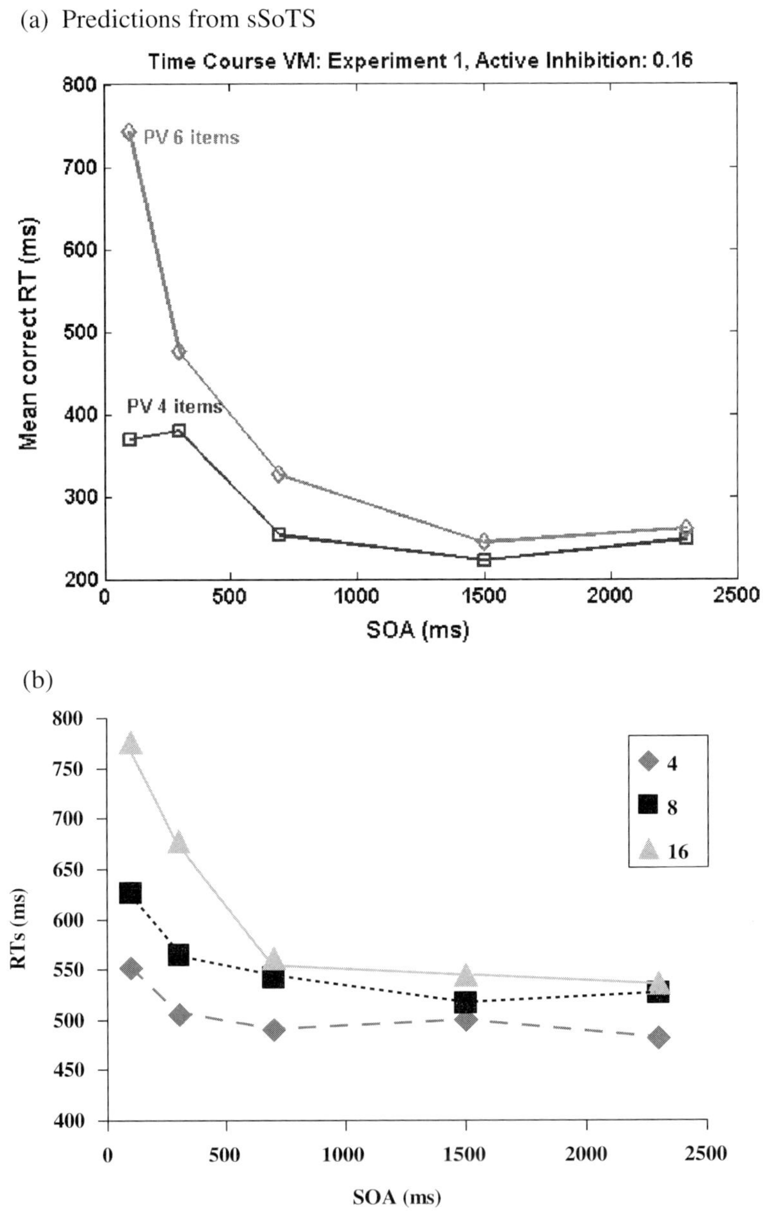

Fig. 5. (a) RTs generated by sSoTS as the duration of the preview is varied prior to the search display. (b) Data from human search as the preview duration is varied (adapted from Humphreys et al., 2004b). The model simulates the slow time course found in studies of preview search.

activation for targets and inhibitory suppression of distractors. Preview search provides good evidence for both processes operating in search. There are at least two pieces of evidence pointing to a role of inhibition. One comes from studies using probe-dot detection to measure where attention is allocated during search. In Humphreys et al's. (2004a) study participants saw a set of distractors as a preview (e.g. green horizontal lines) followed by a search display (red vertical distractors and a red horizontal target, when present). On a majority of trials participants carried out the search task. On a minority of trials they were cued to stop search and to try and detect a small probe that could appear either within an old distractor, a new distractor or the background. When the preview was presented for 800 ms before the new items, search for the new target was efficient. However, probes that fell on old items were difficult to detect, with detection levels in this case being lower than those found for probes presented on the background (see also Agter and Donk, 2005; Olivers and Humphreys, 2002; Watson and Humphreys, 1998). This is consistent with the spatial locations of the old items being inhibited.

A second piece of evidence indicating that there is inhibition of distractors comes from work on 'carry-over effects' in preview search. Braithwaite and Humphreys (2003) and Olivers and Humphreys (2003) presented a preview display of distractors in one colour followed by targets that either did or did not carry the colour of the to-be-ignored old distractors. Targets carrying the colour of the old distractor were difficult to detect — strikingly, this occurred even when the target had a singleton colour relative to the other new items being presented, and even when the old items were removed at the onset of the new displays. Normally such a colour singleton should pop out in search. The problem in detecting such a singleton target provides strong evidence against the view that preview search is simply based on automatic detection of the new items or on the temporal segmentation of the old and new displays (cf. Donk and Theeuwes, 2001; Jiang et al., 2002) — if that were the case, then the singleton should have popped out. This negative colour carry-over effect is consistent with the inhibition of the features as well as the locations of the old items (cf. Humphreys et al., 2004a). The result also fits with the idea of spreading suppression, as put forward by Duncan and Humphreys (1989); in this case, there is a spread of suppression from inhibited old distractors to new items carrying the inhibited properties — the result is that reaction times are slowed targets with these properties. This inhibition is maintained for at least some period even after the old distractors have been removed.

In addition to presenting evidence for the inhibition of old distractors, Braithwaite and Humphreys (2003) also reported data indicating effects of a positive expectancy for targets. In particular, Braithwaite and Humphreys showed that the negative colour carry-over effect could be reduced if participants had advanced knowledge of the target's colour. These authors propose that participants can independently set a top-down positive expectancy for a target along with adopting a negative bias against the properties of irrelevant distractors. The data indicating both positive and negative top-down effects in search match the top-down excitatory and inhibitory components operating in sSoTS.

The neural basis of inhibitory and excitatory biases

When people engage in visual search a range of brain areas are very often activated, most notably there is a conjunction of activity in posterior parietal and frontal cortices which increases as search becomes more difficult (see Corbetta and Shulman, 2002). However, as we have noted, search involves multiple processes (positive activation for targets, inhibitory suppression of distractors, the maintenance of target templates and so forth), so it is useful to explore paradigms such as preview search which can enable different processes to be isolated. There have now been several studies of preview search using functional brain imaging, and it has been consistently found that, relative to search when all the items appear together, preview search is associated with increased activation of several regions of posterior parietal cortex (the superior parietal lobe [SPL] and the precuneus; Allen et al., 2008; Humphreys et al., 2004b; Olivers et al., 2005; Pollmann et al., 2003). This is interesting because preview search can be more efficient than baseline search conditions when all the items appear simultaneously, so the increased activation does not simply reflect the general difficulty of search. Both Allen et al. (2008) and Pollmann et al. (2003) also included some 'dummy preview' trials where only the preview appeared although participants expected a search display to follow the preview (and so participants should engage in the same processing of the preview as on search trials). In these studies there was increased activation of SPL and precuneus when it was a dummy preview trial compared to trials which used equivalent visual displays but where the previews were unlikely to be ignored. This indicates that the SPL/precuneus activation is not tied to the search operation but it is consistent with these brain regions being linked to the inhibitory processing of distractors. The activation of the SPL/precuneus may reflect the operations of inhibitory neurons or some initial attention being paid to the old distractors in order to then inhibit them (see Humphreys et al., 2004a, for evidence consistent with this from probe-dot procedures).

The data on functional brain imaging are supported by neuropsychological studies on selective disorders of search in patients with brain lesions. It is well established that patients with damage to posterior parietal cortex (PPC) are impaired at serial search tasks (Eglin et al., 1989; Riddoch and Humphreys, 1987). Olivers and Humphreys (2004) found that PPC patients were also impaired at preview search, even though normal participants perform preview search efficiently. This again points to effects on particular processes rather there being an exaggerated influence of search difficulty in the patients. Humphreys et al. (2006) found that PPC patients impaired at preview search were nevertheless able to prioritize their attention to new onset stimuli. If prioritized attention to new onsets was sufficient to generate efficient preview search (Donk and Theeuwes, 2001), then the patients should have shown efficient preview search. In contrast to this the impairments indicate that additional processes (such as inhibition of the old distractors) determine the efficiency of preview search, and these additional processes may be disrupted by PPC damage. Further evidence consistent with this was noted by Olivers and Humphreys (2004). They found that the PPC patients were particularly impaired when the new and old stimuli were spatially overlapping compared with when the items did not overlap and could be spatially segmented. These data suggest that the PPC was critical for the segmentation and rejection of the old distractors, and was recruited particularly when spatial segmentation was difficult.

As we have noted (Fig. 4), sSoTS also predicts that damage to the location units within the model, putatively representing PPC, generates problems in search. Within the model this not only affects the detection of conjunction targets, but also the detection of targets in preview search. Humphreys et al. (2009) simulated effects of PPC damage by removing units from one side of the location map. This disrupted both conjunction and preview search. Similarly to the patients, the problems in preview search were most pronounced when the old and new items overlapped spatially, when temporal segmentation would normally augment spatial segmentation (Fig. 6). When lesioned, sSoTS has a reduced ability to implement temporal separation of new targets from old distractors on one side of space; this leads to problems in separating distractors within the same area of field. It is of interest that poor performance of the model was observed even when both the old and new items fell within the undamaged (ipsilesional) field. This is because poor inhibition of old ipsilesional distractors means that they stay as competitors for new items appearing in the same locations. The result is that target detection becomes inefficient.

These simulations of the effects of PPC damage provide important converging evidence linking specific brain regions to functional processes within the model.

Modelling brain imaging data

Another way to link the operation of an abstract computational model to brain function is to simulate data from functional imaging. Activity within a model such as sSoTS can be convolved with an assumed haemodynamic function to predict the BOLD response (Glover, 1999). This then enables us to take functional processes in the model, such as the operation of excitatory and inhibitory activity during search, and to assess within which brain areas the activity correlates with the different functional processes.

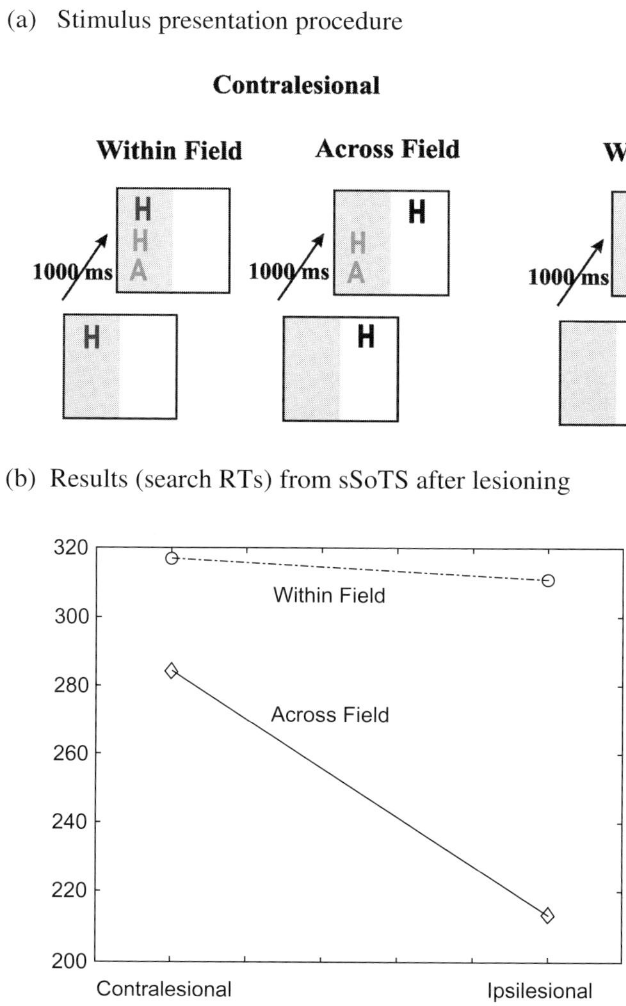

Fig. 6. (a) The stimulus presentation procedure used to examine the effects of spatial and temporal separation in sSoTS. The initial display is the preview and this is followed by the search display. In the across field condition, the preview appears in the opposite field to the new items in the search display. In the within field condition, the new search items appear within the same field as the preview. (b) Mean RTs generated by sSoTS after unilateral lesioning of units on one side of the Location map. Targets were presented either on the contralesional (damaged units) or ipsilesional side of space (undamaged units). The data show that within field discriminations are more difficult than across field discriminations, with the detection of contralesional targets being generally worse. Comparable data from patients with PPC lesions were reported by Olivers and Humphreys (2004).

We (Mavritsaki et al., submitted) have done this by summing activity across different maps in sSoTS according to whether the activity is reflecting top-down excitation of targets or bottom-up inhibition of distractors. The emergent results are shown in Fig. 7. The figure reveals both overlapping and distinct regions within PPC, and also visual processing areas in occipital cortex, where activity is separately correlated with the time course of excitatory and inhibitory activity within the model. This then provides a functional decomposition of the network of areas that is activated during search. As the functional processes of target excitation and distractor inhibition will be involved both in preview search and in search for conjunction stimuli, it follows that damage to the areas supporting the excitation and inhibition processes will disrupt both types of

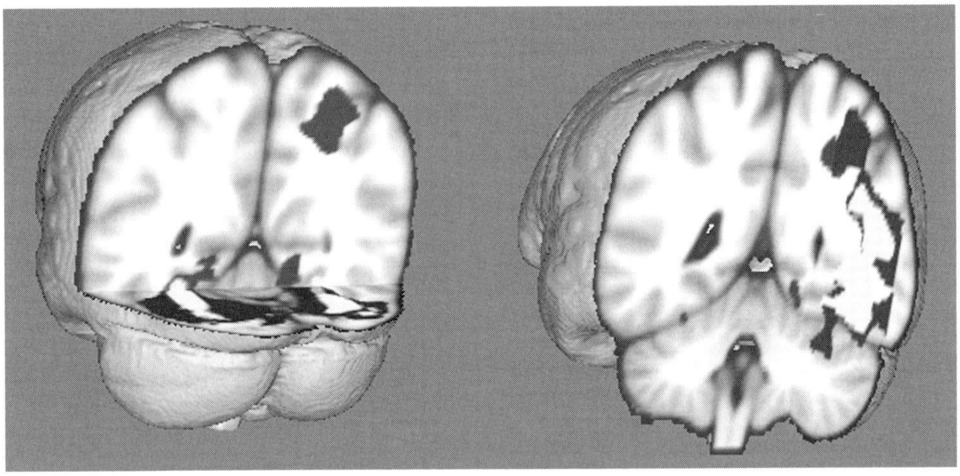

Fig. 7. Images showing brain regions where neural activity in preview search correlates with inhibitory and excitatory activity in sSoTS (when convolved with an assumed haemodynamic function). This model-based analysis pulls apart neurons within overlapping brain regions that perform functionally distinct roles in search. Areas in black correlate with inhibitory activity in the model; areas in white correlate with excitatory activity.

search task. This prediction is supported by the neuropsychological data (Olivers and Humphreys, 2004; Riddoch and Humphreys, 1987).

The importance of multi-level analyses

We have reviewed evidence using behavioural manipulations (such as preview search), neuropsychological analyses (e.g. the effects of PPC lesions), functional brain imaging (e.g. fMRI) and computational modelling, to analyse the processes involved in visual search. Each piece of evidence has its own limitation. For example, behavioural studies reveal 'whole-system' behaviour, but this can make it difficult to analyse the operation of sub-component processes. Studies using functional brain imaging reveal brain areas that correlate with different processes but do not prove that these processes are necessary for a given task. Neuropsychological studies do demonstrate the necessary role of brain regions (since damage to those regions is shown to disrupt performance), but the lesion may affect more than one process which in turn makes it difficult to exactly relate an impaired function to the lesioned area. Given these limitations, it is important to use evidence coming from each approach in order to develop an over-arching framework not subject to one limitation. In the present case, this framework can also be captured at a formal level in terms of the sSoTS model. Models such as sSoTS can help to integrate research in at least two ways. First, it can simulate effects at multiple levels (brain imaging, effects of neuropsychological deficits and effects due to emergent, whole-system behaviour), enabling us to link the different approaches and different types of data together. The weakness inherent in one approach, then, can be compensated for by the strengths of the other. For example, fMRI in normal participants may be able to localise, across the whole brain, processes involved in a given task. The necessary role of these areas would then be addressed through neuropsychological evidence. This relationship should be captured by simulating the effects of lesioning matching regions within the model. Second, the model shows how ideas expressed in different psychological models can be formally linked, enabling us to see how different models relate to one another. For example, distractor similarity influences the amounts of lateral inhibition operating in the model (cf. Duncan and Humphreys, 1989; Wolfe, 1994), while Treisman and Sato's argument for feature-based inhibition is implemented through the top-down inhibition process in the model. Which factors are important, under which conditions, can then be explored. We believe this provides a working framework through which to assess the various factors determining search efficiency. One particularly important point, given the stab at biological plausibility in the model, is that variations in physiological parameters (e.g. the time course over which neurons enter into a refractory state) can generate psychological predictions (e.g. on the time course of visual search); these predictions can be tested and fed back to further inform model development in a (virtual) cycle of modelling and testing. We propose that the formal development of models such as sSoTS will play an important part in the integration of psychological theory and physiological data.

Acknolwedgement

This work was supported by grants from the BBSRC, ESRC and MRC (UK).

References

Agter, , & Donk, M. (2005). Prioritized selection in visual search through onset capture and color inhibition: Evidence from a probe-dot detection-task. *Journal of Experimental Psychology: Human Perception and Performance*, *31*, 722–730.

Allen, H. A., Humphreys, G. W., & Matthews, P. M. (2008). A neural marker for content specific active ignoring. *Journal of Experimental Psychology: Human Perception and Performance*, *34*, 286–297.

Braithwaite, J. J., & Humphreys, G. W. (2003). Inhibition and anticipation in visual search: Evidence from effects of color foreknowledge on preview search. *Perception & Psychophysics*, *65*, 213–237.

Bundesen, C. (1990). A theory of visual-attention. *Psychological Review*, *97*(4), 523–547.

Corbetta, M., & Shulman, G. L. (2002). Control of goal directed and stimulus-driven attention in the brain. *Nature Reviews Neuroscience*, *3*, 201–215.

Courtney, S. M., Ungerleider, L. G., Keil, K., & Haxby, J. V. (1994). Object and spatial visual working memory activate separate neural systems in human cortex. *Cerebral Cortex*, *6*, 39–49.

Donk, M., & Theeuwes, J. (2001). Visual marking beside the mark: Prioritizing selection by abrupt onsets. *Perception & Psychophysics*, *93*, 891–900.

Duncan, J., & Humphreys, G. W. (1989). Visual search and stimulus similarity. *Psychological Review*, *96*, 433–458.

Duncan, J., & Humphreys, G. W. (1992). Beyond the search surface: Visual search and attentional engagement. *Journal of Experimental Psychology: Human Perception and Performance*, *18*, 578–588.

Eglin, M., Robertson, L. C., & Rafal, R. D. (1989). Visual search performance in the neglect syndrome. *Journal of Cognitive Neuroscience*, *1*, 372–385.

Glover, G. (1999). Deconvolution of impulse response in event-related fMRI. *NeuroImage*, *9*, 416–429.

Heinke, D., & Humphreys, G. W. (2003). Attention, spatial representation and visual neglect: Simulating emergent attentional processes in the Selective Attention for Identification Model (SAIM). *Psychological Review*, *110*, 29–87.

Hodsoll, J., & Humphreys, G. W. (2001). Driving attention with the top down: The relative contribution of target templates to the linear separability effect in the size dimension. *Perception & Psychophysics*, *63*, 918–926.

Humphreys, G. W., Jung-Stalmann, B., & Olivers, C. N. L. (2004a). An analysis of the time course of attention in preview search. *Perception & Psychophysics*, *66*, 713–730.

Humphreys, G. W., Kyllingsbæk, S., Watson, D. G., Olivers, C. N. L., Law, I., & Paulson, O. (2004b). Parieto-occipital areas involved in efficient filtering in search: A time course analysis of visual marking using behavioural and functional imaging procedures. *Quarterly Journal of Experimental Psychology*, *57A*, 610–635.

Humphreys, G. W., Mavritsaki, E., Allen, H. A., Heinke, D., & Deco, G. (2009). Modelling visual search in biologically plausible neural networks: Whole-system behaviour, neuropsychological breakdown and BOLD signal activation. In D. Heinke & E. Mavritsaki (Eds.), *Computational modelling in behavioural neuroscience*. London: Psychology Press.

Humphreys, G. W., & Müller, H. M. (1993). SEarch via Recursive Rejection (SERR): A connectionist model of visual search. *Cognitive Psychology*, *25*, 43–110.

Humphreys, G. W., Olivers, C. N. L., & Yoon, E. Y. (2006). An onset advantage without a preview benefit: Neuropsychological evidence separating onset and preview effects in search. *Journal of Cognitive Neuroscience*, *18*, 110–120.

Humphreys, G. W., Quinlan, P. T., & Riddoch, M. J. (1989). Grouping effects in visual search: Effects with single- and combined-feature targets. *Journal of Experimental Psychology: General*, *118*, 258–279.

Jiang, Y., Marks, L. E., & Chun, M. M. (2002). Visual marking: Selective attention to asynchronous temporal groups. *Journal of Experimental Psychology: Human Perception and Performance*, *28*, 717–730.

Mavritsaki, E., Allen, H. A. & Humphreys, G. W. (submitted). Modelling the neural substrate of preview search.

Mavritsaki, E., Heinke, D. G., Humphreys, G. W., & Deco, G. (2006). A computational model of visual marking using an interconnected network of spiking neurons: The spiking Search over Time & Space model (sSoTS). *Journal of Physiology (Paris)*, *100*, 110–124.

Neisser, U. (1967). *Cognitive psychology*. New York: Appleton-Century-Crofts.

Olivers, C., & Humphreys, G. W. (2002). When visual marking meets the attentional blink: More evidence for top-down limited capacity inhibition. *Journal of Experimental Psychology: Human Perception and Performance*, *28*, 22–42.

Olivers, C. N., & Humphreys, G. W. (2004). Spatiotemporal segregation in visual search: Evidence from parietal lesions. *Journal of Experimental Psychology: Human Perception and Performance*, *30*, 667–688.

Olivers, C. N. L., & Humphreys, G. W. (2003). Visual marking and singleton capture: Fractionating the unitary nature of visual selection. *Cognitive Psychology*, *47*, 1–42.

Olivers, C. N. L., Smith, S., Matthews, P., & Humphreys, G. W. (2005). Prioritizing new over old: An fMRI study of the preview search task. *Human Brain Mapping*, *24*, 69–78.

Pollmann, S., Weidner, R., Humphreys, G. W., Olivers, C. N. L., Müller, K., Lohmann, G., et al. (2003). Separating segmentation and target detection in posterior parietal cortex — An event-related fMRI study of visual marking. *NeuroImage*, *18*, 310–323.

Riddoch, M. J., & Humphreys, G. W. (1987). Perceptual and action systems in unilateral neglect. In M. Jeannerod (Ed.), *Neurophysiological and neuropsychological aspects of spatial neglect*. Amsterdam: Elsevier Science.

Treisman, A. (1998). Feature binding, attention and object perception. *Philosophical Transactions of the Royal Society*, *353*, 1295–1306.

Treisman, A., & Gelade, G. (1980). A feature-integration theory of attention. *Cognitive Psychology*, *12*, 97–136.

Treisman, A., & Sato, S. (1990). Conjunction search revisited. *Journal of Experimental Psychology: Human Perception and Performance*, *16*, 459–478.

Watson, D. G., & Humphreys, G. W. (1997). Visual marking: Prioritising selection for new objects by top-down attentional inhibition. *Psychological Review*, *104*, 90–122.

Watson, D. G., & Humphreys, G. W. (1998). Visual marking of moving objects: A role for top-down feature based inhibition in selection. *Journal of Experimental Psychology: Human Perception and Performance*, *24*, 946–962.

Watson, D. G., Humphreys, G. W., & Olivers, C. N. L. (2003). Visual marking: Using time in visual selection. *Trends in Cognitive Sciences*, *7*, 180–186.

Wolfe, J. M. (1994). Guided Search 2.0: A revised model of visual search. *Psychonomic Bulletin and Review*, *1*, 202–238.

N. Srinivasan (Ed.)
Progress in Brain Research, Vol. 176
ISSN 0079-6123

CHAPTER 10

Extinction: a window into attentional competition

M. Jane Riddoch*, Sarah J. Rappaport and Glyn W. Humphreys

Behavioural Brain Sciences, School of Psychology, University of Birmingham, Birmingham, UK

Abstract: Extinction is an example of how stimulus selection may be affected by an imbalance in competition for attentional selection. Patients with extinction are able to process stimuli in either hemispace, but only when presented in isolation. Following brain injury, stimuli will not be processed as efficiently in the damaged hemisphere and so may fail to be detected when other stimuli are competing for selection. In this review we discuss some of the factors that contribute to the recovery from extinction, and consider their implications for functional and neural theories of selection. Work shows that extinction can be modulated by multiple bottom-up factors including: low-level visual grouping (e.g., reflecting Gestalt properties in an array) and grouping based on higher level factors (such as the lexical identity of a stimulus or action relations between objects). Top-down factors (such as holding items in working memory) can also facilitate recovery from extinction. Furthermore, the competition for selection may also be modulated by the programming of action to a given location, consistent with pre-motor feedback to perceptual processes. While often discussed in terms of spatial biases, non-spatial extinction can also be demonstrated (dictated by the coherence of stimuli). In contrast to extinction, a phenomenon of anti-extinction has also been documented where patients are better at report when two items rather than single items are presented. Although superficially distinct, evidence indicates that grouping may be important in both cases, with temporal grouping being important in generating the anti-extinction effect. Overall, the work indicates that the disorder of extinction plays an important role in the understanding of attentional selection.

Keywords: perceptual grouping; competition for selection; action relations; non-spatial extinction; temporal order judgements

Introduction

Extinction is a neurological disorder where the patient is able to detect briefly presented single stimuli on either the ipsilesional or the contralesional side; however, under conditions of double simultaneous stimulation (DSS), there is a failure to report the contralesional stimulus (Karnath, 1988). Extinction can be distinguished from unilateral neglect. Patients with neglect may fail to respond to contralesional stimuli regardless of whether a competing ipsilesional stimulus is present or not. The two conditions often co-occur, and extinction has been considered a mild form of neglect, though the two disorders may have different anatomical substrates (Karnath et al., 2003).

Extinction has been shown in various sensory modalities including vision (di Pellegrino and De Renzi, 1995; Làdavas, 1990), audition (De Renzi et al., 1984) and tactile perception (Vallar et al., 1994). Cross modal extinction has also been

*Corresponding author
Tel.: +44 (0) 121 414 4912; Fax: +44 (0) 121 414 4987;
E-mail: M.J.Riddoch@bham.ac.uk

DOI: 10.1016/S0079-6123(09)17619-7

demonstrated (for a review see Spence and Driver, 2004), as has been extinction purely within the motor domain (Punt et al., 2005; Valenstein and Heilman, 1981). Hence the phenomenon may reflect a rather general property of information processing following brain lesions. In this chapter we will provide an overview of the phenomenon of extinction, discussing what it tells us about how attentional selection operates in both normal and pathological cases.

Accounts of extinction

Spatial inertia and attentional orienting

One early account of extinction was proposed by Birch et al. (1967). These authors argued that extinction arose due to 'cerebral inertia', with slow initiation of processing and delayed return to baseline on the affected side following brain insult. Birch et al. (1967) contrasted the ability of patients to report stimulation of both the affected and the unaffected upper limbs under two conditions: either when there was bilateral DSS or when stimulation of the affected limb preceded that of the unaffected limb. There was significantly better report of the contralesional stimulus in the latter relative to the former condition. Birch et al. argued that the improvement was not due to the successiveness of stimulation in itself, since, if the ipsilesional side was stimulated before the contralesional side, performance did not differ from the DSS condition. They suggested that the significance of stimulating the contralesional side first was in allowing prior activation of the damaged hemisphere to take place, before inhibition and interference from the unaffected hemisphere could develop. More recent evidence for the positive effects of temporally staggering stimuli has been reported by di Pellegrino et al. (1997) who showed that while their patient failed to identify the contralesional item on simultaneous presentation of letters, the problem receded with a 1000-ms SOA when the patient was nearly always able to identify the contralesional stimulus regardless of which came first. The shorter the SOA, the poorer the performance with the contralesional item, but this occurred even when the contralesional item led the ipsilesional stimulus. Poor performance, even when the contralesional item leads, could arise if processing of the contralesional item is particularly slow.

Quite similar effects to the above have been observed in normal participants when attention is manipulated. Stelmach and Herdman (1991) examined the effects of directing attention to either left or right visual hemifield when normal participants had to make temporal order judgements (TOJs). Participants judged the stimuli as occurring at the same time when the stimulus in the unattended hemifield preceded that in the attended hemifield by 40 ms. These findings have been used to argue that reductions in visual attention can increase transmission time of visual signals to central processing mechanisms. TOJs are impaired in patients who show extinction (Rorden et al., 1997). Rorden et al. showed that their patients had a tendency to perceive ipsilesional events as occurring first, and that judgements of simultaneity only occurred if the contralesional event preceded the ipsilesional event by approximately 200 ms. The phenomenon of 'spatial inertia', indicated by these data on TOJs, could thus reflect inherently impaired temporal processing in the patients (see Battelli et al., 2007) or impaired attentional orienting to the affected side (with a knock-on effect on temporal processing). Subsequent imaging work by Davis et al. (2009) is consistent with this first argument. Davis et al. (2009) found that TOJs result in bilateral activation of the temporal parietal junction (TPJ) (see also Battelli et al., 2007). The activation was greater for tasks where the temporal order of stimuli had to be reported relative to when spatial differences had to be reported. Davis et al. conclude that the TPJ, an anatomical area associated with extinction (Karnath et al., 2003), is critical for integrating stimuli over time. In a TMS study with normal participants, Woo et al. (2009) demonstrated that TMS over the right posterior parietal cortex (PCC) affected the performance of TOJs. TMS to the right parietal cortex within a critical time window (50–100 ms after target onset) delayed the detection of a left-side target. The same effects were not shown following

TMS to the left parietal cortex. These results support previous findings indicating a right hemisphere dominance for attention (e.g., Corbetta et al., 1993, 2000) and with right hemisphere lesions being dominant not only for unilateral neglect (Heilman, 1979) but also for extinction (Becker and Karnath, 2007; Stone et al., 1993).

Attentional competition

Rather than attributing deficits associated with extinction to poor temporal processing and/or impaired orienting of attention to the contralesional side, an alternative attentional account is that the normal competition for attention between simultaneously presented stimuli (or simultaneously activated responses in the case of motor extinction; Punt et al., 2005) is unbalanced by the brain lesion. According to this idea, brain lesions lead to stimuli (responses) on the ipsilesional side being assigned greater 'weight' in the competition to be selected (Bundesen, 1990; Duncan and Humphreys, 1989), which results in ipsilesional stimuli (responses) being selected rather than the contralesional stimuli (responses). As the competition for selection operates over time, so effects may emerge between successive as well as simultaneously presented stimuli, at least when the intervals between the stimuli are brief (di Pellegrino et al., 1997).

Related to this biased competition account, it is interesting to note that the processing of single stimuli in the affected field is often not normal, though (by definition) it is better than the processing of contralesional items under conditions of DSS (Gilchrist et al., 1996; Marzi et al., 1997, 2001; Smania et al., 1998). Indeed, the severity of extinction has been shown to correlate with contra–ipsi detection differences on single-item trials (Marzi et al., 1997). Weaker perceptual processing of a contralesional stimulus would naturally give rise to an increased perceptual bias when an ipsilesional stimulus is present, given competition between the stimuli for selection. This has been captured in computational models of visual selection. For instance, in the SAIM model put forward by Heinke and Humphreys (2003), simultaneously presented items compete to gain control of a spatial window of attention, which gates the passage of activation into higher order representations for object recognition. Lesioning units on one side of space can result in units on the spared (ipsilesional) side having a dominant role in this gating process, so that contralesional stimuli are not detected.

Within models such as SAIM, the biased competition for selection does not immediately eliminate bottom-up driven activation from contralesional stimuli. At least during the early stages of selection, this activation can be passed through to higher levels of stimulus representation, enabling higher order effects (grouping, the activation of stored knowledge, etc.) to take place. As we review below, there is mounting evidence that the residual processing of extinguished stimuli can affect performance.

Bottom-up grouping and perceptual recovery

One line of evidence indicating residual processing of extinguished stimuli has already been noted — namely that the grouping between the contralesional and the ipsilesional can mitigate extinction. Here, there is evidence for grouping based on both the shape properties of stimuli (Gilchrist et al., 1996; Ward et al., 1994) and also on their surface properties. For instance, common colour, surface texture and contrast polarity can improve performance in addition to factors such as common shape and collinearity between edges (Humphreys et al., 2000). Mattingley et al. (1997) reported evidence for the coding of surface representations in extinction using Kanisza-type stimuli. Their patient was presented with four circles (two to the left and two to the right of fixation) at the corners of a virtual rectangle. Quarter segments of the circles would then briefly disappear either from the two left or the two right circles, or from all the four circles. When all four quarters disappeared, a Kanisza rectangle was apparent across the centre of the display formed from the 'pacman' wedges at each corner. In addition to the experimental trials, there were control trials where the outer circumference of the circles remained as a rim when the segment disappeared. In the control condition, there was no illusory filling in across the midline.

The patient's ability to detect left-side events was significantly improved in the experimental condition when an illusory figure was present. The same patient was also better able to detect a left-side object which appeared to join a right-side object to represent a solid bar with a central occluding object. If the occluder was not present, the left and right-side objects appeared to be perceptually distinct. Conci et al. (2009) used a similar paradigm, but also introduced intermediate representations so that the degree of contour and surface information across midline could be varied. They showed that surface information can substantially reduce extinction, while contour completions had a relatively small effect.

These data on both shape- and surface-based groupings fit with the idea that basic aspects of perceptual organisation can be spared in patients with extinction. Driver et al. (1992) have further argued that aspects of figure–ground perception can remain intact. These authors showed that a patient with extinction was poor at judging whether single stimuli were symmetrical around a vertical axis. Despite this, judgements about which parts of an ambiguous display were 'figure' and which were 'ground' were influenced by symmetry (there was a bias to choose symmetrical regions as figure). Driver et al. propose that there can be initial coding of figure–ground relations implicitly, although the biased competition for selection leads to ipsilesional regions being more strongly attended, affecting explicit perceptual judgements.

Although bottom-up grouping can take place between stimuli in the contralesional field and stimuli in the ipsilesional field, this does not mean that the grouping process operates normally. Han and Humphreys (2003) examined grouping based on the proximity and similarity of visual elements in two patients with extinction following parietal damage. Using event-related potentials, these authors showed that there was a weakened early perceptual response to grouped elements (in the P1 and N1 components), suggesting that grouping was reduced following parietal damage. There may need to be support from the parietal system to generate a strong perceptual response to grouping signals in early visual cortex.

Top-down perceptual recovery

The above grouping effects between visual elements can be thought of as being driven by bottom-up input, without the involvement of top-down knowledge. There is evidence too for perceptual recovery in extinction based on the use of top-down knowledge. Soto and Humphreys (2006) had patients hold a coloured shape in working memory. Subsequently, a target display appeared which patients had to report. Following this, the patients' memory for the initial item was tested. When the memory cue did not match the target display, there was extinction of the contralesional stimuli under DSS. However, extinction was reduced when the memory item matched the contralesional stimulus. Interestingly, this effect disappeared when the patient did not have to hold the initial cue in memory but merely identified it. Thus, the benefit for reporting the contralesional item was not due to bottom-up priming from the initial visual presentation of the cue; rather it required top-down matching of the cue in working memory to the incoming perceptual input. Since this top-down matching process reduced extinction, the data suggest that top-down processes can modulate the competition for visual selection.

Implicit effects on explicit decisions

In the above cases of bottom-up grouping and top-down matching from memory, evidence of implicit processing of extinguished stimuli is shown by patients being better able to report contralesional items when grouping/matching takes place. Evidence for implicit processing of extinguished stimuli also arises when extinguished items cannot be reported but influence responses to another stimulus. Some initial evidence here was reported by Volpe et al. (1979) who showed that patients with extinction performed well when asked to make same/different matches for two stimuli presented on either side of fixation. Performance dropped, however, when participants were asked to *identify* the contralesional stimulus on different trials (see also Karnath,

1988); however, the drop in performance was not seen if patients were asked to report contralesional stimulus first (e.g., Karnath, 1988). Volpe et al. (1979) argued that contralesional stimuli were processed sufficiently to influence forced-choice decisions though individual identification of contralesional stimuli may be disrupted. Interestingly, these effects on forced-choice decisions may be based not just on primitive coding of visual elements but on the abstraction of shape. Berti et al. (1992) also used same/different match tasks where, on 'same' trials, stimuli could be physically identical or different views of the same item, or physically different exemplars of the same item. The patients performed well on 'same' trials even with physically different stimuli suggesting that visual processing in the contralesional field could extract more than basic image characteristics. Evidence for implicit access to semantic processing has been reported by McGlinchey-Berroth et al. (1993). These authors had participants make a lexical decision to a centrally presented letter string. The string was preceded by the simultaneous brief presentations of drawings, one in the right and the other in the left visual field. The drawings either depicted a common object or consisted of scrambled lines. Judgements of the central word were faster when it was preceded by a semantically related picture, regardless of whether this appeared in the contra- or ipsilesional field (there were no differences between these two conditions), although the contralesional item was extinguished under the DSS condition. This evidence points to high-level processing of extinguished items being possible.

Effects of action

Action relations between objects

Other evidence for high-level processing of extinguished stimuli comes from studies of the action relations between objects. Riddoch et al. (2003) presented patients with extinction pictures of objects that were commonly used together. Pairs of items could be positioned so that the objects were in the locations where they could be used together (e.g., a corkscrew pointing into the top of a wine bottle) or they fell in locations they were not correctly located for action (e.g., a corkscrew going into the bottom of a wine bottle). Extinction was reduced when pairs of objects were correctly positioned for action. The presence of an action relation between the stimuli was critical for recovery from extinction. Humphreys et al. (2006) tested effects with pairs of objects placed in familiar co-locations, which did not appear to be interacting together (e.g., the sun shown above a tree). Despite the stimuli appearing in familiar co-locations, there was no reduction in extinction. Riddoch et al. (2003) also tested for a semantic (but not action-based) relationship between the stimuli. A semantic relationship alone was not sufficient to generate the effect. In additional work, Riddoch et al. (2006) demonstrated recovery from extinction with familiar as well as unfamiliar pairs of objects, provided the objects appeared to interact together – though recovery of extinction was maximised when the objects were both familiar as a pair and positioned to interact together. To account for these data, Riddoch et al. (2003, 2006) proposed that patients remain sensitive to the 'affordance' (Gibson, 1979) based on the possibility of an action occurring between the two stimuli. This affordance could be detected even when patients otherwise show a spatial bias and fail to detect the contralesional item.

Other evidence for the pre-attentive processing of action-related properties of stimuli comes from an examination of performance on trials where patients only report the presence of a single item (i.e., on extinction trials). Riddoch et al. (2003) found that when the objects were positioned for action, the patients tended to report the object in the pair that would be active when the objects were used (e.g., the corkscrew rather than the wine bottle). This occurred irrespective of whether this item fell in the contra- or the ipsilesional field. In contrast, when the objects were not positioned correctly for action, the patients tended to report the ipsilesional stimulus irrespective of whether this was the 'active' or more 'passive' item (the wine bottle). These data from extinction trials indicate that the action relations between the stimuli modulated which object the patient would attend to, even when only one item was explicitly recovered for perceptual report.

Affordance effects from single objects can also be shown. di Pellegrino et al. (2005) asked patients with extinction to report whether a single cup had been presented either to the left or to the right of fixation, or whether two cups had been presented. Report of stimuli on two-item trials was reduced relative to report on single-item trials. However, performance was better on two-item trials when the contralesional cup had a left-oriented handle relative to when it had a right-oriented handle, or if the handle was occluded. The affordance of a left-side grasp, when the left cup was oriented appropriately, facilitated detection even though no hand response was required.

Action programming

In addition to extinction patients being sensitive to action relations between objects, there is also evidence for effects of action programming modulating performance. Kitadono and Humphreys (2007, 2009) tested for extinction when their patient programmed a pointing movement into the ipsi- or contralesional field. They found that extinction was reduced when a contralesional action was programmed. In this study, the targets for perceptual report appeared briefly and disappeared well before any action was initiated; thus the effect was in motor programming rather than the explicit action itself. The effect was not simply due to increased arousal when an action had to be initiated (cf. Robertson et al., 1998). Programming an action to the ipsilesional field *increased* extinction. Also the beneficial effects of programming an action to the contralesional side were maximal when the end position for the movement coincided with where the contralesional stimulus fell. Kitadono and Humphreys proposed that motor programming can be directly coupled to attention so that attention is drawn to the end location of a planned movement. Here, top-down motor enhancement of perception seems to take place, helping remove spatial bias in the competition for selection between contralesional and ipsilesional items (see Deubel and Schneider, 2005).

Kitadono and Humphreys (2009) extended this result by having a patient make sequential movements, first into the good and then into the impaired field. Even though a sequential movement would mean that attention would have to be disengaged from the ipsilesional stimulus, Kitadono and Humphreys reported that performance benefited from programming the movement to the contralesional location. Thus motor programming appears to exert a robust effect on the competition for perceptual selection, sustained over time when sequential actions are made.

Task-based effects on extinction

It is clear that various types of information may still be processed from the extinguished field in order to modulate both the report of the contralesional stimulus itself and its effects on other items (e.g., when priming occurs). There is also evidence that what is extinguished can be modulated by the way a given task is performed. Rafal et al. (2002) presented words (ONE, TWO) or numbers (1, 2) at approximately 3° from fixation to left, right or both visual fields. Patients were asked to respond to both the location and the identity of the stimulus (e.g., 1 on the left, ONE on the right, frequently pointing). Patients were 'trained' to inspect each field. Patients almost always looked and pointed at the ipsilesional field first, then at the contralesional field. Almost twice as much extinction was shown on bilateral trials when the stimuli shared the same meaning. In a second experiment, with one patient, homophones were used (e.g., WON and ONE) — similar effects were shown. Rafal et al. argue that these results indicate that patients with extinction have difficulty in setting up object representations for action selection, and that this level of processing is necessary for explicit detection and awareness.

Another example where task differences seem to be important concerns the effects of stimulus similarity on extinction. We have noted that stimulus similarity can benefit perceptual report, when visual elements group so that contra- and ipsilesional stimuli are selected as a single object (Gilchrist et al., 1996; Humphreys, 1998; Mattingley et al., 1997). However, studies such as those of Rafal et al. (2002) present with what would appear to be the opposite result — namely worse report when patients are presented with two similar rather than dissimilar stimuli. These negative effects of similarity can vary according to the dimensions that have to be reported. In an extinction paradigm, Baylis et al. (1993) presented patients with one (in either the ipsilesional or the contralesional field) or two coloured shapes. On two-item trials, the shapes could be identical (same shape and colour), have the same shape (but different colours), have the same colour (but different shapes), or the items could be different in both shape and colour. In one condition of the experiment, patients had to report the colour and, in the other, the shape of the stimuli. Baylis et al. found that patients showed *more* extinction when the stimuli matched along a response dimension than when the stimuli did not match; thus performance with identical shapes was poor when shapes had to be identified, whereas performance with identical colours was impaired when colours had to be noted (see also Baylis et al., 2001; Ptak and Schnider, 2005). These data are similar to reports of 'repetition blindness' with normal participants (Kanwisher, 1987). One account of repetition blindness effects is that they arise when people register *type* information from different stimuli (the presence of critical features required for task report) but not for *tokens* that indicate the separate representations of the stimuli. Baylis et al. proposed a similar account for their results on extinction, suggesting that extinction patients may find it difficult to establish token representations based on the spatial locations of stimuli.

The disparity between these findings and data showing that extinction reduces with similar items deserves some comment. There are several procedural differences between the studies showing contrasting effects. In studies demonstrating apparent repetition blindness, the patients typically have to localise the stimuli being reported; in contrast this has not been done in studies where positive effects of grouping are reported. The requirement to localise the items may demand that stimuli are bound to their locations, and this may be more difficult when multiple similar stimuli are present. A second difference is that in studies showing repetition blindness, the stimuli have typically appeared at wide eccentricities (e.g., 15° in Baylis et al., 2001). Grouping effects have been found with small distances between items and indeed they decrease as the distance increases (Gilchrist et al., 1996). At very wide eccentricities it seems likely that stimuli are selected consecutively rather than being grouped together as a single perceptual unit. Sequential selection will require that distinct token representations are formed of each stimulus. Poor token representation will generate an extinction effect under these (very different) conditions. We have recently begun to explore these proposals in experiments where eccentricities of the items are varied. Here our preliminary data indicate that a positive effect of similarity when there is a small distance between the stimuli can change to a negative effect at wide eccentricities (see Fig. 1). This fits with the idea that extinction may be modulated by grouping between proximal stimuli, while it is affected by token individuation between more distal items.

The argument that a problem in token individualisation can lead to extinction has also been made by Coslett and Saffran (1991). They studied a patient with bilateral parietal lesions who was better able to report compound (NEWS PAPER) and semantically associated words (HOT COLD) than two unrelated words (e.g., Riddoch et al., 2003). Similarly, the patient was better at identifying two pictures from the same semantic category than two pictures from different categories. To account for this, Coslett and Saffran argued that the patient was unable to maintain independent token representations of multiple stimuli in a visual buffer (VSTM) involved when stimuli are sequentially selected (see Bundesen, 1990). Coslett and Saffran further maintained that semantically related items were better maintained in a visual buffer because there was additional top-down support for their token binding. Though this account of their result remains untested, both this work and that of Baylis and colleagues suggest that extinction may not be a single

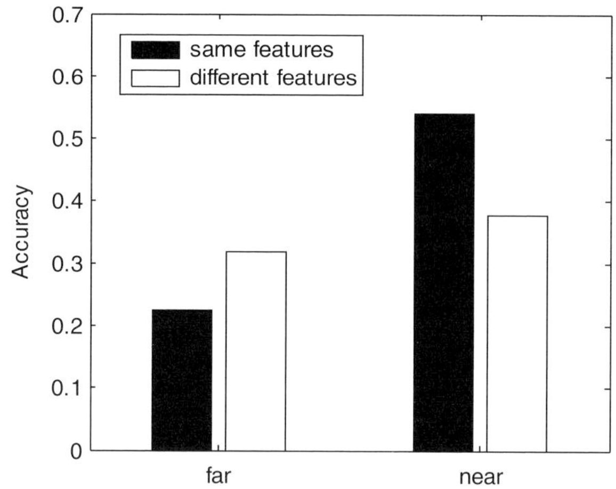

Fig. 1. Performance on two-item trials when stimuli (0.8 × 1.0°) were the same on a feature dimension (shape or colour) versus when they were different. Stimuli are presented either far (12°) or near (3.5°) to fixation.

phenomenon but actually could arise for a number of different reasons in different patients — for example, spatially biased competition in some cases and poor maintenance of token representations in others. Clearly the different forms of extinction need to be distinguished if we are to understand the critical factors determining performance in particular patients.

Non-spatial extinction

Although extinction is typically expressed in spatial terms, there are examples of 'non-spatial' extinction. In these cases, patients only report one object when multiple objects are present, with the item being selected on the basis of its spatial extent and how 'good' a stimulus it is (in terms of how well its parts cohere into a single object). One earlier example of this was reported by Luria (1959). Luria noted a patient who, when presented with a Cross of David, noticed the cross on one occasion, but only one of the component triangles on another. Note that the triangles making up the Cross of David overlap, so the selection of one triangle was not based here on its spatial location but on whether it grouped with or was segmented from the other triangle. Non-spatial extinction was studied more formally by Humphreys et al. (1994). They presented patients with elements that either grouped to form a single closed shape (the corners of a square or a diamond) or did not group together strongly (the corners of a square reflected out so they no longer presented a closed figure). When the closed and non-closed figures were presented, the patients saw the 'good' figure but not the non-closed figure, though the non-closed figure could be reported when presented in isolation. This extinction effect occurred irrespective of the locations of the stimuli. This result can be thought in terms of the patients lacking sufficient processing resources to maintain a representation for the 'weaker' (non-closed) figure when it competes for attention with a strong figure (the closed shape). The stronger figure 'wins' the competition for selection, but the weaker figure cannot then be recovered for perceptual report.

The patients reported by Luria (1959) and Humphreys et al. (1994) suffered bilateral parietal lesions. Other research has pointed to the left intra-parietal sulcus (IPS) for being important in extinction based on the 'strength' of the representation of the stimulus. Mevorach et al. (2008), for example, have reported that the left IPS is significantly activated when a low-saliency target has to be selected and a high-saliency distractor has to be ignored. This activation arose independently of how saliency was defined and even in tasks where targets spatially overlapped with distractors (so emphasising non-spatial selection).

Furthermore, suppressing the left IPS by transcranial magnetic stimulation, or damaging this brain region, leads to poor selection of low-salience targets (Mevorach et al., 2006a, b). This highlights the necessary role of the left IPS in the non-spatial selection of low-saliency stimuli.

The role of the left IPS in non-spatial selection contrasts with the right hemisphere dominance in spatial extinction (Becker and Karnath, 2007; Stone et al., 1993). This may reflect a contrast between the mechanisms of spatial selection and the mechanisms of non-spatial selection, supported through distinct brain regions.

Anti-extinction

Extinction in patients can be thought of as an exaggeration of the normal process of competition for attentional selection, where report can be limited to single objects with briefly presented displays (Duncan, 1984). Reports of anti-extinction thus appear odd. In the phenomenon of anti-extinction, patients can be impaired at attending to a single object while being better able to report two objects under DSS. This was first reported by Goodrich and Ward (1997). Humphreys et al. (2002) found evidence for anti-extinction when displays were briefly presented and then offset, while the report of two stimuli then decreased when displays were exposed for longer. To account for this, Humphreys et al. suggested that anti-extinction reflected temporal grouping between pairs of stimuli which onset and offset together over a short time period (with brief but not more prolonged stimulus exposures). With longer exposures the temporal cues for grouping decreased and patients then tended to select stimuli sequentially. According to this account, anti-extinction is another manifestation of implicit grouping between visual elements that, without grouping, patients would not be able to report. Although the behavioural report indicating anti-extinction differs from that indicating extinction, similar mechanisms of grouping and competition for stimulus selection may be involved.

Another interesting point is that the temporal grouping supporting anti-extinction may not support explicit TOJs. Humphreys et al. (2002) found that anti-extinction depended on the stimuli having common onsets and offsets, but the stimulus on the contralesional side still had to lead the ipsilesional item by a considerable temporal interval for the items to be judged as simultaneous. The temporal signal supporting grouping is not necessarily available to guide conscious TOJs.

Conclusion

Our overview indicates that extinction is a complex phenomenon. While there is evidence for extinction reflecting impairments in attentional competition (due to spatial biases and/or to impaired maintenance of representations of 'weak' stimuli), other work suggests that extinction results from poor binding of elements in short-term memory. Although the processing of extinguished stimuli may not be completely normal, there is evidence for processing continuing to operate. Thus extinguished stimuli can prime responses to non-extinguished items (McGlinchey-Berroth et al., 1993), and extinguished stimuli can enter into grouping relations with ipsilesional items (Gilchrist et al., 1996; Humphreys, 1998; Mattingley et al., 1997; Riddoch et al., 2003). The competition for selection may also be modulated by the programming of action to a given location (Kitadono and Humphreys, 2009), consistent with pre-motor feedback to perceptual processes. Finally, the sensitivity to grouping can lead to the phenomenon of anti-extinction (when temporal grouping operates) and non-spatial extinction. Although extinction can be found after a variety of brain lesions (Duncan et al., 1999), it is associated most strongly with damage to the right inferior parietal lobe (Becker and Karnath, 2007; Stone et al., 1993) consistent with this brain region modulating the competition for selection. In addition, there is evidence for the left IPS being needed for the selection of non-spatial low-saliency targets in the presence of higher saliency distractors (Mevorach et al., 2006a, b). By understanding how extinction may occur, we learn much about attentional selection and its relation to STM and motor processes.

Acknowledgements

This work was supported by grants from the BBSRC, MRC, the EU and the Stroke Association.

References

Battelli, L., Pascual-Leone, A., & Cavanagh, P. (2007). The "when" pathway of the right parietal lobe. *Trends in Cognitive Sciences*, *11*, 204–210.

Baylis, G. C., Driver, J., & Rafal, R. D. (1993). Visual extinction and stimulus repetition. *Journal of Cognitive Neuroscience*, *5*, 453–466.

Baylis, G. C., Gore, C. L., Rodriguez, P. D., & Shisler, R. J. (2001). Visual extinction and awareness: The importance of binding dorsal and ventral visual pathways. *Visual Cognition*, *8*, 359–379.

Becker, E., & Karnath, H. O. (2007). Incidence of visual extinction after left versus right hemisphere stroke. *Stroke*, *38*, 3172–3174.

Berti, A., Allport, A., Driver, J., Dienes, Z., Oxbury, J., & Oxbury, S. (1992). Levels of processing for visual stimuli in an "extinguished" field. *Neuropsychologia*, *30*(5), 403–415.

Birch, H. G., Belmont, I., & Karp, E. (1967). Delayed information processing and extinction following cerebral damage. *Brain*, *90*, 113–130.

Bundesen, C. (1990). A theory of visual attention. *Psychological Review*, *49*, 113–121.

Conci, M., Böbel, E., Matthias, E., Keller, I., Müller, H. J., & Finke, K. (2009). Preattentive surface and contour grouping in Kaniza figures: Evidence from parietal extinction. *Neuropsychologia*, *47*, 726–732.

Corbetta, M., Kincade, M., Ollinger, J. M., McAvoy, M., & Shulman, G. L. (2000). Temporal dynamics of visual attention: Spatial expectancy vs. target detection, as revealed by ANOVA based event-related fMRI. *Neuroimage*, *11*, S8.

Corbetta, M., Miezin, F. M., Shulman, G. L., & Petersen, S. E. (1993). A PET study of visuospatial attention. *The Journal of Neuroscience*, *13*(3), 1202–1226.

Coslett, H. B., & Saffran, E. (1991). Simultanagnosia: To see but not two see. *Brain*, *114*, 1523–1545.

Davis, B., Christie, J., & Rorden, C. (2009). Temporal order judgements activate temporal parietal junction. *The Journal of Neuroscience*, *29*, 3182–3188.

De Renzi, E., Gentilucci, M., & Pattacini, F. (1984). Auditory extinction following right hemisphere damage. *Neuropsychologia*, *22*, 733–744.

Deubel, H., & Schneider, W. X. (2005). Attentional selection in sequential manual movements, movements around an obstacle and in grasping. In G. W. Humphreys & M. J. Riddoch (Eds.), *Attention in action: Advances from cognitive neuroscience*. Hove, UK: Psychology Press.

di Pellegrino, G., Basso, G., & Frassinetti, F. (1997). Visual extinction as a spatio-temporal disorder of attention. *Neuropsychologia*, *35*, 1215–1223.

di Pellegrino, G., & De Renzi, E. (1995). An experimental investigation on the nature of extinction. *Neuropsychologia*, *33*, 153–170.

di Pellegrino, G., Rafal, R., & Tipper, S. P. (2005). Implicitly evoked actions modulate visual selection: Evidence from parietal extinction. *Current Biology*, *15*, 1469–1472.

Driver, J., Baylis, G. C., & Rafal, R. D. (1992). Preserved figure–ground segregation and symmetry perception in visual neglect. *Nature*, *360*, 73–75.

Duncan, J. (1984). Selective attention and the organisation of visual information. *Journal of Experimental Psychology: General*, *10*, 501–517.

Duncan, J., Bundesen, C., Olson, A., Humphreys, G. W., Chavda, S., & Shibuya, H. (1999). Systematic analysis of deficits of visual attention. *Journal of Experimental Psychology: General*, *128*, 450–478.

Duncan, J., & Humphreys, G. W. (1989). Visual search and visual similarity. *Psychological Review*, *96*, 433–458.

Gibson, J. J. (1979). *The ecological approach to visual perception*. Boston, MA: Houghton Mifflin.

Gilchrist, D., Humphreys, G. W., & Riddoch, M. J. (1996). Grouping and extinction: Evidence for low-level modulation of visual selection. *Cognitive Neuropsychology*, *13*, 1223–1249.

Goodrich, S. J., & Ward, R. (1997). Anti-extinction following unilateral parietal damage. *Cognitive Neuropsychology*, *14*, 595–612.

Han, S., & Humphreys, G. W. (2003). Relationship between uniform connectedness and proximity in perceptual grouping. *Science in China*, *46*, 113–126.

Heilman, K. M. (1979). Neglect and related disorders. In K. M. Heilman & E. Valenstein (Eds.), *Clinical neuropsychology*. New York, NY: Oxford University Press.

Heinke, D., & Humphreys, G. W. (2003). Attention, spatial representation and visual neglect. *Psychological Review*, *110*, 29–87.

Humphreys, G. W. (1998). Neural representation of objects in space: A dual coding account. *Philosophical Transactions of the Royal Society*, *B353*, 1341–1352.

Humphreys, G. W., Cinel, C., Wolfe, J., Olson, A., & Klempen, N. (2000). Fractionating the binding process: Neuropsychological evidence distinguishing binding of form from binding of surface features. *Vision Research*, *40*, 1569–1596.

Humphreys, G. W., Riddoch, M. J., & Fortt, H. (2006). Action relations, semantic relations and familiarity of spatial position in Balint's syndrome: Cross-over effects on perceptual report and localisation. *Cognitive, Affective and Behavioural Neuroscience*, *6*, 236–245.

Humphreys, G. W., Riddoch, M. J., Nys, G., & Heinke, D. (2002). Transient binding by time: Neuropsychological evidence from anti-extinction. *Cognitive Neuropsychology*, *19*, 361–380.

Humphreys, G. W., Romani, C., Olson, A., Riddoch, M. J., & Duncan, J. (1994). Non-spatial extinction following lesions of the parietal lobe in humans. *Nature*, *372*, 357–359.

Kanwisher, N. (1987). Repetition blindness: Type recognition without token individuation. *Cognition*, *27*, 117–143.

Karnath, H.-O. (1988). Deficits of attention in acute and recovered visual hemi-neglect. *Neuropsychologia, 26*, 27–43.

Karnath, H.-O., Himmelbach, M., & Küker, W. (2003). The cortical substrate of visual extinction. *NeuroReport, 14*, 437–442.

Kitadono, K., & Humphreys, G. W. (2007). Interactions between perception and action programming: Evidence from visual extinction and optic ataxia. *Cognitive Neuropsychology, 24*, 731–754.

Kitadono, K., & Humphreys, G. W. (2009). Sustained interactions between perception and action in visual extinction: Evidence from sequential pointing. *Neuropsychologia, 47*, 1592–1599.

Làdavas, E. (1990). Selective spatial attention in patients with visual extinction. *Brain, 113*, 1527–1538.

Luria, A. R. (1959). Disorders of "simultaneous perception" in a case of bilateral occipito-parietal brain injury. *Brain, 82*, 437–449.

Marzi, C. A., Girelli, M., Natale, E., & Miniussi, C. (2001). What exactly is extinguished in unilateral visual extinction? Neurophysiological evidence. *Neuropsychologia, 39*, 1354–1366.

Marzi, C. A., Smania, N., Martini, M. C., Gambina, G., Tomelleri, G., & Palamara, A. (1997). Implicit redundant-targets effect in visual extinction. *Neuropsychologia, 34*, 9–22.

Mattingley, J. B., Davis, G., & Driver, J. (1997). Pre-attentive filling in of visual surfaces in parietal extinction. *Science, 275*, 671–674.

McGlinchey-Berroth, R., Milberg, W. P., Verfaellie, M., Alexander, M., & Kilduff, P. (1993). Semantic processing in the neglected visual field: Evidence from a lexical decision task. *Clinical Neuropsychology, 10*, 79–108.

Mevorach, C., Humphreys, G. W., & Shalev, L. (2006a). Effects of saliency, not global dominance, in patients with left parietal damage. *Neuropsychologia, 44*, 307–319.

Mevorach, C., Humphreys, G. W., & Shalev, L. (2006b). Opposite biases in salience-based selection for the left and right posterior parietal cortex. *Nature Neuroscience, 9*, 740–742.

Mevorach, C., Shalev, L., Allen, H. A., & Humphreys, G. W. (2008). The left intraparietal sulcus modulates the selection of low salient stimuli. *Journal of Cognitive Neuroscience, 21*, 303–315.

Ptak, R., & Schnider, A. (2005). Visual extinction of similar and dissimilar stimuli: Evidence for level-dependent attentional competition. *Cognitive Neuropsychology, 22*, 111–127.

Punt, T. D., Riddoch, M. J., & Humphreys, G. W. (2005). Bimanual coordination and perceptual grouping in a patient with motor neglect. *Cognitive Neuropsychology, 22*, 795–815.

Rafal, R., Danziger, S., Grossi, G., Machado, L., & Ward, R. (2002). Visual detection is gated by attending for action: Evidence from hemispatial neglect. *Proceedings of the National Academy of Sciences, 99*, 16371–16375.

Riddoch, M. J., Humphreys, G. W., Edwards, S., Baker, T., & Willson, K. (2003). Seeing the action: Neuropsychological evidence for multiple object selection. *Nature Neuroscience, 6*, 82–89.

Riddoch, M. J., Humphreys, G. W., Hinkman, M., Clift, J., Daly, A., & Collin, J. (2006). I can see what you are doing: Action familiarity and affordance promote recovery from extinction. *Cognitive Neuropsychology, 23*, 583–605.

Robertson, I. H., Mattingley, J. B., Rorden, C., & Driver, J. (1998). Phasic alerting of neglect patient overcomes their spatial deficit in visual awareness. *Nature, 395*, 169–172.

Rorden, C., Mattingley, J. B., Karnath, H. O., & Driver, J. (1997). Visual extinction and prior entry: Impaired perception of temporal order with intact motion perception after unilateral parietal damage. *Neuropsychologia, 35*, 421–433.

Smania, N., Martini, M. C., Gambina, G., Tomelleri, G., Palamara, A., & Natale, E. (1998). The spatial distribution of visual attention in hemineglect and extinction patients. *Brain, 121*, 1759–1770.

Soto, D., & Humphreys, G. W. (2006). Seeing the content of the mind: Enhanced awareness through working memory in patients with visual extinction. *Proceedings of the National Academy of Sciences, 103*, 4789–4792.

Spence, C., & Driver, J. (Eds.). (2004). *Crossmodal space and crossmodal attention*. Oxford, UK: Oxford University Press.

Stelmach, L. B., & Herdman, C. H. (1991). Directed attention and perception of temporal order. *Journal of Experimental Psychology: Human Perception and Performance, 17*, 539–550.

Stone, S. P., Halligan, P. W., & Greenwood, R. (1993). The incidence of neglect phenomena and related disorders in patients with acute right or left hemisphere stroke. *Age and Aging, 22*, 46–52.

Valenstein, E., & Heilman, K. M. (1981). Unilateral hypokinesia and motor extinction. *Neurology, 31*, 445–448.

Vallar, G., Rusconi, M. L., Bignamini, L., Geminiani, G., & Pizzamiglio, D. (1994). Anatomical correlates of visual and tactile extinction in humans: A clinical CT scan study. *Journal of Neurology, Neurosurgery and Psychiatry, 57*, 464–570.

Volpe, B. T., Ledoux, J. F., & Gazzaniga, M. S. (1979). Information processing of visual stimuli in an 'extinguished' field. *Nature, 282*, 722–724.

Ward, R., Goodrich, S. J., & Driver, J. (1994). Grouping reduces visual extinction: Neuropsychological evidence for weight linkage in visual selection. *Visual Cognition, 1*, 101–129.

Woo, S.-H., Kim, K.-H., & Lee, K.-M. (2009). The role of the right posterior parietal cortex in temporal order judgement. *Brain and Cognition, 69*, 337–343.

N. Srinivasan (Ed.)
Progress in Brain Research, Vol. 176
ISSN 0079-6123

CHAPTER 11

An adaptive workspace hypothesis about the neural correlates of consciousness: insights from neuroscience and meditation studies

Antonino Raffone[1,2,*] and Narayanan Srinivasan[3]

[1]*Department of Psychology, "Sapienza" University of Rome, Rome, Italy*
[2]*Perceptual Dynamics Laboratory, BSI RIKEN, Japan*
[3]*Centre of Behavioural and Cognitive Sciences, University of Allahabad, Allahabad, India*

Abstract: While enormous progress has been made to identify neural correlates of consciousness (NCC), crucial NCC aspects are still very controversial. A major hurdle is the lack of an adequate definition and characterization of different aspects of conscious experience and also its relationship to attention and metacognitive processes like monitoring. In this paper, we therefore attempt to develop a unitary theoretical framework for NCC, with an interdependent characterization of endogenous attention, access consciousness, phenomenal awareness, metacognitive consciousness, and a non-referential form of unified consciousness. We advance an adaptive workspace hypothesis about the NCC based on the global workspace model emphasizing transient resonant neurodynamics and prefrontal cortex function, as well as meditation-related characterizations of conscious experiences. In this hypothesis, transient dynamic links within an adaptive coding net in prefrontal cortex, especially in anterior prefrontal cortex, and between it and the rest of the brain, in terms of ongoing intrinsic and long-range signal exchanges, flexibly regulate the interplay between endogenous attention, access consciousness, phenomenal awareness, and metacognitive consciousness processes. Such processes are established in terms of complementary aspects of an ongoing transition between context-sensitive global workspace assemblies, modulated moment-to-moment by body and environment states. Brain regions associated to momentary interoceptive and exteroceptive self-awareness, or first-person experiential perspective as emphasized in open monitoring meditation, play an important modulatory role in adaptive workspace transitions.

Keywords: consciousness; meditation; prefrontal cortex; attention; global workspace; mindfulness; self-awareness

Introduction

As recently remarked by Gaillard et al. (2009, p. 1), "the neural correlates of consciousness (NCC) still remain highly controversial. Indeed, the precise timing, location, and dynamics of neural events causing conscious access are not clearly and unequivocally determined. Do the NCC correspond to late or early brain events? Are they systematically associated with reentrant 'top-down' processing? If so, do they necessarily involve long-range coherent activity, including prefrontal cortex

*Corresponding author.
Tel.: +39 0644427609; Fax: +39 064462449;
E-mail: antonino.raffone@uniroma1.it

DOI: 10.1016/S0079-6123(09)17620-3

as an essential node, or can they be restricted to local patterns of reverberating activity? Is the concept of 'integrated information' relevant, rather than the specific localization of the underlying cerebral network?"

Given this problematic scenario, we suggest a theoretical framework for the NCC, in terms of an *adaptive workspace hypothesis*, based on a set of experimental findings and prior theoretical proposals in the field of neuroscience of consciousness and attention, as well as new insights from meditation studies. We will also consider earlier views on consciousness as developed in Buddhist texts. Apart from aspects of visual awareness and attention, which have been intensively studied over the last years (e.g., Dehaene et al., 2006), we will also include into our theoretical consideration other aspects of consciousness, such as *metacognitive consciousness*, that have been emphasized in earlier philosophical (e.g., Kant, 1781/1996) and contemplative approaches (see Wallace, 1999) as well as in recent neuroscientific theories (e.g., Zeki, 2003; Arenander and Travis, 2004).

In this paper we will first review the main aspects of the influential Baars' global workspace (GW) theory, and then a neural model based on GW with reference to brain structures and processes, including prefrontal cortex, long-range recurrent (reentrant) neural interactions, and oscillatory coherence. We will then pay particular attention to the issue of stability and transience of GW brain dynamics that is crucial for our adaptive workspace hypothesis. We will then consider essential aspects of meditation, as well as recent related neuroscientific and cognitive investigations. Indeed, these aspects are crucial for the development of our adaptive workspace hypothesis. Such a hypothesis will then be outlined in the subsequent sections, with reference to endogenous attention, phenomenal awareness, access consciousness, and metacognitive consciousness as well as their interactions, in a preliminary attempt to develop a unitary neurocognitive framework to characterize human consciousness in its integrity and multiform manifestations. Conclusions and prospective aspects are finally presented.

The global workspace model of conscious access

Baars' global workspace theory

One of the currently most influential theories of human consciousness, with fundamental implications for addressing the NCC, is Bernard Baars' GW theory (1983, 1998, 2002; Baars et al., 2003). In this theory, conscious perception enables access to widespread brain sources, in terms of *broadcasting*, whereas unconscious processing involves brain sources processing information in a substantially segregated or modular fashion. According to Baars, consciousness, although limited in capacity at any given time, provides a gateway to extensive unconscious knowledge sources in the brain, therefore creating the conditions for *global access* in cerebral information processing.

To characterize the functional properties of the GW, Baars (1998) used an interesting theater metaphor, which has been substantially maintained through subsequent versions of the theory (e.g., Baars, 2002; Baars et al., 2003). According to Baars, convergence zones (i.e., the *theater stage*) are needed for the emergence of integrated conscious perceptual information (e.g., in the visual domain). Signals from sensory projection cortical areas, "lit up" by attentional activation, provide consciousness "contents" by converging for integration at the level of more anterior areas. Conscious states can involve a large set of cortical and subcortical brain regions (the *audience*), which can be recruited on an intentional basis for conscious access operations.

A key aspect in Baars' theory is broadcasting of selected contents (i.e., *speaking to the audience*) through which conscious information can be widely disseminated in the brain, in a global access process. Broadcasting occurs through activated distributed (including spatial) maps in the brain, connected through "labeled line" systems, with special reference to intra- and inter-hemispheric cortico-cortical connection systems, and thalamo-cortical connection systems (Baars, 1998). Temporal oscillations would also play a role in coordinating the "speaking to the audience" process in such a distributed brain network for conscious broadcasting. Significant conscious events can be renewed by inner speech, imagery, or conscious emotional feeling states, with a reinitialization of broadcasting-related activity loops.

In GW theory (Baars, 1998, 2002), unconscious or "contextual" brain systems play a role in shaping conscious events, acting as a *backstage* in a theater. Contextual systems in the brain would include the dorsal pathway for visual processing, whereas the ventral visual pathway would produce sensory "contents" (Milner and Goodale, 2008). Interestingly, Baars noticed that parietal neglect, the syndrome characterized by an altered spatial framework for vision, is often accompanied by anosognosia, a loss of awareness about one's body space.

To influence decision making, Baars (1998) hypothesized that conscious events involve self-systems in the brain, with special reference to the "narrative interpreter" in the left hemisphere (with involvement of prefrontal cortex). This interpreter would act as a *stage director* in a theater. Baars particularly refers to Gazzaniga's (1985) findings with split-brain patients, and argues that the two hemispheres might each have an "observer" (or executive interpreter) of the respective conscious flow of visual information.

To address the issue that conscious perception may entail a dialog between specific self-related prefrontal regions ("stage director" or executive interpreter) and sensory cortex (Baars et al., 2003), brain activity patterns produced by a demanding sensory categorization task was compared to those engaged during self-reflective introspection, using similar sensory stimuli (Goldberg et al., 2006). The results showed a complete segregation between the two brain activity patterns arguing against Baars' hypothesis of involvement of self-related observer-like prefrontal regions in perceptual awareness (Goldberg et al., 2006). Moreover, areas characterized by an enhanced activity during introspection exhibited a robust inhibition during the demanding perceptual task, thus suggesting that self-related brain activities are not necessarily implied during sensory perception, and can rather be suppressed.

We will however consider the notions of broadcasting and contextual systems for conscious access, as related to the theater metaphors of "speaking to the audience" and "backstage," in the framework of our model presented in this article. We will also refer to the "observer" or executive interpreter function, as related to the "stage director" metaphor, in terms of a "global backstage" or "consciousness context assembly" in the brain, supporting and constraining in "background" the evolution of transiently stable content-related "core assemblies" characterized by a GW "foreground" dynamics. As will be discussed in section "The adaptive workspace hypothesis" in our model the interplay between global backstage assembly and core assemblies constrains monitoring functions, as related to perceptual conscious access and metacognitive consciousness, in a self-organizing brain-scale dynamics and recursive functional logic.

Neuronal global workspace model

There is wide agreement and remarkable empirical support for GW dynamics in the brain, related to conscious experience (see Dehaene et al., 2006). In particular, Baars' GW theory has been revisited in a neuronal GW framework by Stanislas Dehaene and collaborators, based on a coherent set of psychophysical, neuroimaging, and computational investigations (Dehaene et al., 1998, 2006; Dehaene and Naccache, 2001; Gaillard et al., 2009). This approach postulates that the global availability of information at the whole brain-scale, i.e., in the GW, is what we subjectively experience as a conscious state. With reference to visual awareness, this neuronal GW model proposes that unconscious visual information processing is characterized by the parallel activation of multiple modular brain networks (Dehaene et al., 1998, 2003, 2006; Dehaene and Naccache, 2001; Gaillard et al., 2009).

In the neuronal GW model, three conditions have to be met to enable the access to consciousness of incoming visual information (Dehaene et al., 2006; Gaillard et al., 2009). Condition 1: Information must be *explicitly represented* by neuronal firing in perceptual networks in visual cortical areas coding for the specific features of the conscious percept. Condition 2: This sensory neuronal representation must reach a minimal threshold of *duration* and *intensity* necessary for access to a second stage of processing involving a distributed cortical network, with special reference to parietal and prefrontal cortices. Condition 3: Through joint *bottom-up propagation* and top-down *attentional amplification*, a coherent self-sustained reverberant state must be "ignited" in terms of a brain-scale neural assembly, thus implementing a GW. Neuroscientific findings strongly argue for a major role for prefrontal cortex and anterior cingulate with posterior associative areas that connect to them, in creating the brain-scale workspace dynamics.

The neuronal GW model is characterized by a *winner-take-all* dynamics at higher stages of neural processing (involving prefrontal cortex), a sort of "neural bottleneck" such that only one large-scale reverberating neural assembly is active in the neuronal GW at a given moment. These winner-take-all processes have been highlighted in experimental and related computational settings using the attentional blink (AB) paradigm, in which the subjects have to report the identity of two target stimuli (e.g., digits) in a series of rapidly presented visual stimuli, most of which are distracters (e.g., letters). If the second (T2) of the two target stimuli is presented within 500 ms of the first one (T1) in a rapid sequence of distracters, it is often not detected, thus resulting in an AB (Raymond et al., 1992).

Fries (2005) has proposed a large-scale neural oscillatory mechanism to mediate attention-based broadcasting in the brain, which appears of relevance for the neuronal GW model, in the framework of his "Communication-Through-Coherence" (CTC) hypothesis. In this mechanism, a preferential routing of selected signals takes place by passing the rhythm of a selected sending group onto other groups of neurons, which would be *entrained* to the selected rhythm by membrane potential fluctuations. This set of entrained neuronal groups would then be made sensitive for the selected input, while neuronal "broadcasting centers," such as aspecific thalamic nuclei with widespread reciprocal connections with cortical areas, would distribute the selected rhythm at a whole brain-scale. Thus, as suggested by Fries, the top-down mechanisms reflecting attentional selection are transformed from a spatial to a *temporal code*, as top-down signals reside in the selection-related, entraining-related, and broadcasted temporal information. This hypothesis appears as an interesting specification of Baars' (1998) earlier proposal of "labeled line" systems, coordinated by oscillations, for broadcasting in the GW.

In a very recent intracranial electroencephalographic investigation, Gaillard et al. (2009) have provided an important contribution to characterize the neuronal GW processes, by comparing conscious and nonconscious processing of briefly flashed words. Nonconscious processing of masked words was observed in multiple cortical areas, mostly within an early time window (300 ms), accompanied by induced gamma band activity, but in the absence of coherent long-distance neural activity. In contrast, conscious processing of unmasked words was characterized by the convergence of four distinct neurophysiological markers: sustained voltage changes, particularly in prefrontal cortex, large increases in spectral power in the gamma oscillatory range, increases in long-distance phase synchrony in the beta range, and increases in long-range Granger causality. The analyses of Gaillard et al. (2009) suggest that only late sustained long-distance synchrony and late amplification (after 300 ms) may be causally related to conscious-level processing. In particular, we will focus on the late sustained long-range synchrony reported in the study of Gaillard et al. (2009) for our hypothesis about consciousness presented in this article.

Stability, transience, and adaptive coding in global workspace neurodynamics

As suggested by Maia and Cleeremans (2005) in a connectionist framework, conscious representations involve a distributed network with recurrent connections arriving at an "interpretation" of a given input by settling into a *stable state*. This state is therefore regarded as a function of both the network input and the knowledge embedded in the network's connections, in terms of an interpretation process. Thus, Maia and Cleeremans (2005) suggest that conscious experience reflects stable states corresponding to interpretations that the brain makes of its current inputs, based on a brain-scale *global constraint satisfaction* process. We will however emphasize the importance of transient or metastable neural processes for conscious processes in our adaptive workspace hypothesis (see below).

In line with the GW model, these massive global interactions based on large-scale recurrency are regarded as necessary to reach a stable state supporting a given conscious experience, in terms of a winner-take-all dynamics. As related to Varela and Thompson's (2003) notion of "local-to-global and global-to-local" causality, Maia and Cleeremans (2005) also argue that *strong* and *sustained* neuronal firing at the (global) assembly level makes it more likely that the corresponding representation will reach the conscious level, in a neural competition process at brain-scale level. Conversely, at the local level, neurons characterized by high firing strength and stability are more likely to be inscribed in a winning coalition, and thus receive a higher amount of excitation from the coalition itself. The fact that information that does not enter consciousness tends to decay quickly can be explained in these terms (Dehaene and Naccache, 2001; Maia and Cleeremans, 2005).

Essentially, Maia and Cleeremans' (2005) connectionist framework provides an integrated view of attention, working memory, cognitive control, and consciousness based on a single mechanism: global competition between representations, with top-down biases from prefrontal cortex. A similar integrated approach was proposed earlier by John Duncan (2001), in terms of an *adaptive coding* model of prefrontal cortex function. Based on single-cell recording and neuroimaging data, the central idea of Duncan's adaptive coding model is that, throughout much of prefrontal cortex (with special reference to the lateral areas) the response properties of single cells are highly adaptable, as any given cell has the potential to be driven by many different kinds of input via a dense network of associative synapses. In such a model, prefrontal cortex acts as a GW or working memory onto which the fact needed in a current mental program can be written. Thus, in a particular task context, many neurons become adaptively tuned to code information that is specifically relevant to that task.

Duncan's (2001) approach also refers to the notion of *integrated competition* in visual attention, with emphasis on the problem of *processing coherence*. In this view, despite the fact that the neural representations of an object's features — such as location, color, shape, and motion — are distributed across multiple, partially specialized areas of extrastriate cortex, cognitive experiments show that visual objects are attended as wholes, as directing attention to an object makes its multiple features concurrently available to awareness (Duncan, 1984). According to the integrated competition hypothesis (Desimone and Duncan, 1995; Duncan et al., 1997), objects compete in parallel for representation in multiple extrastriate systems. As an object gains dominance in any one system, its representation is also supported in the other areas, thus resulting in a multiple system convergence. Duncan (2001) suggests that prefrontal cortex plays a guiding role in this integrated competition and convergence with processing coherence, and reflecting the current behavioral significance of objects in terms of adaptive coding and attentional bias. In other words, to achieve processing coherence, multiple brain systems share a strong tendency to converge to represent similar or related information, guided by prefrontal cortex depending on the behavioral or task context. We will also emphasize processing coherence and adaptive coding in prefrontal cortex in the framework of the hypothesis about consciousness presented here.

Global integrative processes for the emergence of consciousness in brain dynamics are also central to Francisco Varela's approach (Varela, 1995; Varela et al., 2001), with a special emphasis on *transient resonant assemblies* and serially established global brain patterns of oscillatory synchronization and desynchronization. In Varela's encompassing view, *for every cognitive act, there is a singular and specific large cell assembly that underlies its emergence and operation.* This approach can also be related to the "dynamic core" model of consciousness (Tononi and Edelman, 1998) that is based on neural complexity and the interplay between integration and differentiation of coherent as well as constantly changing large-scale neural activity patterns.

In Varela's working hypothesis (see also Rodriguez et al., 1999; Le Van Quyen, 2003), the brain-scale endogenous dynamics related to cognitive acts and the emergence of consciousness is characterized by *metastability*, as global activity patterns arise in succession in conditions of dynamical instability, in the absence of settling in any particular state (attractor). In such a framework, global resonant assemblies emerge rapidly in a time frame of 100–300 ms, via widespread cortico-cortical and cortico-thalamic reentrant parallel interactions establishing coherent neural processes. Varela et al. (2001) therefore suggest that large-scale neural integration must implicate not only the establishment of dynamic links, but also their active uncoupling to give way to the next cognitive moment. The integration process is therefore regarded as stemming from the interplay between phase locking and phase scattering across different frequency bands and at different moments in time. In this light, a dynamic neural synchronization model with self-organized spike (burst) synchronization and desynchronization reflecting processing coherence across multiple feature modules was simulated by Raffone and van Leeuwen (2003).

VanRullen and Koch (2003) more recently suggested a mechanism based on neural oscillatory *multiplexing* to account for visual perception and the awareness-related structure of neuronal representations, based on a discrete perception view that may be related to earlier Buddhist texts (von Rospatt, 1995) and psychophysical investigations (see Poeppel, 1997). The mechanism is based on "slow" (especially alpha) and "fast" (gamma) oscillations that are common in the thalamus and visual cortex. In such a mechanism, slow waves would constitute the "context," and fast waves the "content" of neural representations. With reference to an earlier proposal by Llinas et al. (1998), VanRullen and Koch (2003) hypothesized that a nonspecific network or "matrix," a distributed neuronal thalamic network connected with virtually all cortices, in association with the reticular nucleus and modulated by cortical feedback, would be capable of supporting globally coherent oscillations in the alpha range, as a "context" for neuronal representation. Content neural representations would be mediated by specific networks or "core," neuronal subpopulations present in various sensory nuclei of the thalamus, with reentrant connections restricted to the corresponding cortical regions (maps). These specific thalamo-cortical loops would sustain fast (gamma) activity waves. In such a functional logic, the "matrix" and "core" thalamo-cortical networks together would support global and local orchestration of brain activities, implementing a multiplexing oscillatory scheme. We will also refer to the functional and neurodynamical distinction between context (matrix) and core (content) networks in the framework of the hypothesis about consciousness presented in this article.

Finally, the issue of "strength and stability" (as emphasized by Maia and Cleeremans, 2005) versus "transience and rapid integration" (as emphasized by Varela, 1995; Varela et al., 2001) in the dynamics of consciousness, appears also reflected in Block's distinction between *phenomenal consciousness* and *access consciousness* (Block, 1995, 2007). Access conscious content is content information that is "broadcasted" in the GW and phenomenally conscious content is what differs between experiences, say of red and green. Specifically, Block characterizes contents of access consciousness in terms of information that is made available to the brain's "consumer" systems: systems of memory, perceptual categorization, reasoning, planning, evaluation of alternatives, decision making, voluntary direction of attention, and more generally, rational control of action. We will consider these "consumer" systems as an explicit component of our model of metacognitive consciousness presented here.

When we view a complex visual scene, we experience a richness of content that seems to go beyond what we can report, Block proposed a distinct state of "phenomenal consciousness" prior to global access or GW broadcasting. Block's proposal was also based on the report of participants claiming to see the whole array of flashed letters, although they could later report only one subsequently cued row or column in experiments with Sperling's iconic memory paradigm. Along these lines, it has been suggested that access consciousness is related to *more stable* working memory representations, and phenomenal consciousness to a *more transient* iconic memory (Lamme, 2003; Block, 2007). However, this proposal has been criticized by claiming that Block's phenomenal consciousness would just correspond to a preconscious state, and the perceptual awareness experience of viewers in iconic memory experiments to an illusion (Dehaene et al., 2006).

Focused attention and open monitoring in meditation

Meditation can be conceptualized as a family of complex emotional and attentional regulatory practices, in which mental and related somatic events are affected by engaging a specific attentional set. Many recent behavioral, electroencephalographic, and neuroimaging studies have revealed the importance of investigating meditation states and traits to achieve an increased understanding of cognitive and affective neuroplasticity, attention and self-awareness, as well as for relevant clinical implications (Cahn and Polich, 2006; Lutz et al., 2008).

Given that regulation of attention is the central commonality across the many different meditation methods (Davidson and Goleman, 1977), meditation practices can be usefully classified into two main styles — *focused attention* (FA) and *open monitoring* (OM) — depending on how the attentional processes are directed (Cahn and Polich, 2006; Lutz et al., 2008). In the FA ("concentrative") style, attention is focused on an intended object in a sustained fashion. OM ("mindfulness-based") meditation involves the nonreactive monitoring of the content of experience from moment-to-moment, primarily as a means to recognize the nature of emotional and cognitive patterns.

Focused attention meditation

Apart from sustaining the attentional focus on the intended object, FA meditation also entails the regulative skills of monitoring the focus of attention and detecting distraction, disengaging attention from the source of distraction, and (re)directing and engaging attention to the intended object (Lutz et al., 2008). FA meditation techniques involve monitoring of experience fields by allowing other thoughts and sensations to arise and pass without clinging to them, keeping on or bringing attention back to a specific object of concentrative (or focused) awareness to develop an internal "witnessing observer" (Cahn and Polich, 2006). The attentional and monitoring functions in FA meditation have been related to dissociable systems in the brain involved in conflict monitoring, selective and sustained attention (Corbetta and Shulman, 2002; Lutz et al., 2008; Weissman et al., 2006).

It has been observed that FA meditation practice leads to a reduced effort in sustaining the intended focus. The regulative skills of noticing distractions, disengaging from the source of distraction, and redirecting promptly the attentional focus to the chosen object are more frequently involved in novices than in more expert FA meditation practitioner. Expertise in FA meditation would also lead to an attentional focus resting more readily and stably on the intended object, with a more acute monitoring ability to detect any arising distraction or mind wandering, thus implying a reduced cognitive effort in the practice (Lutz et al., 2008).

In Buddhist texts, consciousness is described as a "momentary collection of mental phenomena" or "distinct moments" (von Rospatt, 1995). In such texts it is asserted that the continuum of awareness is characterized by successive *moments*, or *pulses*, of cognition (Wallace, 1999). Based on a view of consciousness as consisting of sequences of discrete events (see also Poeppel, 1997; VanRullen and Koch, 2003), Wallace (1999) argued that the degree of focused *attentional stability* during FA meditation increases in relation to the proportion of ascertaining moments of cognition of the intended object. In this view, in a continuum of perceptual experience, a large amount of moments of awareness consist of non-ascertaining cognition, as objects *appear* to this inattentive awareness, but they are not *ascertained* (Lati Rinbochay, 1981; Wallace, 1999).

As attentional stability increases, a reduced number of moments of ascertaining consciousness are focused on perceptual objects other than the intended object, thus resulting in a homogeneous series of moments of ascertaining perception or perceptual awareness of the chosen object. In this process, the degree of *attentional vividness* corresponds to the ratio of ascertaining to non-ascertaining cognition moments: the higher the frequency of ascertaining perception, the greater the vividness (Wallace, 1999). In FA meditation practice, high attentional stability and vividness are achieved in a mental state of concentrated calm or serene attention, denoted by the word *Samatha* (with the literary meaning of *quiescence*) in the Buddhist contemplative tradition (Wallace, 1999), by FA meditation. By using a telescope analogy, Wallace (1999) observes that the development of attentional stability may be likened to mounting a telescope on a firm platform, while the development of attentional vividness is like highly polishing the lenses and bringing the telescope into clear focus.

Transcendental meditation (TM) can be broadly included in the FA meditation category, as its practice centers on the repetition of a mantra. However, TM primarily emphasizes an absence of concentrative effort and the development of a witnessing, thought-free "transcendental awareness" or "pure consciousness." Maharishi Mahesh Yogi, who brought TM to the West from the Vedic tradition of India, characterized the experience of pure consciousness as follows (see Arenander and Travis, 2004): "When consciousness is flowing out into the field of thoughts and activity it identifies itself with many things, and this is how experience takes place. Consciousness coming back onto itself gains an integrated state … This is pure consciousness." Pure consciousness is thus regarded as "pure" in the sense that it is free from the contents of knowing. It is a state of consciousness in which the individual is fully aware, with the "content" of pure consciousness being awareness itself (Arenander and Travis, 2004). TM meditation practitioners report that the absence of any concentration or effort unfolds experiences of "unboundedness" and the "loss of time, space, and body sense." These "pure consciousness" or "thoughtless awareness" experiences were associated with profound bodily relaxation, marked by spontaneous breath quiescence and global, high amplitude, slow frequency (alpha) EEG patterns which are general highly coherent across frontal leads (Arenander and Travis, 2004; Travis and Wallace, 1999).

Interestingly, Zeki (2003) has recently considered a similar construct of unified or pure consciousness in his hierarchical theory of consciousness. Zeki's theory includes three hierarchical levels of consciousness: the level(s) of micro-consciousness, the level(s) of macro-consciousness, and unified consciousness. With reference to Kant (1996), Zeki regards this pure, unified, or "transcendental consciousness" as consciousness of oneself as the perceiving person, and as amounting to being aware of being aware. Zeki then places it at the apex of the hierarchy of consciousnesses, and remarks that it is the only consciousness that can be described in the singular.

A recent study with a binocular rivalry paradigm showed that Tibetan Buddhist monks were able to perceive a stable, superimposed percept of two dissimilar, competing images presented to separate eyes for a longer duration both during and after FA meditation, but not during and after a form of compassion (emotional OM) meditation (Carter et al., 2005). These extreme increases in perceptual dominance durations suggest that extensive training in FA meditation might improve the abilities to sustain attentional focus on a particular object and to control the flow of items being attended to and accessing consciousness.

A recent functional resonance imaging (fMRI) study investigated the neural correlates of FA meditation in experts (following Tibetan Buddhist traditions) and novices, with meditation focus on an external visual point (Brefczynski-Lewis et al., 2007). FA meditation compared with a rest condition, was associated with activation in multiple brain regions involved in monitoring, such as dorsolateral prefrontal cortex, attentional orienting (e.g., the superior frontal sulcus and intraparietal sulcus), and engaging attention (visual cortex). Srinivasan and Baijal (2007) used the mismatch negativity (MMN) paradigm, that is an indicator of preattentive processing, to investigate the effects of FA (Sudarshan Kriya Yoga) meditation. Meditators were found to have larger MMN amplitudes than non-meditators. The meditators also exhibited significantly increased MMN amplitudes immediately after meditation suggesting transient meditation-related state changes. These findings indicate that FA meditation practice enhances preattentive perceptual processes, enabling better change detection in auditory sensory memory.

Open monitoring meditation

OM meditation involves no explicit attentional focus, and therefore does not seem to be associated to brain areas implicated in sustained or FA, but to brain regions involved in vigilance, monitoring, and disengagement of attention from sources of distraction from the ongoing stream of experience (Lutz et al., 2008). OM practices are based on an attentive set that is characterized by an open presence and a nonjudgmental awareness of sensory, cognitive, and affective fields of experience in the present moment, and involves a higher-order (meta-) awareness of the ongoing mental processes (Cahn and Polich, 2006). The cultivation of this "reflexive" awareness in OM meditation is associated to a more vivid conscious access to the rich features of each experience and enhanced metacognitive and self-regulation skills (Lutz et al., 2008).

Unlike FA meditation, OM meditation does not involve explicit focusing on any object in the field of awareness, and therefore thus not involve attentional selection and de-selection processes. Therefore, in OM meditation cognitive monitoring is reflected in an open-field capacity to detect arising sensory, feeling, and thought events in an unrestricted "background" of awareness, without "grasping" any of these events in an explicitly selected foreground or focus as in FA meditation. In a transition from a FA to an OM meditation state, an object as primary focus is gradually replaced by an "effortless" sustaining of an open background of awareness without explicit attentional selection (Lutz et al., 2008). We will return in next sections of this article to the constructs of OM and awareness background as revealed through OM meditation, in the framework of our model of metacognitive consciousness.

In contemplative practice, as in the Buddhist tradition, attentional stability and vividness (acuity), as developed in FA meditation, are regarded as necessary for deep and reliable introspection to take place, as in the practice of *Vipassana* (insight) OM meditation. As remarked by Wallace (1999), Tsongkhapa (1357–1419), an eminent Tibetan Buddhist contemplative and philosopher, refers to another analogy to highlight the importance of attentional stability and vividness for the cultivation of contemplative insight. If an oil-lamp which is both radiant and unflickering is used at night to light a hanging tapestry, the depicted forms can be vividly observed. By contrast, if the oil-lamp is either dim, or, even if it is bright, flickers due to wind, the depicted images cannot be seen.

In the Buddhist contemplative tradition, introspection, as performed in OM insight meditation, is regarded as a form of metacognition, thus raising the important problem of whether or not it is possible for the mind to observe itself. As Buddhists generally assert that at any given moment consciousness and its concomitant mental processes have coherently the same intentional object, and at any given moment only one consciousness can be produced in a single individual (Vasubandhu, 1991), the problem of whether or not it is possible for the mind to observe itself raises (Wallace, 1999). With this respect, a famous discourse attributed to the Buddha states that the mind cannot observe itself, just as a sword cannot cut itself and a fingertip cannot touch itself; nor can the mind be seen in external sense objects or in the sense organs (*Ratnacutasutra*, cited in Shantideva, 1971 and Wallace, 1999).

To avoid an infinite regress in terms of a noted observer and the one who simultaneously note that observer, the 8th-century Indian Buddhist contemplative Shantideva suggested that instead of such metacognition occurring with respect to a simultaneously existing cognition, a *recollection* of past moments of consciousness would rather take place. In Shantideva's view, when one remembers seeing a certain event, one recalls both the perceived event and oneself perceiving that event. The subject and object are recalled as an integrated, experienced event, from which the subject is retrospectively identified as such; but Shantideva denies that it is possible for a single cognition to take itself as its own object (Dalai Lama, 1994; Shantideva, 1997; Wallace, 1999).

Wallace (1999) suggests an interesting example to clarify Shantideva's view on introspective metacognition: "When one's attention is focused on the color blue, one is not observing one's perception of that color. However, when one's interest shifts to the experience of blue, one is in fact recalling seeing that color just a moment ago. In this process, one conceptually and retrospectively isolates the subjective element from the remembered experienced event, in which the blue and one's experience of it were integrated. Thus, when the attention is shifted back and forth between attending to the color and to remembering seeing the color, it seems as if such a shift is comparable to shifting the attention from the objects at the center of consciousness to those at the periphery. But according to Shantideva, the attention is instead shifted from the perceived object to a short-term recollection of a previous event. And in remembering that event, the subject is isolated and recalled, even though it was not its own object at the time of its own occurrence. When one is recalling a perception of an earlier event, there is still a sense of duality between oneself and the perception that one is recalling. A single cognition does not perceive itself, so the subject/object duality is sustained" (Wallace, 1999, p. 179).

This view of metacognition and conscious access appears converging with a contemporary connectionist approach to metarepresentation, the creation of representations that are then available for reprocessing by the same network, thus implementing a (meta)representational and (recursive) processing cycle that could be regarded as the parallel distributed processing basis of the "stream of thought" (Maia and Cleeremans, 2005). The same operational principles might be underlying a global constraint satisfaction dynamics characterized by sequential (recursive) GW ignitions and transitions, and involving multiple brain systems in parallel. We will return to this aspect in the next section of this article.

Behavioral studies have shown a more distributed attentional focus (Valentine and Sweet, 1999), enhanced conflict monitoring (Tang et al., 2007), and reduced AB or more efficient resource allocation to serially presented targets (Slagter et al., 2007) in OM meditation practitioners. Specifically, Slagter et al. (2007) found that 3 months of intensive OM meditation lead to an observable reduction of elaborative processing of the first of two target stimuli (T1 and T2) presented in a rapid stream of distracters, as indicated by a smaller T1-elicited P3b, a brain potential index of resource allocation. Remarkably, such a reduction in resource allocation to T1 was associated with improved detection of T2. Slagter et al.'s study indeed suggests that an intensive training in OM meditation might result in the development of efficient attentional regulative skills to flexibly engage and disengage from target stimuli in a given task-setting.

Lutz et al. (2004) found a high-amplitude pattern of synchrony in the gamma oscillatory band in expert meditators during an emotional version of OM meditation (non-referential compassion or loving kindness meditation). In that study, compared with a group of novices, the practitioners (with a mental training of 10,000–50,000 h over time periods ranging from 15 to 40 years) self-induced higher-amplitude sustained gamma band oscillations and long-range phase synchrony, especially over lateral fronto-parietal electrodes, during meditation. This pattern of gamma band oscillations and synchrony was also significantly more pronounced in the baseline state of the long-term practitioners compared with the novices, thus suggesting a neuroplasticity-based transformation in the default brain mode of the practitioners.

Lutz et al. (2008) regard that these OM meditation-related neuroelectrical findings suggest the emergence of large-scale coherent neural assemblies which can influence local neuronal processes. They suggest that "some meditation states might not be best understood as top-down influences in a classical neuroanatomical sense but rather as dynamical global states that, in virtue of their dynamical equilibrium, can influence the processing of the brain from moment to moment" (Lutz et al., 2008, p. 5). Indeed, in some versions of OM meditation practitioners drop any explicit effort to control the arising of thoughts or emotions to further stabilize their meditation. Moreover, practitioners with high expertise in FA meditation can sustain attentional focus on the intended object and regulate attention with a low effort. Lutz et al. also argue that: "In this view, the brain goes through a succession of large-scale brain states, with each state becoming the source of top-down influences for the subsequent state. We predict that these large-scale integrative mechanisms participate in the regulatory influence of these meditation states" (Lutz et al., 2008, p. 5).

The adaptive workspace hypothesis

The *adaptive workspace hypothesis* of the interdependent emergence of endogenous attention, access consciousness, phenomenal awareness, and metacognitive consciousness, in a NCC framework, is based on a set of interrelated explanatory and predictive aspects, as characterized in the following subsections.

Adaptive coding net

As considered earlier, firing of a large population of adaptive coding neurons in prefrontal cortex can be driven by different kinds of synaptic input sources which are widespread in the brain (Duncan, 2001; see also Duncan and Miller, 2002). We hypothesize that the *resonant* or recurrent involvement of such adaptive prefrontal neurons is necessary for any form of access-based consciousness to take place, either in terms of access (working memory related) consciousness of specific sensory and thought contents, or content-independent metacognitive consciousness. We refer to the neuronal population of prefrontal adaptive coding neurons as to the *adaptive coding net* (ACN).

We also hypothesize that at any time the number of *dynamic links* that ACN neurons can form is limited. This limitation would be the basis of the notion of limited cognitive resources for conscious access (e.g., Dehaene et al., 2003, 2006). Notice that here we are formulating the limited capacity of conscious access in terms of a neural binding or integration process, centered on the ACN. In biologically and cognitively plausible terms, we further hypothesize that the recurrent signal exchanges involving the ACN are reflected in an enhancement of both neural *co-activation* (neuronal firing rates) and *coherence* (enhanced spatiotemporal correlations in neuronal firing) neural patterns.

The ACN dynamic linking mechanism might involve *short-term plasticity* (von der Malsburg, 1981, 1999; Tononi et al., 1992) or a competitive opening and closing of neural transmission *gates* in hierarchical processing lines, as recently proposed in Bundesen et al's. (2005) Neural Theory of Visual Attention (NTVA). Another possible mechanism can be based on an oscillatory synchrony-based entrainment (Fries, 2005). However, the specification of the hypothesized ACN dynamic linking mechanisms demands further dedicated experimental and modeling investigations.

Consumer systems related to conscious access

In a neurocognitive "working scenario," characterized by a task or performance setting with specified goals and context of stimulus-response mappings, the executive dynamic links of ACN neurons are plausibly established with the *consumer systems* (i.e., systems of explicit memory, perceptual categorization, reasoning, planning, evaluation of alternatives, decision making, voluntary direction of attention, and more generally, rational control of action) for access consciousness (Block, 2007). Such consumer systems are mediated by a large set of anterior and posterior associative (neo)cortical areas, and subcortical regions, such as the hippocampus and related structures in the medial temporal lobe.

In such a scenario, conscious access is oriented by *selective endogenous attention*, thus providing a "working access bias" toward response-relevant (target) stimuli and stimulus-response mappings. A number of ACN dynamic links (e.g., in dorsolateral prefrontal cortex, anterior prefrontal cortex, and anterior cingulate cortex), however, can be allocated to performance *monitoring*.

Metacognitive consciousness and adaptive workspace

We hypothesize that *metacognitive* or *reflexive* consciousness, as reflected in performance monitoring in a cognitive working or goal-based scenario, is mediated by intrinsic ACN dynamic links, i.e., dynamic links established within the population of adaptive coding neurons in PFC.

Therefore, at any given time the degree of metacognitive consciousness can be limited by the number of dynamic links between ACN neurons and consumer system neurons involved in a given task or performance. Note that this aspect of our hypothesis prevents a dual view ultimately leading to an infinite regress argument or homunculus assumption. Indeed, in our hypothesis the neural correlate of what we subjectively experience as an ongoing observation or conscious monitoring of our experience or cognitive performance, would be given by dynamic links within active ACN neurons. The same neurons and dynamic linking mechanism would be involved in a working memory related conscious access to given sensory or thought contents. Rehearsal and active control processes for maintenance in working memory would be activated via dynamic links of rehearsal-related neuronal populations with the ACN, in interaction with dynamic links established recurrently *within* the ACN populations. This aspect relates to the metacognitive (metamemory) character of implemented memory strategies and mnemotechnics (e.g., Sternberg, 2008).

More specifically, we hypothesize that neurons in anterior or rostral prefrontal cortex (Broadmann Area 10, or BA10), especially on the lateral surface, are implied in this reflexive awareness function. Indeed, area BA10 is a very large brain region in humans, is in relative terms twice as large in the human brain as in any of the great apes. Furthermore, this region is possibly the last to achieve myelination, and it has been observed that tardily myelinating areas are implied in complex functions highly related to the organism's experience (Fuster, 1997, p. 37). Such a region does not seem characterized by a primary involvement in standard executive and working memory tasks (Burgess et al., 2005), and has been associated to various aspects of metacognition (e.g., Christoff and Gabrieli, 2000). Moreover, activity or structural changes in Area BA10 have been associated to meditation states and traits (Cahn and Polich, 2006; Lazar et al., 2005).

Distributed endogenous attention and adaptive workspace

We further hypothesize that endogenous (top-down) attention, which is plausibly regarded as crucial for conscious access (e.g., Block, 2007; Dehaene et al., 2006), is generated by dynamic links established between the ACN and a set of prefrontal, parietal posterior, and thalamic nuclei involved in endogenous attentional orienting (Bundesen et al., 2005; Posner and Petersen, 1990). Indeed, it has been recently shown that endogenous attention is guided by prefrontal cortex, with a key role played by oscillatory synchrony in the beta oscillatory range (Buschman and Miller, 2007).

In the adaptive workspace framework, when conscious access demands a high amount of dynamic links with ACN neurons, the otherwise free endogenous attention resources are reduced. This reduction is shown in the inattentional blindness phenomenon, when a perceptually salient stimulus (even if presented within the fovea for a long duration) does not access visual awareness when subjects are engaged in intense mental activity such as in detecting certain stimuli or counting (Simons and Chabris, 1999).

As considered above, Dehaene et al. (2006) stress that top-down attention is necessary for access to consciousness. However, in their neuronal GW model it appears unclear from where this top-down attention derives. In our view, an unfortunate implication of this uncertainty might be a homunculus-like structure or process projecting a top-down attentional bias toward intended representations in perceptual maps. Our proposal here is that endogenous attention is potentially ongoing and open field, i.e., primarily *distributed* (see also Srinivasan et al., 2009, in this volume).

We also hypothesize that when a task is to be performed or some information needs to be accessed, the usually distributed endogenous attention becomes focused or selective (see Srinivasan et al., in this volume). Thus, a goal-based task-setting makes endogenous attention selective (Desimone and Duncan, 1995; Maia and Cleeremans, 2005). Such a task-based setting can be encoded within the ACN population itself (Duncan, 2001). On a trial-by-trial basis, this endogenous attention selectivity can be implemented by transient dynamic links between ACN neurons and fronto-parieto-thalamic neurons (e.g., in parietal posterior cortex and in the pulvinar) which have been shown to be involved in top-down attentional orienting (e.g., Bundesen et al., 2005; Buschman and Miller, 2007; Posner and Petersen, 1990). In such a goal-based selective or biased competition setting, the anterior cingulate cortex would play a crucial role in monitoring and controlling the maintenance of the selective attention (cognitive) focus against arising distractions (Cahn and Polich, 2006; Posner and Petersen, 1990). Conscious access processes based on endogenous attention selectivity can be mediated by dynamic links between ACN neurons and perceptual networks, via fronto-parieto-thalamic neurons (e.g., in parietal posterior cortex and in the pulvinar) involved in endogenous attention orienting. Indeed, a main assumption of the neuronal GW model (Dehaene et al., 2006; Gaillard et al., 2009; see above) is that information in a conscious percept must be *explicitly represented* by neuronal firing in perceptual networks.

The notion of distributed endogenous attention is however different from vigilance (Dehaene et al., 2006). Vigilance mainly refers to a global enabling condition mainly driven by "ascending" projections from mostly "aspecific" subcortical centers. In contrast, in our present hypothesis, endogenous attention is characterized in terms of synaptic signals "descending" by anatomical back-projections from the ACN and dynamically linked cortical areas (e.g., parietal posterior cortex). Our notion of nonselective endogenous attention is also different from an exogenous attention concept, as such a bottom-up form of attention is based on a rapid reactive (automatic) orienting process, likely to be driven by bottom-up neural signaling from posterior cortical areas (e.g., Buschman and Miller, 2007).

Here we endorse Burgess et al.'s (2005) *gateway hypothesis* of rostral prefrontal cortex (Area BA10) function. In the gateway hypothesis, the relative activation of lateral and medial regions of Area BA10 guides the switching of processing resources between stimulus-independent thought (SIT) and stimulus-oriented thought (SOT). Indeed, medial BA10 appears activated in conditions in which subjects attend stimuli in the external world, even when a "shallow" processing of them is required. Lateral BA10, by contrast, shows increased activation when the recollection or manipulation of the products of previous processing is required, and as related to expected targets, even when such targets are not presented (see Burgess et al., 2005). In our present adaptive workspace hypothesis, medial and lateral BA10, in terms of their direct interaction and larger interactions with a set of ACN neurons, would therefore play a key role in governing the ongoing open field or distributed neurodynamics of endogenous attention. When attention becomes selective in a goal-based setting, dynamic links for a sustained selective attention without (before) target appearance can be prominently mediated by lateral Area BA10, and dynamic links for a transient selective enhancement of stimulus-related neural activity by medial Area BA10.

In synthesis, dynamic links within medial and lateral ACN neurons would be differentially associated to two main forms of metacognitive or *second-order* awareness. This would depend on whether endogenous attention and conscious access broadcasting are distributed "externally" or "internally." External metacognitive awareness involves the experience of external stimulus contents (e.g., "I am aware that I am experiencing the sight of a red cherry") and internal metacognitive awareness involves the experience of thought contents (e.g., "I am aware that I am recollecting an episode of my adolescence"). A transcendent form of awareness, characterized as pure "being aware of being aware" (beyond an experiential first or second order "subject–object" awareness), which may be called *third-order*, *non-referential*, or *unified* consciousness, might be characterized by dynamic links and neural activity coherence across medial and lateral Area

BA10, modulated by a background of synaptic signals from "momentary self" or "I" related brain areas (see below).

Adaptive workspace dynamics of first, second, and third-order consciousness

In the adaptive workspace hypothesis each core *refers* to a unique integrated content as explicit in consciousness. As suggested by Baars (1998), however, a larger set of backstage or context neurons can support and modulate the current GW assembly (core). We hypothesize that the phenomenal awareness or the subjective aspect of experience related to a given perceptual object is implicit or contextual when the active core refers to the object features, in terms of explicit mapping to the object neural codes (e.g., in sensory maps). This explicit mapping would be mediated by dynamic links with ACN neurons. However, such implicit (subjective) context would emerge in "foreground" at the decay of the associated object core (assembly): the dynamic links between ACN and explicit object codes in sensory maps would be "released" (e.g., by phase scattering, see Varela et al., 2001), with a complementary enhancement of neural activation and coherence patterns involving neurons formerly in the "backstage," thus resulting in first-person phenomenal awareness of subjective ("I") states (see below). A new implicit context (backstage) could then be formed, which support a subsequent core integrating the perceptual object codes and their implicit (subjective-related) context of experience in the previous consciousness core, by neural broadcasting. Thus, in the second integrated core the subjective aspect of experience as bound to a given perceptual object can become itself an object of conscious access (see the text above about Shantideva and Wallace's views). Alternatively, the second core could just refer to subjective states in backstage during the earlier core, without reference to an external perceptual object, and available for conscious access and possibly "introspective" examination.

This consciousness dynamics can be expressed in terms of global constraint satisfaction and recursive patterns, i.e., sequential (recursive) GW ignitions and transitions, and might plausibly involve multiple brain systems in parallel for broadcasting. Given the emphasis on reversible temporal formation and dissolution of large-scale resonant assemblies, in this view the consciousness-related system neurodynamics would be characterized by transience (metastability), as emphasized in Varela's (1995) approach, rather than in terms of stable attractors (Maia and Cleeremans, 2005; Rumelhart et al., 1986).

In the adaptive workspace hypothesis, the core–context neural assembly transitions are hypothesized to depend on dynamic links with the ACN, linking either with explicit perceptual maps or with *implicit self* neural states. To clarify the last aspect, according to William James (1890) any discussion of the self needs to refer to the distinction between the self as object (the "Me" or explicit self) and the self as subject (the "I" or implicit self) of experience. James stated: "The consciousness of Self involves a stream of thought, each part of which as 'I' can remember those which went before, know the things they knew, and care paramountly for certain ones among them as 'Me,' and appropriate to these the rest." Despite a number of controversial stances, James' distinction between I and Me has been substantially maintained over the decades (see Northoff et al., 2006).

We therefore hypothesize that the awareness of subjective or phenomenal aspects of experience demand the establishment of dynamic links between ACN neurons and neuronal populations involved as neural markers of transient body states, in particular, right lateralized exteroceptive somatic and interoceptive insular cortices (Craig, 2004; Critchley et al., 2004; Damasio, 1999), in a transient neural broadcasting process. Given the transient character of the dynamic links between ACN neurons and somatic marker neurons, it can be assumed that interference on intrinsic ACN dynamic links is limited. By contrast, the consumer systems associated to conscious access broadcasting (see above) are likely to operate on the basis of more sustained reverberations associated to working memory, thus causing larger perturbations of intrinsic ACN patterns.

As remarked by Baars (1998), spatial neglect is often accompanied by anosognosia, a massive loss of awareness about one's body space, thus suggesting a uniform neural context for visuo-spatial awareness and body awareness. Differently, as seen above, brain areas implied in narrative or objective self representation would not participate in perceptual awareness (Goldberg et al., 2006). Finally, somatic marker or momentary self-awareness areas have also been implied in OM meditation (Farb et al., 2007; Lutz et al., 2008).

Our hypothesis is therefore consistent with Thompson and Varela's, (2001) *radical embodiment* view of the neurodynamics of consciousness. Thompson and Varela's approach aims to map the neural substrates of consciousness at the level of large-scale, emergent, and transient dynamical patterns of brain activity, and suggest that the processes crucial for consciousness cut across the *brain–body–world* divisions, rather than being brain-bound neural events. In the adaptive workspace hypothesis, the ongoing coupling between the ACN and body-related and momentary self-related neuronal populations enables the creation of a large set of possible transient resonant modes in the brain, embedded in a rich background of body and environmental states, thus increasing the neurodynamical complexity associated to conscious experience (Le Van Quyen, 2003; Tononi and Edelman, 1998). Based on the high degree of coding adaptivity and intrinsic as well extrinsic synaptic recurrency of its neuronal populations, the ACN would dynamically set the endogenous attention and metacognitive constraints necessary "to go beyond the stimulus given," in the brain–body–environment interplay.

The metacognitive consciousness capabilities related to the ACN would also potentially enable "going beyond the experience given." Indeed, intrinsic links within ACN neurons would mediate the metacognitive consciousness of being aware of either an object or a subjective experience of cognition of an object, where such an object is an external stimulus, an inner thought, or feeling state. In our view, this metacognitive consciousness would be "transversal" to any form of awareness based on the subject–object cognitive duality, whether it refers to an external or internal object per se (first-order consciousness), or to the subjective or phenomenal experience of such an object (second-order consciousness). As such, the intrinsic dynamic links within the ACN can be *non-referential*, and then be related to a third-order (metacognitive) consciousness, going beyond the cognitive subject–object duality, as the "awareness of being aware" (Arenander and Travis, 2004; Zeki, 2003). In our view, this transcendent awareness per se can only be developed as a meditation-based intuition, and can thus also be characterized as an *intuitive awareness* (Sumedho, 2004). This intrinsically unified consciousness would reflect itself as referential and context-dependent metacognition in the processes and neural operations underlying access and phenomenal consciousness. Indeed, in the adaptive workspace hypothesis intrinsic (established within the) ACN and extrinsic (established without the) ACN dynamic links are interacting at any time. We will characterize these interdependent consciousness dynamics with reference to OM and FA meditation in the next subsections, although such interdependent processes are regarded as potentially observable in a wide range of cognitive settings and experiential contexts.

Open monitoring meditation and adaptive workspace

In an OM meditation scenario, attention and monitoring functions are nonselective or open field, inclusive of external sensory fields as well as of internal thoughts and feelings. In light of what has been argued above about the primary nature of endogenous attention in the adaptive workspace model, OM meditation can be regarded as the context for the most direct and natural manifestation of distributed endogenous attention, unbounded to specific goal-related performance (task) settings.

In line with the adaptive coding model of prefrontal cortex function (Duncan, 2001; see also Maia and Cleeremans, 2005), endogenous attention, monitoring, and executive control functions may all be related to underlying ACN dynamics guiding the integration of brain-scale activity patterns. The notion of *mindfulness*, derived from Buddhist texts and always more emphasized in cognitive and clinical psychological contexts (e.g., Cahn and Polich, 2006; Kabat-Zinn, 2003; Sumedho, 2004), might provide a unifying construct for endogenous attention, monitoring, and executive control functions. Indeed, OM meditation is also referred to as *mindfulness meditation* (Cahn and Polich, 2006).

In the adaptive workspace hypothesis, it would therefore be implied that in OM meditation, ACN neurons can dynamically link with multiple brain maps in parallel, such that a large repertoire of ACN firing patterns is available at any time, in resonance with such distributed maps. Related to this aspect, it has been shown that *differentiation*, together with *integration*, is a crucial aspect of large-scale neural processes related to conscious experience, in terms of intrinsic *neural complexity* and in *matching* external stimuli (Tononi et al., 1994, 1996; Tononi and Edelman, 1998). It can therefore be hypothesized that the complexity of ACN resonance repertoires increases with the expertise of OM meditators, probably as a capacity to reflect or match changing patterns of body and environmental states and associated "cognitive" and "affective" responses in distributed brain maps, in ACN firing constellations. Moreover, the fact that endogenous attention more easily rests in an open and nonselective state, i.e., *receptive* state, in (expert) OM meditation practitioners (Cahn and Polich, 2006), can be regarded as in line with a primary nonselective and open field characterization of top-down attention in the adaptive workspace hypothesis.

Specifically, we hypothesize that OM and the related conscious experience reflect the rapid formation and decay of dynamic cores which are assembled via dynamic links with ACN neurons. Metacognitive consciousness would be maintained or enhanced by signal exchanges within the ACN, *induced* by incoming signals from the widespread outer and inner content-related maps. As the goal-related consumer systems (see above) are de-emphasized in meditation, it can be hypothesized that a larger subset of intrinsic dynamic links are available both within the ACN and between the ACN and sensory maps, thus resulting in enhanced metacognitive consciousness as well as phenomenal and a more immediate access awareness of sensory contents. Moreover, any distraction in terms of entrainment in thinking during OM meditation would perturb activity coherence in the ACN, as correlated to distraction awareness and disengagement of ACN (e.g., neurons in medial and lateral BA10) from the distracting process. This monitoring process would be enhanced and more effortless in expert meditators. We will consider this aspect in more depth also with reference to FA meditation in the next subsection.

Related to the findings of Lutz et al. (2004), transient oscillatory coherence in the gamma band might play a crucial role in reversible binding and unbinding of dynamic cores (or proto-cores, see the next subsection) in OM meditation (see also Lutz et al., 2008). A more efficient binding (dynamic linking) and unbinding of neural assemblies encoding for serially presented targets with ACN neurons might explain the reduced AB observed by Slagter et al. (2007).

Focused attention meditation and adaptive workspace

As in OM meditation, goal-based consumer systems for conscious access are likely to be de-emphasized in FA meditation. Given that in FA meditation endogenous attention is focused on a chosen object in a sustained fashion, in our adaptive workspace framework it is hypothesized that driven by the task-setting encoded in the ACN itself, a relatively large subset of ACN neurons are dynamically linked with neurons explicitly encoding for the intended object by long-distance recurrent connectivity. Lateral Area BA10 might plausibly be involved in encoding the FA meditation task-setting and the intentional object-related bias, and medial Area BA10 in the sustained selective conscious access involving the sensory maps for the intended object.

According to our hypothesis, in FA meditation, a relative stability would characterize dynamic core (GW assembly) transitions, possibly in terms of overlapping neurons and ACN dynamic links recruited in subsequent cores and associated contexts (backstage neurons), as related to attentional stability (Wallace, 1999; see above).

Attentional vividness (acuity) could be mediated by the number of ACN neurons and dynamic links recruited in each core, as bound to the neural maps for the intended object.

The monitoring skills of noticing distractions, disengaging from distraction sources, and redirecting attention to the chosen object, would be mediated by dynamic links mostly involving the anterior cingulate cortex in novice practitioners, as related to effortful control, and a more automatic ACN coherence perturbation-based process in expert FA meditation practitioners. It can also be hypothesized that in expert FA meditators, transitions between subsequent dynamic cores are more rapid, with a relative integration between such cores, thus resulting in a reduced occurrence probability of distractions, whereas the anterior cingulate cortex activation would be needed to control transitions between dynamic cores in novices. The slow oscillations observed in the theta and alpha bands during FA meditation (Cahn and Polich, 2006) might then modulate transitions between dynamic cores. In particular, the power of the frontal midline theta, associated to anterior cingulate cortex activation, might correlate with the degree of effort during FA meditation.

During FA meditation, even if the endogenous attention focus is sustained on a chosen object, other events arising in sensory and thought-(feeling) related fields are typically noticed in "background" (Cahn and Polich, 2006; Lutz et al., 2008). In our adaptive workspace framework this can be explained in terms of *proto-cores*, i.e., transient resonant assemblies in the brain which may coexist with a dynamic core or GW assembly for "foreground" conscious access. A similar mechanism has been suggested by Block (2007) for phenomenal consciousness, in terms of loosing neuronal coalitions in posterior associative (e.g., parietal posterior) cortex, maintained in parallel with a winner-take-all GW coalition for access consciousness.

In adaptive workspace terms, such phenomenal consciousness proto-cores would be created as transient perturbations of ongoing ACN-linked activation and coherence patterns. The transience would be emphasized by the "switched-off" state of the consumer systems, and by endogenous attention and conscious access being oriented toward the intended meditation object by reentrant signaling between ACN neurons, selectively accessed sensory maps, and endogenous attention orienting neurons. A mechanism based on proto-cores might also be involved in OM meditation, as related to an open "background" awareness of rapidly changing experience contents (Lutz et al., 2008). Joint FA and OM meditation expertise, as in the practice of the Buddhist *insight meditation* integrating *Samatha* (FA) and *Vipassana* (OM) practice aspects, could then be reflected in an enhanced allocation of neural activity patterns (ACN dynamic links) to metacognitive consciousness (by intrinsic ACN signal exchanges), with a joint reduction of neural signaling (long-range signal exchanges with object-related maps) related to endogenous attention. A transition from phenomenal consciousness proto-cores to access consciousness cores would however be possible at any time, by endogenous attention driving an extended brain broadcasting process, as related to a complementary introspective or investigative stance in the (insight) meditation. Cognitive control processes (e.g., by inner speech or imagery) can be used to reiterate broadcasting in such a sustained conscious access (Baars, 1998).

Finally, as seen above "pure consciousness" (or third-order consciousness in our terminology) experiences during TM meditation are associated to global EEG patterns in the alpha band, which are characterized by a high coherence across frontal leads (see Arenander and Travis, 2004). In an adaptive workspace framework, such a frontal coherence can be related to an intrinsic ACN coherence, i.e., to the brain process that we have associated to the ongoing "awareness of being aware" in the present moment. The observed slow rhythm coherence might be supported by recurrent interactions between the ACN and aspecific thalamic nuclei (such as the intralaminar thalamic nucleus) with widespread cortical projections, possibly in terms of broadcasting of frequency and phase of a selected or emerging rhythm (Fries, 2005; VanRullen and Koch, 2003; Varela et al., 2001). The last idea would nicely fit with the earlier Luria's (1980) integrated working brain model, with reference to interactions between a first subcortical functional unit and a third prefrontal functional unit.

Conclusions and perspectives

The adaptive workspace hypothesis proposed in this paper emphasizes the role of neuronal populations with adaptive coding properties in prefrontal cortex for the emergence of consciousness. The adaptive coding model of prefrontal cortex function (Duncan, 2001; see also Duncan and Miller, 2002), however, needs adequate specification and more supporting evidence. We have also drawn ideas from the gateway hypothesis of anterior prefrontal cortex function (Burgess et al., 2005). A precise functional characterization of such a region, however, is still lacking. On the other hand, given the still relatively unspecified roles of adaptive prefrontal coding and of medial and lateral areas of anterior prefrontal cortex, the primary involvement of prefrontal cortex and its top-down connections in conscious access seems quite well established (Block, 2007; Dehaene et al., 2006).

It also remains to be seen how brain areas involved in momentary self-awareness (Damasio, 1999; Farb et al., 2007) interact with anterior prefrontal cortex and other areas showing adaptive coding responses. Moreover, in light of our adaptive workspace hypothesis it appears interesting to conduct a variant of Goldberg et al's. (2006) fMRI paradigm. In such a variant brain activations in a demanding visual categorization task and a momentary self-awareness (rather than narrative self-awareness) task condition (see Farb et al., 2007), are contrasted. Possibly, a group of OM meditators and a group of control subjects can be involved, to shed light on the role of interoceptive and exteroceptive brain areas in visual awareness.

The neural correlates of different aspects of phenomenal, access, and metacognitive consciousness as characterized in the adaptive workspace hypothesis might be evidenced in the context of FA and especially OM meditation settings. To this aim, the *neurophenomenology approach* (Lutz and Thompson, 2003; Varela, 1996) can be endorsed, with the participation of highly trained meditators capable of switching between different consciousness modes, and to direct attention to sensory or "internal" fields of experience. In the neurophenomenology approach quantitative measures of neural activity are combined with first-person data about the subject's inner experience. Participants' reports can thus be useful in identifying variability in brain activity from moment-to-moment; this unique information might guide the detection and interpretation of neural processes correlated to different aspects of conscious experience. Finally, large-scale computational models with biological and cognitive constraints (see Dehaene et al., 2003; Tononi et al., 1992) would also contribute to shed light on the neural mechanisms implied by the adaptive workspace hypothesis.

Acknowledgment

We would like to thank Dr. Gianluca Baldassarre and Prof. Gezinus Wolters for their very useful comments and suggestions.

References

Arenander, A., & Travis, F. T. (2004). Brain patterns of self-awareness. In B. Beitman & J. Nair (Eds.), *Self-awareness deficits*. New York: W.W. Norton.

Baars, B. J. (1983). Conscious contents provide the nervous system with coherent, global information. In R. J. Davidson, G. E. Schwartz, & D. Shapiro (Eds.), *Consciousness and self-regulation* (p. 41). New York: Plenum Press.

Baars, B. (1998). Metaphors of consciousness and attention in the brain. *Trends in Neurosciences*, *21*, 58–62.

Baars, B. J. (2002). The conscious access hypothesis: Origins and recent evidence. *Trends in Cognitive Science*, *6*, 47–52.

Baars, B. J., Ramsoy, T. Z., & Laureys, S. (2003). Brain, conscious experience and the observing self. *Trends in Neuroscience*, *26*, 671–675.

Block, N. (1995). On a confusion about a function of consciousness. *Behavioural and Brain Sciences*, *18*, 227–287.

Block, N. (2007). Consciousness, accessibility, and the mesh between psychology and neuroscience. *Behavioural and Brains Sciences*, *30*, 481–548.

Brefczynski-Lewis, J. A., Lutz, A., Schaefer, H. S., Levinson, D. B., & Davidson, R. J. (2007). Neural correlates of attentional expertise in long-term meditation practitioners. *Proceedings of the National Academy of Sciences of the United States of America*, *104*, 11483–11488.

Bundesen, C., Habekost, T., & Kyllingsbaek, S. (2005). A neural theory of visual attention: Bridging cognition and neurophysiology. *Psychological Review*, *112*, 291–328.

Burgess, P. W., Simons, J. S., Dumontheil, I., & Gilbert, S. J. (2005). The gateway hypothesis of rostral prefrontal cortex (area 10) function. In J. Duncan, P. McLeod, & L. Phillips (Eds.), *Measuring the mind: Speed, control and age* (pp. 215–246). Oxford: Oxford University Press.

Buschman, T. J., & Miller, E. K. (2007). Top-down versus bottom-up control of attention in the prefrontal and posterior parietal cortices. *Science*, *315*, 1860–1862.

Cahn, B. R., & Polich, J. (2006). Meditation states and traits: EEG, ERP, and neuroimaging studies. *Psychological Bulletin*, *132*, 180–211.

Carter, O. L. Presti, D., Callistemon, C., Liu, G. B., Ungerer, Y., & Pettigrew, J. D. (2005). Meditation alters perceptual rivalry in Tibetan Buddhist monks. *Current Biology*, *15*, R412–R413.

Christoff, K., & Gabrieli, J. D. E. (2000). The frontopolar cortex and human cognition: Evidence for a rostrocaudal hierarchical organization within the human prefrontal cortex. *Psychobiology*, *28*, 168–186.

Corbetta, M., & Shulman, G. L. (2002). Control of goal-directed and stimulus-driven attention in the brain. *Nature Review Neuroscience*, *3*, 201–215.

Craig, A. D. (2004). Human feelings: Why are some more aware than others? *Trends in Cognitive Sciences*, *8*, 239–241.

Critchley, H. D., Wiens, S., Rotshtein, P., Ohman, A., & Dolan, R. J. (2004). Neural systems supporting interoceptive awareness. *Nature Neuroscience*, *7*, 189–195.

Dalai Lama. (1994). *Transcendent wisdom: A teaching on the wisdom section of Shantideva's guide to the Bodhisattva way of life*. In: B. Alan Wallace (Trans., Ed., and Annot.). Ithaca, NY: Snow Lion.

Damasio, A. (1999). *The feeling of what happens: Body and emotion in the making of consciousness*. New York: Harcourt, Brace.

Davidson, R. J., & Goleman, D. J. (1977). The role of attention in meditation and hypnosis: A psychobiological perspective on transformations of consciousness. *International Journal of Clinical and Experimental Hypnosis*, *25*, 291–308.

Dehaene, S., Changeux, J. P., Naccache, L., Sackur, J., & Sergent, C. (2006). Conscious, preconscious, and subliminal processing: A testable taxonomy. *Trends in Cognitive Sciences*, *10*, 204–211.

Dehaene, S., Kerszberg, M., & Changeux, J. P. (1998). A neuronal model of a global workspace in effortful cognitive tasks. *Proceedings of the National Academy of Sciences of the United States of America*, *95*, 14529–14534.

Dehaene, S., & Naccache, L. (2001). Towards a cognitive neuroscience of consciousness: Basic evidence and a workspace framework. *Cognition*, *79*, 1–37.

Dehaene, S., Sergent, C., & Changeux, J.-P. (2003). A neuronal network model linking subjective reports and objective physiological data during conscious perception. *Proceedings of the National Academy of Sciences (USA)*, *100*, 8520–8525.

Desimone, R., & Duncan, J. (1995). Neural mechanisms of selective visual attention. *Annual Review of Neuroscience*, *18*, 193–222.

Duncan, J. (1984). Selective attention and the organization of visual information. *Journal of Experimental Psychology: General*, *113*, 501–517.

Duncan, J. (2001). An adaptive coding model of neural function in prefrontal cortex. *Nature Reviews Neuroscience*, *2*, 820–829.

Duncan, J., Humphreys, G. W., & Ward, R. (1997). Competitive brain activity in visual attention. *Current Opinion in Neurobiology*, *7*, 255–261.

Duncan, J., & Miller, E. K. (2002). Cognitive focusing through adaptive neural coding in the primate prefrontal cortex. In D. Stuss & R. T. Knight (Eds.), *Principles of frontal lobe function*. Oxford University Press.

Farb, N. A. S., Segal, Z. V., Mayberg, H., Bean, J., McKeon, D., Fatima, Z., & Anderson, A. K. (2007). Attending to the present: meditation reveals distinct neural modes of self-reference. *Social Cognitive and Affective Neuroscience*, *2*, 313–322.

Fries, P. (2005). A mechanism for cognitive dynamics: Neuronal communication through neuronal coherence. *Trends in Cognitive Sciences*, *9*, 474–480.

Fuster, J. M. (1997). *The prefrontal cortex. Anatomy, physiology, and neuropsychology of the frontal lobe*. Philadelphia, PA: Lippincott-Raven.

Gaillard, R., Dehaene, S., Adam, C., Clémenceau, S., Hasboun, D., et al. (2009). Converging intracranial markers of conscious access. *PLoS Biology*, *7*, e1000061. 10.1371/journal.pbio.1000061.

Gazzaniga, M. S. (1985). *The social brain*. New York: Basic Books.

Goldberg, I. I., Harel, M., & Malach, R. (2006). When the brain loses its self: Prefrontal inactivation during sensorimotor processing. *Neuron*, *50*, 329–339.

James, W. (1890). *The principles of psychology* (Vol. I). New York: Dover.

Kabat-Zinn, J. (2003). Mindfulness-based interventions in context: Past, present, and future. *Clinical Psychology: Science and Practice*, *10*, 144–156.

Kant, I. (1781/1996). Kritik der reinen Vernunft. In W. S. Pluhar (Trans.), *Critique of pure reason*. Indianapolis, IN: Hackett.

Lamme, V. (2003). Why attention and awareness are different. *Trends in Cognitive Sciences*, *7*, 12–18.

Lati Rinbochay. (1981). *Mind in Tibetan Buddhism*. In E. Napper (Trans. and Ed.). Valois: Gabriel/Snow Lion.

Lazar, S. W., Kerr, C. E., Wasserman, R. H., et al. (2005). Meditation experience is associated with increased cortical thickness. *Neuroreport*, *16*, 1893–1897.

Le Van Quyen, M. (2003). Disentangling the dynamic core: A research program for neurodynamics at the large scale. *Biological Research*, *36*, 67–88.

Llinas, R., et al. (1998). The neuronal basis for consciousness. *Philosophical Transactions of the Royal Society of London B: Biological Science*, *353*, 1841–1849.

Luria, A. R. (1980). *Higher cortical functions in man*. Dordrecht: Kluwer.

Lutz, A., Greischar, L., Rawlings, N. B., Ricard, M., & Davidson, R. J. (2004). Long-term meditators self-induce high-amplitude synchrony during mental practice. *Proceedings of the National Academy of Sciences*, *101*, 16369–16373.

Lutz, A., Slagter, H. A., Dunne, J. D., & Davidson, R. J. (2008). Attention regulation and monitoring in meditation. *Trends in Cognitive Sciences*, *12*, 163–169.

Lutz, A., & Thompson, E. (2003). Neurophenomenology. *Journal of Consciousness Studies*, *10*, 31–52.

Maia, T. V., & Cleeremans, A. (2005). Consciousness: Converging insights from connectionist modeling and neuroscience. *Trends in Cognitive Sciences*, *9*, 397–404.

Milner, A. D., & Goodale, M. A. (2008). Two visual systems re-viewed. *Neuropsychologia*, *46*, 774–785.

Northoff, G., Heinzel, A., de Greck, M., Bermpohl, F., Dobrowolny, H., & Panksepp, J. (2006). Self-referential processing in our brain: A meta-analysis of imaging studies on the self. *NeuroImage*, *31*, 440–457.

Poeppel, E. (1997). A hierarchical model of temporal perception. *Trends in Cognitive Sciences*, *1*, 56–61.

Posner, M. I., & Petersen, S. E. (1990). The attention system of the human brain. *Annual Review of Neuroscience*, *13*, 25–42.

Raffone, A., & van Leeuwen, C. (2003). Dynamic synchronization and chaos in an associative neural network with multiple active memories. *Chaos*, *13*, 1090–1104.

Raymond, J. E., Shapiro, K. L., & Arnell, K. M. (1992). Temporary suppression of visual processing in an RSVP task: An attentional blink? *Journal of Experimental Psychology: Human Perception and Performance*, *18*, 849–860.

Rodriguez, E., George, N., Lachaux, J. P., Martinerie, J., Renault, B., & Varela, F. (1999). Perception's shadow: Long-distance synchronization in the human brain. *Nature*, *397*, 340–343.

Rumelhart, D. E., et al. (1986). Schemata and sequential thought processes in PDP models. In J. L. McClelland, et al. (Eds.), *Parallel distributed processing: Explorations in the microstructure of cognition* (Vol. 2, pp. 7–57). MIT Press.

Shantideva. (1971). *Siksa-samuccaya: A compendium of Buddhist doctrine*. In C. Bendall & W. H. D. Rouse (Trans. from Sanskrit). Delhi: Motilal Banarsidass.

Shantideva. (1997). *A guide to the Bodhisattva way of life*. In Vesna A. Wallace & B. Alan Wallace (Trans.). Ithaca, NY: Snow Lion.

Simons, D. J., & Chabris, C. F. (1999). Gorillas in our midst: Sustained inattentional blindness for dynamic events. *Perception*, *28*, 1059–1074.

Slagter, H. A., Lutz, A., Greischar, L. L., Francis, A. D., Nieuwenhuis, S., Davis, J. M., et al. (2007). Mental training affects distribution of limited brain resources. *PLoS Biology*, *5*, e138.

Srinivasan, N., & Baijal, S. (2007). Concentrative meditation enhances pre-attentive processing: A mismatch negativity study. *Neuroreport*, *18*, 1709–1712.

Srinivasan, N., Srivastava, P., Lohani, M., & Baijal, S. (2009). Focused and distributed attention. In N. Srinivasan (Ed.), *Progress in brain research: Attention* (Vol. 176). Amsterdam: Elsevier.

Sternberg, R. J. (2008). *Cognitive psychology* (5th ed.). Belmont, CA: Cengage.

Sumedho, A. (2004). *Intuitive awareness*. Hemel Hempstead, UK: Amaravati Buddhist Monastery.

Tang, Y., Ma, Y., Wang, J., Fan, Y., Feng, S., Lu, Q., et al. (2007). Short-term meditation training improves attention and self-regulation. *Proceedings of the National Academy of Sciences of the United States of America*, *104*, 17152–17156.

Thompson, E., & Varela, F. (2001). Radical embodiment: Neural dynamics and consciousness. *Trends in Cognitive Sciences*, *5*, 418–425.

Tononi, G., & Edelman, G. M. (1998). Consciousness and complexity. *Science*, *282*, 1846–1851.

Tononi, G., Sporns, O., & Edelman, G. M. (1992). Reentry and the problem of integrating multiple cortical areas: Simulation of dynamic integration in the visual system. *Cerebral Cortex*, *2*, 310–335.

Tononi, G., Sporns, O., & Edelman, G. M. (1994). A measure for brain complexity: Relating functional segregation and integration in the nervous system. *Proceedings of the National Academy of Sciences of the United States of America*, *91*, 5033–5037.

Tononi, G., Sporns, O., & Edelman, G. M. (1996). A measure for the selective matching of signals by the brain. *Proceedings of the National Academy of Sciences of the United States of America*, *93*, 3422–3427.

Travis, F., & Wallace, R. K. (1999). Autonomic and EEG patterns during eyes-closed rest and transcendental meditation (TM) practice: The basis for a neural model of TM practice. *Consciousness and Cognition*, *8*, 302–318.

Valentine, E. R., & Sweet, P. L. G. (1999). Meditation and attention: A comparison of the effects of concentrative and mindfulness meditation on sustained attention. *Mental Health, Religion and Culture*, *2*, 59–70.

VanRullen, R., & Koch, C. (2003). Is perception discrete or continuous? *Trends in Cognitive Sciences*, *7*, 207–213.

Varela, F. (1995). Resonant cell assemblies: A new approach to cognitive functioning and neuronal synchrony. *Biological Research*, *28*, 81–95.

Varela, F. (1996). Neurophenomenology: A methodological remedy for the hard problem. *Journal of Consciousness Studies*, *3*, 330–349.

Varela, F. J., Lachaux, J.-P., Rodriguez, E., & Martinerie, J. (2001). The brainweb: Phase synchronization and large-scale integration. *Nature Reviews Neuroscience*, *2*, 229–239.

Varela, F. J., & Thompson, E. (2003). Neural synchrony and the unity of mind: A neurophenomenological perspective. In A. Cleeremans (Ed.), *The unity of consciousness: Binding, integration, and dissociation* (pp. 266–287). Oxford University Press.

Vasubandhu. (1991). *Abhidharma Kosabhasyam*. Louis de La Vallée Poussin (French Trans.), Leo M. Pruden (English Trans.). Berkeley, CA: Asian Humanities Press.

Von der Malsburg, C. (1981). *The correlation theory of brain function*. Internal Report 81-2. MPI für Biophysikalische Chemie, Göttingen.

Von der Malsburg, C. (1999). The what and why of binding: The modeler's perspective. *Neuron*, *24*, 95–104.

Von Rospatt, A. (1995). *The Buddhist doctrine of momentariness: A survey of the origins and early phase of this doctrine up to Vasubandhu*. Stuttgart: Franz Steiner Verlag.

Wallace, A. (1999). The Buddhist tradition of Samatha: Methods for refining and examining consciousness. *Journal of Consciousness Studies*, *6*, 175–187.

Weissman, D. H., Roberts, K. C., Visscher, K. M., & Woldorff, M. G. (2006). The neural bases of momentary lapses in attention. *Nature Neuroscience*, *9*, 971–978.

Zeki, S. (2003). The disunity of consciousness. *Trends in Cognitive Sciences*, *7*, 214–218.

N. Srinivasan (Ed.)
Progress in Brain Research, Vol. 176
ISSN 0079-6123

CHAPTER 12

Cognitive maps and attention

Oliver Hardt[1,*] and Lynn Nadel[2]

[1]*Department of Psychology, McGill University, Montreal, Quebec, Canada*
[2]*Department of Psychology, University of Arizona, Tucson, AZ, USA*

Abstract: Cognitive map theory suggested that exploring an environment and attending to a stimulus should lead to its integration into an allocentric environmental representation. We here report that directed attention in the form of exploration serves to gather information needed to determine an optimal spatial strategy, given task demands and characteristics of the environment. Attended environmental features may integrate into spatial representations if they meet the requirements of the optimal spatial strategy: when learning involves a cognitive mapping strategy, cues with high codability (e.g., concrete objects) will be incorporated into a map, but cues with low codability (e.g., abstract paintings) will not. However, instructions encouraging map learning can lead to the incorporation of cues with low codability. On the other hand, if spatial learning is not map-based, abstract cues can and will be used to encode locations. Since exploration appears to determine what strategy to apply and whether or not to encode a cue, recognition memory for environmental features is independent of whether or not a cue is part of a spatial representation. In fact, when abstract cues were used in a way that was not map-based, or when they were not used for spatial navigation at all, they were nevertheless recognized as familiar. Thus, the relation between exploratory activity on the one hand and spatial strategy and memory on the other appears more complex than initially suggested by cognitive map theory.

Keywords: visual extinction; visual attention; biased competition; visual grouping; action

Traditional learning theory (e.g., Rescorla and Wagner, 1972; Mackintosh, 1975) supposes that all forms of learning obey the same fundamental laws of association. A basic principle underlying these laws is that learning depends upon predictability — when a given outcome is already fully predicted by stimuli currently available in the environment, learning about stimuli newly added to the environment is said to be "blocked" (Kamin, 1969), i.e., the new stimuli will not acquire control over the animal's behavior. A challenge to this assertion comes from cognitive map theory (O'Keefe and Nadel, 1978), which puts emphasis on the idea that organisms automatically update their internal representations of environments (their "cognitive maps") when something novel is observed. Cognitive map theory posits two major forms of learning: "locale" learning, which is thought to be dependent on cognitive maps and the neural systems underpinning these maps, while "taxon" learning is dependent on other, non-maplike, representations and their associated neural systems, which are different from the brain areas supporting the

*Corresponding author.
Tel.: +1 514 398 3167; Fax: +1 514 398 4896;
E-mail: oliver.hardt@mac.com

DOI: 10.1016/S0079-6123(09)17610-0

representation of cognitive maps. The updating assumption of cognitive map theory predicts that when an organism recruits locale learning, blocking will not occur. On the other hand, when taxon systems are involved, in which learning is assumed to obey standard associative rules, blocking should be observed in spatial learning.

This prediction from cognitive map theory raises issues about possible interactions between spatial learning and attention. While it can be readily assumed that one cannot learn about something that has not been attended, it is an open question as to whether one learns about something that has been attended. The blocking phenomenon seems to raise the possibility that an animal could attend to a given stimulus, yet not learn what it would have learned about it had the animal been naïve. For example, the stimulus might not acquire control over the animal's subsequent behavior if the animal encounters it later in its learning history.

The question of whether or not blocking is observed in spatial learning is thus of some interest in considering the relation between attention and learning. This question has been the focus of a substantial number of studies in animals, but with inconclusive results to date (Biegler and Morris, 1999; Brown et al., 2002; Chamizo et al., 1985; Hayward et al., 2003; Pearce et al., 2001; Rodrigo et al., 1997; Sanchez-Moreno et al., 1999). One factor that appears to influence the presence or absence of blocking is the nature of the cues: geometrical cues do not seem to yield the same outcome as landmarks (Doeller and Burgess, 2008; Graham et al., 2006; Gray et al., 2005; Hayward et al., 2004; Pearce et al., 2004; Wall et al., 2004; but see Horne and Pearce, 2009, for an important caveat).

We have recently approached this question in a series of studies in humans, using a computer-generated version of the well-known watermaze task for rodents (Jacobs et al., 1997; Jacobs, et al., 1998). The watermaze task, first developed by Morris (1981), is particularly useful in assessing spatial learning because it requires the animal to locate a hidden target (a platform below the surface of a large tank filled with water made opaque with milk-like powder). The only cues available to help locate this platform are typically at a considerable distance, thereby forcing the animal to use these distal cues to navigate to an otherwise unmarked place. In rodents this task typically calls into play the development of a cognitive map, and ought to provide a situation in which one can test the assertion that blocking will (or will not) be observed in this kind of spatial learning. Studying spatial learning in the watermaze task in humans has the advantage that one can ask questions about how instructions and cue salience affect attention. This chapter will review a series of experiments using our computer implementation of the watermaze, which demonstrate that spatial learning has special properties and suggest that a complex bidirectional relationship exists between spatial learning strategy choice and attention.

A challenge to cognitive map theory

We begin with what appears to be a direct test of the very question we raise above: is blocking observed in human spatial learning? Hamilton and Sutherland (1999) developed a computer version of the watermaze that was similar in many respects to the one developed by Jacobs et al. (1997, 1998). They used this maze to test human subjects in a standard blocking experiment modeled carefully after what one does in the animal lab: virtually no instructions were given to the subjects, beyond being told to search for a hidden platform, which would become visible only when the subject arrived at its location in the maze. Their subjects received a series of training trials in which the platform's location was fixed relative to various distal cues presented on the four walls of the computer-generated maze. During the initial 20 training trials four distal cues were presented (cues 1, 2, 3, 4; one cue on each wall), and during the subsequent 12 compound trials an additional four cues (5, 6, 7, 8) were added to the walls of the computer maze (one cue to each wall). Following these compound trials, a "probe" trial was given on which only the added cues (5–8) were available, and the target platform was removed unknown to the subjects — if subjects learned the location of the target, they should spend more time looking for it in the area where it was located than elsewhere in the maze. It is assumed that if blocking has occurred during the compound trials, the relationship between the newly added cues and the target platform was not acquired and subjects will not be able to use them to find the hidden platform. And that indeed is what Hamilton and Sutherland observed, leading them to conclude that blocking does occur in allocentric spatial learning, i.e., when subjects are using cognitive maps, contrary to the prediction of cognitive map theory.

We were surprised by this result because we had already been using our own computer-generated watermaze to ask essentially the same question regarding what type of learning governs the acquisition and use of cognitive maps, and our preliminary results suggested that blocking would not be observed in this task. Given the differences between our computer implementation and the one used by Hamilton and Sutherland, we decided to explore their maze in some detail, asking whether subjects had indeed engaged in creating and using cognitive maps in their computer-generated environment and training protocol. Their assertions about the rules governing cognitive map learning only apply if their apparatus indeed invoked such learning. This is a relevant issue because subjects (both rats and humans) can solve this task in many ways, only some of which involve the formation and use of a cognitive map (Nadel and Hardt, 2004). For example, rats in some cases have been shown to swim around the water tank at a relatively fixed distance from the wall, thereby maximizing their chances of finding a platform located at that distance from the wall.

One way to determine if subjects are using a cognitive map is to study how flexibly they use the cues present in the environment. O'Keefe and Nadel (1978) pointed out that maps have the property of allowing for the interchangeable use of all the cues in the environment: no specific cues are essential, but some minimal subset is necessary. This property of map use predicts that after an animal (or human) has learned about a space, any subset of the available cues could be deleted with little or no impact on performance, so long as the minimal number remained. Jacobs et al. (1998) tested this prediction in their computer maze and confirmed that, in that maze, performance could be supported by any two of the cues.

Critically, Hamilton and Sutherland (1999) did not run these kinds of cue elimination probes to ask whether or not their subjects were actually forming and using cognitive maps in their apparatus. To check on this possibility we replicated their maze using our software (original and our replication could not be distinguished by raters blind to the maze origin) and ran the critical elimination experiment using the same instructions and procedures Hamilton and Sutherland did in the original study. As seen in Fig. 1, our subjects were trained with all eight cues on the walls of the apparatus (i.e., like Hamilton and Sutherland's subjects during compound training). After the training trials, we composed a control group and four different experimental groups for the critical probe trial. No cues were removed for the control group, which thus was tested with all eight cues that had been present throughout all of training. However, each of the experimental groups had four of the cues removed, as seen in the figure: if the subjects were using a cognitive map, their performance in all these conditions would be good, as an adequate number of cues was always available.

On the critical probe trial the control group, as expected, showed a strong preference for searching in the target quadrant. As Fig. 2 shows, a difference emerged among the various experimental groups. The subjects tested after deletion of the cues on the walls nearest the target quadrant performed at chance levels. That is, they seemed incapable of determining where the platform had been located. In contrast, the subjects tested after deletion of any of the other cue subsets were well above chance in searching the target quadrant. This pattern of results strongly suggests that the subjects in this apparatus, with minimal instructions, were not forming and using cognitive maps to solve the task. Instead, they seemed to be relying mostly on the cues on the walls closest to the target. Given this, the results reported by Hamilton and Sutherland had no bearing on whether blocking would be seen in the context of cognitive map learning, as subjects most likely did not create such a representation

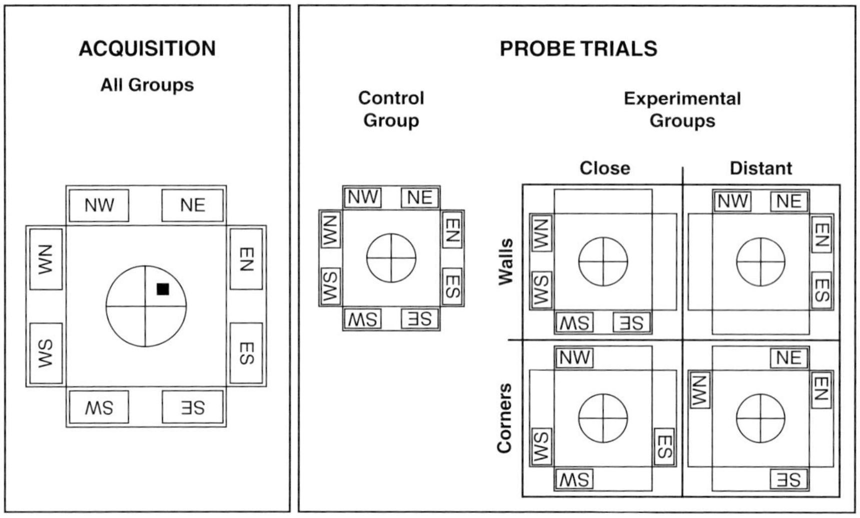

Fig. 1. Protocol used in the cue deletion experiment. During all 32 training trials, two cues were present on each wall. Rats were then divided into five groups, and two probe trials were administered. The Control group was presented with the same cues as during training. The other four groups lost half of the cues as shown in the figure.

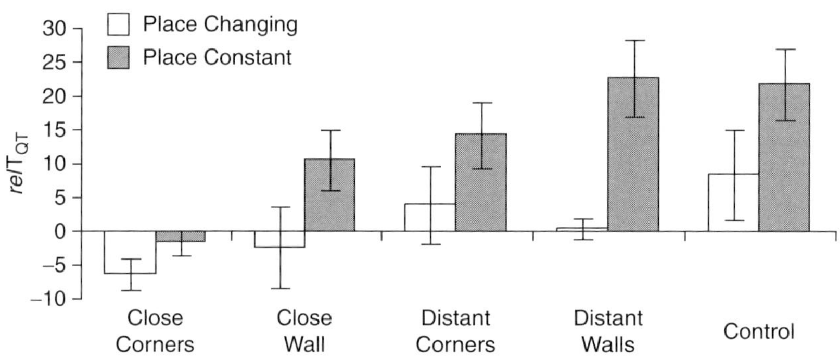

Fig. 2. Search preference for the target in the correct quadrant. Performance plotted for each group by the "Place Hypothesis" subjects formed during training. Each subject's Place Hypothesis was assessed after the last probe trial by asking whether the target remained in the same position or whether the position changed during training. Those who believed that the target remained stationary were classified as "Place Constant," while those who came to (wrongly) believe that the target moved between trials were classified as "Place Changing." In all studies reported here, Place Hypothesis explains a large amount of variance; only those subjects who reported that the target was maintaining a constant location from trial to trial showed knowledge of the target location on the probe trials, but perhaps more interesting, only the "Place Constant" subjects showed learning at all. In this and subsequent figures, performance in the standard probe trial, in which the target is removed without informing subjects about its absence, is represented in a single score. This score to index the relative time in the target quadrant (relT_{qT}), was calculated by subtracting 25% (which accounts for chance) from the percent of time spent in the target quadrant during the entire probe trial:

$$\mathrm{rel}T_{qt} = \left(\frac{t_{qT}}{L} \times 100\right) - 25 \quad \text{with } L = t_{qT} + t_{q\mathrm{Left}} + t_{q\mathrm{Right}} + t_{q\mathrm{Opposite}}$$

in their implementation of the watermaze task. That some cues were less likely to be associated with the target location in this experiment could reflect overshadowing. Overshadowing is a well-documented associative effect of learning suppression, which can be observed when stimuli of different salience are presented together in conditions trials. It is diagnosed when less salient stimuli fail to acquire behavioral control, while the salient ones do. It is possible that the cues in the corners overshadowed the ones further away so that removal of the former reveals the behavioral deficit. If this were the case, it further corroborates our conclusion, that the human subjects in this specific watermaze implementation did not recruit cognitive map-based strategies, and that is the reason why blocking was observed. Jacobs et al. (1997) failed to detect overshadowing in their implementation of the watermaze task. These divergent results can be explained by cognitive map theory, which assumes that two classes of spatial behavior are available. Cognitive map learning represents an instance of the class of locale strategies alone, which employs learning mechanism different from associative learning, so that blocking, overshadowing, and similar associative effects will not be observed. Spatial behaviors supported by the taxon system, however, recruit these kinds of learning mechanisms, and, consequently, associative learning effects can be observed. The question now is why cognitive maps are not generated in the watermaze task variant used by Hamilton and Sutherland, a basic task that subjects (rats and humans) normally solve by using locale learning and memory?

To establish that indeed our computer-generated watermaze leads to cognitive map learning, we asked whether we would observe blocking as Hamilton and Sutherland did. We trained subjects in this maze, using instructions that encouraged exploration of the environment. There were basically two kinds of groups (see Table 1): one group received training with cues on two of the four walls for the first eight trials (i.e., Cue-Set S_1), which were followed by eight compound training trials with cues on all four walls (i.e., Cue-Set S_1+Cue-Set S_2). During the subsequent probe trial only the cues added during compound training were available (i.e., Cue-Set S_2). If learning about these cues had been "blocked" then the performance of these groups should be at chance. A control group received no initial training but the same compound training with cues on all four walls (i.e., Cue-Set S_1+S_2 for eight trials), as well as the probe trial with cues on only two of the four walls (i.e., Cue-Set S_2). We expected good performance for all groups, i.e., the absence of a blocking effect.

Table 1. Design of blocking study

	Acquisition		
Group	Phase I	Phase II	Probes
Blocking	S_1[a]	$S_1 \cup S_2$[b]	S_2
Control		$S_1 \cup S_2$	S_2

Notes: Six different cue-sets were used, each consisting of the cues from two walls. All possible set-combinations were used for Blocking and Control groups. The cues in each set could be in one of two spatial relations: They were either on walls (a) opposite each other (e.g., North–South), or (b) adjacent to each other (e.g., Northwest). To obtain the same number of observations for each of these two cases, the number of subjects for the conditions in which the walls were adjacent to each other was half the number of subjects for walls opposite each other (i.e., four and eight, respectively).

[a] $S_1 \in \{(N, S), (W, E), (N, E), (N, W), (S, E), (S, W)\}$.

[b] $S_2 \in \{N, E, S, W\} \cap S_1$.

As Fig. 3 shows, subjects in all our groups indeed performed well on the probe trial: they searched for the target where it had been located throughout training using either only the cues available from the beginning or only the cues added during compound training. In other words, we did not find a blocking effect.

These results suggest that when subjects engage in cognitive map learning, they will not show blocking when new cues are introduced after original learning. As such, the data support the premise of cognitive map theory (O'Keefe and Nadel, 1978) that when new cues are added to familiar situations subjects will learn about them. They also raise the question of why such learning does not occur when subjects do not form a cognitive map, as in the Hamilton and Sutherland (1999) study? Is it possible that in this study they failed to notice, or attend to, the cues that were added after initial training?

To test this possibility, we used their computer maze, but with the instructions that we had used in our watermaze instead, in which subjects are encouraged to explore the environment more thoroughly. Specifically, subjects were told: "The target will always be in the same location. The pictures on the wall of the room will help you to find the target. So have a good look around each time you find the target — that may help you to

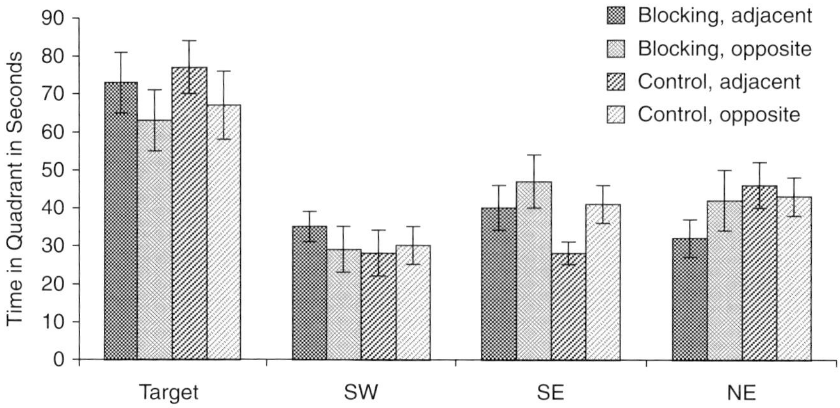

Fig. 3. Performance in the search persistence probe trial for control and blocking groups. Persistence of search for the target was greater in the target quadrant than in any other quadrant for all groups, indicating target location acquisition in all groups, i.e., the absence of a blocking effect.

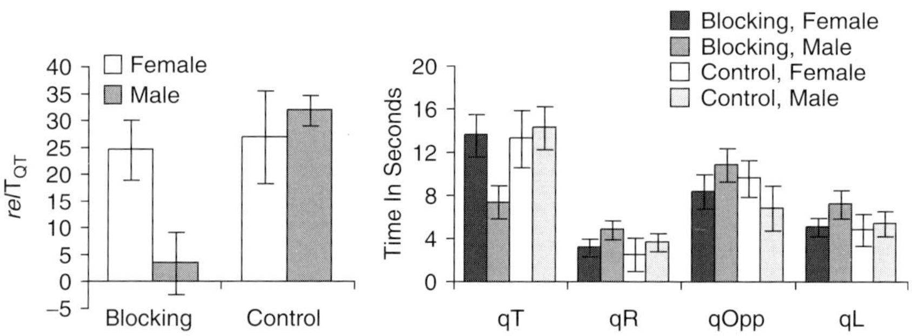

Fig. 4. Performance on the search persistence probe trial. Only males show a profound blocking effect, which is absent in females. The right graph shows time spent in all quadrants during this probe trials and reveals that males in the blocking group search almost all quadrants equally long for the target.

quickly find the target again in the next trial.'' Once again we used two groups, replicating exactly the original study by Hamilton and Sutherland (1999): a control group that only received compound training and was only trained with two cues available on all four walls, and an experimental group that was initially trained with half the cues available, and then with the full set of four cues during compound training. Both groups were then given two kinds of probe trials after training: we started with the standard probe trial as described above, which was followed by a new spatial assessment procedure, the ''location accuracy probe trial''. Before this probe trial, subjects were debriefed, i.e., told that the target had been removed and that the task was simply to go to the place in the maze where they thought the target had been located. This kind of probe trial is a more objective measure of spatial learning than the standard ''search'' probe trial. On the standard search probe one uses search persistence as the measure of spatial knowledge, but such persistence can reflect either the strength of the subject's spatial knowledge or the speed with which it extinguishes once it is discovered that the target is not where the subject thought.

On the standard probe trial we observed a different result in males and females: males showed blocking but females did not (see Fig. 4). However, on the more objective location probe neither group showed blocking (Fig. 5).

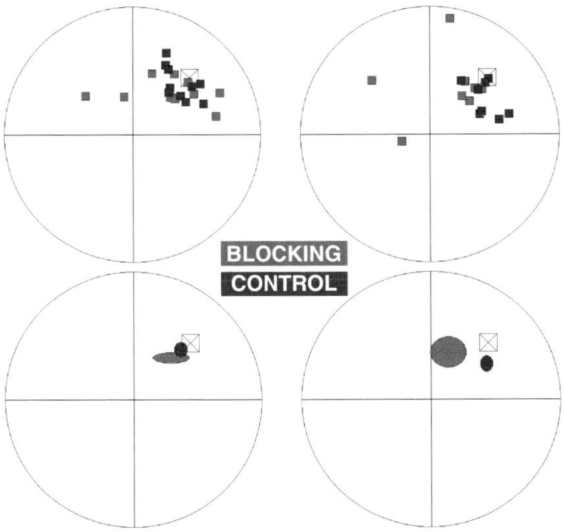

Fig. 5. Performance on the location accuracy probe trial. The blocking effect for males found in the search persistence probe trial is absent. Both males and females located the target in the actual quadrant, close to its actual location. Although females were more accurate than males in recalling the target location, both males and females knew in which quadrant the target was located and in about what area within this quadrant. Since the two probe trials were administered in succession, and since the actual target position was never revealed during the probe trials nor between them, male subjects must have known the target location during the search persistence probe trial as well, however, for yet unknown reasons they did not continue to exclusively search the target quadrant for it. Upper graphs show location recall for each subject, while lower graphs represent the average of the recalled locations for each group in form of an ellipse, the center of which denotes the arithmetic mean, while major and minor axis represent the standard deviation of the *x* and *y* coordinates of the recalled target positions, respectively. (See Color Plate 12.5 in color plate section.)

These results suggest that when subjects are encouraged to explore their environment, they are more likely to form cognitive maps, and that having done so, they will incorporate cues added to the environment after initial training. They also hint at interesting differences between males and females, although both showed an absence of blocking in at least one of the probe conditions. Exactly why the males showed blocking in the search probe but not in the location probe is unclear, but their performance in the search probe shows that they clearly learned about the added cues, as predicted by cognitive map theory. Note that there were no interpolated learning trials between the search persistence probe trial, in which they showed blocking, and the location accuracy probe trial, in which such blocking was absent. Possibly, males were more confident than females about the platform location, and terminated their search shortly after they could not find it, thus searching for it in the other quadrants. Although we cannot corroborate this hypothesis empirically by, for example, analyses of the initial path during this probe trial (our software did not permit this), the learning curve of male subjects lends some support for this possible interpretation. Males reached asymptote during learning much earlier than females, and thus were able to quickly and reliably locate the target from early on (data not shown). As a consequence, they might have started to search for the target elsewhere, when they were not able to locate it rapidly during the first probe trial.

To further assess what knowledge subjects acquired during training we added a recognition test in which subjects were asked if they recognized the cues, that is, the ones added during compound training as well as the ones available from the beginning. Both males and females showed hit rates that were well above chance for both types of cues, showing once again that the apparent blocking observed in the males on the search probe did not reflect a failure to attend to and even learn about the added cues.

Cue codability

As noted earlier, cognitive map theory makes assumptions about learning rates. It assumes that maps are formed rapidly, often in a single trial. In contrast, many associative learning accounts assume that such spatial learning is incremental, developing slowly over trials. This presumed difference might be another factor explaining why we obtained different results in the two computer mazes. Rapid learning follows from the ability of the subject to quickly form a representation of the to-be-learned cues. When this is not possible, rapid learning might prove difficult, and map formation could suffer. This matters in our experiments because of the nature of the cues placed on the walls of the two mazes. Hamilton and Sutherland (1999) used cues that were, by and large, rather abstract, and difficult if not impossible to name. This might have worked against map formation. Our maze, in contrast, used cues that were mostly nameable, hence easier to code and represent.

We decided to explore this factor of cue codability directly to see how it interacted with exploration and attention in the computer-generated maze. We started by asking if subjects would form maps, and hence show an absence of blocking, if tested with highly concrete cues but no explicit instructions to explore. This latter condition seems to have worked against map formation in the original Hamilton and Sutherland study, which employed largely abstract cues.

Figure 6 shows the kinds of cues we used in our first study exploring this issue. Each cue depicted a line drawing of a familiar object. The design of the study was straightforward: Subjects in the Blocking group were initially trained with four of the eight cues for 20 trials (one cue on each wall, just as in Hamilton & Sutherland's experiment, i.e., eiher Cue-Set A or B), then four more cues were added during the 12 compound training trials (such that two cues were now on each wall, i.e., subjects were now trained with Cue-Set A+B). The Control group was trained for 32 trials with the full set of cues (i.e., with Cue-Set A+B). Then both groups received a search probe trial first and a location probe during which only the cues added during the Blocking group's

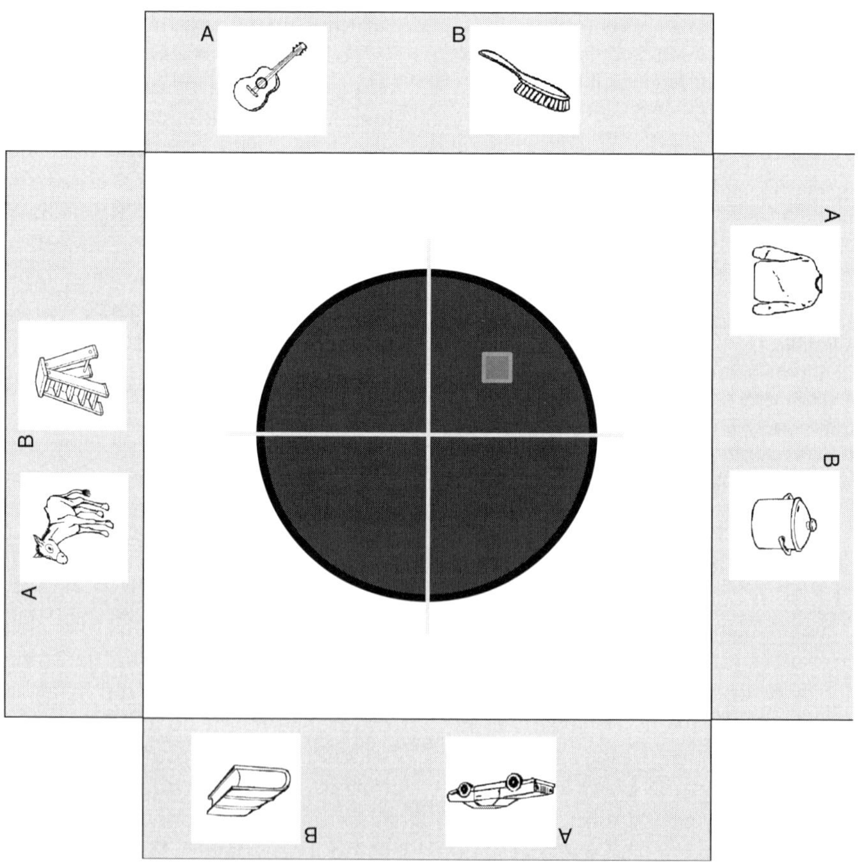

Fig. 6. Schematic of the computer-generated environment (from a bird's-eye perspective). Note that subjects perceived the maze from a first-person perspective, that is, simulated three-dimensional space, and that this two-dimensional schematic depicts only the layout and the relative positions of the cues at the walls. The letters (A and B) indicate to which cue-set the cues belong.

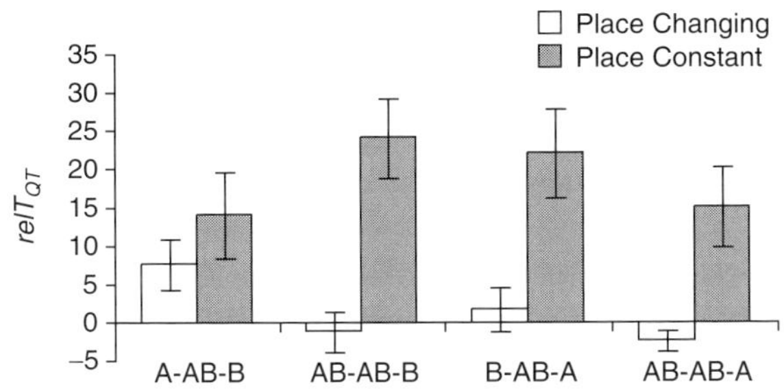

Fig. 7. Search persistence probe trial shows that blocking (A-AB-B and B-AB-A) and control groups (AB-AB-B and AB-AB-A) predominantly search the target quadrant. Group differences are absent, indicating that the preference was the same for all groups. As before, Place Hypothesis predicts spatial knowledge, as subjects who believed that the target's position changed during training performed worse than subjects who believed otherwise.

compound training were available (i.e., either Cue-Set A or B, depending on which set was added during compound training).

In both the search persistence and the location accuracy probes, an absence of blocking was observed. Subjects in both the Blocking and Control groups persisted in searching the target quadrant during the search probe trial. They also located the target in the correct quadrant in both groups (see Figs. 7 and 8).

In this study, as in most of our other studies, we saw two kinds of subjects: those who reported that the target was changing location from trial to trial and those who reported that it was maintaining a constant location. Only the latter subjects showed knowledge of the target location

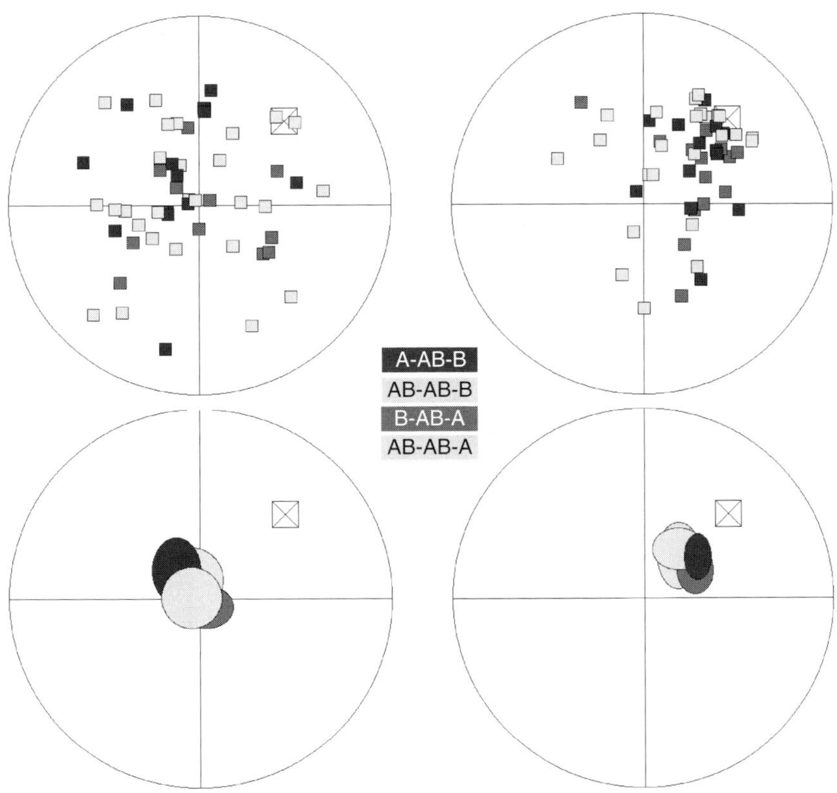

Fig. 8. Performance during the location accuracy probe trial reveals that location knowledge is the same for all groups (see panels on right hand, showing recall performance of subjects who believed that the target remained in the same location during training). The panels on the left hand show location recall of subjects who believed that the target location changed during training: for them, all quadrants were equally plausible locations for the target. Upper panels show recalled locations for all individuals, lower panels the group averages. (See Color Plate 12.8 in color plate section.)

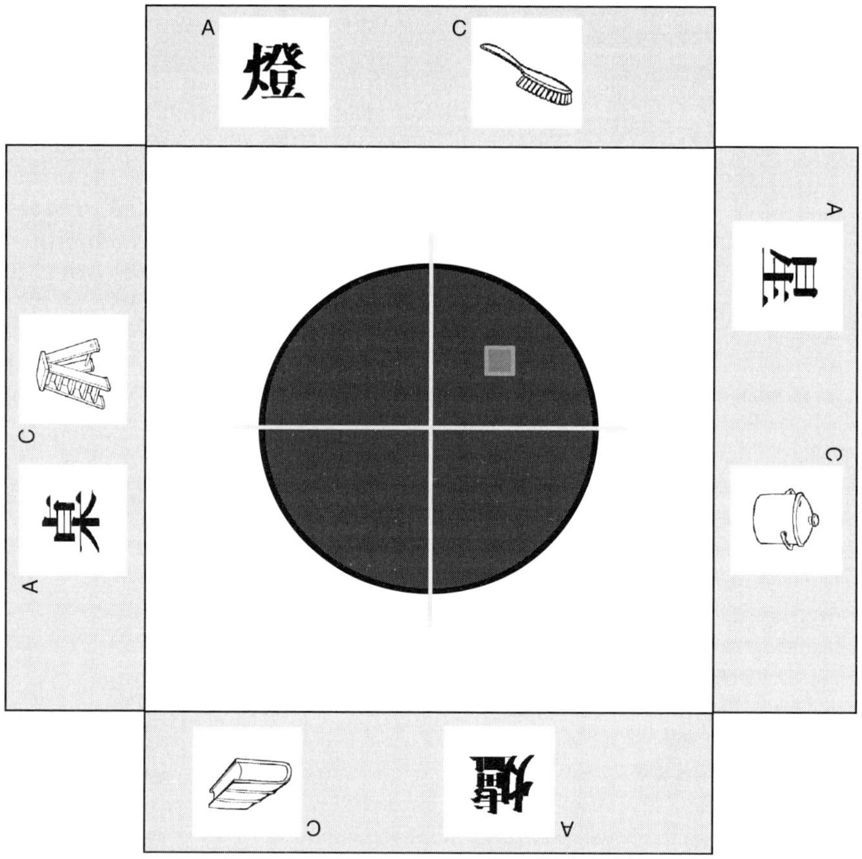

Fig. 9. Schematic of the computer-generated environment (from a bird's-eye perspective), in which blocking of concrete (cues labeled with "C") and abstract cues (labeled with "A") was tested.

on the probe trials, but perhaps more interesting, only the "Constant Location" subjects showed learning at all.

These data suggest that the use of concrete and readily nameable cues leads to the formation and use of maps, at least as indexed by the absence of blocking. We next tested whether abstract cues can induce a blocking effect, which might provide an explanation as to why Hamilton and Sutherland found a blocking effect in their maze using abstract paintings as distal cues. In our study subjects in the Blocking groups were initially trained with one cue on each wall: The Concrete group was trained with cues like the ones used in our previous study, that is, line drawings of familiar objects (Cue Set C), while the Abstract group was trained with cues based on Japanese characters (Cue Set A, see Fig. 9). During

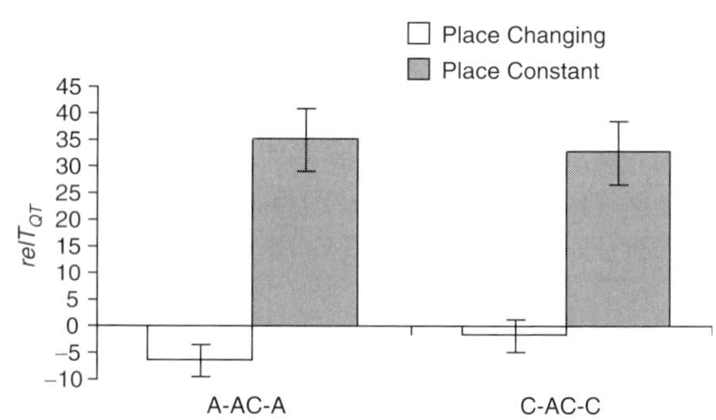

Fig. 10. Search persistence probe trial. Blocking was observed only when abstract cues were introduced to the environment. When only the added concrete cues were available, subjects searched the correct quadrant for the target more so than any other. However, when abstract cues were available, search preference for the target quadrant was at chance level. Once again we observed that subjects who thought the target was shifting location across trials were not able to learn the target location.

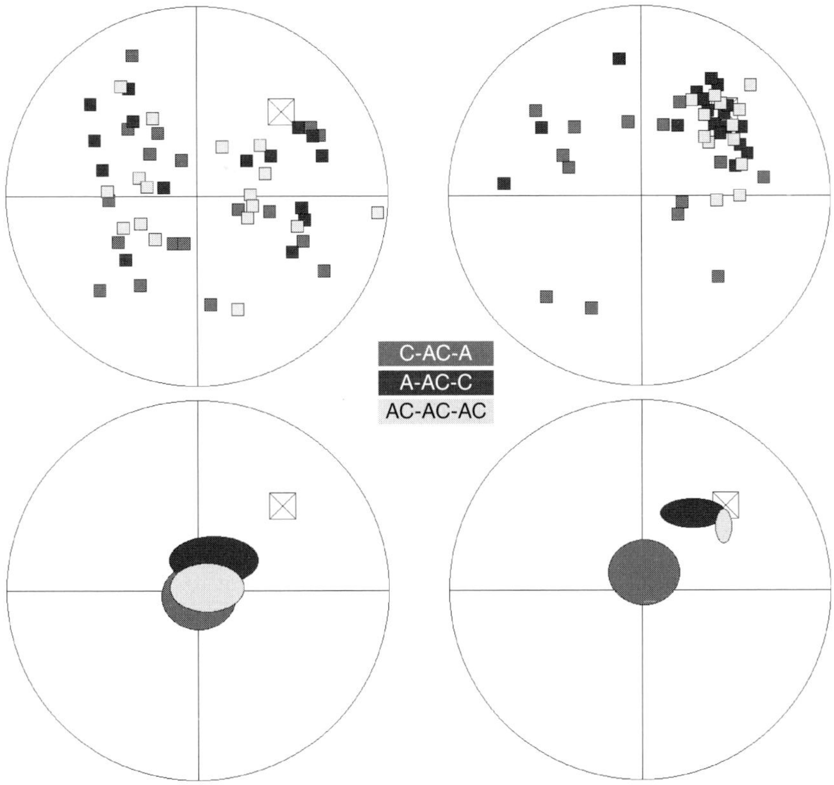

Fig. 11. Location accuracy probe trial. When only the added abstract cues were available, subjects were not able to accurately recall the target location. This was not the case when concrete cues were present (see right-hand panels). Subjects who believed that the target location was changing recalled the target location equally often in every quadrant (left-hand panels). Upper panels show recall locations of all individuals, lower panels the group averages of these positions. (See Color Plate 12.11 in color plate section.)

compound training, subjects were further trained with all eight cues (Cue-Set A+C), half of them concrete and half of them abstract. Then subjects received the two types of probe trials, with either only the added Abstract cues (in the Concrete Blocking group) or the added Concrete cues (in the Abstract Blocking group) or both sets of cues (in the Control group). On the crucial probe trials performance in the blocking groups depended on which cues the subjects had been trained with and which cues were added during compound training. When subjects were initially trained with abstract cues and had concrete cues added, we did not observe blocking. That is, subjects were able to perform quite well with just the added concrete cues when tested on the probe trials. However, when subjects were initially trained with concrete cues, and abstract cues were added, we did observe blocking (see Figs. 10 and 11).[1]

These results highlight the role that cue codability plays in determining whether subjects will show blocking or not. When concrete cues are used in original training, subjects clearly form and use maps. When abstract cues are subsequently added, they are apparently not integrated into these maps, since the subjects cannot use them to navigate to the target. When abstract cues are used in original

[1]In a pilot study we ensured that the abstract cues alone suffice to learn the target location. In this study, only the abstract cues were used, and the results from both types of probe trials clearly showed that subjects learned the target location at the same level of precision as when only the concrete cues were used.

training, subjects can also form and use maps, given adequate instructions. When concrete cues are now added, these are integrated into the maps, and can be used to navigate to the target. Given the failure to integrate abstract cues into existing maps, we went on to ask whether subjects had attended to the abstract cues at all. Perhaps, having been trained with concrete cues and performing quite well on the basis of these cues, subjects simply ignored the hard-to-encode abstract cues that were added during compound training. We addressed this question in a four-alternative forced choice recognition test. During this assessment, subjects were presented with four stimuli: the original cue that was in the maze (e.g., the image of a donkey), a modified version of this cue (e.g., a different looking donkey), a new image conceptually related to the cue (e.g., a horse), and a different instance of this cue (e.g., a different horse). Subjects were asked to select the image that they had seen during training. While recognition rates were lower for the abstract cues as compared to the concrete cues, they were nevertheless well above chance for all the cues. The presence of blocking did not imply that the subjects had learned nothing about the "blocked" abstract cues. It was instead the case that these cues had not been incorporated into their maps of the maze.

Conclusions

The presence or absence of blocking reflects the behavioral strategies humans recruit in solving a spatial task. In our computer-generated maze, when subjects explore the space adequately and when they conclude that the target remains in a fixed location, they are highly likely both to form a map of the space and to use this map for navigating to the hidden target. Under these conditions, as cognitive map theory predicted, blocking will not be observed.

We isolated factors upon which these strategic choices depend. Our results suggest that exploration is a function both of knowledge about the nature of the spatial task and the codability of the distal cues found in the environment. When subjects are informed that the target will remain in a fixed location, that distal cues will be powerful place predictors, and that intensive exploration will be beneficial, they are likely to invoke strategies that lead to formation of a cognitive map. This is true even when the available cues are predominantly abstract and hard to encode. In the absence of explicit knowledge about task demands, the presence of abstract cues biases human subjects to engage in a strategy of associating the target location with a small subset of cues; a cognitive map encoding the cues and their spatial relations is not created. Furthermore, when subjects initially encounter only easily encodable concrete cues, added abstract cues are not incorporated into the map. However, if only abstract cues are initially available, added concrete cues will be integrated into a map.

Even though abstract cues are not all associated with the target location and hence do not become part of a representation of space, they are nevertheless attended to and encoded. Our subjects recognized these cues in a demanding cue discrimination task well above chance. What fails to be encoded are the spatial relations among these cues in the environment. This is a very important point: Subjects apparently form one kind of representation involving these cues — sufficient to permit accurate recognition — but fail to form the kind of map-like representation required for flexible navigation.

Map learning seems to favor strategies involving broad attention to all cues and, more importantly, to their complex spatial relations. Thus, the relation between attention and learning in our paradigm is rather complex. Attended cues can be learned in the sense that they will be subsequently recognized, but will still not be usable for purposes of accurate navigation (cf. Doeller and Burgess, 2008). Our results add to the growing body of evidence showing that object recognition is not a function of the mapping system, which mainly allows for place navigation. For example, Save et al. (1992) dissociated hippocampus-mediated detection of spatial novelty from hippocampus-independent detection of object novelty. Animals with lesions in the hippocampus did not explore displaced objects, but readily explored new objects (see also Mumby et al., 2002). In a similar task,

Hardt et al. (2008) showed that blocking the atypical protein-kinase isoform M zeta (PKMζ) in the dorsal hippocampus leads to a loss of knowledge for *where* objects had been located one day previously, while knowledge about *what* objects had been encountered remained unaffected. The knowledge for object identity appears to be mediated by perirhinal cortex (for a review see Winters et al., 2008). Immediate-early gene studies have shown differential expression patterns for *c-fos* following exposure to object novelty as compared to spatial novelty (Aggleton and Brown, 2005). When familiar objects were regrouped, thereby changing their spatial relations without introducing new objects, hippocampus, but not perirhinal cortex responded with increased *c-fos* expression. However, presentation of novel objects led to increases in *c-fos* levels only in perirhinal cortex, but not hippocampus. Thus, spatial knowledge and the detection of spatial novelty appears a function exclusive to hippocampus, while knowledge about objects relies on perirhinal cortex, which explains why our subjects could correctly recognize distal cues that they had not incorporated into a cognitive map.

A cognitive map, which encodes spatial relations among distal cues, is computationally more complex than associating the target location with one or two discrete cues. Therefore, it is possible that when the available distal cues cannot be quickly encoded, a comprehensive map of the environment will not be created, although the stimuli might have been attended to and encoded. Instead, if the spatial task permits, a different system will be recruited, and comparably less complex spatial representations will be generated. According to this account, spatial behavior is the organism's answer to a specific adaptive problem. An organism will apply the most efficient solution to a given spatial problem, and it will use available environmental information and its knowledge of the task demands to decide which strategy is most likely to result in successful behavior. Attention may thus be dynamically allocated and not an automatic process (Fig. 10): The detection of certain environmental features or preexisting knowledge about the task lead to recruitment of a specific spatial strategy, i.e., either map learning or the formation of rather simple cue-place associations. If the environment changes through the introduction of new objects, these changes will be noticed, but whether the spatial strategy changes as a response to these cues depends on the nature of the cues and the organism's knowledge about the situational demands: If a mapping strategy is evoked as a response to the new cues, then the spatial relationships that these cues form within the space are explored and encoded. Otherwise, this type of spatial information will be ignored. In any case, the identity of the cues will be retained. The blocking effect thus is a function of what spatial strategy best fits the newly added cues, and how exploration of spatial relations is allocated as a consequence.

Thus, strategy choice affects attention and exploration, which affects what is learned, and what is learned affects what subsequently will be more likely noticed, attended, and explored. In this sense, our results suggest that cognitive map theory needs to be revised in a subtle way. According to the theory, organisms in a novel spatial environment will always generate a cognitive map. In light of our results, it appears instead that after a consideration of task demands as well as analysis of the features in the environment, the system "decides" to create a map or not. Cognitive map theory linked exploratory behavior to map creation, but it appears that exploration simply allows for the gathering of knowledge necessary to decide on the most effective spatial strategy, which might or might not entail map creation. This strategy then will influence subsequent attention to, and exploration of, distal cues and spatial features.

Acknowledgment

This paper was part of the first author's dissertation at The University of Arizona and was supported by grants to the Cognitive Neuroscience Center from the Flinn Foundation and the McDonnell-Pew Program.

Oliver Hardt is currently a post-doctoral fellow at McGill University supported by the Deutsche Forschungsgemeinschaft (DFG, German Research Foundation).

Bibliography

Aggleton, J. P., & Brown, M. W. (2005). Contrasting hippocampal and perirhinal cortex function using immediate early gene imaging. *The Quarterly Journal of Experimental Psychology*, *58B*(3/4), 218–233.

Biegler, R., & Morris, R. G. M. (1999). Blocking in the spatial domain with arrays of discrete landmarks. *Journal of Experimental Psychology: Animal Behavioral Processes*, *25*, 334–351.

Brown, M. F., Yang, S. Y., & DiGian, K. A. (2002). No evidence for overshadowing of facilitation of spatial pattern learning by visual cues. *Animal Learning & Behavior*, *30*(4), 363–375.

Chamizo, V. D., Sterio, D., & Mackintosh, N. J. (1985). Blocking and overshadowing between intra-maze and extra-maze cues: A test of the independence of locale and guidance learning. *Quarterly Journal of Experimental Psychology*, *37B*, 235–253.

Doeller, C. F., & Burgess, N. (2008). Distinct error-correcting and incidental learning of location relative to landmarks and boundaries. *Proceedings of the National Academy of Sciences of the USA*, *105*, 5909–5914.

Graham, M., Good, M. A., McGregor, A., & Pearce, J. M. (2006). Spatial learning based on the shape of the environment is influenced by properties of the objects forming the shape. *Journal of Experimental Psychology: Animal Behavior Processes*, *32*, 44–59.

Gray, E. R., Bloomfield, L. L., Ferrey, A., Spetch, M. L., & Sturdy, C. B. (2005). Spatial encoding in mountain chickadees: Features overshadow geometry. *Biology Letters*, *1*, 314–317.

Hamilton, D. A., & Sutherland, R. J. (1999). Blocking in human place learning: Evidence from virtual navigation. *Psychobiology*, *27*, 453–461.

Hardt, O., Hastings, M., Wong, J., Migues, P. V., & Nader, K. (2008). Long-term memory for objects and object location: Effects of ZIP infusions into dorsal hippocampus. Poster presented at the annual meeting of the Society for Neuroscience, November, Washington, DC (online: http://web.mac.com/oliver.hardt/Job/Pubs_&_Peer-Review_files/hardt%20hastings%20wong%20migues%20nader.sfn%202008.pdf).

Hayward, A., Good, M. A., & Pearce, J. M. (2004). Failure of a landmark to restrict spatial learning based on the shape of the environment. *Quarterly Journal of Experimental Psychology: Comparative and Physiological Psychology*, *57*(B), 289–314.

Hayward, A., McGregor, A., Good, M. A., & Pearce, J. M. (2003). Absence of overshadowing and blocking between landmarks and the geometric cues provided by the shape of a test arena. *The Quarterly Journal of Experimental Psychology B, Comparative and Physiological Psychology*, *56*(1), 114–126.

Horne, M. R., & Pearce, J. M. (2009). A landmark blocks searching for a hidden platform in an environment with a distinctive shape after extended pretraining. *Learning & Behavior*, *37*, 167–178.

Jacobs, J. W., Laurance, H. E., & Thomas, K. G. F. (1997). Place learning in virtual space I: Acquisition, overshadowing, and transfer. *Learning and Motivation*, *28*, 521–541.

Jacobs, J. W., Thomas, K. G. F., Laurance, H. E., & Nadel, L. (1998). Place learning in virtual space II: Topographical relations as one dimension of stimulus control. *Learning and Motivation*, *29*, 288–308.

Kamin, L. J. (1969). Predictability, surprise, attention, and conditioning. In B. A. Campbell & R. M. Church (Eds.), *Punishment and aversive behavior* (pp. 276–296). New York: Appleton-Century-Crofts.

Mackintosh, N. J. (1975). A theory of attention: Variations in the associability of stimuli with reinforcement. *Psychological Review*, *82*, 276–298.

Morris, R. G. M. (1981). Spatial localisation does not require the presence of local cues. *Learning and Motivation*, *12*, 239–260.

Mumby, D. G., Gaskin, S., Glenn, M. J., Schramek, T. E., & Lehmann, H. (2002). Hippocampal damage and exploratory preferences in rats: Memory for objects, places, and contexts. *Learning & Memory*, *9*, 49–57.

Nadel, L., & Hardt, O. (2004). The spatial brain. *Neuropsychology*, *18*(3), 473–476.

O'Keefe, J., & Nadel, L. (1978). *The hippocampus as a cognitive map*. Oxford: Clarendon Press.

Pearce, J. M., Graham, M., Good, M., Jones, P. M., & McGregor, A. (2004). Transfer of spatial behavior between different environments: Implications for theories of spatial learning and for the role of the hippocampus in spatial learning. *Journal of Experimental Psychology: Animal Behavior Processes*, *30*, 135–147.

Pearce, J. M., Ward-Robinson, J., Good, M., Fussell, C., & Aydin, A. (2001). Influence of a beacon on spatial learning based on the shape of the test environment. *Journal of Experimental Psychology: Animal Behavior Processes*, *27*(4), 329–344.

Rescorla, R. A., & Wagner, A. W. (1972). A theory of Pavlovian conditioning: Variations in the effectiveness of reinforcement and nonreinforcement. In A. H. Black & W. F. Proksay (Eds.), *Classical conditioning II: Current research and theory* (pp. 64–99). New York: Appleton-Century-Crofts.

Rodrigo, T., Chamizo, V. D., McLaren, I. P. I., & Mackintosh, N. J. (1997). Blocking in the spatial domain. *Journal of Experimental Psychology: Animal Behavior Processes*, *25*, 225–235.

Sanchez-Moreno, J., Rodrigo, T., Chamizo, V. D., & Mackintosh, N. J. (1999). Overshadowing in the spatial domain. *Animal Learning & Behavior*, *27*(4), 391–398.

Save, E., Poucet, B., Foreman, N., & Buhot, M.-C. (1992). Object exploration and reactions to spatial and nonspatial changes in hooded rats following damage to parietal cortex or hippocampal formation. *Behavioral Neuroscience*, *106*(3), 447–456.

Wall, P. L., Botly, L. C. P., Black, C. K., & Shettleworth, S. J. (2004). The geometric module in the rat: Independence of shape and feature learning in a food finding task. *Learning & Behavior*, *32*, 289–298.

Winters, B. D., Saksida, L. M., & Bussey, T. J. (2008). Object recognition memory: Neurobiological mechanisms of encoding, consolidation and retrieval. *Neuroscience and Biobehavioral Reviews*, *32*, 1055–1070.

N. Srinivasan (Ed.)
Progress in Brain Research, Vol. 176
ISSN 0079-6123

CHAPTER 13

The remains of the trial: goal-determined inter-trial suppression of selective attention

Alejandro Lleras[1,*], Brian R. Levinthal[1] and Jun Kawahara[2]

[1]*Department of Psychology, University of Illinois at Urbana-Champaign, Champaign, IL, USA*
[2]*National Institute of Advanced Industrial Science and Technology, Tsukuba, Japan*

Abstract: When an observer is searching through the environment for a target, what are the consequences of not finding a target in a given environment? We examine this issue in detail and propose that the visual system systematically tags environmental information during a search, in an effort to improve performance in future search events. Information that led to search successes is positively tagged, so as to favor future deployments of attention toward that type of information, whereas information that led to search failures is negatively tagged, so as to discourage future deployments of attention toward such failed information. To study this, we use an oddball-search task, where participants search for one item that differs from all others along one feature or belongs to a different visual category, from the other stimuli in the display. We find that when participants perform oddball-search tasks, the absence of a target delays identification of future targets containing the feature or category that was shared by all distractors in the target-absent trial. We interpret this effect as reflecting an implicit assessment of performance: target-absent trials can be viewed as processing "failures" insofar as they do not provide the visual system with the information needed to complete the task. Here, we study the goal-oriented nature of this bias in three ways. First, we show that the direction of the bias is determined by the experimental task. Second, we show that the effect is independent of the mode of presentation of stimuli: it happens with both serial and simultaneous stimuli presentation. Third, we show that, when using categorically defined oddballs as the search stimuli (find the face among houses or vice versa), the bias generalizes to unseen members of the "failed" category. Together, these findings support the idea that this inter-trial attentional biases arise from high-level, task-constrained, implicit assessments of performance, involving categorical associations between classes of stimuli and behavioral outcomes (success/failure), which are independent of attentional modality (temporal vs. spatial attention).

Keywords: adaptive workspace; consciousness; attention; meditation; neural mechanisms

Whenever we face the world, we do so tainted by our prior experience. In experimental psychology, there are countless examples of how prior experience affects our current behavior. For example, we tend to slow down after making an erroneous response (e.g., Jentzsch and Dudschig, 2009; Laming, 1968; Rabbitt and Rogers, 1977), and we tend to speed-up too before we make an error (e.g., Ridderinkhof et al., 2003). Whenever

*Corresponding author.
Tel.: +1 217 265 6709; Fax: +1 217 244 5876;
E-mail: Alejandro.Lleras@gmail.com

DOI: 10.1016/S0079-6123(09)17611-2

we have to do the same thing twice, we are faster the second time around (Kraut et al., 1981; Pashler and Baylis, 1991). In perceptual discrimination tasks, *visual priming* is commonplace: we are better at identifying a stimulus that we have already identified in the past (e.g., Bar and Biederman, 1998; James et al., 2000), one similar to it (e.g., Fiser and Biederman, 2001; Forster et al., 1987) or even one semantically related to it (e.g., Marcel, 1983; Potter, 1999). And cognitive phenomena such as the Flanker effect, the Stroop effect, and the Simon effect are also all influenced by our recent experience with these tasks (e.g., Gratton et al., 1992; Kems et al., 2004; Sturmer et al., 2002, respectively). So, it should come as no surprise that the manner in which we attend to a scene is also heavily influenced by our prior experience with that same (or similar scene) scene (e.g., Chun and Jiang, 1998; Frings and Wühr, 2007; Huang and Pashler, 2005; Lamy et al., 2006; Olivers and Humphreys, 2003; Tipper, 1985). In other words, our attentional system receives strong inputs from our prior experience, and the manner in which we view any subsequent visual scene must therefore partly reflect these memory-based inputs.

One well-controlled and well-documented example of experience-driven effects on visual attention can be found in the phenomenon known as priming of pop-out (POP, e.g., Fecteau, 2007; Kristjánsson et al., 2007; Maljkovic and Nakayama, 1994, 1996, 2000). A pop-out effect in vision occurs whenever a highly salient item is presented in a scene among a more homogeneous set of distractors. The salient object seems to grab the observer's attention, as if the object were popping-out of the scene. In terms of behavior, the time to find a pop-out target is typically independent of the number of distractors in the scene. This pop-out effect is a cornerstone of many theories of vision and visual search (e.g., Itti and Koch, 2001; Treisman and Gelade, 1980; Wolfe, 1994) as it illustrates the powerful guiding force of bottom-up scene characteristics on attention. That is, our visual system is biased toward attending — as a matter of first priority — to the big, to the bright, to the most visually unique region in a scene. That said, the work of Maljkovic and Nakayama (and others since) has documented that even this most basic and apparently automatic form of attentional deployment is heavily influenced by prior events. For instance, when looking for a color-oddball target (say either a green target among red distractors or a red target among green distractors), we are faster at deploying our attention to a red target on the current trial, if on the previous trial, the target was also a red oddball. This history effect of past experience on current attentional deployment is very robust (e.g., Hillstrom, 2000; Kristjánsson, 2006; Kristjánsson et al., 2002) and has also been documented in monkeys (e.g., Bichot and Schall, 2002). Current theories of POP account for this effect in terms of a processing benefit from having to attend to the same target feature twice in a row (e.g., Huang et al., 2004; Maljkovic and Nakayama, 1994; Wolfe et al., 2003). That is to say, we more readily attend to a feature that we have recently attended to than to any other feature. Further, there is recent evidence that POP is not a passive automatic modulation of attentional biases, but rather that it reflects the current goals of the observer (Fecteau, 2007). In other words, if the observer is looking for color oddballs, the repetition of a color will produce a POP, not the repetition of a shape, and vice versa.

One way of interpreting the POP phenomenon is to characterize it within the context of the observer–environment interaction and the task that the observer must accomplish within this environment. Observers are asked to find a color oddball, for example, and on the upcoming trial there is total uncertainty as to what the color of the target will be. Observers do not approach this upcoming trial in a vacuum: they carry with them their recent experience with this task. Imagine, then, a case in which the observer on the most recent trial found a red target. All else being equal, why should not the attentional system bet on this recent success and bias future attentional deployments toward red objects? The same would be true for shape-defined targets. Further, imagine an experiment in which the search task randomly changed from trial to trial, such that sometimes observers were asked to find color oddballs, sometimes shape oddballs. As Fecteau

(2007) found the repetition of a given feature across trials should only benefit performance if that feature was also relevant to the observer on the previous trial (i.e., only if the attentional system was tuned to that feature on the previous trial and that feature *produced* a target). In sum, POP might be interpreted as a case of the attentional system *betting on past success* in the search task as a way to dynamically adjust to the uncertainty in the environment.

If POP is a case of the attentional system betting on past success, is there an analogous situation when attention *bets against past failures*? We believe the phenomenon known as the distractor preview effect (DPE) represents such a scenario. Like POP, the DPE arises in the context of oddball-search tasks, but unlike POP, in DPE experiments target-absent trials are intermixed with target-present trials (Ariga and Kawahara, 2004; Goolsby et al., 2005; Levinthal and Lleras, 2008a, b; Lleras et al., 2008; Shin et al., 2008). For example, imagine a case when observers are asked to find a color oddball in a display, and on a large proportion of trials, there is no oddball in the display (i.e., all items are of the same color). The DPE refers to the attentional modulation that is observed on target-present trials that immediately follow a target-absent trial. In the case of search for a color oddball, it represents the cost of selecting a target that has the same color as the distractors in the preceding target-absent trial. For example, if on trial *N*, the target is green (among red distractors, creating a color pop-out effect), observers would be 60–100 ms slower in selecting that green target if on the preceding trial all distractors had been green (the target-color previewed condition), than to select the same green target if the distractors on the preceding trial had all been red (the distractor-color previewed condition).

Thus far, several interesting characteristics of the DPE have been described. Like POP, the DPE has been shown to be sensitive to the observers' task set (Levinthal and Lleras, 2008a), such that only the repetition of features that are relevant to the search task creates an inter-trial effect, whereas repetition of features outside of the task set does not. The effect is not specific to color and can be obtained with a number of basic visual features such as shape oddball tasks (Levinthal and Lleras, 2008a) and motion-direction oddball (Ariga and Kawahara, 2004). Furthermore, Shin et al. (2008) demonstrated that the slowing of reaction times (RTs) in the target-color previewed condition was also correlated with an analogous delay in the onset of the N2pc (a component of the ERP associated with lateralized shifts of attention; see Eimer, 1996; Luck and Hillyard, 1994). That is, lateralized shifts of attention toward a color-oddball target, as indexed by the N2pc, occurred later for targets that were of the same color as the distractors in the preceding target-absent trial, than for targets in a different color. Finally, the DPE only arises when participants are asked to identify the color oddball, and it is absent when participants merely have to detect the presence or absence of the color oddball (Lleras et al., 2008).

We would like to argue that the DPE represents, in terms of the attentional system, the flip side of the POP bias. When observers are asked to find a color oddball, one of two things can happen. Either the observer finds an oddball, or he or she does not. When he or she does, the attentional system notes this "search success" and associates the feature that produced the target (i.e., the target-defining feature present in the target) with that success, creating a bias toward selecting that particular feature in upcoming trials. This bias is responsible for the POP effect. On the other hand, when he or she fails to find a target in the display, the attentional system notes this "search failure" and associates the feature that failed to produce a target (i.e., the target-defining feature common to all the distractors in the target-absent display) with that failure, creating a bias against selecting that particular feature in upcoming trials. This bias is responsible for the DPE. In terms of the DPE, this bias would: (a) allow observers to more accurately stay away from future distractors containing that failed feature and (b) prevent observers from quickly selecting future targets containing that failed feature. Both of these effects have been documented when a non-previewed color condition is included in the design (Lleras et al., 2008).

If this interpretation of POP and the DPE is correct, that would imply that the attentional system is sensitive to an implicit assessment of performance that takes into account not only the failure or success of its search effort, but also the features involved in that success or failure.[1] Further, that this assessment is responsible for producing a very strong attentional bias that affects where we will direct our attention in the future. That said, it may yet be premature to argue that this implicit assessment of performance is actually taking place and that these effects are not arising out of more automatic processes, perhaps at early levels of the visual system. Here, we describe three approaches we have taken in the context of the DPE to show that this assessment is indeed taking place and may be responsible for these inter-trial effects. The rationale is as follows. First, we will show that, using the very same displays, one can produce different biases, depending on what observers are asked to judge about these displays. That is, it is what observers do with the information on target-absent trials that determines the direction of the effect, and not the repetition itself of any task-relevant feature. Second, we will show that an identical assessment of performance is produced when observers search for oddball targets in time (using RSVP displays), and that the bias produced by these performance assessments is independent of the search context in which it was produced. That is, failing to find a color oddball in a temporal RSVP task is disruptive to future color searches both in space and in time (the same being true for failures to find oddball in spatial searches tasks). Finally, we will show that when the oddball task is defined at a categorical level (find the house among faces or vice versa), failure to find a categorical oddball produces a bias against focusing on any-and-all items belonging to the "failed" category, even to previously unseen members of the failed category. In sum, this is evidence that the assessment of performance producing the DPE bias occurs at the same level of discrimination that defines the search task, and not at some lower, perhaps feature-based level.

[1]This notion of implicit assessment of performance has also been discussed within the literature on "error processing", where it has been demonstrated that RTs slow down following error trials, as well as speed-up on trials preceding error trials, suggesting that there are subtle threshold adjustments of response-related processes that squarely depend on whether a task has been successfully completed or an error was produced (see Jentzsch and Dudschig, 2009; Riddenrinkhof et al., 2003).

Part 1: what is "failure"?

The failure that we argue is responsible for the DPE is defined relative to the search task observers are asked to perform. On target-absent trials, a search "failure" arises only because there is no target for participants to select in the display. We argue that it is this search failure that underlies the DPE, and not the mere act of looking at a target-absent display. In other words, seeing a group of homogeneously colored red distractors should not automatically produce a bias against selecting red items in the future. It does so only when observers are actually looking for a color target in that display.

Experiment 1: subitizing and oddball-search tasks

Here, we tested the "failure is relative to the task" assumption by manipulating the task participants performed with our displays while otherwise using identical displays and experimental procedures across tasks. One group of participants was asked to perform a color-oddball search task (as in previous DPE experiments), whereas a second group of subjects was asked to subitize the smaller set of items in the display (see Fig. 1 for an illustration of the displays and the task).

Our displays consisted of eight items and contained: all items of the same color (homogeneous displays), one item of a different color from the remaining distractors, or two items in a color different from that of the remaining distractors. We were most interested in the effects that trials with homogeneous displays would have on subsequent trials containing one color oddball. We expected that, for the participants in the subitizing group, trials with homogeneous displays

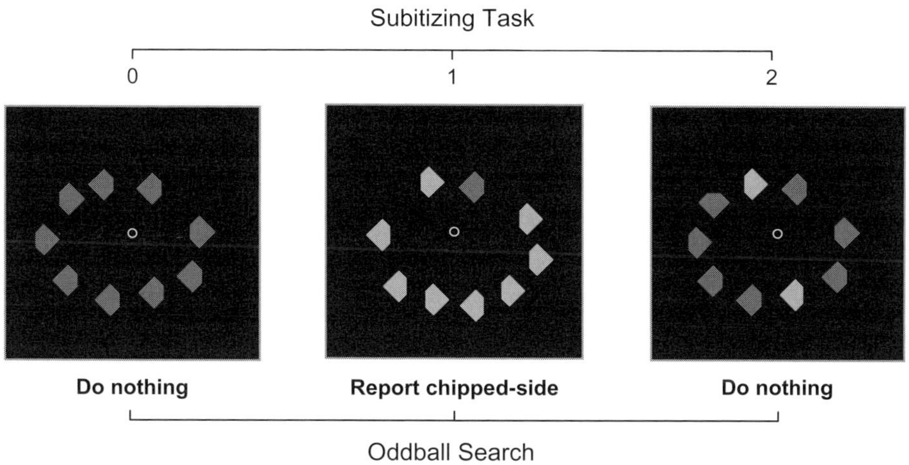

Fig. 1. Schematic representation of stimuli in Experiment 1. Observers in the subitizing group reported the number of the minority-color diamonds. Observers in the oddball-search group reported the chipped side when there was one color oddball.

would not be considered "failures" because these displays are easily associated with a correct behavioral response in the subitizing task ("there are 0 items in the small set"). As a result, we expected to find no DPE in the subitizing group, whereas we expect to replicate the DPE in the oddball-search group.

Methods

Participants. Seventeen and 18 observers, recruited from the AIST subject pool, participated in the subitizing and color-oddball search groups, respectively.

Stimuli and apparatus. Stimuli were presented on a 17-in. Sony monitor, at a refresh rate of 60 Hz, at a screen resolution of 1024×768, controlled by a Windows PC. The experiments were programmed using the Psychophysics Toolbox (Brainard, 1997; Pelli, 1997). Displays consisted of 9–12 diamond shapes, each subtending $1.3^\circ \times 1.3^\circ$ in visual angle. The diamonds were each missing either the right or the left corner (a 0.22° chip). They were presented at 10 of 12 possible locations on an approximate imaginary iso-acuity ellipse (Rovano and Virsu, 1979) with a vertical axis of 8.2° and horizontal axis of 10.1°. The locations of the diamonds were determined randomly trial by trial. The diamonds could be either red or green. On every trial, 0, 1, or 2 odd-colored items were presented. For example, when the majority of items were red, 0, 1, or 2 of them could be colored green (0 odd-colored items corresponds to a homogeneous display). Participants viewed the displays in a dimly illuminated room, at a distance of approximately 57 cm.

Task and design. A between-subject design was used. Half of the participants completed a "subitizing" task in which they counted the number of oddball items in the display by pressing the 0, 1, or 2 keys, respectively, in the 10-keypad. The other half of participants completed an "oddball-search" task in which they indicated the missing corner of a unique color oddball (right or left) in the display by pressing the right or left arrow keys, respectively, on the keyboard, when there was one color oddball. In trials where no color oddball was present or where two color oddballs were presented, participants were to withhold their responses. Each participant completed 768 trials (256 trials for each target number). All participants were encouraged to respond as quickly as possible, while maintaining a high level of accuracy. Importantly, the displays were identical across these two groups of participants. The only differences between groups were the experimental task and the time-out of displays: in the oddball-search task, 0 and 2 odd-colored displays were only displayed for 600 ms. All other displays remained on the screen until the participant responded.

For every participant, a sequence of trials was determined to ensure that the color of items in zero-target trials was equally likely to become the color of the oddball(s) or the color of the majority items in the immediately following one- or two-target trial. When one- or two-target display was preceded by a zero-target display, two types of trials were defined: target-color previewed trials (when the current target color was the same as the color of the items in the preceding zero-target trial) and distractor-color previewed trials (when the color of the current majority items was the same as the color of the items in the preceding zero-target trial). We chose these condition labels in the interest of clarity and consistency labels across tasks. The DPE is the RT difference between target-previewed RTs and distractor-previewed RTs.

Procedure. Trials began with the presentation of a fixation display with a random duration between 800 and 2300 ms. The search display was then presented until participants' response. On one-target trials, the odd color diamond was chipped either on the left or on the right side. On two-target trials, one of the minority diamonds had the chipped corner on the right side and the other on the left side. In all trials, the majority diamonds also had its left or right corner chipped.

Results

Subitizing group. Participants' mean accuracy was 97.5%. The critical comparison here is between RTs in the distractor-color previewed and the target-color previewed conditions on trials in which one odd-colored item was presented (Fig. 2, Left). The respective RTs in these conditions were 764.6 and 726.0 ms, which were not significantly different from each other, $t(16) = .80$, n.s. If anything, the difference between these RTs was in the opposite direction than the normal DPE.

Oddball-search group. Participants' mean accuracy was 93.0%. The RTs in the distractor-color previewed condition was 615.3 ms, which was significantly smaller than the 646.0-ms RT in the

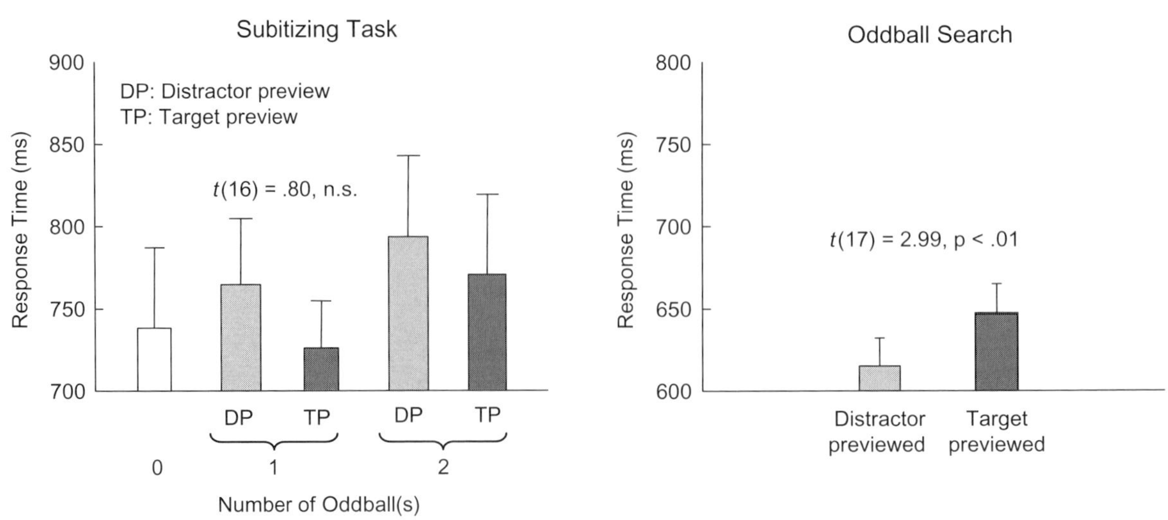

Fig. 2. Results of Experiment 1. Left panel: subitizing group; right panel: oddball-search group. The error bars indicate standard errors for each condition.

target-color previewed conditions, $t(17) = 2.99$, $p < .01$ (Fig. 2, Right). Note that the RTs in this task had smaller variances than in the subitizing task, likely due to the smaller number of response alternatives in this task (Hick, 1952).

Discussion

As expected, we obtained a DPE in the oddball-search group, albeit the magnitude of this effect was smaller than what we have observed in the past. This reduction in magnitude might be ascribed to the larger than usual set size that we used in this experiment (an analogous reduction in the magnitude of POP with increasing set size has been documented). Crucially, however, we failed to obtain any evidence of a DPE in the subitizing group. If anything, there was a trend for shorter (rather than longer) RTs in the target-color previewed condition in this group, almost as if the color of the zero-target display had been now associated with behavioral success at this task. Although this reversal did not reach significance, the absence of a DPE in the subitizing group can be taken as partial evidence that the DPE does not automatically arise as a consequence of being exposed to a homogeneous display (as has been proposed elsewhere, Goolsby et al., 2005), but rather that the task observers perform on these displays determine the inter-trial effect (if any) that they give rise to. As a further test of this hypothesis, we designed a second experiment, aimed at more strongly associating the color of the homogeneous displays with behavioral success.

Experiment 2: contingent number identification task

We wanted to create a situation in which participants would search the display for oddballs, while providing them with a positive outcome on oddball-absent trials. We modified the typical oddball-search task by introducing a character at fixation on every trial. On trials in which participants do not find a color target, they were instructed to judge whether the character at fixation was a number higher or smaller than five. In this manner, we wished to create a positive assessment of these target-absent trials, in which the inspected color (of the distractors) might now be associated with a successful behavioral outcome. If so, we would expect to now find a reversed DPE, with target-color previewed trials producing shorter RTs than distractor-color previewed trials. Note that this is not a dual-task paradigm: on oddball-present trials, participants only identified the oddball and on oddball-absent trials, participants only identified the character at fixation. Two responses were never produced on a single trial.

Methods

Participants. Seventeen subjects from the AIST subject pool participated in this experiment.

Stimuli and apparatus. The stimuli and the procedures were the same as those used by Lleras et al. (2008), and those used in oddball-search condition of Experiment 1. The differences are as follows. Displays contained only three chipped red or green diamonds, as well as a letter or a digit, which subtended 0.8° in height, presented at fixation. The locations of the diamonds were randomly determined at the beginning of a trial with the constraint that all three were equally spaced. On oddball-absent trials, the central character was a digit, whereas on oddball-present trials, the central character was a letter (Fig. 3).

Procedure. The sequence of the displays was as follows. A fixation display was presented for 2000–2500 ms followed by a search display. The display remained until response. The primary task of the observer was to identify the chipped side of the color oddball (left/right) by pressing the left or right arrow key on the keyboard. When there was no color oddball in the periphery, observers indicated whether the central digit was greater or less than five by pressing the up or down arrow keys on the keyboard.

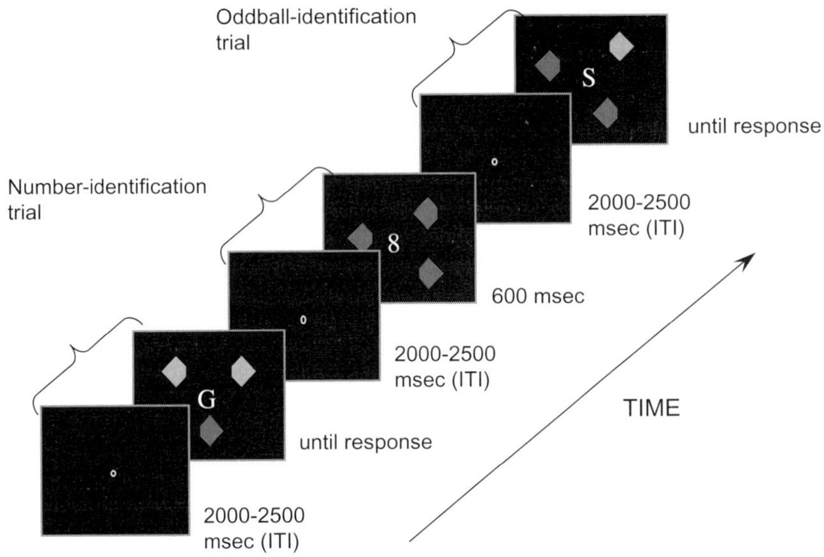

Fig. 3. Schematic representation of stimuli in Experiment 2.

Results

The mean accuracies on oddball-present trials and oddball-absent trials were 94.2 and 99.2%, respectively. To remind the reader, we assessed the DPE on oddball-present trials, when participants were finding and identifying the color oddball in the periphery. In these trials, RTs in target-color previewed condition were 992.0 ms, significantly shorter than RTs in the distractor-color previewed condition 1043.6 ms, $t(16) = 2.44$, $p < .05$ (Fig. 4). The RT in the non-search trials, when participants were doing the number identification task, was 1284.2 ms. We should note that the "reversal" of the DPE cannot be attributed to the overall longer RTs in this task, as we have found DPEs with longer RTs in the past (see also Experiment 5 and its follow-up).

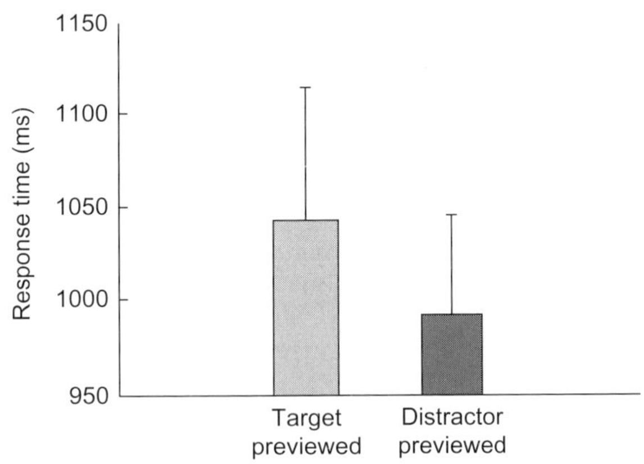

Fig. 4. Results of Experiment 2. The error bars indicate standard errors for each condition.

Discussion

In Experiment 2, we were able to produce a significant reversal of the DPE. Whereas on previous published experiments on the DPE, oddball-absent displays always produced a negative or suppressive bias aimed at keeping attention away from focusing on the "failed" feature that defined that display, here oddball-absent displays produced the reverse: a positive bias aimed at directing attention toward items containing the color in the oddball-absent display. We take this result as further converging evidence that the presence and the direction of the biases that are created by homogeneous displays are in fact related to the behavioral success (or failure) that comes to be associated with the features in those homogeneous displays. It is also further evidence against the view that the DPE arises automatically whenever observers are engaged in a search for color oddballs and target-absent trials are included in the experiment.

Together, the results of Experiments 1 and 2 provide good support for the proposal that the DPE bias arises as a consequence of an implicit assessment of behavior, in which current success or failure at the task-at-hand is associated with the visual features responsible for those outcomes. In the case of the "failed" searches that occur on target-absent trials, this assessment produces an attentional bias to try to keep attention away from selecting items that contain that "failed" feature in the future. In Part 2, we show that this rationale can also be applied to other tasks and that the bias created in these implicit assessments is somewhat independent from the search context that produced it.

Part 2: looking for oddballs in time (and space again)

We begin this section with an extension of our logic to a different search context; rather than asking participants to look for a color oddball among spatially distributed distractors, we now ask participants to search for a color oddball among temporally distributed distractors: participants view RSVP displays and identify the case of the oddly colored letter. If our logic holds, then one would expect that after viewing an entire RSVP sequence containing all distractors of the same color, a bias against selecting letters of that "failed" color would be produced. As a result, on the following trial, selection of an odd-colored target should be impaired if the target happens to be of that "failed" color, whereas there should be no particular impairment if the odd-colored target is shown on a different (non-failed) color.

Experiment 3: oddball color search in RSVP sequences

Methods

Participants. Fifteen undergraduates (age range 18–23 years) at the University of Illinois at Urbana-Champaign participated in the experiment.

Apparatus and stimuli. All stimuli were presented on 17-in. Samsung monitors, at a refresh rate of 60 Hz, a resolution of 1280 × 1024, and driven by 3.4 GHz Pentium-4 Dell Optiplex GX620 PCs with 3.5 gb RAM. The experiments were programmed using the Psychophysics Toolbox for Matlab, and stimuli were presented against a black background. Letters were presented in either red, green, or white.

Task and procedure. Participants performed a color-oddball temporal search task. Each trial consisted of a 12-item sequence of English letters presented randomly in uppercase or lowercase (Fig. 5). On oddball-absent trials, all letters were of the same color and participants refrained from responding. On oddball-present trials, one letter was of a different color from all others and participants had to report that its case using the keyboard (up-arrow: uppercase; down-arrow: lowercase). Each character was presented for two monitor refreshes (~33 ms), followed by a blank screen for five refreshes (~83 ms). Targets appeared in approximately half of all trials, and could appear at any temporal position within the trial sequence, except for the first, second, and final positions. A 1000-ms blank interval followed each target-absent trial, during which time no response was required, and a response window followed each target-present trial that lasted until either a response was made or 2500 ms had elapsed.

The experiment consisted of five blocks of 102 trials. The trial sequence was determined such that 82% of all pairs of trials consisted of a target-absent trial followed by a target-present trial. Of these, 25% of trial pairs were distractor-color previewed pairs, 25% were target-color previewed pairs, and 50% were neither-color previewed pairs (when the color of the items in target-absent trial was neither the color of the target nor the color of the distractors on the subsequent target-present trial). The remaining trial pairs consisted of two consecutive target-absent or target-present trials. Responses were marked as incorrect if no response was made during the response interval following a target-present trial.

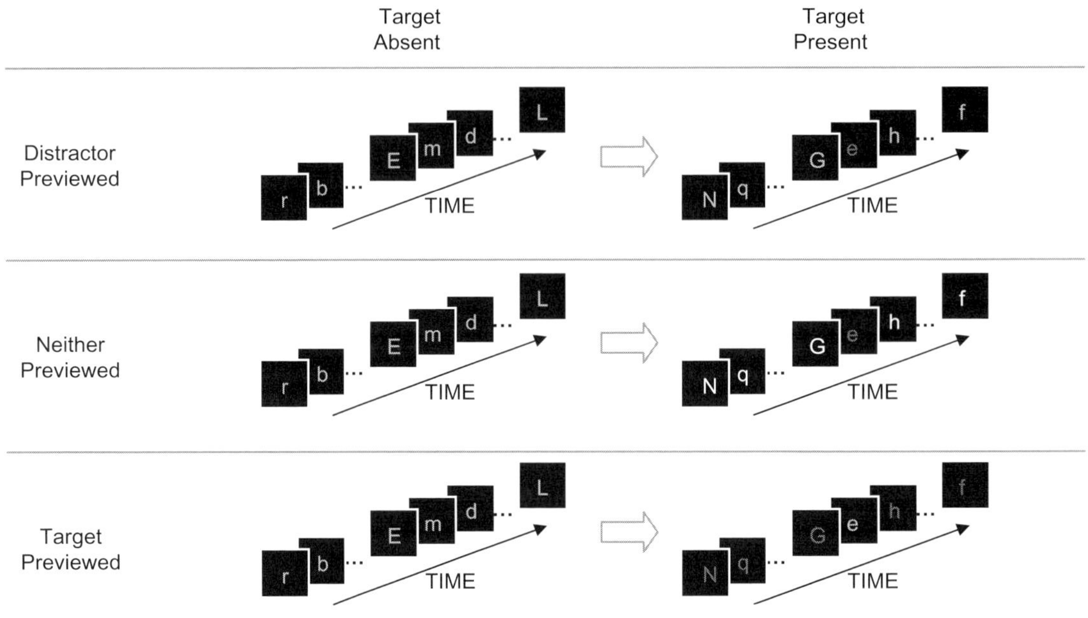

Fig. 5. Schematic representation of trial sequences in Experiment 3. The participants' task is to report the case (upper or lower) of an odd-colored letter in the RSVP stream. Top: distractor-color previewed condition; middle: neither-color previewed condition; bottom: target-color previewed condition (adapted from Lleras et al., 2009).

Results

Participants' data were analyzed by repeated-measures ANOVA, including as factors the preview condition (distractor, target, and neither-color previewed) and the serial position of the target within the RSVP stream. Due to the relatively low number of observations per position ($\sim$11.6 observations per cell), we performed the analysis on the full design with each position (3–11) by preview condition, and also performed a more conservative analysis in which positions were pooled as early (positions 3–5), intermediate (positions 6–8), and late (positions 9–11). Both analyses revealed the same patterns in the data, therefore only the results of the full design will be reported.

The full repeated-measures ANOVA revealed main effects of both target position, $F(8,112) = 8.59$, $p < .001$, and preview condition, $F(2,28) = 18.98$, $p < .001$. Furthermore, the two-way interaction was also significant, $F(16,224) = 5.52$, $p < .001$. Post-hoc comparisons revealed a significantly lower accuracy in the target-color previewed condition compared to the distractor-color previewed condition at position 3 (28.3%), $t(14) = 6.68$, $p < .001$, position 4 (12.2%), $t(14) = 3.38$, $p < .004$, and position 5 (9.1%), $t(14) = 3.81$, $p < .002$, and the difference in accuracy approached significance at position 6 (6.2%), $t(14) = 1.82$, $p > .09$. None of the comparisons at later positions approached significance. Although accuracy for the neither-color previewed condition was consistently lower than that of the distractor-color previewed condition at all positions, none of these differences approached significance (all t's < 1.17, all p's $> .05$). These results are illustrated in Fig. 6.

Discussion

The results of Experiment 3 provide early evidence that the logic we used to account for the DPE can also be extended to temporal search tasks. That is, oddball-absent trials produce an attentional bias that impairs future selection of visual objects containing the "failed" feature. This attentional bias is apparent not only in spatial

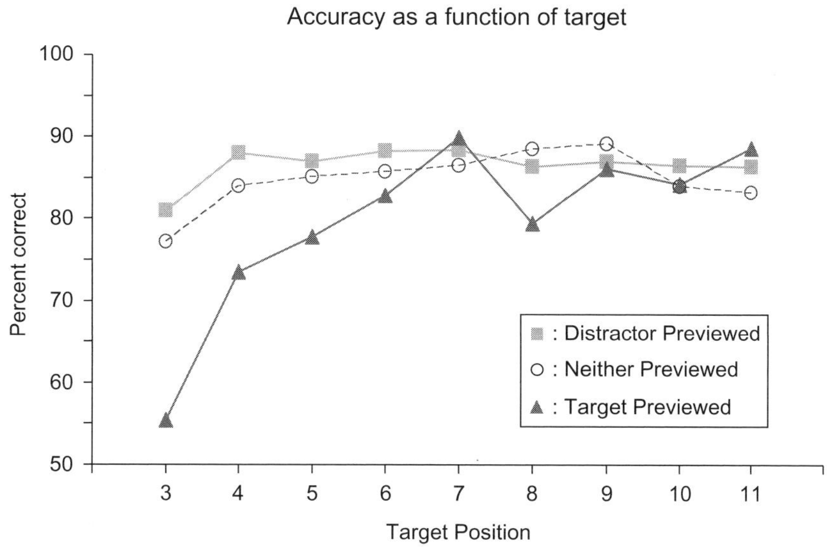

Fig. 6. Results of Experiment 3.

search tasks (as in Experiments 1 and 2) but also in temporal search tasks, when attention has to select a brief target event out of a rapid sequence of distractor stimuli. It should be noted that we have replicated the existence of this attentional bias in temporal searches several times (see Lleras et al., 2009), so we can be confident about the observed pattern of results. The current results also tell us something new: it takes participants about 600 ms to counteract the effects of these failure biases; after position 6 in the RSVP stream, there is no longer an inter-trial effect observed in the target-color previewed condition, possibly signaling the fact that the attentional system has determined that other colors (such as the distractor color on the current trial) are more worthy of being suppressed. Finally, we should point out that the results of Experiment 3, on their own, could conceivably be once again explained in terms of low-level perceptual modulations due to the repeated exposure to a given color in these RSVP streams. However, when put in the context of Experiments 1 and 2, as well as Experiment 5 and a follow-up experiment described in the Discussion section of Experiment 5, it becomes clear that it is not the repetition of a given low-level feature that underlies this effect, but rather that it is driven at a high-level categorical evaluation of behavior, in which a type of stimuli (here, a given color) is associated with a type of behavioral outcome (here, failure). This issue will be further discussed in Part 3 and in the Conclusion.

Experiment 4: intermixing temporal and spatial search tasks

In Experiment 3, we documented that analogous attentional biasing effects are observed in temporal search tasks as in spatial search tasks. A new question now arises: are these biases specific to the task (or environment) that produce them? Or are they specific to the feature itself that they are attached to, and somewhat independent of the search context that produced them? Experiment 4 was designed to address this question. We simply intermixed spatial and temporal search task trials. We asked participants to find color oddballs that could be present either in a sequence of briefly presented letters or in a display of spatially distributed items. The search context was changed randomly from trial to trial, and we looked for crossover biasing effects. That is, does a target-absent RSVP trial produce a bias that will affect the spatial deployment of attention (and vice versa)?

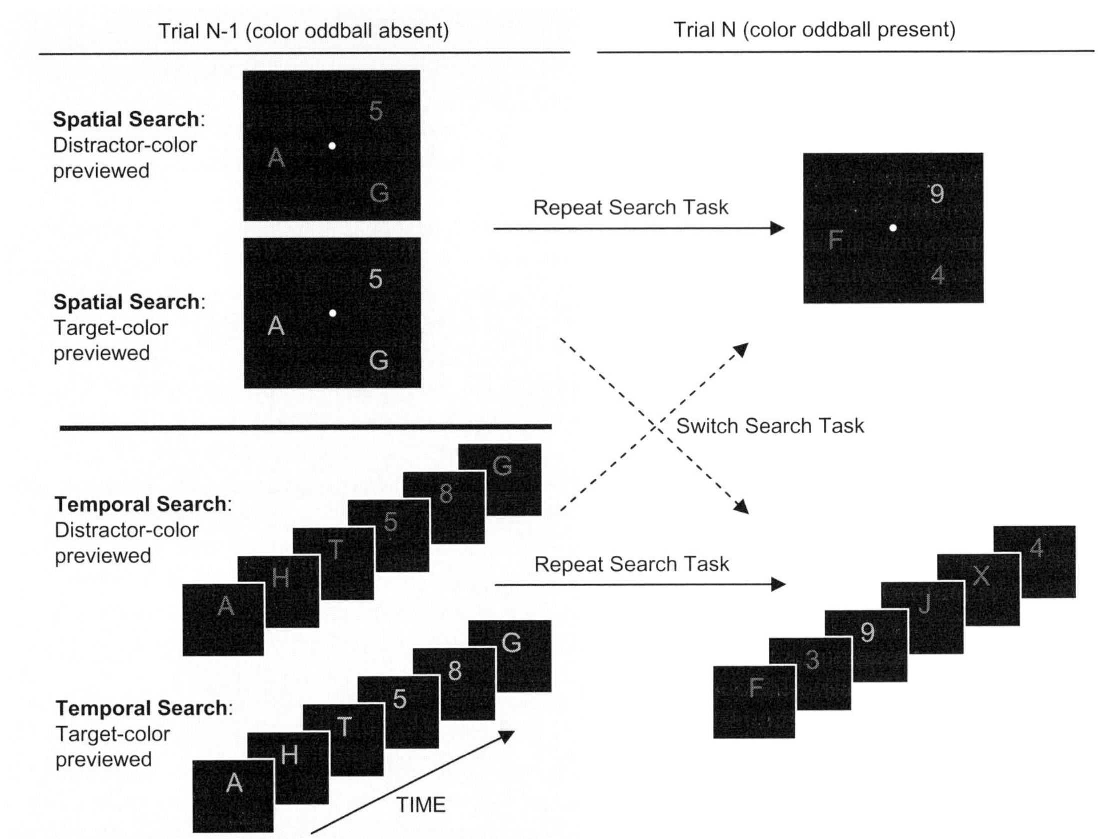

Fig. 7. Schematic representation of trial sequences in Experiment 4. Irrespective of the search modality (space or time), participants were always asked to find the odd-colored item and report its category (letter or number) (adapted from Levinthal and Lleras, 2008b).

Methods

Participants. Fourteen students from the University of Illinois at Urbana-Champaign (age range 21–29 years) participated in the experiment.

Stimuli and apparatus. The stimuli and apparatus were identical to those used in Experiment 3, except for the following. On temporal search tasks, RSVP sequences contained only six characters (English letters and Arabic numerals). Participants were asked to find (if any) the odd-colored character in the sequence and report whether the character was a letter or a number. The target could appear at positions 3, 4, or 5 in the stream. On spatial search tasks, the search stimuli were three colored characters (76 pt Arial font), presented in the periphery, equally spaced from each other along an iso-acuity ellipse surrounding the center of the screen with a horizontal axis of approximately 10° and a vertical axis of approximately 8°. Sample displays are illustrated in Fig. 7.

Procedure. Participants were presented with either a temporal search display or a spatial search display; the type of search was randomly determined prior to each trial. Inter-trial relations were determined along two orthogonal dimensions: color-preview condition (target-color vs. distractor-color previewed conditions), irrespective of search context; and task repetition: a target-present trial was labeled as "repeat", if the preceding target-absent trial had been presented in the same search context (e.g., two consecutive temporal searches), and was labeled a "switch" if the preceding trial had been presented in a different context (e.g., a temporal search followed by a spatial search).

Target-absent displays in the spatial context were presented for 600 ms. There was a variable 800–1300 ms blank screen interval between trials.

Results

In the spatial search task, the DPE was measured as the difference in response time between the target and the distractor-feature previewed trials. In the temporal search task, inter-trial inhibition was measured as the difference in accuracy between these same trials, estimated when the target was at position 3 in the stream (which produced the strongest inter-trial relation in Experiment 3).

Spatial search task. We performed a repeated-measures ANOVA including task repetition and color-preview condition. The analysis revealed a main effect of preview condition, $F(1,13) = 80.12$, $p < .001$, but no main effect of task repetition and no interaction. Paired comparisons indicated that performance on target-present trials in the spatial search context was significantly modulated by the color-preview condition, irrespective of the task context on the preceding trial. That is, a significant DPE occurred following both spatial target-absent trials (60 ms, $t(13) = 4.67$, $p < .001$) as well as temporal target-absent trials (44 ms, $t(13) = 3.04$, $p < .009$), as illustrated in Fig. 8 (Left).

Temporal search task. An analogous ANOVA on accuracy revealed a main effect of preview condition, $F(1,13) = 20.84$, $p < .001$, but no main effect of task repetition or an interaction. Paired comparisons indicated that performance on target-present trials in the RSVP task was significantly modulated by the color-preview condition, irrespective of the task context on the preceding trial. That is, a significant DPE occurred following both temporal target-absent trials (25%, $t(13) = 6.47$, $p < .001$) as well as spatial target-absent trials (15%, $t(13) = 5.69$, $p < .001$), as illustrated in Fig. 8 (Right).

Discussion

The results of Experiment 4 are important because they show a type of inter-task transfer that is seldom observed in cognitive psychology.

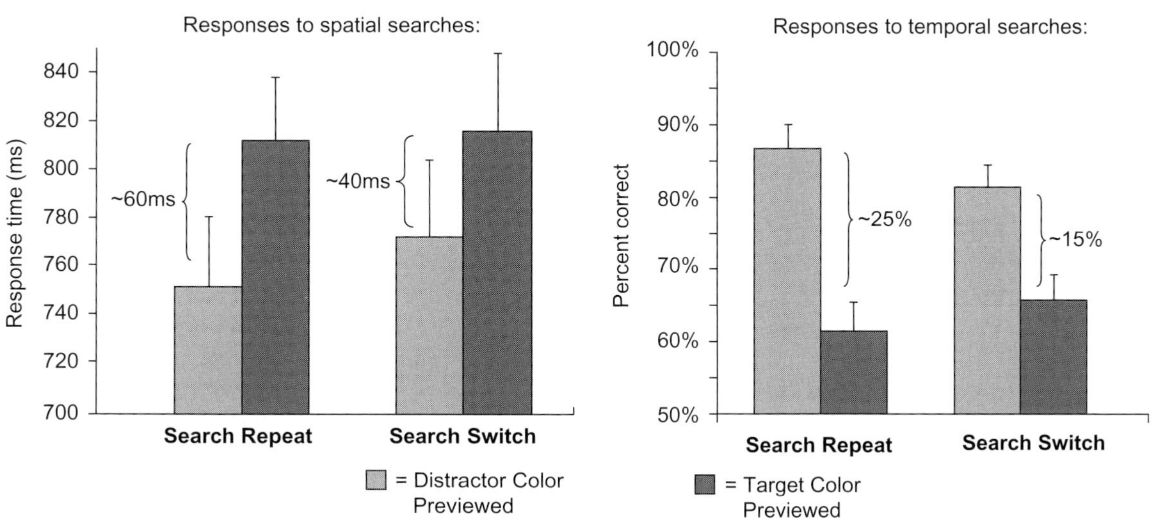

Fig. 8. Results of Experiment 4. Left panel: reaction times in the spatial search task; right panel: accuracies in the temporal search task. The error bars indicate standard errors for each condition.

Specifically, we showed that participants' performance in a temporal search task can have large effects on their performance in a subsequent spatial search task (and vice versa). These results suggest that there is a common attentional mechanism that is being used in both temporal and spatial tasks. Further, our results show that attentional biases can be "context-free": regardless of the search context in which the attentional system judges the value of a feature, the bias against that feature will be present whenever attention is directed to a scene in the future, irrespective of the specific way in which information is presented in that scene (Levinthal and Lleras, 2008b).

Part 3: categorical oddball search

So far, we have shown various demonstrations of how our attentional system evaluates the usefulness of information in the environment, and creates biases based upon this assessment that affect our future deployments of attention (either in space or in time). However, the biases may reside at some fairly low level in vision and, therefore, their use may be restricted only to situations when low-level features can actually drive performance. Instead, we would like to argue that the biases are, in fact, created at a task-relative level: if a task requires a color discrimination, color biases will be produced, but if higher level discriminations are required in the task, we believe similar attentional biases ought to be observed at that higher level. There is already some evidence that this may be the case. Ariga and Kawahara (2004) obtained a DPE when the task required participants to find the odd-gender face in a display containing three faces. However, the experiment used only a very small set of faces. Further, on target-absent trials, all faces were identical pictures of the same person, and on target-present trials, two faces were of the same person and the odd-gender face belonged to a second person. As a result, it is difficult to gauge the extent to which the inter-trial biases observed in that study reflected a true categorical-level bias, that is, a bias that is generalized to any-and-all faces of a given gender. Below we present one final experiment in which we test for category-wide level inter-trial effects.

Experiment 5: faces versus houses

In this experiment, we simply asked participants to look at displays containing three images and find the image that did not belong to the same category as the other two (house among faces or vice versa). All images contained a superimposed red dot that was slightly to the left or right of the picture, and participants were asked to report the location of this dot on the category-oddball image. Most importantly, any given image was only used once in the experiment (no images were ever repeated). Once again, we were interested in whether the category viewed on a target-absent trial would affect the deployment of attention on the subsequent target-present trial.

Methods

Participants. Eleven students from the University of Illinois at Urbana-Champaign (age range 18–21 years) participated in the experiment.

Stimuli, apparatus, task, and procedure. Methods were identical to the ones used in the spatial search task of Experiment 4, except that the stimuli used were square grayscale images of houses or faces (300 px square), approximately matched in overall luminance.

Results

A repeated-measures ANOVA with preview condition and target category (house or face target) as conditions revealed a significant main effect of the preview condition ($\sim$85 ms), $F(1,10) = 21.605$, $p < .001$, but no main effect of the target category, $F(1,10) = .921$, $p > .360$. Furthermore, an interaction between the target category and preview condition did not approach significance, $F(1,10) = 1.14$, $p > .264$. The data are shown in Fig. 9(A), separately for trials in which a house was a target and trials in which a face was a target.

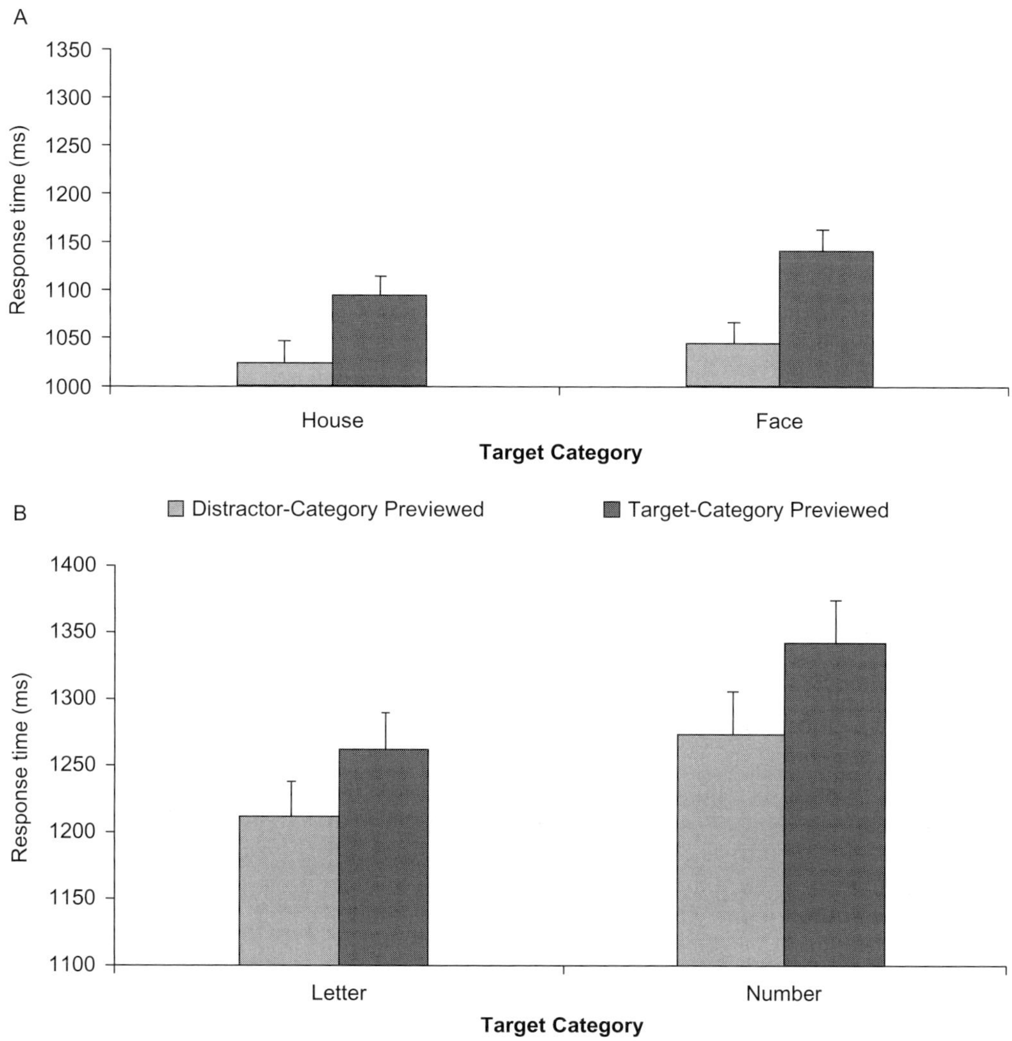

Fig. 9. Results of Experiment 5. (A) Mean reaction times for trials in which a house was a target and trials in which a face was a target. (B) Mean reaction times for trials in which a letter was a target and trials in which a number was a target. The error bars indicate standard errors for each condition.

Discussion

The results show that a category-wide level bias was at play in the experiment: because no face or house was ever shown twice in the experiment, whatever bias was formed on a given target-absent trial must have been associated with the entire category of stimulus present on that trial, and not with any specific member of that category. Skeptical readers may argue that our categorical effect may in fact be a low-level effect: because houses and faces differ dramatically in their low-level visual features, it is possible that the effect we observed here was a bias against some predominantly face-type or house-type low-level features. To rule out this possibility, we reran Experiment 5 using two categories of stimuli that are more closely matched in low-level features: letters and digits. Everything was identical to Experiment 5, except that now participants were asked to identify the odd-category character (and report the position of a red dot that appeared slightly to right or left of that character). Fig. 9(B) shows the results of that experiment. Once again we observed a category-level bias occurring between trials: when letters are targets, RTs are slower if on the preceding trial all items were letters, and RTs are faster if on the preceding trial all items were digits (51 ms; $t(13) = 2.069$, $p < .029$). The same was true for digit targets (67 ms; $t(13) = 2.257$, $p < .021$). This inter-trial effect must really be occurring at a category level, because the letters and digits that we used in the experiment are made up of the same group of low-level features. We should note too that we have also replicated this letter/digit inter-trial category effect in temporal search tasks (see Lleras et al., 2009). These results are remarkable because they suggest that attentional biases can be quite flexible: irrespective of the featural variability within a given category of stimuli, attention can be biased away from focusing on any one member of that category, in very broad fashion.

In sum, our results strongly suggest that the biases we have been observing in these DPE experiments are truly defined at a task-relative level of information processing: when the oddball task is defined at a low-level feature level of scene analysis, attention is biased away from a specific low-level feature, whereas when the oddball task is defined at a higher level, object category level of scene analysis, attention can be biased away from a whole entire range of potential stimuli belonging to a given category of objects. This fits well with our account of the DPE because it emphasizes that the attentional system is sensitive to the implicit assessment of performance that occurs at a task-level of behavioral analysis: was the task completed or not? Was the information present in the environment related to the success or failure in this task? We argue that it is the answers to these questions that determine the type and level of attentional biasing that will be put in place in anticipation of future events.

Finally, we want to draw a link (and a distinction) between this work and the work on "error processing" that has been recently gathering a lot of attention. The work of Jentzsch and Dudschig (2009), among others, discusses the consequences of making an erroneous response on future performance (typically a slowing of RTs). Jentzsch and Dudschig propose that there are at least two mechanisms underlying these RT modulations: a fast error-monitoring process and a slower response threshold adjustment. It is possible that whatever executive control center is responsible for the error-monitoring process in their task may also be involved in the behavioral assessment that we assume is happening in our tasks. More research looking at the similarities between these two effects, in that sense, might be very valuable, perhaps looking at the overlap (or not) of the neural substrates at play. That said, it also seems clear to us that the effects of the assessment are quite different in the two scenarios. In our case, participants are faced with an environment that prevents them from completing a task, and this "processing failure" results in a bias in the attentional system against focusing on similar features/categories in future encounters of similar environments. These biases affect processes occurring prior to response selection and execution (for evidence on this, see Shin et al., 2008). In Jentzsch and Dudschig's case, participants have access to the correct information, but they make an incorrect response to this information, and this "execution failure" results in the stiffing of response-related thresholds. Whereas it is clear that in both cases the cognitive system is undergoing adjustments to improve performance in the near future, the adjustments are taking place at different levels of processing.

Conclusion

Through various demonstrations, we have shown how attention can be biased by our recent behavioral success or failure at a task, in a fairly straightforward manner: information that afforded successful completion of a task is judged positively by the attentional system whereas information that prevented the completion of a task is judged negatively by the attentional system. These assessments are aimed at biasing the system in the presence of environmental uncertainty, probably to increase the chances of success on upcoming events: a bias to direct attention toward a recently "successful" feature is akin to betting on past success (as in POP), and a bias to direct attention away from a recently "failed" feature is akin to betting against recent failures (as in the DPE). From a larger perspective, this account of inter-trial priming might be telling us something more general about the way our cognitive systems interact with our surroundings. These experience-based attentional biases can be viewed, in a way, as an act of tagging environmental information with internally adjudicated valences to direct future behavior. Within the attentional system, positive valence may be like a "look-at-me" tag and negative valence like a "stay-away-from-me" tag.

We should note that these ideas are not new but parallel other proposals in the field. For example, the theoretical work of Aston-Jones and Cohen (2005) describes two modes of attentional interaction with the environment: an "exploration" mode, where attention is encouraged to wander from one stimulus type to the next, and an "exploitation" mode, where attention is encouraged to continue to inspect the same type of information. Similar ideas also underlie some accounts of negative priming (e.g., Tipper, 2001). One can also see a clear parallel with the phenomenon of inhibition of return (IOR; Posner and Cohen, 1984). IOR can be viewed as associating a negative tag with a location that, when inspected, failed to produce a target. That is, IOR may not be an automatic mechanism that prevents re-visitations to previously fixated locations in a scene in general, but rather a task-dependent mechanism aimed at improving performance specifically in those tasks when observers are looking for a spatially uncertain target, as a way to differentiate locations that recently produced at a target, from those that failed to produce one, when last inspected. Dodd et al. (2009) have evidence supporting this view: when participants viewed identical scenes, IOR was only observed when their task was to find a target, but not when participants were memorizing the scene, rating its pleasantness or free-viewing it (when *facilitation* of return was observed). These results nicely parallel our ideas that internally adjudicated valences of environmental information (whether positive or negative) are contextual to the behavioral task that participants are performing. As such, the same information can lead to different valences (positive or negative) depending on the behavioral task.

More intriguing still is the parallel with recent work by Raymond and colleagues on *distractor devaluation* studies, which suggests strong links between attention and emotional valences (e.g., Goolsby et al., 2009, in press; Raymond et al., 2003). This work shows how previously ignored stimuli tend to be consistently judged with a more negative emotional valence (e.g., "less attractive", "less trustworthy") than those selected as targets. In the context of our current results, this work suggests what could be further ramifications for "failed" information in the environment: not only will it be less likely to summon attention (as the current work suggests), but it may also be emotionally devalued.

In sum, we think that there are strong commonalities between all these phenomena and theoretical perspectives on the workings of attention, which together frame a picture of attention-by-experience effects as influencing not only the future workings of attention but also other cognitive systems, such as emotional evaluation and categorization processes. Future work will explore these intriguing new avenues of research.

Acknowledgement

This work was partly supported by a grant from the National Science Foundation to Alejandro Lleras, award number BCS 07-46586 CAR.

References

Ariga, A., & Kawahara, J. (2004). The perceptual and cognitive distractor-previewing effect. *Journal of Vision*, *4*, 891–903.

Aston-Jones, G., & Cohen, J. (2005). An integrative theory of locus coeruleus-norepinephrine function: Adaptive gain and optimal performance. *Annual Review of Neuroscience*, *28*, 403–450.

Bar, M., & Biederman, I. (1998). Subliminal visual priming. *Psychological Science*, *9*, 464–469.

Bichot, N. P., & Schall, J. D. (2002). Priming in macaque frontal cortex during popout visual search: Feature-based facilitation and location-based inhibition of return. *Journal of Neuroscience*, *22*, 4675–4685.

Brainard, D. H. (1997). The Psychophysics Toolbox. *Spatial Vision*, *10*, 433–436.

Chun, M. M., & Jiang, Y. (1998). Contextual cueing: Implicit learning and memory of visual context guides spatial attention. *Cognitive Psychology*, *36*, 28–71.

Dodd, M. D., Van der Stigchel, S., & Hollingworth, A. (2009). Novelty is not always the best policy: Inhibition of return and facilitation of return as a function of visual task. *Psychological Science*, *20*, 333–339.

Eimer, M. (1996). The N2pc component as an indicator of attentional selectivity. *Electroencephalography and Clinical Neurophysiology*, *99*(3), 225–234.

Fecteau, J. H. (2007). Priming of pop-out depends upon the current goals of observers. *Journal of Vision*, *7*, 1–11.

Fiser, J., & Biederman, I. (2001). Invariance of long-term visual priming to scale, reflection, translation, and hemisphere. *Vision Research*, *41*, 221–234.

Forster, K. I., Davis, C., Schoknecht, C., & Carter, R. (1987). Masked priming with graphemically related forms: Repetition or partial activation? *The Quarterly Journal of Experimental Psychology*, *39*, 211–251.

Frings, C., & Wühr, P. (2007). Prime display offset modulates negative priming only for easy-selection tasks. *Memory & Cognition*, *35*, 504–513.

Goolsby, B. A., Grabowecky, M., & Suzuki, S. (2005). Adaptive modulation of color salience contingent upon global form coding and task relevance. *Vision Research*, *45*, 910–930.

Goolsby, B., Raymond, J. E., & Shapiro, K. L. (in press). Affective consequences of effortful attention. *Visual Cognition*.

Goolsby, B., Shapiro, K. L., & Raymond, J. E. (2009). Distractor devaluation requires working memory. *Psychonomic Bulletin & Review*, *16*, 133–138.

Gratton, G., Coles, M. G. H., & Donchin, E. (1992). Optimizing the use of information: Strategic control of activation of responses. *Journal of Experimental Psychology. General*, *121*, 480–506.

Hick, W. E. (1952). On the rate of gain of information. *The Quarterly Journal of Experimental Psychology*, *4*, 11–26.

Hillstrom, A. P. (2000). Repetition effects in visual search. *Perception & Psychophysics*, *62*, 800–817.

Huang, L., Holcombe, A. O., & Pashler, H. (2004). Repetition priming in visual search: Episodic retrieval, not feature priming. *Memory & Cognition*, *32*, 12–20.

Huang, L., & Pashler, H. (2005). Expectation and repetition effects in searching for featural singletons in very brief displays. *Perception & Psychophysics*, *67*, 150–157.

Itti, L., & Koch, C. (2001). Computational modeling of visual attention. *Nature Reviews. Neuroscience*, *2*, 194–203.

James, T. W., Humphrey, G. K., Gati, J. S., Menon, R. S., & Goodale, M. A. (2000). The effects of visual object priming on brain activation before and after recognition. *Current Biology*, *10*, 1017–1024.

Jentzsch, I., & Dudschig, C. (2009). Why do we slowdown after an error? Mechanisms underlying the effects of posterror slowing. *The Quarterly Journal of Experimental Psychology*, *62*, 209–218.

Kems, J. G., Cohen, J. D., MacDonald, A. W., Cho, R. Y., Stenger, V. A., & Carter, C. S. (2004). Anterior cingulate conflict monitoring and adjustments in control. *Science*, *303*, 1023–1026.

Kraut, A. G., Smothergill, D. W., & Farkas, M. S. (1981). Stimulus repetition effects on attention to words and colors. *Journal of Experimental Psychology. Human Perception and Performance*, *7*, 1303–1311.

Kristjánsson, A. (2006). Simultaneous priming along multiple feature dimensions in a visual search task. *Vision Research*, *46*, 2554–2570.

Kristjánsson, Á., Vuilleumier, P., Schwartz, S., Macaluso, E., & Driver, J. (2007). Neural basis for priming of pop-out revealed with fMRI. *Cerebral Cortex*, *17*, 1612–1624.

Kristjánsson, Á., Wang, D., & Nakayama, K. (2002). The role of priming in conjunctive visual search. *Cognition*, *85*, 37–52.

Laming, D. R. J. (1968). *Information theory of choice reaction times*. London: Academic Press.

Lamy, D., Carmel, T., Egeth, H. E., & Leber, A. B. (2006). Effects of search mode and inter-trial priming on singleton search. *Perception & Psychophysics*, *68*(6), 919–932.

Levinthal, B. R., & Lleras, A. (2008a). Inter-trial inhibition of attention to features is modulated by task relevance. *Journal of Vision*, *8*(15), 12.1–12.15.

Levinthal, B. R., & Lleras, A. (2008b). Context-free inhibition: Attentional biases transfer strongly across temporal and spatial search tasks. *Visual Cognition*, *16*, 1119–1123.

Lleras, A., Kawahara, J., & Levinthal, B. R. (2009). Past rejections lead to future misses: Selection-related inhibition produces blink-like misses of future (easily detectable) events. *Journal of Vision*, *9*(3), 26.1–26.12.

Lleras, A., Kawahara, J., Wan, X. I., & Ariga, A. (2008). Inter-trial inhibition of focused attention in pop-out search. *Perception & Psychophysics*, *70*, 114–131.

Luck, S. J., & Hillyard, S. A. (1994). Spatial filtering during visual search: Evidence from human electrophysiology. *Journal of Experimental Psychology. Human Perception and Performance*, *20*, 1000–1014.

Maljkovic, V., & Nakayama, K. (1994). Priming of popout: I. Role of features. *Memory & Cognition*, *22*, 657–672.

Maljkovic, V., & Nakayama, K. (1996). Priming of popout: II. Role of position. *Perception & Psychophysics*, *58*, 977–991.

Maljkovic, V., & Nakayama, K. (2000). Priming of popout: III. A short term implicit memory system beneficial for rapid target selection. *Visual Cognition*, *7*, 571–595.

Marcel, A. J. (1983). Conscious and unconscious perception: An approach to the relations between phenomenal experience and perceptual processes. *Cognitive Psychology*, *15*, 238–300.

Olivers, C. N. L., & Humphreys, G. W. (2003). Visual marking inhibits singleton capture. *Cognitive Psychology*, *47*, 1–42.

Pashler, H., & Baylis, G. (1991). Procedural learning: 2. Intertrial repetition effects in speeded choice tasks. *Journal of Experimental Psychology. Learning, Memory, and Cognition*, *17*, 33–48.

Pelli, D. G. (1997). The VideoToolbox software for visual psychophysics: Transforming numbers into movies. *Spatial Vision*, *10*, 437–442.

Posner, M. I., & Cohen, Y. (1984). Components of visual orienting. In H. Bouma & D. Bonwhuis (Eds.), *Attention and performance X: Control of language processes* (pp. 551–556). Hillsdale, NJ: Erlbaum.

Potter, M. C. (1999). Understanding sentences and scenes: The role of conceptual short term memory. In V. Coltheart (Ed.), *Fleeting memories* (pp. 13–46). Cambridge, MA: MIT Press.

Rabbitt, P. M. A., & Rogers, B. (1977). What does man do after he makes an error? An analysis of response programming. *The Quarterly Journal of Experimental Psychology*, *29*, 232–240.

Raymond, J. E., Fenske, M. J., & Tavassoli, N. T. (2003). Selective attention determines emotional responses to novel visual stimuli. *Psychological Science*, *14*, 537–542.

Ridderinkhof, K. R., Nieuwenhuis, S., & Bashore, T. R. (2003). Errors are foreshadowed in brain potentials associated with action monitoring in cingulate cortex in humans. *Neuroscience Letters*, *348*, 1–4.

Rovano, J., & Virsu, V. (1979). Visual resolution, contrast sensitivity, and the cortical magnification factor. *Experimental Brain Research*, *37*, 475–494.

Shin, E., Wan, X. I., Fabiani, M., Gratton, G., & Lleras, A. (2008). Electrophysiological evidence of feature-based inhibition of focused attention across consecutive trials. *Psychophysiology*, *45*, 804–811.

Sturmer, B., Leuthold, H., Soetens, E., Schroter, H., & Sommer, W. (2002). Control over location-based response activation in the Simon task: Behavioral and electrophysiological evidence. *Journal of Experimental Psychology. Human Perception and Performance*, *28*, 1345–1363.

Tipper, S. P. (1985). The negative priming effect: Inhibitory priming by ignored objects. *The Quarterly Journal of Experimental Psychology*, *37A*, 571–590.

Tipper, S. P. (2001). Does negative priming reflect inhibitory mechanisms? A review and integration of conflicting views. *The Quarterly Journal of Experimental Psychology*, *54A*, 321–343.

Treisman, A. M., & Gelade, G. (1980). A feature-integration theory of attention. *Cognitive Psychology*, *12*, 97–136.

Wolfe, J. M. (1994). Guided search 2.0: A revised model of visual search. *Psychonomic Bulletin & Review*, *1*, 202–238.

Wolfe, J. M., Butcher, S. J., Lee, C., & Hyle, M. (2003). Changing your mind: on the contributions of top-down and bottom-up guidance in visual search for feature singletons. *Journal of Experimental Psychology. Human Perception and Performance*, *29*, 483–502.

N. Srinivasan (Ed.)
Progress in Brain Research, Vol. 176
ISSN 0079-6123

CHAPTER 14

Attentional limits and freedom in visually guided action

James T. Enns* and Geniva Liu

Department of Psychology, University of British Columbia, Vancouver, BC, Canada

Abstract: Human vision supports both the conscious perception of objects (e.g., identifying a tea cup) and the control of visually guided action on objects (e.g., reaching for a tea cup). The distinction between these two functions is supported by neuroanatomy, neuropsychology, animal lesion studies, psychophysics and kinematics in healthy humans, as well as in overarching theories of the visual brain. Yet research on visual attention, which concerns limitations in processing information concurrently from multiple objects, has so far not fully exploited this functional distinction. Attention research has focused primarily on the conscious aspects of vision, to the relative neglect of the unconscious control of limb actions and whether we can perceive some objects concurrently while acting on others. Here we review research from our lab leading to the conclusions that (1) the finger is guided visually by an *automatic pilot* that uses different information from that of conscious vision, (2) conscious object identification interferes with concurrent planning of pointing to a second object, though not with the online control needed to complete the pointing action, (3) concurrent perception and action sometimes lead to benefits in motor performance, and (4) the automatic pilot is itself capacity limited in processing information concurrently from multiple locations. These findings help clarify the conditions under which interference-free multitasking is possible and point to new challenges for research on the attentional limits of unconscious visual processing.

Keywords: cognitive map; blocking; associative learning; spatial learning

Introduction

It is time for research on attention to catch up with recent developments in our understanding of how the human brain uses visual information to perform distinctly different functions. Although there is only one stream of light that impinges on our eyes at any moment in time (hereafter referred to as *the image*), the information contained in that image can be used to serve one of two functions. On the one hand, the image can be used to construct a visual experience of the environment around us, allowing us to recognize a colleague or to discern whether we are viewing scissors or a spoon. Yet the same image can also guide our limb actions so that we interact appropriately with that environment, allowing us to shake our colleague's hand when we greet them and to pick up scissors using a different grasp than we use when picking up a spoon. Research on human vision has demonstrated in numerous ways

*Corresponding author.
Tel.: 604-822-6634; Fax: 604-822-6923;
E-mail: jenns@psych.ubc.ca

DOI: 10.1016/S0079-6123(09)17612-4

over the past 25 years that these two functions are distinct (Goodale and Milner, 2004; Goodale et al., 2004; Milner and Goodale, 1995; Norman, 2002; Ungerleider and Mishkin, 1982). Specifically, our visual experience in the absence of overt action (the conscious experience of objects at a distance, subserved by the *ventral* visual stream) is governed by brain regions, neural tracts, and mechanisms that are distinct from those that guide our interactions with objects (unconscious control of visually guided action, subserved by the *dorsal* stream). It is not our purpose to review those arguments here, as they have been made extensively in many reviews. Our purpose is, instead, to consider some consequences of this understanding of the visual brain for research on visual attention.

Research on visual attention over the past 150 years can be characterized, at a first approximation, as the study of channel limitations of the human brain as it concerns conscious experience (see Pashler, 1994, 1998; Shapiro et al., 1997 for reviews). We seem to be aware, and to become aware, of little more than one discrete event at a time. If we are expecting imminent visual information from one location, we will be delayed in processing the expected visual information at another location (e.g., location orienting, as in Posner, 1980). If we are engaged in identifying one person, the identification of a second person must wait (e.g., two object cost, as in Duncan, 1984). Even when we are focused solely on viewing a single person's face, if we are engaged in the process of identifying that person, we will be impaired in making judgments of the emotional expression currently displayed by that person (e.g., we are blind to changes that are deemed unlikely, as in Austen and Enns, 2003). If, during the performance of any of these visual tasks, we temporarily allow our thoughts to drift among the ideas constantly being offered to our awareness by our long-term memories, then our ability to identify and localize visual information becomes impaired (e.g., mind wandering, as in Giambra, 1995).

Relative to this intense interest in attentional limits on conscious aspects of seeing, much less effort has been devoted to developing methods for exploring limits of the visually guided action system, and for exploring the conditions under which conscious vision and visually guided action might interfere with one another, might operate independently of one another, or perhaps might even mutually enhance one another. Among the obstacles encountered in the design of experiments to probe unconscious visually guided action is that it is often difficult to provide input to the action system that does not itself have to pass through the bottleneck of consciousness before it can be used to inform the visually guided action system. Take, as a case in point, dual-task studies already in the literature that have combined tasks of action and perception in an effort to measure possible interference when these systems must use the same visual image. Some of these studies required participants to point to one colored shape while simultaneously trying to identify a symbol in a separate location (Deubel et al., 1998). Other studies required participants to grasp a target object while simultaneously monitoring for changes in the luminance of a second object (Castiello, 1996). Since pointing and grasping are thought to be under *dorsal* stream control and object identification under *ventral* stream control, these could be construed as proper tests of cross-stream interference. And since significant task interference was observed in both studies, one might conclude that efficient multitasking is not possible between the visual streams. However, we do not consider these results to be strong tests, mainly because in order to carry out the limb action required in both types of studies, the color or the shape of an item has to be processed (*ventral* stream function) before the appropriate action (*dorsal* stream function) can be initiated.

A second difficulty encountered when designing experiments to probe unconscious visually guided action separately from conscious vision concerns the important distinction between action planning (or preparation) versus the online control of that action once it is already underway (Henry and Rogers, 1960). Action planning is generally considered to involve processes that occur prior to action initiation, and therefore may be influenced by the *ventral* stream as well as other consciously accessible brain regions such as the *frontal lobe* functions implicated in executive task control. It is typically indexed by measuring the period from target onset to movement onset (response initiation time), and as such, is potentially influenced by the mental processes of target identification, response selection, and movement planning (or preprogramming). In contrast, action execution consists of processes involved in bringing the action to completion and is usually considered more uniquely a function of the *dorsal* stream system. It is indexed by the time that elapses between action onset and action completion (movement time, MT) and is thus uniquely influenced by processes that occur only once an initiated action is already underway.

An example of research that ignores this distinction comes from recent report by Kunde et al. (2007). These authors paired an auditory tone discrimination task (*ventral* stream) with either a visual size discrimination task (*ventral* stream) or a visually guided grasping task (ostensibly *dorsal* stream). Their results indicated that response time in both visual tasks was interfered with when these tasks were paired with the auditory task. Although, the dependent measures taken in the grasping task included both response time (preparation) and MT (online control) only the response time data were reported in detail, leaving open the possibility that interference only occurred during the planning phases of the action. Consistent with this possibility was a terse one-line report by the authors that they found no reliable influences of dual-task performance on MT. But this null result was not explored in any greater detail, even though it suggested the possibility that online control of grasping was immune from the task interference measured for action initiation. Our reading of the existing literature on dual-task performance involving ventral and dorsal stream functions therefore suggests that much remains to be done. In particular, there is a need for research involving *dorsal* stream tasks that (1) does not rely on consciously processed input for their initiation, and (2) allow for the separate measurement of action planning (or preparation) from the online control of already initiated actions.

A model task for studying online control: the finger's automatic pilot

Pressing an elevator button while we are walking toward the door, reaching to grasp someone's hand as we both move toward one another, and striking a tennis ball that has just tipped the top of the net, are all highly sophisticated visual-motor skills that we as healthy humans take for granted. Many complex computations are involved, though they remain hidden from our ability to access them through conscious introspection. This human ability to make rapid, online adjustments in pointing and reaching in response to unexpected changes in the task environment, is often referred to the body's *automatic pilot* (Bard et al., 1999; Brenner and Smeets, 2003; Castiello et al., 1991; Desmurget et al., 1999; Paulignan et al., 1991; Pisella et al., 2000; Prablanc and Martin, 1992). By this, researchers mean that limb movements are modified rapidly and in flight (often called *online*) to changes in the location or shape of a target object, and that these limb modifications occur before or in the absence of individuals being consciously aware of the changing task environment.

A laboratory model for studying the finger's automatic pilot is the *double-step task* pioneered by Goodale et al. (1986). In this task, participants move their finger quickly and accurately from a centrally fixated home position to a suddenly appearing target in the visual periphery. At the onset of the target, participants typically make both an initial saccadic eye movement and a finger pointing movement in its direction. The first saccade is followed rapidly by a corrective saccade, allowing the higher-resolution foveal information to guide the finger precisely to the target location. The actions guiding the eye and the finger to the target location are referred to as the *first step*. The critical manipulation in this task is an unexpected displacement in the target location on a random half of the trials, occurring near the time the initial saccade reaches peak velocity. This displacement of the target is referred to as the *second step* (sometimes also the *jump*). Although this displacement is large enough to be visible when viewed without eye movements (up to 10% of the total movement), participants seem unaware of the target jumps when they occur during a saccade. However, the finger demonstrably takes the jump into account in coming to its final resting place, while at the same time, the conscious brain seems blissfully unaware that a double-step has occurred.

Direct evidence that the *dorsal* stream is involved in the online control of double-step pointing comes from studies of patients with dorsal stream brain damage, who are able to point much more accurately on single-step trials than on double-step trials (Pisella et al., 2000), and from similar results in studies using transcranial magnetic stimulation to inhibit the same brain region on otherwise healthy individuals (Desmurget et al., 1999). The dorsal stream, therefore, appears critical for rapid online adjustments to actions that are already underway, and it can effect these changes in as little time as 200 ms.

Decoupling action and awareness

In our first exploration of double-step pointing, we asked whether the claims made by Goodale et al. (1986) for a complete dissociation between the finger's automatic pilot and conscious awareness of target displacements was really justified. In the original study, Goodale et al. had only questioned their participants about seeing the target jump after a series of many pointing trials. Their study had also confounded the magnitude of the jump with visual awareness (and therefore also the extent of online action correction), such that large jumps ($>10\%$ of the distance from home) were both visible to participants and resulted in measurable kinematic corrections of the finger, whereas small jumps ($<10\%$ of distance) were both invisible to conscious perception and yielded no measurable kinematic differences between single- and double-step trials.

The novel features of our study (Fecteau et al., 2001) included giving participants full knowledge that single- and double-step trials would occur in equal proportions and asking them to indicate on each trial whether they were able to discriminate a jump (double-step) from a stationary (single-step) target. This is the strictest possible test of conscious awareness, since it is based on an objective threshold (Merikle and Cheesman, 1987). If jump sensitivity was greater than could be guessed by chance among our participants, it would mean that some signal concerning the jump was accessible to awareness even though participants may not have had the expected subjective experience of seeing a jump (i.e., proprioception of the limb, extra-retinal signals from the eyes).

The second feature of our study was a factorial combination of target jumps that were either small or large (space) and timing of the jumps that occurred either near maximum saccadic velocity or 100 ms later (time). Our intent was to compare small jumps that were visible (because their timing was delayed) with small jumps that were invisible (because they occurred when the eye was moving most rapidly away from the home position. If movement correction on jump trials was unaffected by awareness, then movement kinematics should be unaffected by jump visibility. We measured the activity of the eye using electrooculography (EOG), sampled at 500 Hz (every 2 ms) and we sampled the movement of the limb using a handheld stylus at a rate of 118 Hz (every 8.5 ms).

The results were clear. Our examination of the kinematic data indicated that every one of the stimulus factors had an influence on movement parameters. For example, the trajectory of the moving limb was influenced in predictable ways by whether or not the target jumped (double-step), whether the jump was small or large (space), and whether the jump was immediate or delayed by 100 ms (time). These kinematic results point to an automatic pilot in the finger that is exquisitely responsive, in an online fashion, to the details of the relationship between the visual target and the moving limb.

The results for the trial-by-trial measures of awareness of the target jump were equally clear. In no case did conscious awareness of the jump have any influence on the kinematics of finger movement. This was true when we compared movement on small-immediate jump trials on which participants reported seeing the jump with the same type of trials on which they reported not seeing any jump. It was also true when we compared the clearly visible delayed-small jump trials with the invisible immediate small-jump trials. In none of the conditions were we able to detect an influence on kinematics that was related to participants' reports of whether or not they detected a jump.

We interpreted these results as supporting two conclusions. First, the results pointed to a more sophisticated form of online control and feedback than some researchers had thought necessary for this task (Jeannerod, 1988; Keele, 1981). If single- and double-step trials were equivalent in that they each began with the same ballistic eye movement and initial ballistic pointing action to the general region (coarse coding), followed up by a corrective saccade and finer control in the limb action to complete the point based on the final target position (fine coding), then the kinematics of small single- and double-step trials should not differ. The fact that the kinematics differed suggests that there is an important form of control that occurs while the eye and the finger are in flight; a form of visual control that uses visual information that is not necessarily accessible to conscious awareness.

Our second conclusion was that there was perhaps an even sharper degree of visual stream separation in the online control of action than some researchers had proposed (Goodale et al., 1986). In particular, we argued that these data rule out the possibility that kinematic features of the movement are being influenced by conscious awareness of where the target appears to be. When a kinematic correction occurs, it does so in the same way regardless of whether or not the change in the environment that caused the correction has also been seen by the conscious brain.

Can the finger reveal its secrets?

In another study of the automatic pilot (Chua and Enns, 2005), we asked whether spatial movements of the finger, either through our direct visual perception of them or perhaps through our inner proprioception of them, could inform the conscious brain when they were being guided by information that was not otherwise accessible to a participant. An important arena in which the action systems (i.e., eye, hand) and conscious awareness are guided by different aspects of the same visual scene occurs with regard to perception of an object's location in space.

When it comes to conscious awareness of whether an object is stationary or has moved, the conscious judgment is based on perceived continuity between views of the same scene. For example, if the target object of the saccade is displaced a large distance while the eye is in motion, its displacement is readily detected (Bridgeman et al., 1975; Fecteau et al., 2001). Also, if the target object is extinguished briefly (blanked) during the saccade and then redisplayed after the saccade is complete, even small image displacements can be detected (Deubel and Schneider, 1994; Deubel et al., 1996). Taken together, these results suggest that when objects are deemed continuously present during a saccade, they are judged to be stationary (within some bound of spatial tolerance that is estimated as large as 50% of saccade distance by some, e.g., Deubel et al., 1998) and when they are deemed to be discontinuous during the saccade, they are judged to have moved.

Chua and Enns (2005) asked whether these rules also apply when participants were engaged in finger pointing. We thought that finger pointing, unlike eye movements, might reveal to participants that a target was actually displaced during the arm motion. This is because, unlike the profound insensitivity we have for our own eye movements, we are fully conscious that we are pointing with our finger. We can see them move and can even monitor movements through proprioception. Our question was whether this information could be made available for conscious report.

To test this possibility, we conducted what we think must be the simplest possible *change blindness* experiment (Rensink, 2002). As in the previous experiment, each trial began with the participant's eyes and finger resting at the home position on a display table. The sudden appearance of two objects on the right side of the panel was a signal for the participant to point as rapidly as possible to the lower of the two objects (the lower one was always the pointing target; the upper object served only as a reference point and was itself never a target). We monitored both the eye movement (using electrooculography) and the hand movement made to the target (using an Optotrak motion analysis system) at 500 Hz. During the first eye movement away from the home position, one of the two objects was spatially displaced by 2 cm (a "jump") and, independently either one, both, or neither of the two objects was extinguished for 100 ms (a "blank") before being redisplayed. At the end of the pointing action, participants were forced to choose which object had *jumped* during the trial. This allowed us to associate a measure of visual awareness of object stability with kinematic measures of hand position on every trial.

The results were very dramatic in showing that the movement of the finger was governed by completely different visual information than the perception of object stability. Although the finger's position was ruled very reliably by the final position of the target (i.e., pointing accurately regardless of whether the target was stationary or jumped), the perceptual judgment of which object had remained stationary was governed entirely by which object had remained continuously visible (i.e., if an object was not blanked, it was judged to be stationary). Conversely, if an object had blanked briefly while the eyes and finger were in flight, that is, the object blanked while staying in the same place or while jumping, then it appeared as though the blanked object had moved.

We conducted an even more stringent test of this dissociation by repeating the experiment, but this time asking participants to *stop* (to withdraw their fingers from the table entirely) whenever they detected that the target they were moving toward had jumped. These results showed that the finger tracked the target position just as reliably as before, with very few finger withdrawals occurring prior to completion of the point. However, when the finger was withdrawn on the roughly 50% of trials on which participants thought that the target had jumped, the withdrawal was based on whether target visibility had been interrupted (blanked) briefly during the saccade to the target, rather than on whether the target had actually jumped. This meant that finger withdrawals occurred equally for jumping targets and for completely stationary targets; the only criterion for withdrawal was that these targets had blanked during the flight of the eye and finger.

Unconscious does not mean inflexible

Establishing the existence of an unconscious influence on visually guided behavior does not in itself imply that the influence is occurring online, that is, during the recruitment and implementation of the appropriate motor routines to complete a task. An unconscious influence could, in principle, come about by the initiation of a response that had already been consciously prepared at an earlier time. For example, as a consequence of the instructions provided by the experimenter at the beginning of some tasks, an action plan may be established, or preprogrammed. If this plan is in place prior to the onset of a trial, events processed subliminally during the trial may be sufficient to initiate the preprogrammed response. Indeed, this is what many researchers believe underlies many unconscious perceptual priming effects. Thus, to firmly establish that ongoing visually guided actions can be controlled by subliminal visual events in a dynamic way, it is necessary to show that the priming itself occurs while the motor actions are ongoing. Moreover, once online subliminal control has been established, it opens up the possibility of exploring the intelligence (i.e., the adaptability or flexibility to changing circumstances) of the unconscious action system.

These were the challenges we set upon in Cressman et al. (2007). In this study, we asked whether the shape of an unseen object could influence an action after a different action had already been initiated. Figure 1 illustrates the spatial layout of the stimuli and the trial sequence. Participants initiated pointing to a center target location, which triggered the appearance of a stimulus shape (left arrow, right arrow, or composite pattern of superimposed left and right arrows) in that location. If the central shape was the composite pattern (75% of the time), they were to continue pointing to the central pattern, but if the central shape was a left- or right-pointing arrow (25% of the time), they were to

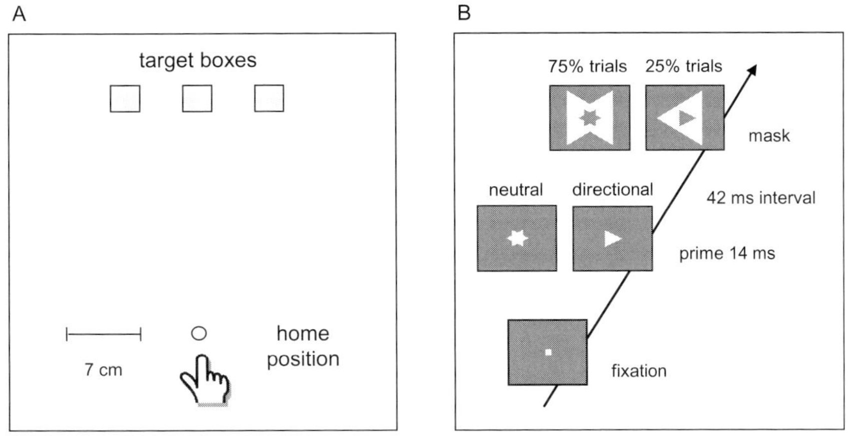

Fig. 1. The spatial layout of the experiment (A) and the sequence of trial events (B) as given in Cressman et al. (2007).

point instead to a left or a right target location as rapidly as possible.

Unbeknownst to participants, upon pointing initiation, the large visible shapes that appeared in the central pointing location were preceded by smaller and briefly flashed prime shapes that were made invisible by the large mask shapes that followed. We reasoned that if the masked prime shapes could influence the online control of a goal-directed action, we would observe deviations in movement trajectories that corresponded to the identity of the unseen primes. Furthermore, there should be a difference in the pointing trajectories observed for congruent (small and large shapes point in the same direction) and incongruent (small and large shapes point in different directions) shape sequences. On the other hand, if the invisible prime shapes are only able to influence the initiation of an already-prepared action and not online control, then deviations in the pointing trajectories should only be measured in response to the direction indicated by the visible mask, and congruent versus incongruent prime-mask sequences should not exert any influence on pointing trajectories.

After the pointing task, we ensured that the small shapes in our procedure were inaccessible to participants' consciousness by measuring their visibility when presented immediately before the large shapes. When participants were fully informed of the presence of the small shapes and encouraged to guess their identity, their accuracy was quite low. However, pointing was influenced by the small unseen shapes, replicating numerous previous studies indicating *masked priming* despite a lack of consciousness of the stimuli governing the priming. Additionally, the kinematic measures of pointing indicated that invisible shapes can not only alter the speed with which a goal-directed action can be initiated, as in many other previous studies (Ansorge et al., 2002; Breitmeyer et al., 2004; Neumann and Klotz, 1994; Schmidt, 2002), but that invisible shapes also influence the control of an ongoing action.

The main evidence for this conclusion came from a detailed look at the pointing trajectories for the different shape sequences. These showed that initial deviations of the finger, away from the center target, were consistently in the direction indicated by the first small shape, regardless of the direction of the subsequent larger visible shape. Thus, *congruent* shape sequences gave the pointing action a head start in the correct direction, with these modifications occurring within 277 ms of movement onset. By way of comparison, *incongruent* shape sequences resulted in initial trajectory deviations toward the wrong target, with the consequence that additional MT was required for participants to complete the action. Pointing trajectories were not directed to the correct target until approximately 330 ms into the movement, a delay that was roughly equivalent to difference in the onset of the first and second shapes (56 ms). This suggests that pointing relevant visual information was being incorporated into control of the pointing almost as soon as it was available.

We concluded that the automatic pilot is indeed flexible enough to respond to the differences in shape between left- and right-pointing arrows and that it can respond to changes in this information even while an already programmed action is underway. This is consistent with other recent findings of unconscious action responding to visual size (Binsted et al., 2007). The question we turn to next is whether the visually guided control of actions such as this can be accomplished simultaneously with tasks of conscious perception, or whether the usual interference that we have come to expect when two visual tasks are combined will occur.

Attention sharing between visual identification and the automatic pilot

The tasks we chose to combine in a dual-task setting involved participants monitoring a rapid-fire sequence of digits for the presence of a central letter target (*ventral* task) while simultaneously pointing to a second letter target that appeared with variable temporal lag in the visual periphery (*dorsal* task) (Liu et al., 2008). Additionally, in a single-task condition, participants ignored the central target while pointing to the peripheral target. This is a version of the well-established method for obtaining an *attentional blink* (AB), which refers to a reduction in accuracy for the second of two targets presented in rapid succession (Shapiro et al., 1997). We chose to use this methodology because it allowed us to make a direct comparison between the conscious perceptions of the second target (through participant reports of target identity) with their ability to point rapidly and accurately to it (a visual-motor function governed by the *dorsal* stream).

A second key feature of our method was that we included both single- and double-step pointing trials. Measuring pointing in only the single-step condition would leave our results vulnerable to the interpretation that pointing could be based solely on the initial feed-forward activation of a response. Double-step pointing trials ensure that error-correcting feedback is also involved in the action that is measured (Desmurget et al., 1999).

A third key feature of the methodology in Liu et al. (2008) was that we measured both action planning (response initiation time) and action execution (MT and accuracy). This is important because, as we mentioned earlier, planning is generally considered to involve processes that occur prior to action initiation, whereas execution consists of processes involved in online control. Studies measuring the real-time kinematics of limb movement have shown that total MT is comprised of two distinct phases, an initial, ballistic phase based on previously programmed movement characteristics (e.g., movement direction, amplitude), and a later feedback-sensitive phase in which the movement is refined (Desmurget et al., 1999; Elliott et al., 2001).

Measurements of initiation time (IT) and MT in Liu et al. (2008) were made using a touch screen display, rather than with a three-dimensional limb tracker as in previous studies. This meant we were unable to measure the fine-grained details of movement trajectories, but these were not required to answer our main question, which was whether planning, execution, or both of these components of visually guided pointing to a target were influenced by a concurrent perception task. Guided by the dual vision systems framework (Goodale and Milner, 2004; Norman, 2002), we predicted that the task of letter identification (*ventral* stream) would interfere with IT (movement planning), but not with MT or pointing accuracy (both measures of online action execution).

The results clearly showed that, compared to when the central letter was ignored, successful central letter identification interfered with identifying a second letter in a peripheral location, replicating the standard attentional blink for the conscious perception of two targets presented in rapid succession (Shapiro et al., 1997). Results also showed that central letter identification interfered with the initiation of the pointing response to this peripheral letter (by slowing down IT). This is the expected result, one that is consistent with the hypothesis that conscious action planning shares cognitive resources with conscious letter identification (Goodale and Milner, 2004). The truly novel finding was that there was no evidence of interference from central letter identification in either the speed or accuracy of the pointing action to the peripheral letter. That is, MT and pointing accuracy were sensitive to the distance the finger had travel and to whether online feedback was required or not (double- vs. single-step pointing), but there was no hint of any difference in these pointing measures as a function of whether a central letter was being identified at the same time.

One unexpected finding of this study was the emergence of significant dual-task *facilitation* in MT when the pointing target was presented in temporal proximity to the central letter target. This result is surprising from the perspective of typical dual-task costs in performance, although we note that it concurs with other research showing dual-task benefits in visually guided action. For example, researchers have reported benefits in bimanual over manual pointing (Diedrichsen et al., 2004) and in manual peg placement by Parkinson's patients when combined with a tapping task (Brown and Jahanshahi, 1998). Finding MT facilitation in the Liu et al. (2008) study concurs with these researchers who remind us that focused, singular attention on an automatic task can interfere with fluent performance (e.g., in sports, performance arts). Attention to a second task can ensure that such overfocusing does not occur, leading to better performance (Arend et al., 2006).

Capacity limits of the automatic pilot?

In the study of attentional limitations on conscious perception, it is conventional to refer to a task as *automatic* when it can be accomplished without interference from concurrent tasks, when it can be done with little or no cognitive effort, and/or when it is not influenced by increases in the sensory information available during performance of the task. Tasks of conscious perception can sometimes meet these criteria for automaticity, either because they involved innately privileged processing (e.g., biological relevance or a high degree of stimulus saliency) or because an individual has devoted a great deal of learning and practice to their performance (e.g., some aspects of driving, over-learned visual search involving a consistent mapping of targets and distractors).

From this perceptive, the Liu et al. (2008) result, showing no interference between central letter identification and online control of visually guided pointing, makes it tempting to surmise that the finger's automatic pilot is also capacity unlimited. However, we think caution is in order when extrapolating from studies of conscious perception to questions about unconscious visual processing. To begin exploring these possibilities for the automatic pilot, we designed a study to explicitly test for capacity limitations in visually guided pointing. Specifically, we measured the finger's ability to modify movements in a double-step pointing task under dual-target conditions. This laboratory task is directly relevant to the many actions we make every day involving multiple targets and multiple movement components. Any time we key in telephone numbers or place sugar cubes into a cup of tea, multiple targets are involved in the movement sequence. The question of whether the visuomotor system, like the conscious perception system, suffers interference under multiple target conditions was the focus of a study by Cameron et al. (2007).

Before summarizing Cameron et al. (2007), it is important to note that there have been previous studies of sequential action. For example, participants in some studies tap two or more targets in rapid sequence. When the kinematics of a single tap was compared to those for a tap to the same first location in a series of taps, movement times were typically longer in the multiple target condition (i.e., there is a "one-target advantage"). This in itself would appear to implicate a capacity limit for dorsal stream visual processing. However, there is reason to be cautious in this interpretation because all studies showing the one-target advantage have emphasized speed of response and very low accuracy criteria, with no requirement to correct the action online in response to changing display characteristics (see review by Adam et al., 2000). As such, the actions measured may have been memorized and thus have little connection to the online dorsal stream processes of interest in the present discussion.

Participants in Cameron et al. (2007) made fast sequential aiming movements to either one (single target) or two targets in succession (targets in different locations), with a goal of reaching the first target within 200 ms of target onset. The main measure of interest was the extent to which the finger deviated in its movement trajectory in response to a double-step in either or both targets (a target jump). The prediction was that if the online control processes of the dorsal stream are capacity limited, then there should be less responsiveness of the automatic pilot to the double-step in the two-target than in the single-target conditions.

The participants were also tested under both the standard "go" instructions (i.e., point rapidly and accurately to the target(s) when they appear) and the "stop" instructions (i.e., the same instructions as in the "go" condition, with the caveat to withdraw the finger immediately from the table if a target jump is detected). As in previous studies, the stop instructions are useful in assessing whether the results are influenced in any way by the conscious intent to modify the action in response to a consciously perceived event.

The results of Cameron et al. (2007) revealed that the finger's automatic pilot suffers interference from multiple targets in much the same way that conscious perception is reduced when more than one object must be attended in rapid succession. Although the actions of the finger took the double-step of a single target into account, these corrections were much less in evidence (both in frequency and in magnitude) when the same double-step target was the first of two targets to be pointed to in quick succession. Moreover, detailed kinematic analyses indicated that this interference came about specifically because the preparation of the second target action interfered with the online control of pointing to the first target. Similarly, examination of pointing accuracy to the first target revealed that it was specifically impaired on double-step trials when a second target was present. Simply having to take that target into account, even before it moved to a second possible location, reduced the automatic pilot's ability to correct an action in response to a change in the first target's location.

Our confidence that we were actually measuring the dorsal stream's online abilities grew when the results also showed that the "stop" instructions led to the same conclusions as the "go" instructions. In short, participants following "stop" instructions were unable to alter the actions of their finger in response to consciously perceived events (i.e., jumps) before their finger touched down and the results we have already described had taken place. Their finger withdrawals occurred after the first touchdown, implicating conscious rather than unconscious control. We have now also replicated this finding in the context of an experiment in which the double-step occurs entirely during the intra-saccadic period (i.e., while the eye is in transit between two fixations) (Cameron et al., 2009). That experiment revealed that the online control of actions could be based on target jumps that began and ended entirely within the span of what is normally referred to as *saccadic suppression*. This implies that while the ventral stream of visual processing may indeed be suppressed during a saccade, the dorsal stream governing the online control of action is not. This finding further boosts our confidence that we are studying the unconscious control system of the dorsal stream, because participants are typically unaware of visual events that occur while a saccade is being made.

Future research on attentional interactions involving seeing and acting

Lurking beneath the surface of our normally smooth visual-motor interactions with the environment is the fact that there are two independent streams of visual processing that make use of the same pattern of light. One of these streams enables us to consciously identify objects and to apprehend their layout in space; the other serves to unconsciously facilitate our motor interactions with these objects. Although most previous research on the attentional limitations of vision have focused on the conscious stream to date, the purpose of this chapter has been to emphasize that we should be focusing equal effort in our understanding on the attentional capabilities — both limits and freedoms — of the unconscious action stream.

But doing so confronts us with new challenges. One of these challenges is that care must be taken to measure visual functions that are uniquely governed by the unconscious action system. As we pointed out, studies in the past that have been aimed at this question have often inadvertently posed tasks that involved conscious attentional processing prior to the unconscious processes that were the focus of the study. Another problem of past research in this area is that adequate care has not been taken to distinguish between the planning (or preparation) of visually guided action and the online control of those actions.

In our own beginning efforts to overcome these challenges with the research we highlighted here, we have come to the conclusions that (1) the finger is guided visually by an *automatic pilot* that uses different information from that of conscious vision (Chua and Enns, 2005; Cressman et al., 2007; Fecteau et al., 2001); (2) conscious object identification interferes with concurrent planning of pointing to a second object, though not with the online control needed to complete the pointing action (Liu et al., 2008); (3) concurrent perception and action sometimes lead to benefits in motor performance because the action task is not over-managed by conscious processes (Liu et al., 2008); and (4) the automatic pilot is itself capacity limited in processing information concurrently from multiple locations (Cameron et al., 2007, 2009). The challenge for future research will be to find methods for exploring the attentional capabilities of unconscious visual processes that take us beyond what can be learned from studying the action of the visually guided finger (i.e., the automatic pilot), its responsiveness to changes in spatial position (i.e., the double-step paradigm), and the participant's intent to either execute or halt an action (i.e., the go-stop paradigm) in response to simple visual targets (i.e., luminance defined discs in an otherwise empty display).

References

Adam, J. J., Nieuwenstein, J. H., Huys, R., Paas, F. G. W. C., Kingma, H., Willems, P., et al. (2000). Control of rapid aimed hand movements: The one target advantage. *Journal of Experimental Psychology: Human Perception and Performance*, *26*, 295–312.

Ansorge, U., Heumann, M., & Scharlau, I. (2002). Influences of visibility, intentions, and probability in a peripheral cuing task. *Consciousness and Cognition*, *11*, 528–545.

Arend, I., Johnston, S., & Shapiro, K. (2006). Task-irrelevant motion and flicker attenuate the attentional blink. *Psychonomic Bulletin and Review*, *13*, 600–607.

Austen, E., & Enns, J. T. (2003). Change detection in an attended face depends on the expectation of the observer. *Journal of Vision*, *3*, 64–74.

Bard, C., Turell, Y., Fleury, M., Teasdale, N., Lamarre, Y., & Martin, O. (1999). Deafferentation and pointing with visual double step perturbations. *Experimental Brain Research*, *125*, 410–416.

Binsted, G., Brownell, K., Vorontsova, Z., Heath, M., & Saucier, D. (2007). Visuomotor system uses target features unavailable to conscious awareness. *Proceedings of the National Academy of Sciences of the United States of America*, *104*, 12669–12672.

Breitmeyer, B. G., Ro, T., & Singhal, N. S. (2004). Unconscious color priming occurs at stimulus- not percept-dependent levels of processing. *Psychological Science*, *15*, 198–202.

Brenner, E., & Smeets, J. B. (2003). Perceptual requirements for fast manual responses. *Experimental Brain Research*, *153*, 246–252.

Bridgeman, B., Hendry, D., & Stark, L. (1975). Failure to detect displacement of the visual world during saccadic eye movements. *Vision Research*, *15*, 719–722.

Brown, R. G., & Jahanshahi, M. (1998). An unusual enhancement of motor performance during bimanual movement in Parkinson's disease. *Journal of Neurological Neurosurgery and Psychiatry*, *64*, 813–816.

Cameron, B. D., Enns, J. T., Franks, I. M., & Chua, R. (2009). The hand's automatic pilot can update visual information while the eye is in motion. *Experimental Brain Research*, *195*(3), 445–454.

Cameron, B. D., Franks, I. M., Enns, J. T., & Chua, R. (2007). Dual-target interference for the 'automatic pilot' in the dorsal stream. *Experimental Brain Research*, *181*, 297–305.

Castiello, U. (1996). Grasping a fruit: Selection for action. *Journal of Experimental Psychology: Human Perception and Performance*, *22*, 582–603.

Castiello, U., Paulignan, Y., & Jeannerod, M. (1991). Temporal dissociation of motor responses and subjective awareness. *Brain, 114*, 2639–2655.

Chua, R., & Enns, J. T. (2005). What the hand can't tell the eye: Illusion of space constancy during accurate pointing. *Experimental Brain Research, 162*, 109–114.

Cressman, E. K., Chua, R., Franks, I. M., & Enns, J. T. (2007). On-line control of pointing is modified by unseen visual shapes. *Consciousness and Cognition, 16*, 265–275.

Desmurget, M., Epstein, C. M., Turner, R. S., Prablanc, C., Alexander, G. E., & Grafton, S. T. (1999). Role of the posterior cortex in updating reaching movements to a visual target. *Nature Neuroscience, 2*, 563–567.

Deubel, H., & Schneider, W. X. (1994). Can man bridge a gap? *Behavioral and Brain Sciences, 17*, 259–260.

Deubel, H., Schneider, W. X., & Bridgeman, B. (1996). Postsaccadic target blanking prevents saccadic suppression of image displacement. *Vision Research, 36*, 985–996.

Deubel, H., Schneider, W. X., & Paprotta, I. (1998). Selective dorsal and ventral processing. Evidence for a common attentional mechanism in reaching and perception. *Visual Cognition, 5*, 81–107.

Diedrichsen, J., Nambisan, R., Kennerley, S. W., & Ivry, R. B. (2004). Independent on-line control of the two hands during bimanual reaching. *European Journal of Neuroscience, 19*, 1643–1652.

Duncan, J. (1984). Selective attention and the organization of visual information. *Journal of Experimental Psychology: General, 113*(4), 501–517.

Elliott, D., Helsen, W. F., & Chua, R. (2001). A century later: Woodworth's two-component model of goal-directed aiming. *Psychological Bulletin, 127*, 342–357.

Fecteau, J. H., Chua, R., Franks, I., & Enns, J. T. (2001). Visual awareness and the on-line modification of action. *Canadian Journal of Experimental Psychology, 55*, 106–112.

Giambra, L. M. (1995). A laboratory based method for investigating influences on switching attention to task unrelated imagery and thought. *Consciousness and Cognition, 4*, 1–21.

Goodale, M. A., & Milner, A. D. (2004). *Sight unseen: An exploration of conscious and unconscious vision*. Oxford, England: Oxford University Press.

Goodale, M. A., Pélisson, D., & Prablanc, C. (1986). Large adjustments in visually guided reaching do not depend on vision of the hand or perception of target displacement. *Nature, 320*, 748–750.

Goodale, M. A., Westwood, D. A., & Milner, A. D. (2004). Two distinct modes of control for object-directed action. *Progress in Brain Research, 144*, 131–144.

Henry, F. M., & Rogers, D. E. (1960). Increased response latency for complicated movements and a 'memory drum' theory of neuromotor action. *Research Quarterly, 31*, 448–458.

Jeannerod, M. (1988). *The neural and behavioural organization of goal-directed movements. Motor control: Concepts and issues*. New York: Wiley.

Keele, S. W. (1981). Motor control. In W. B. Brooks (Ed.), *Handbook of physiology, Section 1: The nervous system* Vol. 2, (pp. 1391–1414). Baltimore, MD: Williams and Wilkins.

Kunde, W., Landgraf, F., Paelecke, M., & Kiesel, A. (2007). Dorsal and ventral processing under dual-task conditions. *Psychological Science, 18*(2), 100–104.

Liu, G., Chua, R., & Enns, J. T. (2008). Attention for perception and action: Task interference for action planning, but not for online control. *Experimental Brain Research, 185*, 709–717.

Merikle, P. M., & Cheesman, J. (1987). Current status of research on subliminal perception. In M. Wallendorf & P. F. Anderson (Eds.), *Advances in consumer research* (Vol. XIV), Provo, UT: Association for Consumer Research.

Milner, A. D., & Goodale, M. A. (1995). *The visual brain in action*. London: Oxford University Press.

Neumann, O., & Klotz, W. (1994). Motor responses to nonreportable, masked stimuli: Where is the limit of direct parameter specification. In C. Umilta & M. Moscovitch (Eds.), *Attention and performance XV: Conscious and non-conscious information processing* (pp. 123–150). Cambridge, MA: MIT Press.

Norman, J. (2002). Two visual systems and two theories of perception: An attempt to reconcile the constructivist and ecological approaches. *Behavioral and Brain Sciences, 25*, 73–144.

Pashler, H. E. (1994). Dual-task interference in simple tasks: Data and theory. *Psychological Bulletin, 116*, 220–244.

Pashler, H. E. (1998). *The psychology of attention*. Cambridge: MIT Press.

Paulignan, Y., MacKenzie, C., Marteniuk, R., & Jeannerod, M. (1991). Selective perturbation of visual input during prehension movements 1. The effects of changing object position. *Experimental Brain Research, 83*, 502–512.

Pisella, L., Grea, H., Tilikete, C., Vighetto, A., Desmurget, M., Rode, G., et al. (2000). An "automatic pilot" for the hand in human posterior parietal cortex: toward reinterpreting optic ataxia. *Nature Neuroscience, 3*, 729–736.

Posner, M. I. (1980). The orienting of attention. *Quarterly Journal of Experimental Psychology, 32*, 3–25.

Prablanc, C., & Martin, O. (1992). Automatic control during hand reaching at undetected two-dimensional target displacements. *Journal of Neurophysiology, 67*, 455–469.

Rensink, R. A. (2002). Change detection. *Annual Review of Psychology, 53*, 245–277.

Schmidt, T. (2002). The finger in flight: Real-time motor control by visually masked color stimuli. *Psychological Science, 13*, 112–118.

Shapiro, K. L., Arnell, K. M., & Raymond, J. E. (1997). The attentional blink. *Trends in Cognitive Science, 1*, 291–296.

Ungerleider, L. G., & Mishkin, M. (1982). Two cortical visual systems. In D. J. Ingle, M. A. Goodale, & R. J. W. Mansfield (Eds.), *Analysis of visual behavior* (pp. 549–586). Cambridge: MIT Press.

N. Srinivasan (Ed.)
Progress in Brain Research, Vol. 176
ISSN 0079-6123

CHAPTER 15

Attention for action during error correction

K.M. Sharika, Supriya Ray[a] and Aditya Murthy*

National Brain Research Centre, Nainwal More, Manesar, Haryana, India

Abstract: While the role of attention in selecting visual attributes is well acknowledged, relatively less is known about the mechanisms that facilitate the selection of actions during goal-directed behaviors. The notion of an executive attention has provided a particularly fruitful framework to understand how the brain coordinates the selection of appropriate modules in a sequence that optimizes behavior. However, to do this, theorists have recognized the need to parcel out this unitary system into subcomponents. Two modules that have been commonly invoked are performance monitoring and response inhibition. Visuomotor control of eye movements provides an elegant model system to investigate these mechanisms of selection and control specially occurring during "double-step" tasks in which goals are suddenly changed, demanding inhibition and error detection/correction. Here, we describe our work that has focused on the executive mechanisms that regulate the production of saccadic movements during double-step tasks in different cognitive contexts and target-shift double-step tasks. By examining the pattern of response in the context of quantitative models of saccadic reaction times, we provide behavioral evidence of predictive error correction that produces fast, corrective responses. The predictions from these behavioral experiments were also tested and supported by analyzing neural data from the frontal cortex of monkeys performing similar tasks. Finally, we present data that tested the possibility of an interaction between the inhibitory control and error correction and suggest a model in which predictive error correction may be engaged when the likelihood of error is high. We propose that these results when used in conjunction with electrophysiological recordings, may provide an important approach to understand how error detection/correction and inhibition, two vital cogs in the functioning of executive control, may interact to govern goal-directed behaviors.

Keywords: intertrial effects; selective attention; inhibition; category; distractors; task switching

To move things is all that mankind can do … for such the sole executant is muscle, whether in whispering a syllable or in felling a forest.

C.S. Sherrington

Introduction

Nervous systems have evolved to transform sensory information into meaningful motor acts. The early theories, largely motivated by Sherrington's seminal work on the reflex arc, assumed that if

*Corresponding author.
Tel.: +91 124 233 8922 26; Fax: +91 124 233 8910/28;
E-mail: aditya@nbrc.ac.in

[a]Current affiliation: Department of Psychology, Vanderbilt University, Nashville, TN, USA

DOI: 10.1016/S0079-6123(09)17613-6

appropriately conditioned, our brains had the capacity to implement a complex "lookup table" that linked stimuli to appropriate responses without the need for an explicit representation of goals. However, the need for an executive control system that actively represents current goals and guides behavior has been recognized by cognitive theorists (e.g., Logan, 1985; Norman and Shallice, 1986; Cohen et al., 1990; Allport et al., 1994; Baddeley and Della Sala, 1996). This system is presumably called into action particularly when the sensory environment is ambiguous or presents competing demands, or the mapping of stimulus onto response is complex or contrary to some learned pattern, making performance prone to errors. In general, executive control can be thought of as a set of higher-level cognitive control processes that coordinates the selection of the appropriate sequence of modules that optimizes behavior. Such executive control has also been likened to an "attention for action" that places the selection of particular events or stimuli from the environment that are relevant to the current plan of action at the heart of goal-directed behavior (Allport, 1987).

The hallmark of executive control over our behavior is the ability to respond to changes that make the present goals inappropriate. The same can be tested in the laboratory by means of a task that typically entails inhibition of the ongoing response and programming of the one, more appropriate to the new context. A stimulus paradigm that can be used to access such executive control using the oculomotor system is the double-step task. Here, a single target is displaced to successive locations called "target steps" and subjects are asked to rapidly follow the targets and fixate afresh (Wheeless et al., 1966; Komoda et al., 1973; Hallett and Lightstone, 1976a, b; Becker and Jürgens, 1979; Hou and Fender, 1979; Findlay and Harris, 1984). If the interval between the target displacements, called the target step delay, is short, subjects typically inhibit the planned saccade to the initial target location and respond with a single saccade directed toward the second target location. However, if the target step delay is long, subjects often respond with a sequence of two saccades; an initial erroneous saccade toward the initial target location and a second corrective saccade directed at the final target location.

Critical insights into the ability of the oculomotor system to concurrently program two saccades in a sequence have been obtained by analyzing the pause between the two saccades generated in the above sequence as a function of reprocessing time (RPT), i.e., the duration between the appearance of the second target and the beginning of the first saccade, which is the time available to the saccadic system to reprogram the second saccade. Parallel programming occurs when the preparation of the second saccade, following the appearance of the second target, begins while the first saccade is being programmed and executed. More specifically, it is now well established that as the RPT increases, the interval between the two saccades decreases, and may even become shorter than the normal saccadic reaction time, indicating that although the generation of saccades is usually sequential, the processing of preceding saccades can occur in parallel (Becker and Jürgens, 1979; McPeek et al., 2000). Here, we describe our work that has focused on the mechanisms that regulate the production of sequential eye movements, particularly in the context of error detection/correction. We also present data from a modified double-step task that attempt to describe the nature and time course of motor preparation during error correction. Because inferences of brain mechanisms from behavioral data alone are indirect, we also tested the predictions from these behavioral experiments by analyzing neural data from the frontal cortex of monkeys performing similar tasks. Collectively, these data suggest how error detection/correction and inhibitory control, two modules that have been suggested by cognitive theorists to underlie executive function, may interact to govern goal-directed behaviors.

Modulation of saccade programming by cognitive control

To explore the role of cognitive control in modulating the extent of concurrent saccade preparation, Ray et al. (2004) recorded eye movements of subjects performing a modified version of

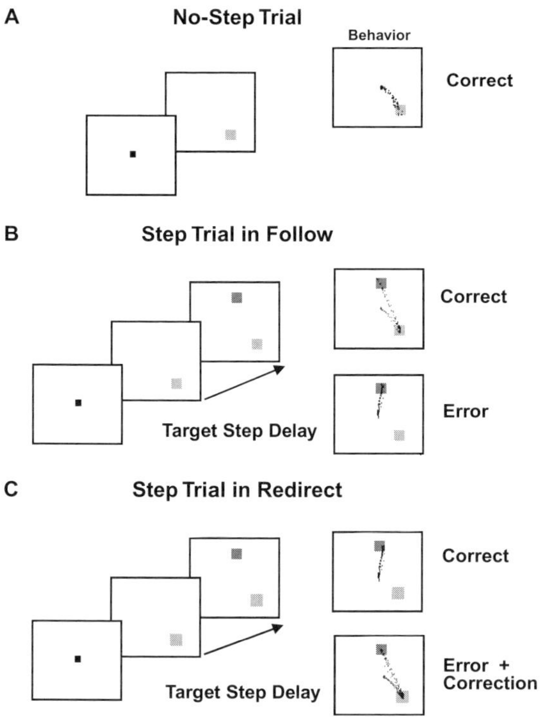

Fig. 1. Schematic representation of events in different types of trials in the *FOLLOW* and *REDIRECT* conditions. (A) *No-step* trial. Common to both conditions, in these trials following fixation at the centre of the screen, a green target (light gray square) appears at any one of the eight evenly spaced locations centered on an imaginary circle of radius of 10°. Subjects are instructed to make a saccade to the target as soon as possible. (B) *Step* trial in the *FOLLOW* condition. Following fixation, a green target (light gray square) is presented just as in a *no-step* trial. After a variable target step delay (~50–200 ms), a final red target (dark gray square) appears and subjects are instructed to make an eye movement first to the initial target and then to the final target. (Right panel) Eye position traces are shown for a few trials producing the correct sequence of saccades to the initial and final target (top) and the incorrect saccadic response to final target (bottom). (C) *Step* trial in the *REDIRECT* condition. Following fixation, an initial green (light gray square) and final red target (dark gray square) are presented just as in the *step* trial of the *FOLLOW* condition. However, subjects are instructed to cancel the saccade to the initial target and look directly at the final target. (Right panel) Eye position traces for a few trials depicting the reprogrammed correct saccade to the final target (top), and the response consisting of an incorrect saccade to the initial target and second corrective saccade to the final target (bottom). Step trials in each condition were randomly interleaved with no-step trials such that subjects could not predict or anticipate the appearance of the targets. Adapted with permission from Ray et al. (2004).

the classic double-step task under two different conditions (Fig. 1). In the *FOLLOW* condition (Fig. 1B), subjects were explicitly instructed to follow target steps with successive saccades while in the *REDIRECT* condition (Fig. 1C) they were asked to cancel the initial saccade and straightaway direct gaze to the final target. In cases of failure to cancel the pre-planned saccade in the *REDIRECT* condition, subjects were found to make a sequence of two saccades: an initial erroneous saccade to the location of the first target, followed by a second corrective saccade to the final target. Although in both instances a sequence of two saccades were elicited, the second saccade in the *FOLLOW* condition constituted part of the correct response, whereas in the *REDIRECT* condition, the same was a corrective response following an error. As the stimuli and responses engendered by the *FOLLOW* and *REDIRECT* instructions were the same, this paradigm allowed Ray et al. (2004) to examine cognitive control during saccade programming.

Using the framework described by Becker and Jürgens (1979) to understand the inverse relation between the inter-saccadic interval (ISI) and the RPT, Ray et al. (2004) estimated the extent to which parallel processing ensues in a given interval of the RPT (Fig. 2). Their rationale was that if the planning of the second saccade occurred independently and in parallel with that of the first saccade during the RPT, then an increase in RPTs should decrease the ISIs by a proportionate amount, depending on the extent of parallel programming occurring for these two saccades. For example, if entirely independent parallel processing of the second saccade occurred during the RPT, then for a trial in which RPT is longer by an interval of $RPT_{2\text{-}1}$, the corresponding inter-saccadic interval (ISI_2) should be shorter by a duration of $RPT_{2\text{-}1}$, resulting in an ISI versus RPT slope of −1. However, if processing of the second saccade is slowed during the interval $RPT_{2\text{-}1}$ then the extent to which ISI is reduced should be less than the duration of $RPT_{2\text{-}1}$, i.e., the ISI versus RPT slope during this interval should be, in absolute terms, less than one. Thus, the slope of the plot of ISI as a function of RPT can be construed as a metric that describes the extent to which processing of the

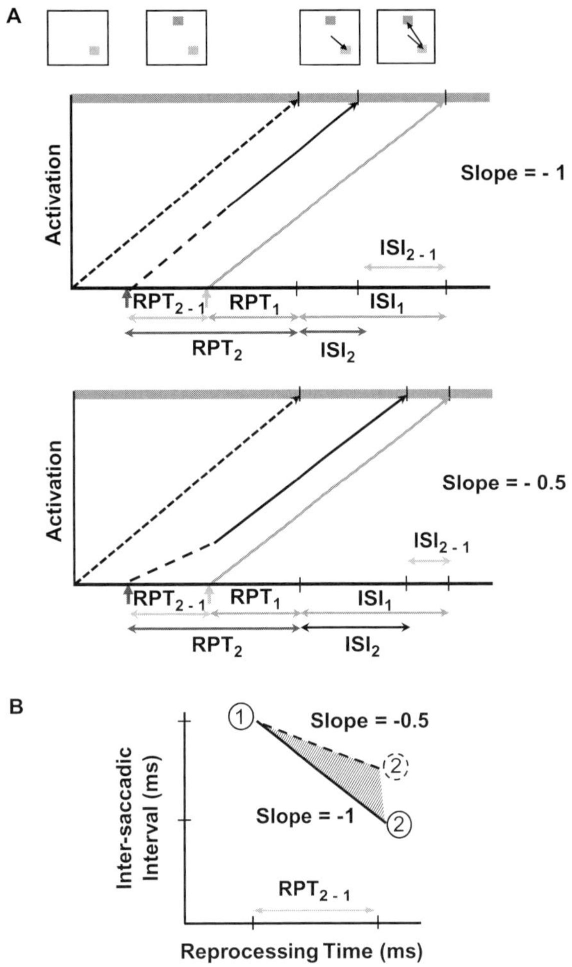

Fig. 2. Schematic diagram of how the slope between ISI and RPT can represent the rate of parallel processing occurring in case of two sequential saccades. (A) Vertical arrows (dark and light gray) on the x axis of upper and lower panels indicate the time of final target presentation in two representative trials associated with longer (RPT_2) and shorter (RPT_1) RPTs, respectively. If entirely independent parallel processing of the second saccade occurs during the reprocessing time, then for a trial in which RPT is longer by $RPT_{2\text{-}1}$, the ISI_2 should decrease by a duration equal to $RPT_{2\text{-}1}$, resulting in an ISI versus RPT slope of -1 (upper panel). However, if processing is slowed during the interval (represented by the relatively shallow accumulation of activation; compare broken lines in the upper and lower panels) such that on average, the rate of processing the second saccade is only half as fast, then the corresponding decrease in ISI should be equal to half the duration of $RPT_{2\text{-}1}$, giving rise to an ISI versus RPT slope of -0.5 (lower panel). (B) Thus the local slope between the ISI and RPT can be considered a metric representing the rate of processing during the RPT interval. Adapted with permission from Ray et al. (2004).

second saccade occurs while the first saccade is being programmed.

Figure 3 summarizes the essential results of this study. Amid variability across subjects, three important trends were seen in the population. First, ISIs decreased with RPTs indicating some degree of parallel programming in both *FOLLOW* and *REDIRECT* conditions. Second, the slope of the function characterizing the ISI versus RPT relation tended to decrease progressively with increasing RPT up to a minimum. Because the slope is thought to represent the rate of processing of the second saccade during the RPT interval, such a pattern indicates that the degree of parallel processing diminishes with increasing RPTs up to a point, beyond which processing appears to be serial. Stated differently this implies that because ISI does not decrease, there is little or no processing of the second saccade at the longest RPT intervals. Third, and the most important from the perspective of cognitive control, was that processing rates were faster in the *REDIRECT* condition as compared to the *FOLLOW* condition, particularly during the shorter RPTs, wherein the inverse relationship between the ISIs and RPTs was invariably a slope of more than negative one. This, together with the fact that the reaction times associated with the "corrective" second saccades in the *REDIRECT* condition were faster than the "correct" second saccades of the *FOLLOW* condition, highlight the possible role of supervisory control in modulating the parallel processing of sequential saccades.

Parallel programming during error correction

Concurrent processing of corrective saccades has also been reported in the visual search paradigm (McPeek et al., 2000), in which the oddball target and distractor colors were randomly switched across trials leading to saccade errors when gaze was directed toward the color of the "target" in the previous trial (McPeek et al., 1999). Such incorrect saccades were then usually followed by a second saccade to the actual target. To test if the preparation of corrective saccades occurred in parallel with the initial ones McPeek et al. (2000) employed a saccade-contingent search task.

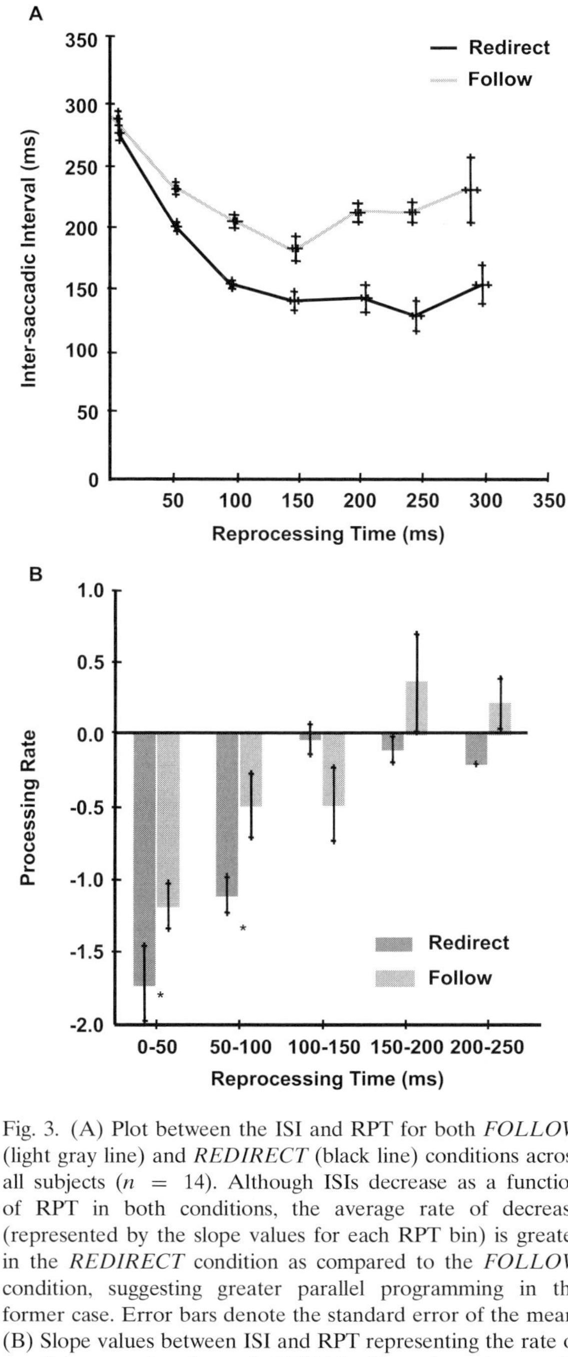

Fig. 3. (A) Plot between the ISI and RPT for both *FOLLOW* (light gray line) and *REDIRECT* (black line) conditions across all subjects ($n = 14$). Although ISIs decrease as a function of RPT in both conditions, the average rate of decrease (represented by the slope values for each RPT bin) is greater in the *REDIRECT* condition as compared to the *FOLLOW* condition, suggesting greater parallel programming in the former case. Error bars denote the standard error of the mean. (B) Slope values between ISI and RPT representing the rate of parallel processing for each RPT bin. Error bars denote the standard error of the mean. Asterisks denote significant (pair wise t-test $p < 0.05$) differences between the conditions. Adapted with permission from Ray et al. (2004).

The task involved switching target and distractor positions as soon as the initial incorrect saccade began in some trials. Consistent with the notion that corrective saccades could be prepared in parallel with the erroneous saccade, McPeek et al. (2000) observed that on a number of occasions, subjects directed their corrective saccades at the original position of the target even when the associated ISIs were as large as the normal latency of visually guided saccades. This suggests that the preparation of the second saccade was well underway even before the first saccade was executed. However, as saccade programming comprises of at least two stages, a visual stage that specifies the location of a target, and a motor stage that prepares and executes the oculomotor command (Viviani, 1990; Hooge and Erkelens, 1996; Thompson et al., 1996; Ludwig et al., 2005), the extent to which parallel processing of the corrective saccade occurs in anticipation of an error is not evident from these results. Additionally, because both the original target and the distractor were presented simultaneously before the first saccade in a visual-search task, and the corrective saccade preparation was internally triggered, the time course of error correction could not be investigated. By modifying the *REDIRECT* task to include what are called the *target-shift step* trials, Sharika et al. (2008), were able to overcome these difficulties and examine the extent of concurrent preparation during error correction, as well as study the time course of planning the second saccade relative to the error onset.

Figure 4 illustrates the task paradigm used to test our hypothesis. *Target-shift step* trials occurred with *no-shift step* trials and were randomized with *no-step* trials such that subjects could not predict or anticipate the appearance of the targets. Although in a *no-shift step* trial, the initial green and final red targets were presented as in the standard step trials of the *REDIRECT* task (for details see Sharika et al., 2008), in a *target-shift step* trial, after the presentation of initial and final targets following fixation, the final red target was stepped to a new position during the execution of the first saccade (Fig. 4). The "shifted" position of the final target (referred, hereafter, as the new position of the final target) was either to

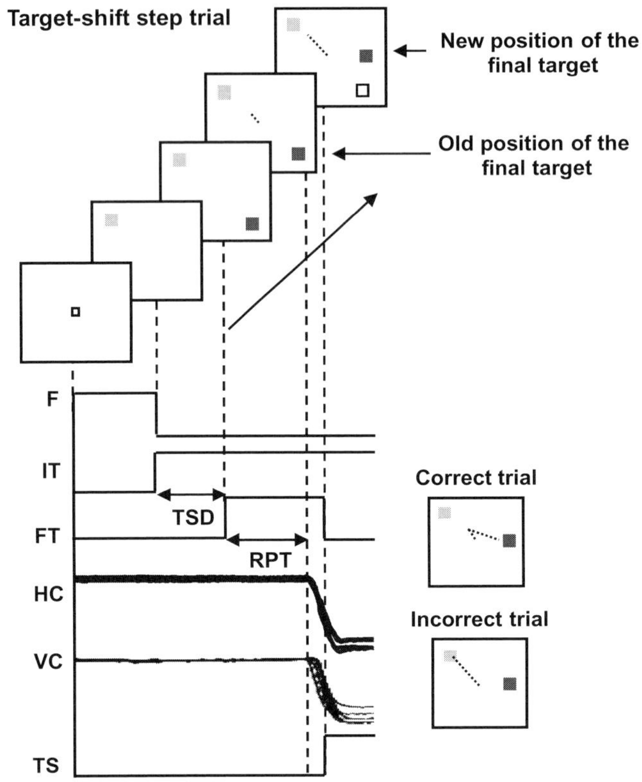

Fig. 4. Temporal sequence of events in a *target-shift step* trial. (Top half) Following fixation, an initial green target (light gray square) is presented at any of the four diagonal locations of the screen at a radius of 21 degrees. After a variable target step delay (~50–200 ms), the final red target (dark gray square) is presented at any of the three remaining locations and subjects are instructed to cancel the saccade to the green target and make a saccade directly to the red target. During the execution of the first saccade, the red target is shifted vertically to the left or right of the fixation point depending on whether the original position of the final target (unfilled square) was on the left or right hemi-field, respectively. (Bottom half-left) Solid vertical line denotes the beginning of the trial while broken vertical lines align the following events to their representative time of occurrence: the time of presentation of the fixation box (F), Initial Target (IT), Final Target (FT), the occurrence of horizontal (HC) and vertical (VC) components of the first saccade, and the shift of the final target (TS). TSD – target step delay. RPT – reprocessing time. (Bottom half-right) Only those trials in which the subjects made the first saccade directly to the new position of the final target were scored as correct, while those in which the first saccade reached the initial green target were considered incorrect. Adapted with permission from Sharika et al. (2008).

the right or left of the fixation point depending on whether the original position of the final target (referred, hereafter, as the old position of the final target) was on the right or left hemi-field, respectively.

Target-shift step trials were used to test whether the brain can begin the motor preparation for the second corrective saccade in parallel with the first erroneous saccade. The logic was that if motor preparation cannot occur in parallel and commences only after the first saccade, the second corrective saccades should always be directed to the new, shifted position of the final target. However, if motor preparation of the second corrective saccade can occur in parallel and begins before the final target shifts, one should find instances when such saccades end up at the old position of the final target. Because we found examples of the latter in some trials (Fig. 5B), these results appear to be consistent with the hypothesis that motor preparation of the second corrective saccade may begin during the preparation of the first erroneous saccade itself.

If directing gaze to the old position of the final target is a consequence of parallel motor preparation, the propensity for such behavior should be

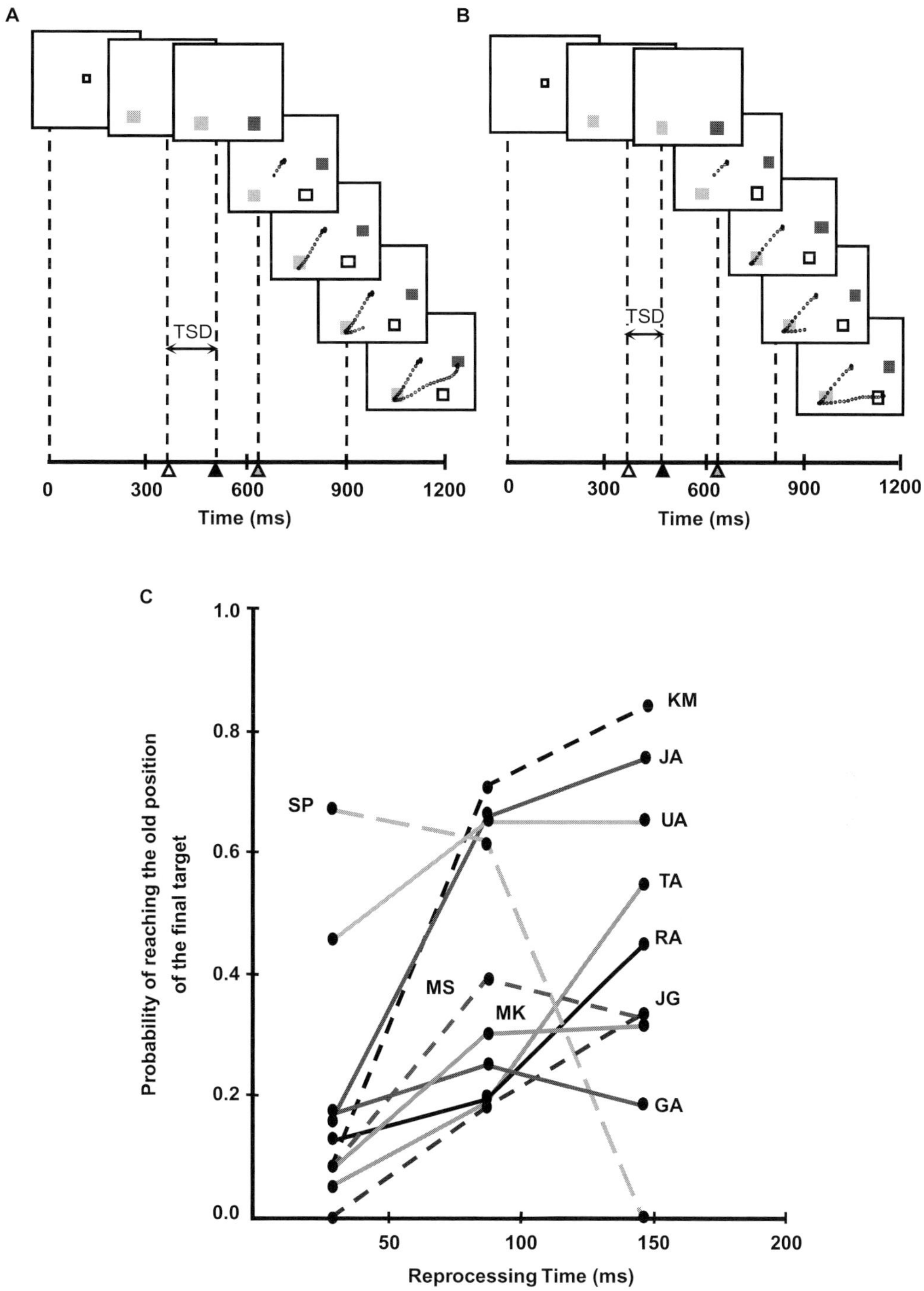

Fig. 5. Schematic representation of events in a *target-shift step* trial when the corrective saccade is made to the new (A) and old (B) position of the final target. White, black, and gray triangles on the time axis denote the presentation of the initial target (light gray square), final target (dark gray square) and target-shift, respectively. For the convenience of the reader, the old position of the final target is represented by an unfilled square. TSD — target step delay. (C) Plot shows the probability of the corrective saccades to the old position of the final target increasing as a function of RPT for seven out of ten subjects. Adapted with permission from Sharika et al. (2008).

related to the RPT. Because at longer RPTs, the extent of motor preparation is likely to be more advanced, the tendency to execute the second corrective saccade to the old final target position should be higher. We tested this by plotting the probability of reaching the old position of the final target as a function of RPT. Figure 5C shows that this probability increased as a function of RPT for most subjects indicating that motor preparation of the corrective saccade may indeed occur concurrently with the processing of the first erroneous saccade.

Are corrective saccades to the old final target location produced as a result of insufficient time for saccade reprogramming? In other words, if the shifted target is presented within the so-called saccadic dead time, which is the point in time during saccadic preparation at which no new visual information can change the upcoming movement, then saccades directed at the old final target position are to be expected. Saccadic dead time is typically around 80 ms prior to the movement onset (Findlay and Harris, 1984) and is assumed to be the result of afferent and efferent delays in the transmission of information between the eye and the brain regions responsible for generating the oculomotor commands (Becker, 1991; Van Loon et al., 2002; Ludwig et al., 2007). However, the mean ISI for corrective saccades to the old final target location across subjects was estimated to be about 143 ms, indicating adequate time for visual information to modify the decision process. Hence, these trials must represent instances in which the preparation for the corrective saccade has already reached some "point of no return" in decision-making (e.g., Osman et al., 1986).

Time course of concurrent corrective saccade preparation

The time at which the second corrective saccade's preparation begins for each subject was also estimated by Sharika et al. (2008) using the *LATER* model (Carpenter and Williams, 1995; Hanes and Schall, 1996; Hanes and Carpenter, 1999; Reddi et al., 2003). This model envisions the delay responsible in generating a saccade as the time taken by the decision variable to linearly accumulate information from the environment till it reaches a criterion level of activation at which time the saccade is executed. In their analysis, Sharika et al. (2008) also assumed the rates at which the decision variables responsible for generating a correct saccade in the *no-step* trial and a corrective saccade in a *target-shift step* trial, rise toward the activation threshold to vary with the same distribution and thus, give rise to the same latency profiles. The essential logic of the estimation is graphically described in Fig. 6.

If the second corrective saccade preparation to the old final target position begins assuming an afferent delay of ~50 ms (Ludwig et al., 2007) (Fig. 6B), then the predicted reaction times of these saccades fail to match the observed longer reaction times. Thus, it is likely that either the $GO_{Corrective}$ process itself started with a delay from the time of final target presentation or it began early but was slowed down later in the process. However, as the *LATER* model assumes the rate of accumulation to vary in a Gaussian fashion across trials but to be invariant during the latency of any one trial, the observed longer reaction times of corrective saccades was assumed to be caused by the delay in the onset of the $GO_{Corrective}$ process alone. By shifting the onset of the $GO_{Corrective}$ process iteratively, the minimum delay (denoted by d in Fig. 6C) required for the predicted reaction times to be the same as the observed reaction times of the corrective saccades for any set of trials was computed.

Figure 7 shows the frequency distribution of the onset of corrective saccade preparation with respect to the end of the error (denoted by o in Fig. 6B and 6C) obtained by adding the mean error/saccadic duration across all subjects (54 ms) to the onsets of corrective saccade preparation computed across all RPTs. Consistent with the notion of parallel programming, for 47% of the times, the planning for correction was estimated to have begun before the onset of the first erroneous saccade itself, while in 97% of the cases it was estimated to have begun before or during the execution of this erroneous saccade, i.e., in the absence of any sensory feedback.

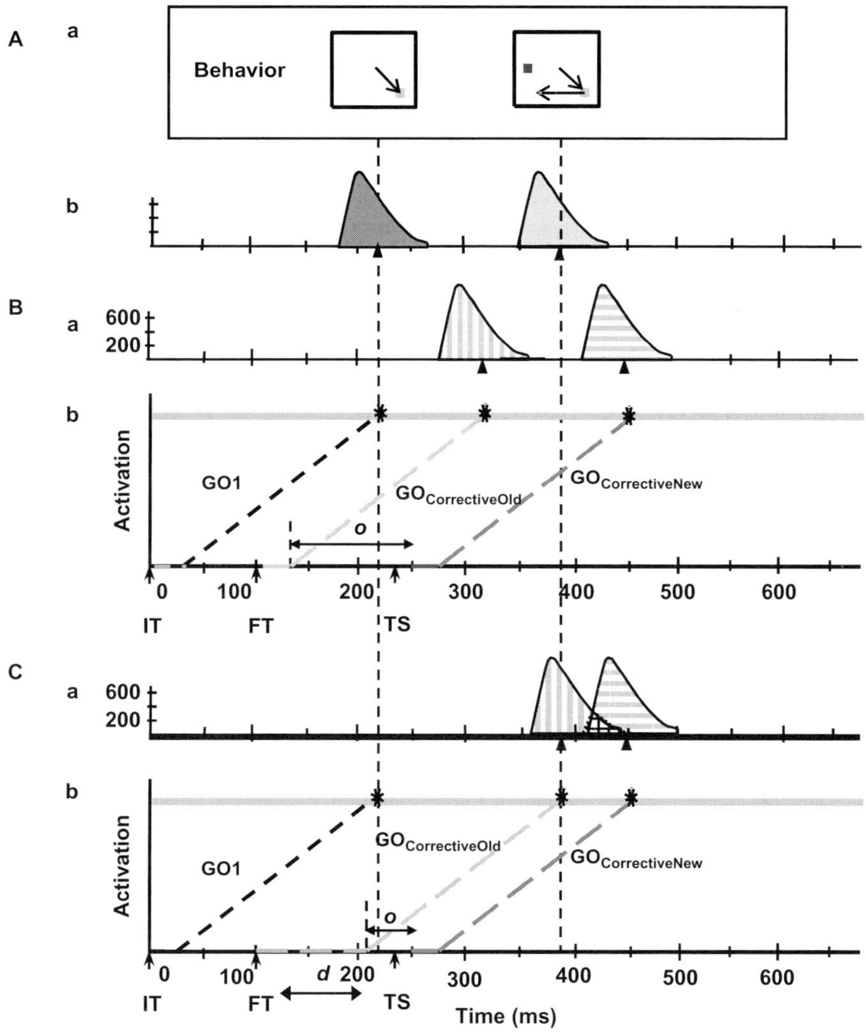

Fig. 6. *LATER* model-based estimation of the onset of corrective saccade preparation. A (a) Behavioral representation of a saccade to the green target (light gray square) in a *no-step* trial and a corrective saccade to the old position of the final target (dark gray square) aligned to their actual mean onset times (for a representative subject) on the x axis of (B) and (C). In A (b) the light gray, skewed reaction time distribution represents the latency profile of corrective saccades to the old position of the final target assumed to be the same as that of saccades made to single targets in *no-step* trials (represented by the dark gray, skewed reaction time distribution). Black triangles on the x-axis mark the mean of the reaction time distributions. In (B) and (C), panels (a) show the corrective saccade reaction time distributions with respect to the final target presentation (distribution with vertical lines) and that starting from the end of first saccade (distribution with horizontal lines) assumed to be responsible for corrective saccades directed at the old and new location of the final target, respectively. Vertical arrows on the x-axis mark the typical time stamps of the following events: initial target presentation (IT), final target presentation (FT) and target-shift (TS). (B) In panel (b), the horizontal gray band represents the activation threshold while GO_1 (black short-dashed line), $GO_{CorrectiveOld}$ (light gray long-dashed line) and $GO_{CorrectiveNew}$ (dark gray long-dashed line) represent the rise to threshold of the decision variables responsible for generating a correct saccade in the *no-step* trial and a corrective saccade to the old and new final target positions in a *target-shift step* trial, respectively. If the $GO_{CorrectiveOld}$ process were to begin as soon as the final target was perceived, then the predicted reaction times of these corrective saccades (distribution with vertical lines) fail to match the longer reaction times observed in case of these saccades [light gray distribution of A (b)]. o represents the onset of corrective saccade preparation with respect to the end of the error. (C) By shifting the onset of the $GO_{CorrectiveOld}$ process iteratively, the minimum delay (denoted by d in Fig. 6C) required for the predicted reaction times to be the same as the observed reaction times of these corrective saccades is calculated. Note that during the iteration any overlap of these two distributions [grid region in panel C(a)] was assumed to correspond to corrective saccades landing midway between the old and new positions of the final target and hence, and hence were not included for predicting the latency of corrective saccades to the old final target position. Adapted with permission from Sharika et al. (2008).

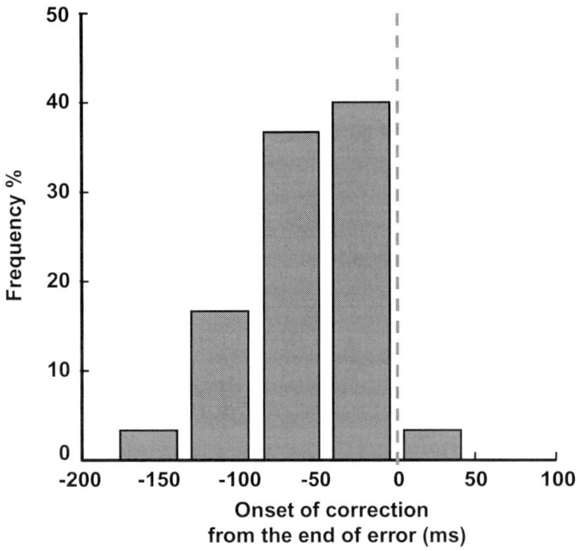

Fig. 7. Frequency distribution of the onsets of corrective saccade preparation with respect to the end of the erroneous saccade for all subjects, across all RPTs. Bars on the negative scale of the *x* axis represent onsets of corrective saccade preparation before the end of the error while bars on the positive scale of the *x* axis denote onsets after the end of the error. Adapted with permission from Sharika et al. (2008).

Neurophysiological evidence for predictive error correction in frontal eye field (FEF)

If subjects are capable of generating accurate corrective saccades with latencies less than the latency of visual feedback, there must be neural mechanisms present in the oculomotor network, comprising of areas such as the frontal eye field (FEF), the lateral intraparietal cortex (LIP) and the superior colliculus (SC), which underlie such behavior. One mechanism that can account for fast online error correction involves a comparison of the spatial location of the goal with the current eye displacement, for which a specific role for a class of neurons called postsaccadic neurons in FEF has been invoked (Goldberg and Bruce, 1990). However, within this framework error correction can begin only after the erroneous saccade has occurred. Alternatively, some motor theorists have suggested that fast online error correction may involve a comparison of the spatial location of the goal with the anticipated movement displacement (Kawato et al., 1987; Wolpert et al., 1995; Wolpert and Kawato, 1998). As a consequence, programming the corrective saccade can begin even before the occurrence of the error and in parallel with the erroneous saccade. Evidence in support of this view is derived from the finding of a characteristic modulation of some visual neurons across the brain areas constituting oculomotor network which has been described as "remapping" (Duhamel et al., 1992; Walker et al., 1995; Umeno and Goldberg, 1997; reviewed by Colby and Goldberg, 1999; Umeno and Goldberg, 2001). These neurons respond to stimuli that will be brought into their receptive fields by saccades, even before the saccades actually take place. More recently, McPeek and Keller (2002a, b) have shown that the activity of visually responsive neurons in SC can represent the location of the salient object in a search array if monkeys shift gaze to another location before producing a corrective saccade to the overlooked target. They suggested that the sustained representation by these visual neurons remapped the location of the target in the reference frame of the errant saccade to afford rapid error correction. If remapping the target location enables the brain to rapidly and accurately correct saccade errors (Vaziri et al., 2006), several implications of this hypothesis can be assessed. First, if the remapping is established early enough, parallel programming for rapid error correction can occur. Second, neurons that are representing the target location as well as neurons associated with saccade programming should be activated before afferent visual signals occur. Third, the timing of activation of these neurons should predict the time of initiation of the corrective saccade. Because FEF is known to contribute to the selection of targets and preparation of saccadic eye movements (Thompson et al., 2001; reviewed by Schall, 2002; Schall et al., 2003), Murthy et al. (2007) investigated how remapped visual information before a saccade in the FEF can be used to rapidly correct errors.

In this experiment, monkeys were trained to perform a task that combines color singleton search with the *REDIRECT* task described before (see Fig. 8). On *no-step* trials monkeys were

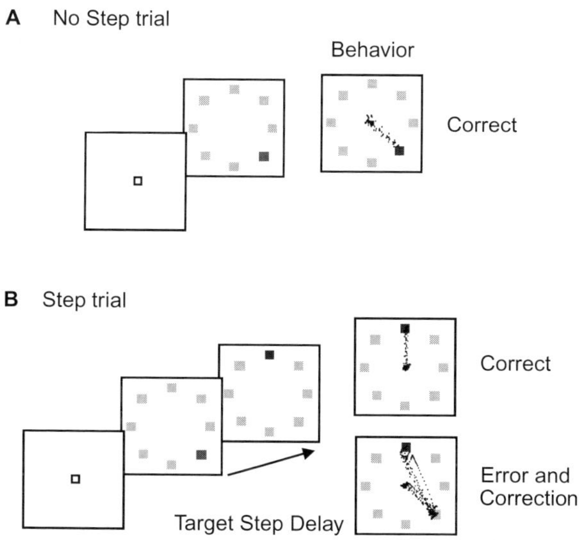

Fig. 8. Schematic representation of trials in the search step task (A) No-step trials: Following fixation, an odd-colored target was presented amongst many distractors and monkeys were rewarded if they made a saccade to the target. (B) Step trials: After a variable target step delay from the presentation of target amongst distractors but before any saccade was initiated, the target swapped positions with a distractor. Monkeys were rewarded if they looked directly to the new target position (correct) and not when they made a saccade to the old target position (error). Erroneous saccades were often followed by a corrective saccade to the new target location. Adapted with permission from Murthy et al. (2007).

rewarded for making a saccade to a color singleton target among distractors. On random *step* trials, the target and a distractor swapped locations through an isoluminant color change. In other words, the target stepped to a new location in the search array with a variable delay after presentation of the search array but before any saccade was produced. Targets steps were spatially constrained so that targets stepped into or out of response fields but never stepped within a response field. Because the analysis of data for this study required matching the direction and amplitude of correct and corrective saccades, only those combinations of target steps were chosen that resulted in subsequent corrective saccades falling into the response field of the cell in question (for details, see Murthy et al., 2007). Here, we restrict our discussion to only movement-related neurons in the FEF that serve as a more proximal basis of saccade preparation.

The goal of this experiment was to determine if the timing of corrective saccade production could be accounted for by the timing of FEF activity after the erroneous saccade. If monkeys produce corrective saccades before the visual feedback about the error could be encoded, then movement-related neurons should modulate earlier than a typical ISI after execution of the error. In contrast, if programming corrective saccades can begin only after visual processing of the new image is finished, then movement-related neurons should increase their activity later and at a time that does not vary with RPT. To identify the neural activity associated with error correction, the activity associated with corrective saccades into the movement field after errors outside of the movement field (Fig. 9) was compared with the activity associated with correct *no-step* saccades directed to the same location outside of the movement field. Trials were matched for saccade latency and direction so that any difference in activity between these trials should arise from the process of producing the corrective saccade. Figure 9 illustrates an exemplar neuron that showed the activity producing the corrective saccade began 29 ms before the end of the erroneous saccade. Qualitatively similar modulation was observed in double-step trials. Considering there were no distractors, these double-step data indicate that the early movement-related activity observed in the search-step trials could not just be a consequence of the distractor appearing in the response field before the target step.

The time of modulation measured relative to the time of the erroneous saccade provides the main evidence for the parallel programming of the corrective saccade. Effectively, all of the movement-related neurons programming the corrective saccade became active within 100 ms after the error saccade was terminated. As the mean visual latency in FEF is about 70 ms (Thompson et al., 1996; Schmolesky et al., 1998; Pouget et al., 2005) and the mean latency of explicit error signals in the medial frontal lobe is 110–180 ms (Stuphorn et al., 2000; Ito et al., 2003), the data indicate that

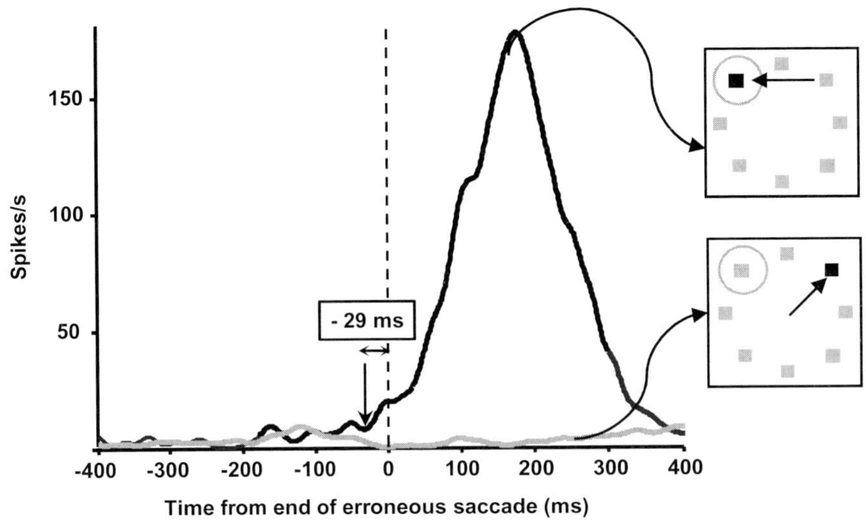

Fig. 9. Activity of a typical movement-related neuron of the FEF during the search-step task. Activity associated with corrective saccades (black) into the movement field after noncompensated errors outside of the movement field are compared with that of the correct no-step saccades (gray) directed to the same location outside of the movement field. Trials are matched for saccade latency and direction, and the activity aligned on the termination of the respective saccades (represented by a vertical broken line at zero on the time scale). Right panel illustrates the arrangement of stimuli and saccadic behavior used for the comparison; a circle around the stimulus marks the movement field location. The arrow represents the time of significant difference in activity (corresponding to start of corrective activity), which begins just before the occurrence of the error. Adapted with permission from Murthy et al. (2007).

the activity producing corrective saccades began occasionally before the errant saccade is accomplished, often before postsaccadic visual input about the error can be registered, and almost always before the brain registers that an error was made through endogenous error monitoring presumed to occur in the medial frontal lobes (Gehring and Fenscik, 2001). This constitutes the first clear physiological evidence for how corrective saccades can be produced faster than serial processing would permit.

Further compelling evidence that movement-related activity in FEF contributed to concurrent saccade programming was the systematic relationship between the time of the neural activation and the RPT that is expected to echo the well-known tendency of the latency of the corrective saccade relative to the error saccade, i.e., one expects the ISI to decrease as the latency of the error saccade relative to the target step (RPT) increases. Specifically, if movement-related activity in FEF contributes to programming the corrective saccade, that activity must begin earlier after longer RPTs. In contrast, if the onset of the movement-related activity arose from error feedback of some kind, then it should depend on the timing of the errant saccade alone and not RPT. To evaluate these alternatives, corrective saccades were divided into those trials associated with longer RPTs (greater than the mean RPT) and those trials associated with the shorter RPTs (less than the mean RPT). Figure 10 illustrates the activity for the movement neuron described in Fig. 9. Figure 10A (lower panel) shows that corrective activity began later with respect to errors in trials with shorter RPTs (and longer ISIs) and earlier before errors associated with longer RPTs (and shorter ISIs; Fig. 10A upper panel). To quantify this relation, the mean time of the beginning of corrective activity was related to the mean RPT from the groups of shorter and longer RPT trials. For this neuron, the two measures were inversely related (slope $= -0.89$), consistent with the hypothesis that the timing of the corrective saccade is dictated by the timing of movement-related activity. The same analysis was performed for movement-related neurons that provided sufficient and reliable data for both long and

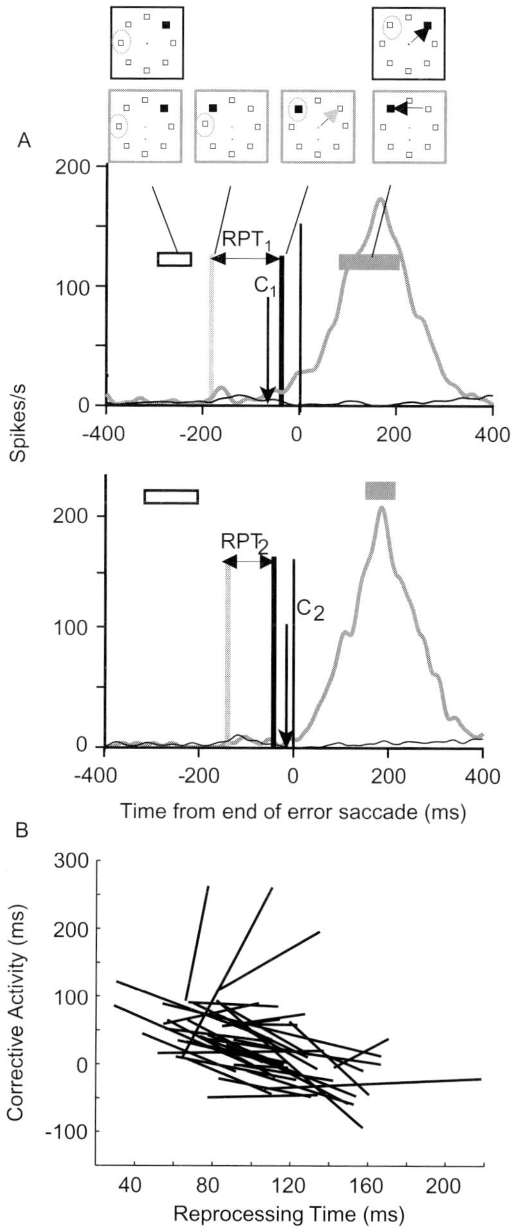

Fig. 10. Relation between movement-related activity and RPT for the neuron in the same conditions as Fig. 9. (A) Arrows representing the time when corrective activity became significantly different shows earlier onsets of correction for errors associated with longer RPTs (upper panel) than those associated with shorter RPTs (lower panel); −59 ms earlier as against −10 ms before the occurrence of the error. The white horizontal bar indicates time of presentation of the search array, and the black bar indicates the range of corrective saccade latencies. (B) Plot between time of corrective activity and RPT for movement-related neurons that provided sufficient data. Each line represents data from a single neuron, with 81% of neurons showing an inverse relation of earlier corrective activity with increasing RPT. Adapted with permission from Murthy et al. (2007).

short RPTs during the search-step task. A significant inverse relationship between the beginning of corrective activity and RPT was observed across this sample; the average slope of −0.72 was significantly less than zero (one-tailed t-test, $t(35) = -5.1$, $P<0.001$) with 81% (29/36) of neurons exhibiting negative slopes (Fig. 10B). Furthermore, the mean time of corrective neural activity was inversely related to the mean RPT in each search-step session (slope = − 0.8, $R^2 = 0.16$; $F(43) = 7.9$, $P = 0.007$). The same inverse relationship was also observed in five of seven double-step sessions. Together these data provide additional evidence that the latency of the corrective saccades depends on the timing of the movement-related activity in FEF and provide the neural basis for rapid error correction.

Performance monitoring, inhibitory control, and error correction

While error correction in choice reaction time tasks may be construed as a delayed correct response, predictive error correction in a motor task raises a fundamental question about the nature of control. This is particularly borne out by the results in the study by Sharika et al. (2008), in which subjects explicitly attempted to inhibit a potential error. More specifically, their results suggest that if correction can start before the occurrence of the error, the brain must have the capacity to predict the likelihood of an error as it is trying to inhibit an unwanted movement. Although the mechanism underlying such ability is not clear as yet, the well known finding about the reaction times of subjects being much slower in trials following errors (Rabbitt, 1966; Rabbitt and Phillips, 1967) implies that the brain can maintain a representation of past performance. In line with this view, neuronal representations of past performance have been recently recorded in the activity of single neurons in the prefrontal cortex of awake, behaving monkeys (Hasegawa et al., 2000). More recently, Brown and Braver (2005) used modeling and imaging studies to hypothesize and describe the role of anterior cingulate cortex (ACC), another executive control area of the brain, while subjects performed a high

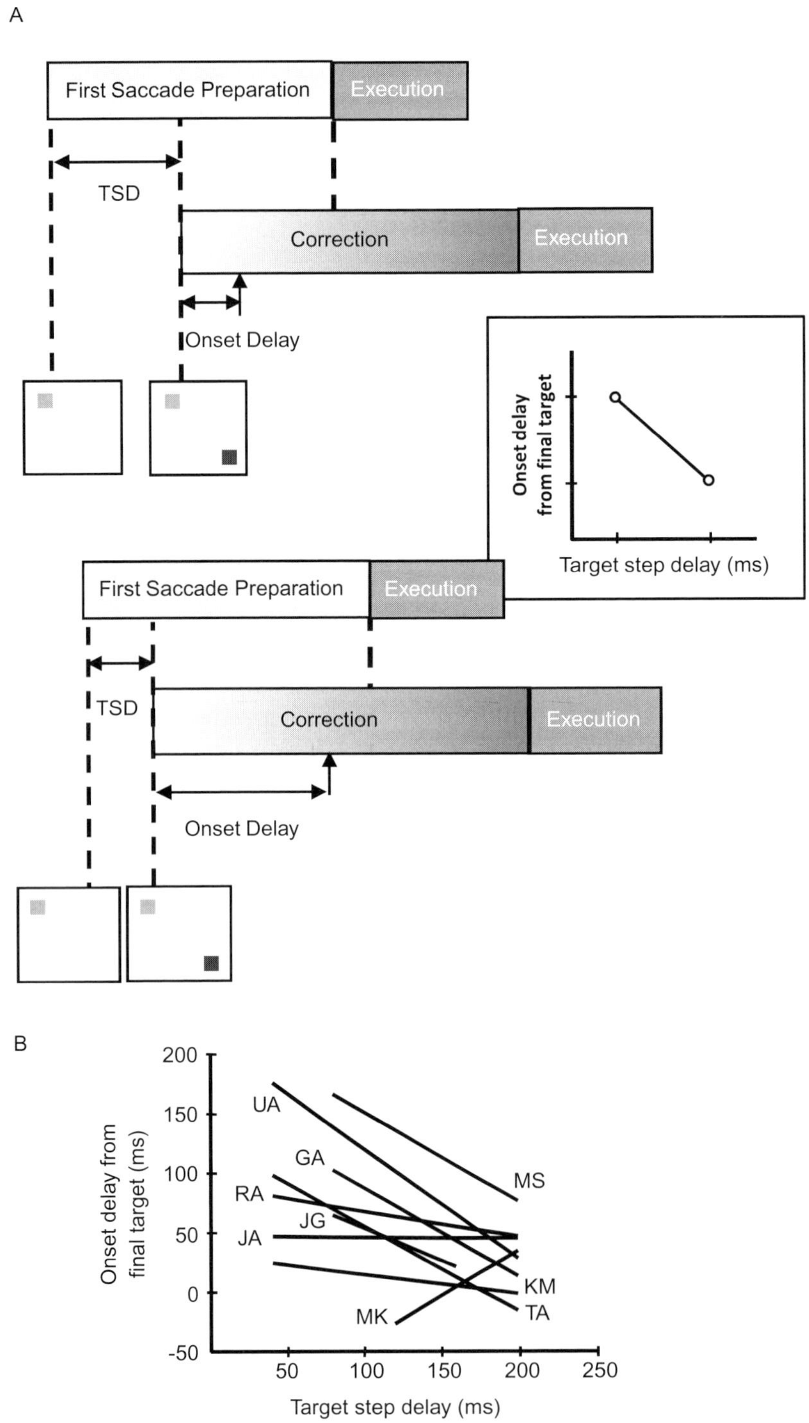

Fig. 11. (A) Schematic model for parallel processing of error correction. According to the model, at longer target step delays, when the likelihood of error is high, the onset of correction should be associated with a shorter delay (top panel). On the other hand, in trials with shorter target step delays, where the likelihood of error is low, the delay in the onset of correction should be longer (bottom panel). Thus, this model predicts an inverse relation between the delay in the onset of error correction and target step delays (inset). (B) Plot showing the estimated delay in the onset of corrective saccade preparation from the final target presentation (after a visual delay of ~50 ms) as a function of target step delay. Negative slopes obtained by linear fitting of data show that for eight out of nine subjects the delay in correction onsets decreases with increasing target step delays. Adapted with permission from Sharika et al. (2008).

versus low conflict task. They obtained an increased metabolic activity in ACC during trials with no response conflict and high rate of error probability, even when subjects performed correctly in them, leading them to propose that subjects learn to predict the likelihood of error on the basis of the stimuli–outcome relationship of previous trials.

Because in the *REDIRECT* task error likelihood is related to the degree of inhibitory control, Sharika et al. (2008) had the opportunity to examine the hypothesis that predictive error correction depends on error likelihood, and thereby study the potential interaction between inhibition and error correction. Exploring such an interaction, of which little is known to date, is an important step toward understanding the workings of an executive control system that has been hypothesized to flexibly coordinate goal-directed behavior (Norman and Shallice, 1980; Logan and Cowan, 1984; Baddeley and Della Sala, 1996). Sharika et al. (2008) examined these issues by using a *target-shift* double-step task in the context of a race model widely used to interpret inhibitory control in saccadic reaction time tasks (Logan and Cowan, 1984; Hanes and Schall, 1996; Hanes and Carpenter, 1999).

The logic of the argument rests on the basis of the race model (Logan and Cowan, 1984; Camalier et al., 2007; Kapoor and Murthy, 2008) that successfully describes saccade production or cancellation in a double-step task as an outcome of a race between a GO and a STOP process (initiated following the presentation of the initial and final target, respectively). In the context of such a race model, it is plausible that at some point in time after the initiation of the STOP process, when its likelihood of finishing first becomes very low, it becomes prudent for the oculomotor system to program a corrective saccade in parallel. If this idea is correct and the onset of correction actually takes into account the likelihood of error, then one ought to expect the time of correction onsets to reflect such performance-based stimulus–response associations. A more specific test is the hypothesized relation of the onset of correction and target step delay; while at longer delays, correction should begin sooner, presumably reflecting the condition of high error probability; at shorter delays correction ought to start later, presumably reflecting the condition of lower error probability. The data in Fig. 11 bear out this prediction. In eight out of nine subjects, the onset of correction was increasingly delayed as target step delays became shorter and the likelihood of an error presumably reduced. These results are in accordance with the idea of the brain estimating error likelihood for performance monitoring (Brown and Braver, 2005) and suggests how such predictive estimation may influence the time course of concurrent corrective saccade processing as well.

However, alternative hypotheses cannot be ruled out. It is possible, as a general consequence of there being a bottleneck at some stage of saccade processing, that at shorter delays the preparation of the corrective saccade may be attenuated by either the motor preparation of the first erroneous saccade and/or the preparation of the correct saccade that was never executed. If the attenuation due to such a bottleneck varies with target step delay such that interference increases as the temporal overlap between two saccades increases, then a similar relation between target step delay and onset of correction should occur. Although at one level the error likelihood and bottleneck hypotheses may be distinct, models of executive control (Botvinick et al., 2001) postulating error/conflict detection in the brain to be a consequence of simultaneous programming of mutually incompatible motor programs, suggest that these two hypotheses need not be necessarily incompatible. The data of Sharika et al. (2008) suggests a model of how inhibitory control — generated as a consequence of either performance monitoring or a general bottleneck — and error detection/correction may interact for successful production of voluntary action.

Conclusions

In this article we have described our work that has focused on the mechanisms that regulate the production of sequential eye movements engaged by double-step tasks, particularly in the context of error detection/correction. The behavioral data from the *target shift* double-step task in our laboratory and those of others have provided fairly convincing evidence of the neural networks that may implement a form of predictive error control to facilitate fast error correction under difficult task conditions. The data from FEF and the SC also provide converging neurophysiological evidence for the existence of such predictive control, and although a bit speculative at this juncture, the data from Sharika et al. (2008) may nevertheless provide a starting conceptual basis to understand the core of this predictive control by suggesting that error correction may derive in part from the capacity of the brain to estimate the likelihood of an error as it is trying to inhibit an unwanted movement. Collectively these results, used in conjunction with electrophysiological recordings, may provide an important approach to understand how error detection/correction and inhibition, two vital cogs in the functioning of executive control, may interact to govern goal-directed behaviors.

Acknowledgments

We would like to thank our long-time collaborator Jeffrey D. Schall for his constant encouragement and critical intellectual input; Sheldon Hoffmann of Reflective Computing (USA) for help in software-related issues and Jitender Ahlawat for manuscript preparation. This work was generously supported by grants from the Department of Science and Technology (DST) and the Department of Biotechnology (DBT), Government of India, and from core funding from the National Brain Research Centre.

References

Allport, D. A. (1987). Selection for action: Some behavioral and neurophysiological considerations of attention and action. In H. Heuer & A. F. Sanders (Eds.), *Perspectives on perception and action*. Hillsdale, NJ: Lawrence Erlbaum Associates.

Allport, D. A., Styles, E. A., & Hsieh, S. (1994). Shifting intentional set: Exploring the dynamic control of tasks. In C. Umilta & M. Moscovitch (Eds.), *Attention and performance XV: Conscious and nonconscious information processing* (pp. 396–419). Cambridge, MA: MIT Press.

Baddeley, A., & Della Sala, S. (1996). Working memory and executive control. *Philosophical Transactions of the Royal Society of London Series B Biological Science*, *351*, 1397–1403.

Becker, W. (1991). Saccades. In R. H. S. Carpenter (Ed.), *Vision and visual dysfunction* (pp. 95–137). Boca Raton, FL: CRC Press.

Becker, W., & Jürgens, R. (1979). An analysis of the saccadic system by means of double step stimuli. *Vision Research*, *19*, 967–983.

Botvinick, M. M., Braver, T. S., Barch, D. M., Carter, C. S., & Cohen, J. D. (2001). Conflict monitoring and cognitive control. *Psychological Reviews*, *108*, 624–652.

Brown, J., & Braver, T. (2005). Learned predictions of error likelihood in the anterior cingulate cortex. *Science*, *307*, 1118–1121.

Camalier, C. R., Gotler, A., Murthy, A., Thompson, K. G., Logan, G. D., Palmeri, T. J., et al. (2007). Dynamics of saccade target selection: Race model analysis of double step and search step saccade production in human and macaque. *Vision Research*, *47*, 2187–2211.

Carpenter, R. H., & Williams, M. L. (1995). Neural computation of log likelihood in control of saccadic eye movements. *Nature*, *377*, 59–62.

Cohen, J. D., Dunbar, K., & McClelland, J. L. (1990). On the control of automatic processes: A parallel distributed processing account of the Stroop effect. *Psychological Reviews*, *97*, 332–361.

Colby, C. L., & Goldberg, M. E. (1999). Space and attention in the parietal cortex. *Annual Review of Neuroscience*, *22*, 319–349.

Duhamel, J. R., Colby, C. L., & Goldberg, M. E. (1992). The updating of the representation of visual space in parietal cortex by intended eye movements. *Science*, *255*, 90–92.

Findlay, J., & Harris, L. (1984). Small saccades to double-stepped targets moving in two dimensions. In A. G. Gale & F. Johnson (Eds.), *Theoretical and applied aspects of eye movement research* (pp. 71–78). Amsterdam: Elsevier.

Gehring, W. J., & Fenscik, D. E. (2001). Functions of medial frontal cortex in the processing of conflict and errors. *The Journal of Neuroscience*, *21*, 9430–9437.

Goldberg, M. E., & Bruce, C. J. (1990). Primate frontal eye fields. III. Maintenance of a spatially accurate saccade signal. *Journal of Neurophysiology*, *64*, 489–508.

Hallett, P. E., & Lightstone, A. D. (1976a). Saccadic eye movements towards stimuli triggered by prior saccades. *Vision Research*, *16*, 99–106.

Hallett, P. E., & Lightstone, A. D. (1976b). Saccadic eye movements to flashed targets. *Vision Research*, *16*, 107–114.

Hanes, D. P., & Carpenter, R. H. (1999). Countermanding saccades in humans. *Vision Research*, *39*, 2777–2791.

Hanes, D. P., & Schall, J. D. (1996). Neural control of voluntary movement initiation. *Science*, *274*, 427–430.

Hasegawa, R., Blitz, A., Geller, N., & Goldberg, M. (2000). Neurons in monkey prefrontal cortex that track past or predict future performance. *Science*, *290*, 1786–1789.

Hooge, I., & Erkelens, C. (1996). Control of fixation duration in a simple search task. *Perception & Psychophysics*, *58*, 969–976.

Hou, R., & Fender, D. (1979). Processing of direction and magnitude by the saccadic eye-movement system. *Vision Research*, *19*, 1421–1426.

Ito, S., Stuphorn, V., Brown, J. W., & Schall, J. D. (2003). Performance monitoring by the anterior cingulate cortex during saccade countermanding. *Science*, *302*, 120–122.

Kapoor, V., & Murthy, A. (2008). Covert inhibition potentiates online control in a double-step task. *Journal of Vision*, *8*, 1–16.

Kawato, M., Furukawa, K., & Suzuki, R. (1987). A hierarchical neural-network model for control and learning of voluntary movement. *Biological Cybernetics*, *57*, 169–185.

Komoda, M., Festinger, L., Phillips, L., Duckman, R., & Young, R. (1973). Some observations concerning saccadic eye movements. *Vision Research*, *13*, 1009–1020.

Logan, G., & Cowan, W. (1984). On the ability to inhibit thought and action: A theory of an act of control. *Psychological Reviews*, *91*, 295–327.

Logan, G. D. (1985). Executive control of thought and action. *Acta Psychologica*, *60*, 193–210.

Ludwig, C. J., Gilchrist, I. D., McSorley, E., & Baddeley, R. J. (2005). The temporal impulse response underlying saccadic decisions. *The Journal of Neuroscience*, *25*, 9907–9912.

Ludwig, C. J., Mildinhall, J. W., & Gilchrist, I. D. (2007). A population coding account for systematic variation in saccadic dead time. *Journal of Neurophysiology*, *97*, 795–805.

McPeek, R., & Keller, E. (2002). Saccade target selection in the superior colliculus during a visual search task. *Journal of Neurophysiology*, *88*, 2019–2034.

McPeek, R., Maljkovic, V., & Nakayama, K. (1999). Saccades require focal attention and are facilitated by a short-term memory system. *Vision Research*, *39*, 1555–1566.

McPeek, R., Skavenski, A., & Nakayama, K. (2000). Concurrent processing of saccades in visual search. *Vision Research*, *40*, 2499–2516.

McPeek, R. M., & Keller, E. L. (2002). Superior colliculus activity related to concurrent processing of saccade goals in a visual search task. *Journal of Neurophysiology*, *87*, 1805–1815.

Murthy, A., Ray, S., Shorter, S. M., Priddy, E. G., Schall, J. D., & Thompson, K. G. (2007). Frontal eye field contributions to rapid corrective saccades. *Journal of Neurophysiology*, *97*, 1457–1469.

Norman, D., & Shallice, T. (1980). *Attention to action: Willed and automatic control of behavior*. San Diego, CA: Center for Human Information Processing, University of California.

Norman, D. A., & Shallice, T. (1986). Attention to action: Willed and automatic control of behavior. In R. J. Davidson, G. E. Schwartz, & D. Shapiro (Eds.), *Consciousness and self-regulation: Advances in research and theory* (Vol. 4, pp. 1–18). New York: Plenum Press.

Osman, A., Kornblum, S., & Meyer, D. E. (1986). The point of no return in choice reaction time: Controlled and ballistic stages of response preparation. *Journal of Experimental Psychology: Human Perception and Performance*, *12*, 243–258.

Pouget, P., Emeric, E. E., Stuphorn, V., Reis, K., & Schall, J. D. (2005). Chronometry of visual responses in frontal eye field, supplementary eye field, and anterior cingulated cortex. *Journal of Neurophysiology*, *94*, 2086–2092.

Rabbitt, P. M. (1966). Errors and error correction in choice-response tasks. *Journal of Experimental Psychology*, *71*, 264–272.

Rabbitt, P. M., & Phillips, S. (1967). Error-detection and correction latencies as a function of S–R compatibility. *Journal of Experimental Psychology*, *19*, 37–42.

Ray, S., Schall, J., & Murthy, A. (2004). Programming of double-step saccade sequences: Modulation by cognitive control. *Vision Research*, *44*, 2707–2718.

Reddi, B. A. J., Asrress, K. N., & Carpenter, R. H. S. (2003). Accuracy, information and response time in a saccadic decision task. *Journal of Neurophysiology*, *90*, 3538–3546.

Schall, J. D. (2002). The neural selection and control of saccades by the frontal eye field. *Philosophical Transactions of the Royal Society of London Series B Biological Science*, *357*, 1073–1082.

Schall, J. D., Thompson, K. G., Bichot, N. P., Murthy, A., & Sato, T. R. (2003). Visual processing in the frontal eye field. In J. Kaas & C. Collins (Eds.), *The primate visual system* (pp. 205–230). Boca Raton, FL: CRC Press.

Schmolesky, M. T., Wang, Y., Hanes, D. P., Thompson, K. G., Leutgeb, S., Schall, J. D., et al. (1998). Signal timing across the macaque visual system. *Journal of Neurophysiology*, *79*, 3272–3278.

Sharika, K. M., Ramakrishnan, A., & Murthy, A. (2008). Control of predictive error correction during a saccadic double-step task. *Journal of Neurophysiology*, *100*, 2757–2770.

Stuphorn, V., Taylor, T. L., & Schall, J. D. (2000). Performance monitoring by supplementary eye field. *Nature*, *408*, 857–860.

Thompson, K., Hanes, D., Bichot, N., & Schall, J. (1996). Perceptual and motor processing stages identified in the activity of macaque frontal eye field neurons during visual search. *Journal of Neurophysiology*, *76*, 4040–4055.

Thompson, K. G., Bichot, N. P., & Schall, J. D. (2001). From attention to action in frontal cortex. In J. Braun & C. Koch (Eds.), *Visual attention and cortical circuits* (pp. 137–157). Cambridge, MA: MIT Press.

Umeno, M. M., & Goldberg, M. E. (1997). Spatial processing in the monkey frontal eye field. I. Predictive visual responses. *Journal of Neurophysiology*, *78*, 1373–1383.

Umeno, M. M., & Goldberg, M. E. (2001). Spatial processing in the monkey frontal eye field. II. Memory responses. *Journal of Neurophysiology*, *86*, 2344–2352.

Van Loon, E. M., Hooge, I. T., & Van den Berg, A. V. (2002). The timing of sequences of saccades in visual search. *Proceedings of the Royal Society B: Biological Sciences*, *269*, 1571–1579.

Vaziri, S., Diedrichsen, J., & Shadmehr, R. (2006). Why does the brain predict sensory consequences of oculomotor commands? Optimal integration of the predicted and the actual sensory feedback. *The Journal of Neuroscience*, *26*, 4188–4197.

Viviani, P. (1990). Eye movements in visual search: Cognitive, perceptual and motor control aspects. *Reviews of Oculomotor Research*, *4*, 353–393.

Walker, M. F., Fitzgibbon, E. J., & Goldberg, M. E. (1995). Neurons in the monkey superior colliculus predict the visual result of impending saccadic eye movements. *Journal of Neurophysiology*, *73*, 1988–2003.

Wolpert, D. M., Ghahramani, Z., & Jordan, M. I. (1995). An internal model for sensorimotor integration. *Science*, *269*, 1880–1882.

Wolpert, D. M., & Kawato, M. (1998). Multiple paired forward and inverse models for motor control. *Neural Networks*, *11*, 1317–1329.

Wheeless, L., Jr., Boynton, R., & Cohen, G. (1966). Eye-movement responses to step and pulse-step stimuli. *Journal of Optical Society of America*, *56*, 956–960.

N. Srinivasan (Ed.)
Progress in Brain Research, Vol. 176
ISSN 0079-6123

CHAPTER 16

Explaining the Colavita visual dominance effect

Charles Spence*

Crossmodal Research Laboratory, Department of Experimental Psychology, University of Oxford, Oxford, UK

Abstract: The last couple of years have seen a resurgence of interest in the Colavita visual dominance effect. In the basic experimental paradigm, a random series of auditory, visual, and audiovisual stimuli are presented to participants who are instructed to make one response whenever they see a visual target and another response whenever they hear an auditory target. Many studies have now shown that participants sometimes fail to respond to auditory targets when they are presented at the same time as visual targets (i.e., on the bimodal trials), despite the fact that they have no problems in responding to the auditory and visual stimuli when they are presented individually. The existence of the Colavita visual dominance effect provides an intriguing contrast with the results of the many other recent studies showing the superiority of multisensory (over unisensory) information processing in humans. Various accounts have been put forward over the years in order to try and explain the effect, including the suggestion that it reflects nothing more than an underlying bias to attend to the visual modality. Here, the empirical literature on the Colavita visual dominance effect is reviewed and some of the key factors modulating the effect highlighted. The available research has now provided evidence against all previous accounts of the Colavita effect. A novel explanation of the Colavita effect is therefore put forward here, one that is based on the latest findings highlighting the asymmetrical effect that auditory and visual stimuli exert on people's responses to stimuli presented in the other modality.

Keywords: Colavita effect; sensory dominance; extinction; vision; auditory; attention

Introduction

One of the most intriguing examples of visual dominance was popularized by the research of Frank B. Colavita more than 30 years ago (see Colavita, 1974, 1982; Colavita et al., 1976; Colavita and Weisberg, 1979; see also Osborn et al., 1963). In his original 1974 study, Colavita described a remarkably simple experiment that gave rise to the most surprising of results. A random sequence of 30 clearly suprathreshold auditory and visual targets were presented to the participants, one every 15 s.[1]

*Corresponding author.
Tel.: +44-1865-271364; Fax: +44-1865-310447;
E-mail: charles.spence@psy.ox.ac.uk

[1]It is perhaps worth noting that all of the participants in Colavita's (1974) experiments initially matched the intensity of the auditory and visual stimulus. They were then presented with 30 unimodal target trials in which the upcoming target modality was announced at the start of each trial. Finally, they were presented with the 35 critical trials (30 unimodal and 5 bimodal) in which the target modality on each trial was unpredictable. After each modality discrimination response, the participant was asked by the experimenter whether s/he had pressed the correct response key or not. Note that the auditory and visual stimuli were presented until the participant responded, with their manual response terminating any stimuli that were being presented.

DOI: 10.1016/S0079-6123(09)17615-X

The participants had to discriminate the modality of each target by making a speeded discrimination response, pressing one button in response to the auditory targets and another button in response to the visual targets. As one might expect, the participants responded rapidly and accurately on these unimodal target trials. However, unbeknownst to the 10 participants, a few trials (5) in which both the light and sound were presented at the same time were interleaved into the sequence of unimodal target trials.

Colavita (1974) was interested in finding out how his participants would respond on these "unexpected" bimodal target trials: Would they respond to the light (thus showing visual dominance over behavior), would they respond to the sound, or would they press both response keys? In fact, the participants in Colavita's original study responded to the visual stimulus on 49 out of the 50 bimodal trials (5 bimodal trials were presented to each of the 10 participants in the study).[2] Even more dramatically, the participants reported being unaware that a sound had even been presented on 16 of those 50 bimodal trials. Given the intriguing nature of the Colavita effect (as this phenomenon is now known), and given the simplicity of the experimental design used to elicit it, it is rather surprising to find that relatively little research has been published on this form of visual "prepotency" in the 35 years since Colavita first reported the phenomenon in either humans (for exceptions, see Shapiro et al., 1984; Johnson and Shapiro, 1989; Zahn et al., 1994; Quinlan, 2000; Van Damme et al., 2009) or, for that matter, in other species (Randich et al., 1978). This apparent lack of interest in the Colavita effect stands in marked contrast to the recent resurgence of interest in the topic of multisensory information processing more generally (e.g., see Calvert et al., 2004; Spence, 2007, for reviews).

Looking back with hindsight, one cannot though help but be struck now by the somewhat idiosyncratic nature of the experimental procedure Colavita used originally. For example, the participants in Colavita's (1974), Experiment 1, were not informed (prior to their taking part in the study) that any bimodal trials would be presented, nor were they informed as to how they should respond on such trials. What is more, when, on verbal questioning after having made their response, participants mentioned that they had perceived both an auditory and a visual stimulus, the experimenter took the rather unconventional step of apologizing for this "error" (of presenting both stimuli at the same time), and explaining how/why it had occurred, before continuing on with the rest of the experiment!

Why, then, has this fascinating empirical phenomenon fallen out of favor among researchers? While many of the methodological shortcomings in Colavita's original research were soon corrected, it would appear that certain researchers perhaps still believe that the Colavita effect is simply such a surprising (unbelievable even) phenomenon that it must reflect some kind of artifact in the design of Colavita's experiments. This skepticism may well have been compounded by the results of an early series of experiments reported by Egeth and Sager (1977) that, at first glance, seemingly failed to replicate the Colavita visual dominance effect; That is, the participants in their study always managed to respond correctly to both the auditory and visual target on the bimodal target trials.[3]

[2]In fact, when subsequently questioned, the participant claimed to have responded erroneously on the one trial in which they had pressed the auditory response key.

[3]It is important to note that in contrast to Colavita's studies, where the detection of a participant's manual response resulted in the termination of all stimuli, only the target that participants responded to was terminated in Egeth and Sager's (1977) studies. Thus, if participants initially made a visual response on the bimodal target trials that would only result in the visual target being turned off. The auditory target would, however, continue to be presented. This minor methodological modification essentially ensured that all of Egeth and Sager's participants would eventually make a response to each and every sound. Nevertheless, Egeth and Sager did observe some evidence of visual dominance over audition when they analyzed the relative latencies of their participants' responses to the unimodal and bimodal target stimuli: While their participants responded more rapidly to the auditory targets than to the visual targets on the unimodal target trials, this pattern of results was reversed on the bimodal trials (cf. Colavita, 1974, 1982; Colavita and Weisberg, 1979; Cooper, 1998; Zahn et al., 1994; Koppen and Spence, 2007a; Sinnett

Here at the Crossmodal Research Laboratory in Oxford, we wondered just how reproducible the Colavita effect really was. We set out to resolve once-and-for-all whether it constitutes a robust empirical phenomenon; if so, we thought that researching it would provide a stark contrast to the innumerable studies currently showing that multisensory stimuli lead to enhanced perception/information processing relative to their unimodal counterparts. A second potentially important reason for studying the Colavita visual dominance effect relates to the claim that it might provide an analog in normal participants to the phenomenon of crossmodal extinction that has been reported in certain clinical patients (see Rapp and Hendel, 2003; Sarri et al., 2006; see also Egeth and Sager, 1977; Hartcher-O'Brien et al., 2008).

Recent research on the Colavita visual dominance effect

In 2007, Camille Koppen and I conducted a series of studies in order to investigate whether the Colavita visual dominance effect could be reproduced under rather more stringent methodological conditions than those described in Colavita's (1974) original research. In our first series of experiments (see Koppen and Spence, 2007a), the participants had to make speeded detection/discrimination responses[4] to a random sequence of briefly presented (for 50 ms) auditory (40% of targets), visual (40%), and audiovisual (20%) targets. Note that we approximately matched the stimulus probabilities used by Colavita in his early research. However, in contrast to Colavita's original study, all of our participants were explicitly informed that the auditory and visual targets would be presented together on some proportion of the trials. Furthermore, all of our participants were instructed by the experimenter to respond to such bimodal targets by pressing both the auditory and visual response keys. Successive targets were separated by an intertarget interval of 1800 ms, thus providing the participants with sufficient time in which to make an additional response should they have realized that their initial response had been incomplete (e.g., pressing only the visual key when they should have pressed both response keys on a bimodal target trial).

Under these conditions, Koppen and Spence (2007a, Experiment 1) observed a small, but significant, Colavita visual dominance effect; that is, the participants made significantly more visual-only than auditory-only responses on the bimodal trials (see Fig. 1A). More recently, Hartcher-O'Brien et al. (2008, Experiment 1) have shown that visual stimuli also dominate over tactile stimuli in much the same way (see Fig. 1B). Importantly, these effects still occur when the probability of each trial type is equalized (i.e., when 33% auditory, visual, and audiovisual targets are presented; Koppen and Spence, 2007a, Experiment 2; see also Hartcher-O'Brien et al., 2008, Experiment 2). For some reason though, when presented with a bimodal audio-tactile target, people do not appear to have a strong bias toward either auditory-only or tactile-only responses (see Hecht and Reiner, 2009; Occelli et al., submitted). The latter result has been taken (perhaps prematurely) to suggest that the Colavita effect reflects a specifically visual form of dominance (or prepotency) effect. Interestingly, Hecht and Reiner have also shown that modality (in particular, visual) dominance effects are not found under conditions where auditory, visual, and tactile stimuli are all presented at the same time, and where the participants are given a separate key to respond to the target stimuli presented in each of the three target modalities.

Stimulus intensity and the Colavita visual dominance effect

Many researchers have wondered what role, if any, stimulus intensity might play in modulating

et al., 2007, for similar results). Egeth and Sager argued that this result was consistent with their participants exhibiting visually dominant behavior as a result of the shifting of their attention toward the visual modality on the bimodal target trials (see also Falkenstein et al., 1991; Jaśkowski, 1996).

[4]Note that this task has elements of both detection and discrimination tasks.

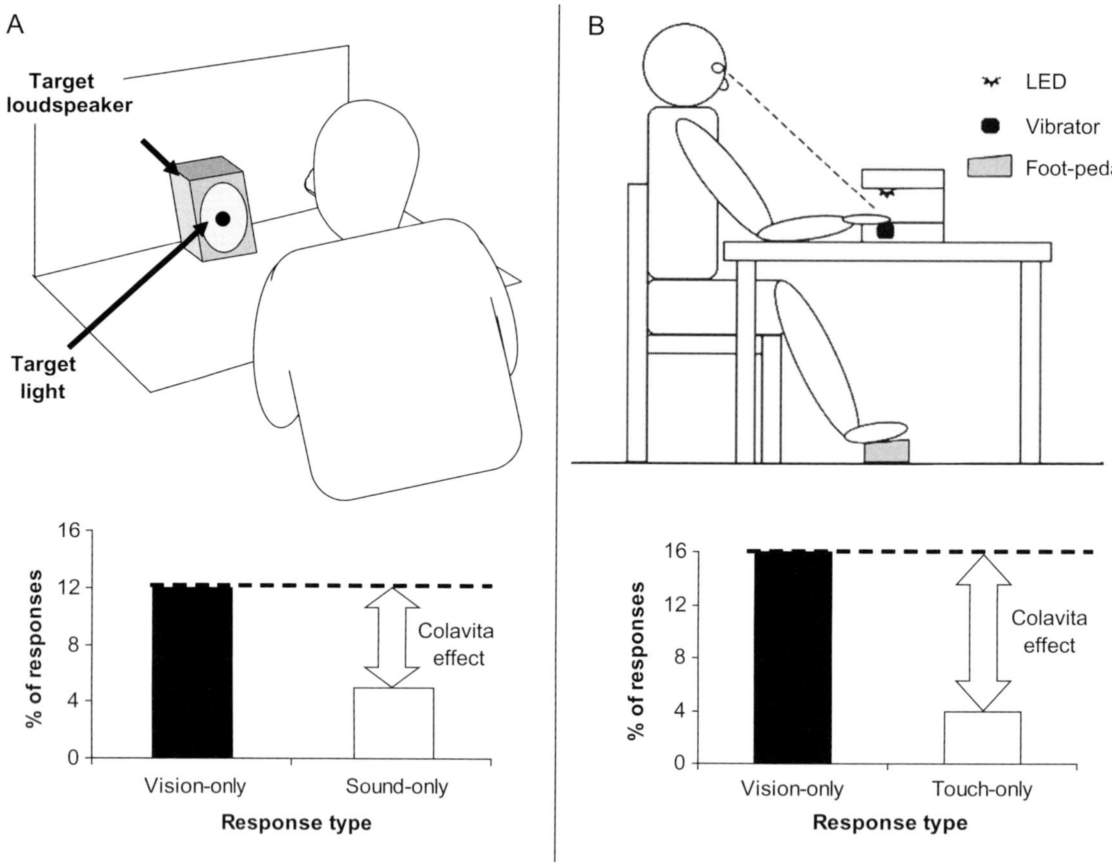

Fig. 1. (A) Schematic figure showing the experimental set-up used in Koppen and Spence's (2007a) studies of the audiovisual Colavita effect, and (B) Hartcher-O'Brien's et al.'s (2008) studies of the visuotactile Colavita effect. The visual stimulus consisted of the illumination of the loudspeaker cone used to present the auditory stimuli in Koppen and Spence's study and the illumination of the finger where the vibration was presented in Hartcher-O'Brien et al.'s study. Note that the target stimuli were presented from exactly the same spatial location in both studies (cf. Koppen and Spence, 2007c; Hartcher-O'Brien et al., 2008, Experiment 3). The results of both studies highlight a significant Colavita visual dominance effect (see the graphs at the lower part of the figure). The values reported in the graphs refer to the percentage of the bimodal target trials in which the participants made either a visual-only or an auditory- (tactile-) only response. Thus, in both of the studies shown here, the participants actually responded correctly on the majority of the bimodal target trials (i.e., the magnitude of the Colavita effect is much smaller than that reported in Colavita's early studies).

the Colavita visual dominance effect. Could it simply be, for example, that the more intense stimulus dominates over the less intense one (cf. Shapiro and Johnson, 1987)? Such an explanation seems unlikely given that Colavita had his participants match the subjective intensity of the auditory and visual stimuli at the start of the experiment (see also Zahn et al., 1994). Colavita (1974, Experiment 2) even showed that the effect that bears his name remains essentially undiminished when participants adjust the loudness of the auditory stimulus until it is subjectively twice as intense as the visual stimulus (cf. O'Connor and Hermelin, 1963). Furthermore, Hartcher-O'Brien et al. (2008, Experiment 4) recently demonstrated a robust visual dominance over touch when the intensity of the tactile stimulus was matched to that of the visual target at the start of each participant's experimental session (and where the intensity of the visual stimulus had been set at the 75%-detection threshold). Under these conditions, participants made 29% visual-only responses as compared to only 11% tactile-only responses on the bimodal visuotactile target trials.

Attention and the Colavita visual dominance effect

Having ruled out an intensity-based explanation of the Colavita visual dominance effect, the next suggestion made by researchers was that it might reflect the consequences of some kind of attentional bias toward the visual modality instead (Posner et al., 1976). Indeed, consistent with this claim, the only manipulation that Colavita himself found to effectively reduce the magnitude of visual dominance effect was explicitly to instruct his participants to respond to the sound on bimodal trials (Colavita, 1974, Experiment 4). Even under these conditions, though, it is worth noting that Colavita's participants still responded to the light on 36 out of 60 of the bimodal trials. Subsequently, other researchers have used a variety of different methods in order to try and bias their participants' endogenous attention toward one sensory modality (usually audition) or the other. One popular method has been to vary the ratio of unimodal trials presented in each modality (see Egeth and Sager, 1977; Quinlan, 2000; Koppen and Spence, 2007b; Sinnett et al., 2007). For example, Sinnett et al. demonstrated an increased Colavita visual dominance effect by increasing the proportion of unimodal visual targets to 60%, while at the same time reducing the proportion of auditory targets to just 20% (the remaining 20% of targets being bimodal). Sinnett and his colleagues were also able to eliminate (but crucially not to reverse) the Colavita effect by presenting more unimodal auditory targets than unimodal visual targets (i.e., 60% auditory, 20% visual). These manipulations should have led to a sustained shift of participants' attention toward the more likely target modality.

Koppen and Spence (2007a, Experiment 4) have also investigated the consequences for the Colavita effect of transiently directing their participants' exogenous attention toward either the auditory or visual modality, by presenting a non-predictive auditory or visual cue prior to the onset of the target on each trial [at a cue-target stimulus onset asynchrony (SOA) of 200 ms; cf. Spence et al., 2001; Turatto et al., 2002; Rodway, 2005]. Although the modality of the cue was completely non-predictive with respect to the likely target type, Koppen and Spence nevertheless still found that the magnitude (but not the direction) of the Colavita visual dominance effect was influenced by the modality of the cue: That is, the presentation of a visual cue shortly before the onset of the audiovisual targets gave rise to a significantly larger Colavita visual dominance effect than seen following the presentation of an auditory cue.

It is important to point out here that while these attentional manipulations (of both endogenous and exogenous attention) have been relatively successful in modulating the size of the Colavita visual dominance effect, not one of them has proved effective in reversing the effect (i.e., to the extent that participants start behaving in an auditorily dominant way — making more auditory-only than visual-only responses). These results therefore suggest that, contrary to Posner et al.'s (1976) early suggestion, the Colavita effect cannot be explained simply in terms of participants having a predisposition to attend to the visual modality.

Response demands and the Colavita visual dominance effect

Across a number of studies of the Colavita effect, it has been reported that reaction times (RTs) on bimodal trials tend to be significantly slower than on unimodal trials (e.g., Koppen and Spence, 2007a; Hartcher-O'Brien et al., 2008; Hecht and Reiner, 2009). This result has led researchers to suggest that the Colavita visual dominance effect might be caused, at least in part, by a difficulty (or cost) associated with participants having to make two responses on the bimodal target trials, as opposed to just a single response on the unimodal trials (cf. Egeth and Sager, 1977; Sinnett et al., 2007). One might think here of the difficulty of making two responses in rapid succession that has been highlighted by research into the

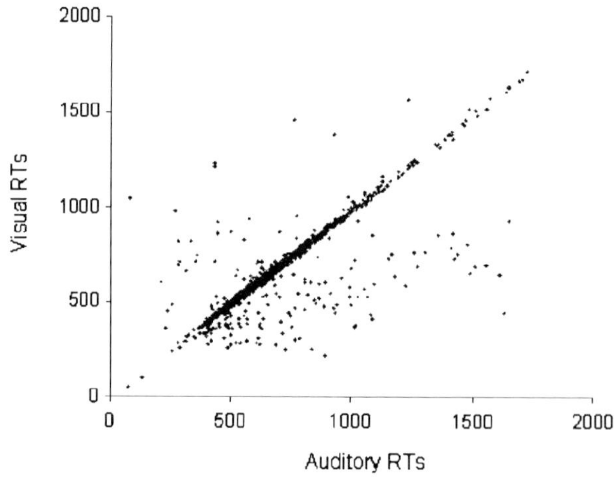

Fig. 2. Scatter plot highlighting the responses to the visual and auditory components of the bimodal targets, for those trials in which the participants made a correct response, in Koppen and Spence's (2007a, Experiment 1) study of the Colavita visual dominance effect. Note that in this study the participants had two response keys, one for auditory targets and the other for visual targets. The participants were explicitly instructed to make both responses on the relatively infrequent bimodal target trials. Each dot represents an individual trial from one of the participants in which they responded correctly to the bimodal audiovisual target. The graph shows that the majority of participants' responses fell on the identity line, meaning that the participants had coupled their auditory and visual responses into a single bi-finger response (cf. Fagot and Pashler, 1992) Adapted with permission from Koppen and Spence (2007a).

psychological refractory period (see Pashler, 1994; Spence, 2008, for reviews). However, it should be noted that an analysis of the RT data from those bimodal target trials in which the participants responded correctly (by pressing both the auditory and visual target response keys) revealed that participants typically make the two responses at more or less the same time (see Fig. 2). This result suggests that when the participants responded on the bimodal target trials it may only have required a single act of response selection (i.e., some form of response coupling may have occurred; cf. Fagot and Pashler, 1992; Schumacher et al., 2001).

Direct evidence against the "difficulty of bimodal responding" account of the Colavita visual dominance effect comes from studies showing that the audiovisual and visuotactile Colavita effects still occur even when participants are given three separate response keys (auditory, visual, and audiovisual; or tactile, visual, and visuotactile), thus equating the response requirements for the unimodal and bimodal target trials (see Koppen and Spence, 2007a; Sinnett et al., 2007; Hartcher-O'Brien et al., 2008; Koppen et al., 2008; Hecht and Reiner, 2009).

That said, the precise nature of the participants' task does seem to influence whether response facilitation or response inhibition is observed. In particular, Sinnett et al. (2008, Experiment 1) have recently shown that when participants simply have to detect the presence of any target (i.e., regardless of the target modality) in a simple speeded detection version of the Colavita task, then response facilitation (over-and-above that expected merely due to statistical facilitation; see Miller, 1982, 1991) is observed. By contrast, as soon as participants make a separate response to targets presented in each modality (by pressing one response key for auditory targets, and another for visual targets), the typical Colavita visual dominance effect is once again found. These results therefore show that the nature of the response (and task) can have a profound effect on the pattern of results (i.e., multisensory facilitation vs. inhibition) that will be observed. In particular, the Colavita effect only seems to occur under those conditions in which participants have to detect/discriminate the modalities (or identities; see Koppen et al., 2008) of the targets that have been presented.

Does the Colavita effect occur with complex stimuli?

It is important to note that the Colavita visual dominance effect is not restricted to the processing of (or responses to) simple sensory stimuli, such as beeps, taps, and flashes of light (cf. Rapp and Hendel, 2003). Researchers have now shown that the Colavita effect also occurs when participants have to respond to more complex auditory and visual stimuli as well (see Sinnett et al., 2007, 2008; Koppen et al., 2008). For example, Sinnett et al. (2007) conducted several experiments demonstrating the dominance of vision over audition when the visual target consisted of the outline picture of a set of traffic lights (taken from the Snodgrass and Vanderwart, 1980 database), while the auditory target consisted of the sound of a cat meowing (i.e., a naturalistic sound). A significant Colavita visual dominance effect was demonstrated both when these stimuli were presented in isolation (as in a typical Colavita study), and when the targets were presented among a rapidly presented stream of auditory and visual distractors (consisting of 50 different objects presented in each modality) instead. Koppen et al. also observed a Colavita visual dominance effect when the participants had to detect/discriminate the presence of cats and dogs (either seen or heard; that is, regardless of whether the animals were presented in the auditory and/or visual modalities).

Koppen et al. (2008) have recently extended these results by showing that the semantic congruency of the auditory and visual stimuli has absolutely no effect on the magnitude of the Colavita visual dominance effect. This null result, which was replicated in three separate experiments (using different stimulus materials), did not, however, simply reflect the fact that semantic congruency failed to affect participants' performance at all. It did. Specifically, there were significant differences in terms of the speed with which the participants were able to respond on the semantically congruent versus incongruent trials. This latter result is important because it shows that the semantic congruency of the stimuli had been coded at some level by the participants in Koppen et al.'s studies, it is just that this factor did not modulate the Colavita visual dominance effect.

Summary of recent findings on the Colavita effect

In conclusion, research published over the last couple of years has confirmed the robustness of the Colavita visual dominance effect. The available empirical evidence now convincingly shows that people's responses to visual stimuli tend to dominate over (and sometimes even eclipse) their responses to both auditory and tactile stimuli (Koppen and Spence, 2007a; Hartcher-O'Brien et al., 2008). It should, however, be pointed out that the magnitude of the Colavita effect reported recently has tended to be much smaller than that highlighted by Colavita's early work (e.g., see Fig. 1). To date, there is no evidence of either auditory or tactile stimuli dominating when they are presented at the same time (Hecht and Reiner, 2009; Occelli et al., submitted). The Colavita effect cannot simply be accounted for by differences in the intensity of the stimuli used (Colavita, 1974; Zahn et al., 1994; Hartcher-O'Brien et al., 2008), nor can it be accounted for solely in terms of any bias that participants might have to attend preferentially to the visual modality (Koppen and Spence, 2007a, b, d; Sinnett et al., 2007). Furthermore, given that the Colavita effect occurs regardless of whether or not the participants are given a separate response key with which to respond to the bimodal targets, it would appear that the effect is not caused by any difficulty that participants may have in making two responses at once on the bimodal target trials either (though note that participants do have to detect/discriminate each target modality in order for the Colavita visual dominance effect to be observed; see Sinnett et al., 2008, Experiment 1). Thus, the latest research has now effectively ruled out all of the explanations that have been put forward previously to account for the Colavita visual dominance effect.[5]

[5]That is not to say that these various factors (stimulus intensity, attention, response demands) do not influence the magnitude of the Colavita effect, they clearly do; rather, the point is that in-and-of-themselves none of them can account for the whole of the Colavita effect.

Explaining the Colavita visual dominance effect

So how then can the Colavita visual dominance effect be explained? One potentially exciting development here comes from the results of a recent study by Sinnett et al. (2008, Experiment 2) in which participants had to make speeded target detection responses to either auditory targets or else to visual targets. Importantly, however, the target stimuli consisted of auditory (40%), visual (40%), and audiovisual targets (20%) just like in a typical study of the Colavita effect (this task can, then, also be thought of as a kind of go/no-go task; see also Egeth and Sager, 1977; Quinlan, 2000). Under these particular experimental conditions, participants responded significantly faster to the visual targets when they were accompanied by an accessory sound than when they were presented in silence. By contrast, participants' responses to the auditory targets were actually slowed by the presentation of a visual accessory stimulus (see Fig. 3A). These results therefore suggest that while the presentation of an auditory accessory stimulus can speed the initiation of a participant's visual detection response, the presentation of a visual stimulus may actually slow their responses to auditory stimuli (cf. Egeth and Sager, 1977).

How does this result help to resolve what might be going on in the Colavita effect? Well, let us imagine that participants set themselves one internal threshold for initiating their responses to the relatively common (i.e., typically appearing on around 40% of trials) unimodal auditory stimuli, and another threshold for initiating their responses to the equally common unimodal visual stimuli. Sinnett et al.'s (2008) results suggest that on the relatively infrequent bimodal trials, the threshold for responding to visual targets would actually be reached sooner than on the unimodal visual trials, whereas it would be reached more

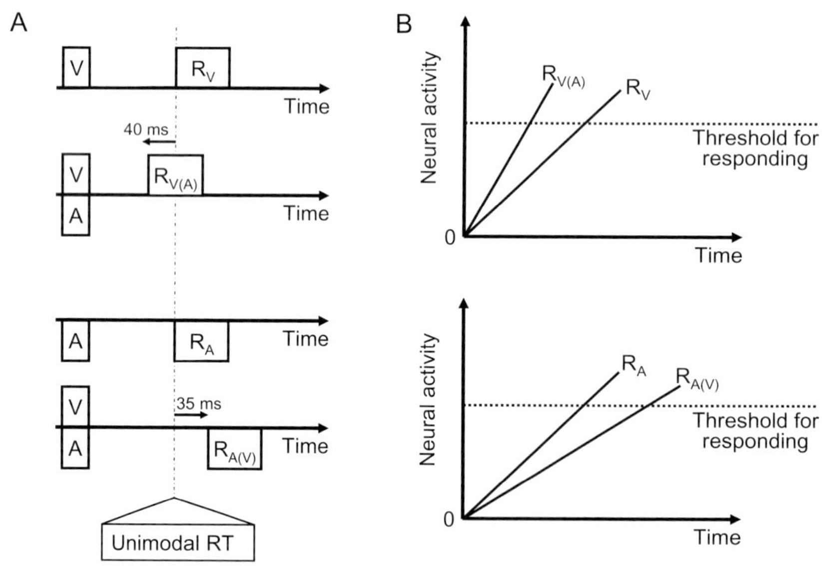

Fig. 3. (A) Schematic illustration of the results of Sinnett et al.'s (2008) study. The figure shows how the presentation of an accessory sound facilitates visual RTs [$R_{V(A)}$] whereas the presentation of a light delays auditory RTs [$R_{A(V)}$]. Note that the unimodal auditory and visual response latencies (R_A and R_V, respectively) were matched in this study (V = visual target; A = auditory target; AV = bimodal audiovisual target). (B) Schematic diagram showing how these asymmetric accessory stimulus effects might lead to more (and more rapid) vision-only than auditory-only responses on bimodal trials. Note the assumption implicit in these two graphs is that the threshold for responding remains constant, while the rate of information accrual changes as a function of the crossmodal presentation of the accessory stimulus. However, a logically plausible alternative account of these results would be to suggest that the rate of information accrual does not actually change, but rather the threshold for responding is changed by the presence of the accessory stimulus in the other modality. It will be for future research to discriminate between these two possibilities.

slowly for auditory targets (see Fig. 3B). Under time pressure (e.g., when successive targets are presented in rapid succession, and hence when participants have little opportunity to delay their responses), vision-only responses on bimodal trials would therefore be expected to occur more frequently than auditory-only responses.

It is important to note here that these differential effects on the timing of a participant's (visual and auditory) responses need not necessarily have a concomitant perceptual correlate (such as the "prior entry" of the visual stimulus to a participant's awareness; see Titchener, 1908; Spence, in press). In fact, they seem not to; Koppen and Spence (2008d) actually found that their participants perceived the auditory stimulus to have been presented slightly ahead of a simultaneously presented visual stimulus (their results suggested that the visual stimulus would have had to have been presented 12 ms before the auditory stimulus in order for the two stimuli to have been judged as simultaneous). This result is the opposite direction to what would be expected according to a prior entry account of the Colavita visual dominance effect, according to which the visual stimulus dominates a participant's responding because it is perceived first.[6]

While Sinnett et al.'s (2008) results may help to explain why a participant might make more vision-only than auditory-only responses, they do not explain why participants do not quickly recognize the error of their ways (after making a vision-only response, say), and then quickly initiate an additional auditory-only response. The participants certainly had sufficient time in which to make a response before the next trial started in Koppen and Spence's research, where the intertarget interval was in the region of 1500–1800 ms (Koppen and Spence, 2007a–c).[7] One possibility here is that a participant's awareness of the target stimuli may actually be determined by the responses that they happen to make (or initiate). That is, in order to explain what is going on in the Colavita visual dominance effect one needs to break with the intuitive view that there is a causal link between conscious perception and action. Instead, one needs to adopt the view that the only causal link is the one that exists between stimulus and response (i.e., bypassing conscious perception and relying on unconscious stimulus processing instead).

Neumann (1990) refers to this direct route from stimulus to response as "direct parameter specification"; he specifically argues that perception should not be conceptualized as a necessary stage in the chain of human information processing, but rather as "*a class of actions that serve to establish and update an internal representation of the environment*" (Neumann, 1990, p. 207; see also Johnson and Haggard, 2003; Banks and Isham, 2009). Support for Neumann's view that stimuli can control responses in the absence of perception comes from studies showing that participants can execute rapid and accurate discrimination responses to masked stimuli that they are subjectively unaware of (e.g., Taylor and McCloskey, 1996; see also Fehrer and Biederman, 1962; Libet et al., 1983; Allport, 1988; Taylor and McCloskey, 1990). Furthermore, other researchers have also demonstrated that the way in which a person responds can influence their subsequent perception of a previously presented stimulus (e.g., Bridgeman, 1990; Müssler and Hommel, 1997): What Müssler and Hommel (p. 861) describe as

[6]Koppen and Spence's (2007d) results therefore add support to the argument that there may be a dissociation between those factors that modulate the perception of temporal order and those that influence the speed of a participant's responding (e.g., Rutschmann and Link, 1964; Jaśkowski, 1996; see also Cardoso-Leite et al., 2007, 2009). It should, however, be noted that the Colavita effect is eliminated when bimodal audiovisual targets are presented too frequently (see Koppen and Spence, 2007b; Sinnett et al., 2007). Therefore, given that an auditory and a visual stimulus was presented in every single trial of Koppen and Spence's (2007d) TOJ study, future research should ideally look for any evidence of the prior entry of the visual stimulus under conditions where the bimodal targets are actually presented as infrequently as when the Colavita effect is demonstrated behaviorally.

[7]Note that according to Koppen and Spence's (2007a) results (see Fig. 2), participants do, on at least a few trials of the typical Colavita paradigm, make a second (typically auditory) response in order to correct for their initial incorrect (or rather incomplete) visual response (each trial on which a participant made such a delayed auditory response appears as a data point falling below the identity line in the figure).

the *"aftereffects of response programming or execution on perception."*

To summarize, the argument here is that when participants try to respond rapidly in the Colavita visual dominance task, they may sometimes end up initiating their response as soon as the threshold is passed, prior to becoming aware of the stimuli eliciting that response. Their awareness of which stimuli have, in fact, been presented is then modulated (or driven) by the response(s) that they actually happen to make: In other words, if (as a participant) I realize that I am going to make a vision-only response, it would seem unsurprising that I only subsequently become aware of the visual target, even if an auditory target had also been presented. Sinnett et al.'s (2008) research is crucial here as it shows that vision-only responses are more likely to occur than auditory-only responses, due to asymmetrical crossmodal accessory stimulus effects.

What happens to the neural representation of the extinguished stimulus on the bimodal trials?

One question that the account of the Colavita visual dominance effect outlined here has not yet answered concerns the fate of the "extinguished" stimulus on the bimodal trials. One possibility here is that the presentation of the visual stimulus simply reduces the salience/perceptibility of the simultaneously presented auditory (or tactile) stimulus. Alternatively, however, the initial neural representation of the auditory stimulus might remain essentially intact, but participants might just "forget" to respond to the auditory (or tactile) target on some proportion of the trials? If a response is not rapidly made to a stimulus (i.e., to the auditory or tactile stimulus) then it may not be consolidated into short-term memory, and will hence be rapidly forgotten. In the future, neuroimaging data might well help to discriminate between these two competing accounts (cf. Sarri et al., 2006).

In the meantime, however, there are a couple of pieces of evidence that argue against the reduced perceptual saliency account: On the one hand, Koppen et al. (2009) recently used signal detection theory to assess the effect of presenting an accessory visual stimulus on the perceptibility of auditory targets. They found only a very small (albeit significant) decrement in auditory sensitivity when an irrelevant accessory visual stimulus was presented at the same time (see also Thompson et al., 1958). No such effect of an auditory accessory stimulus was observed on visual sensitivity. Importantly, however, the decrease in auditory sensitivity was nothing like big enough to account for the failure to respond to sound that is often seen in the Colavita visual dominance effect.

On the other hand, the rapid forgetting account actually fits rather well with one of Colavita's early observations, namely that participants do sometimes appear able to recall the presentation of the stimuli that they have just failed to respond to, if they are quizzed verbally immediately after having made their response (see Colavita, 1974, 1982). As Colavita (1982, p. 412) puts it, on questioning, at least some of his participants appeared able *"to attend to the last vestige of a rapidly fading memory trace."* (This after having initially made a manual response indicating a visual target.) In this case, the experimenter's questioning may therefore have helped to prevent the neural representation of the auditory stimulus from being forgotten too rapidly (the prediction here being that if participants' ability to recall the stimulus that they had just failed to make a manual response to, would decline rapidly with increasing delay). Indeed, subjectively, when one knows about the Colavita visual dominance effect, one is able to check one's memory after every visual-only response in order to determine whether an auditory stimulus had also been presented but had simply not been responded to. In other words, knowledge of the Colavita effect can, under certain conditions, ameliorate the deficit (by encouraging the use of a compensatory strategy).

Theoretical predictions

While the account of the Colavita visual dominance effect put forward here is admittedly somewhat speculative, there are nevertheless a number of relatively clear (and testable) predictions that follow on from it: First, according to the account outlined here, one might predict that if the SOA between the auditory and visual stimuli were to be varied then auditory dominance should be observed whenever the auditory stimulus is presented sufficiently far ahead of the visual stimulus. Indeed, that is precisely what Koppen and Spence (2007d) showed in a Colavita study in which the SOA between the auditory and visual targets on the bimodal target trials was varied randomly between any one of 10 values (±600, ±300, ±150, ±75, and ±35 ms; and where negative values indicate that the auditory stimulus was presented before the visual stimulus). On those trials in which the auditory stimulus was presented 600 ms before the visual target, auditory dominance was, for the first time, observed (although a similar trend was observed at both 300 and 150 ms; see Fig. 4).

A second prediction to emerge from the explanation of the Colavita visual dominance effect outlined here is that the vision-only RTs on the bimodal trials should, on average, be fast (i.e., because the threshold for responding has been reached quickly in these trials); In particular, they should typically be faster than the visual responses observed on those bimodal trials in which the participants actually responded

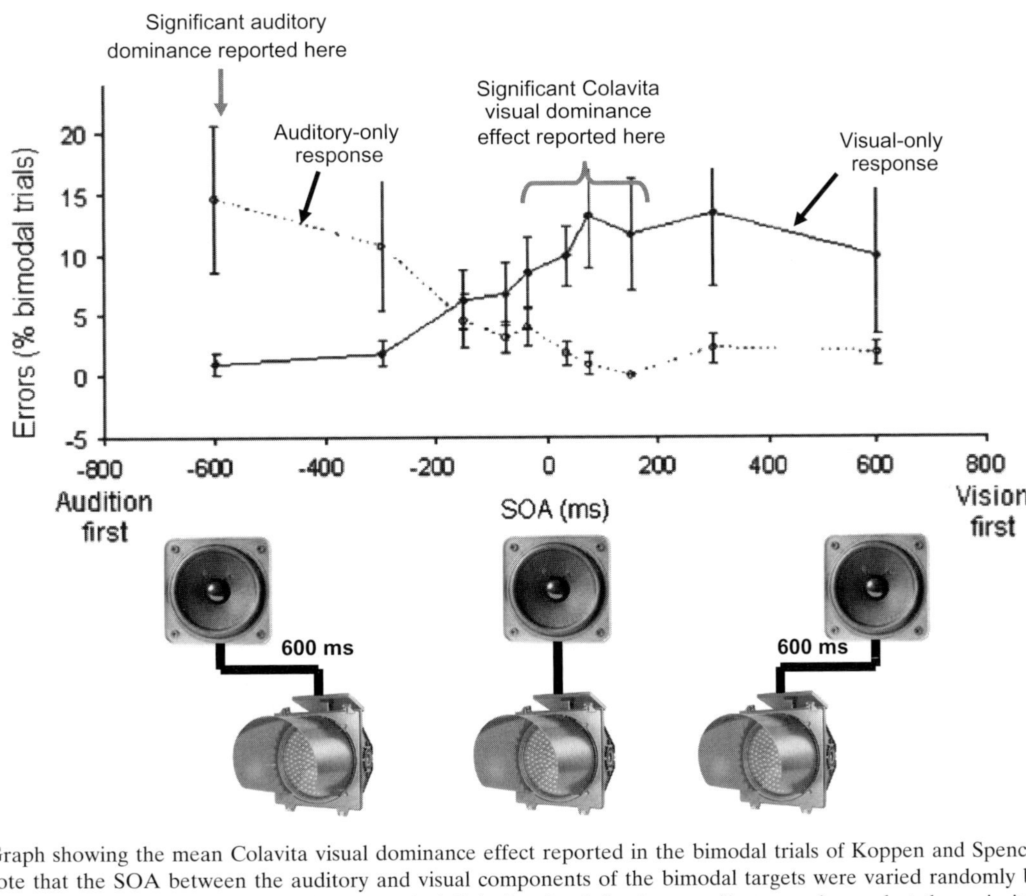

Fig. 4. Graph showing the mean Colavita visual dominance effect reported in the bimodal trials of Koppen and Spence's (2007d) study. Note that the SOA between the auditory and visual components of the bimodal targets were varied randomly between 10 different values. The graph shows that while the typical Colavita visual dominance effect was observed at the majority of SOAs, auditory dominance was observed when the auditory component of the bimodal target was presented 600 ms before the visual component. Note that the numerical trend toward auditory dominance highlighted when the auditory stimulus led by 150 or 300 ms SOAs failed to reach statistical significance.

correctly. Once again, support for this claim comes from another of Koppen and Spence's studies in which the participants had three response keys, one for auditory targets, one for visual targets, and a third for the bimodal targets (see Koppen and Spence, 2007a, Experiment 2; note that the three-response alternative version of the Colavita task needs to be analyzed to test this particular prediction in order to rule out a response difficulty account of any differences that might be observed between the various conditions; see above). In line with the theory's predictions, Koppen and Spence's results showed that the incorrect vision-only responses (mean = 563 ms) were made more rapidly, on average, than correct responses to unimodal visual targets (mean = 582 ms), and crucially, far more rapidly than correct responses were made to the bimodal targets (mean = 641 ms). Thus, it really does seem as though the occurrence of the Colavita visual dominance effect (in particular, the making of a vision-only response on the bimodal audiovisual target trials) is tightly linked to the speed with which a participant initiates his/her response. When the participant responds rapidly they are likely to make an erroneous visual-only response, whereas when they respond more slowly, they are more likely to respond correctly; This result is all the more impressive given the fact that auditory-only responses often tend to be faster than visual-only responses on the unimodal target trials.[8]

Finally, there are at least two further predictions that emerge from the explanation of the Colavita visual dominance effect outlined here that await future research: First, under conditions where the time pressure on a participant to respond rapidly is increased, perhaps using some form of adaptive responding procedure (cf. Terbeck et al., 2008), the incidence of vision dominating over audition and/or touch should increase (since this would likely increase the likelihood of participants making very fast responses, and those seem to be precisely the trials in which the Colavita effect is observed). Furthermore, if Sinnett et al.'s (2008, Experiment 2) study was to be repeated using auditory and tactile stimuli rather than auditory and visual stimuli, then they might well be found to have relatively symmetrical accessory stimulus effects. Should such a result be obtained, it would then help to explain why no Colavita effect has, as yet, been found when participants respond to the combination of auditory and tactile targets when presented together.

[8]Of course, it is in some sense unsurprising to find that participants are more likely to make an error when they respond rapidly. However, what is unique about the Colavita effect is that these errors on the rapid response trials are significantly more likely to be visual-only responses than auditory-only responses.

Abbreviations

RTs	reaction times
SOA	stimulus onset asynchrony

Acknowledgments

This manuscript is based on an invited presentation that I gave at the ASSC13 meeting held in Taiwan in June 2008. I would like to thank Dr. Tim Bayne, for a very stimulating and helpful discussion of these findings on that long flight back from Taiwan. Yi-Chuan Chen and Cesare Parise also provided useful comments on an earlier version of this manuscript.

References

Allport, A. (1988). What concept of consciousness? In A. J. Marcel & E. Bisiach (Eds.), *Consciousness in contemporary science* (pp. 174–175). Oxford: Clarendon.

Banks, W. P., & Isham, E. A. (2009). We infer rather than perceive the moment we decided to act. *Psychological Science*, *20*, 17–21.

Bridgeman, B. (1990). The physiological basis of the act of perceiving. In O. Neumann & W. Prinz (Eds.), *Relationships between perception and action: Current approaches* (pp. 21–42). Berlin: Springer.

Calvert, G. A.,, Spence, C., & Stein, B. E. (Eds.). (2004). *The handbook of multisensory processes*. Cambridge, MA: MIT Press.

Cardoso-Leite, P., Gorea, A., & Mamassian, P. (2007). Temporal order judgment and simple reaction times:

Evidence for a common processing system. *Journal of Vision*, *7*(6), 11. 1–14.
Cardoso-Leite, P., Mamassian, P., & Gorea, A. (2009). Comparison of perceptual and motor latencies via anticipatory and reactive response times. *Attention, Perception & Psychophysics*, *71*, 82–94.
Colavita, F. B. (1974). Human sensory dominance. *Perception & Psychophysics*, *16*, 409–412.
Colavita, F. B. (1982). Visual dominance and attention in space. *Bulletin of the Psychonomic Society*, *19*, 261–262.
Colavita, F. B., Tomko, R., & Weisberg, D. (1976). Visual prepotency and eye orientation. *Bulletin of the Psychonomic Society*, *8*, 25–26.
Colavita, F. B., & Weisberg, D. (1979). A further investigation of visual dominance. *Perception & Psychophysics*, *25*, 345–347.
Cooper, R. (1998). Visual dominance and the control of action. In: M. A. Gernsbacher & S. J. Derry (Eds.), *Proceedings of the 20th Annual Conference of the Cognitive Science Society* (pp. 250–255).
Egeth, H. E., & Sager, L. C. (1977). On the locus of visual dominance. *Perception & Psychophysics*, *22*, 77–86.
Fagot, C., & Pashler, H. (1992). Making two responses to a single object: Exploring the central bottleneck. *Journal of Experimental Psychology: Human Perception and Performance*, *18*, 1058–1079 (see errata, same journal, 19, p. 443).
Falkenstein, M., Hohnsbein, J., Hoormann, J., & Blanke, L. (1991). Effects of crossmodal divided attention on late ERP components. II. Error processing in choice reaction tasks. *Electroencephalogy and Clinical Neurophysiology*, *78*, 447–455.
Fehrer, E., & Biederman, I. (1962). A comparison of reaction time and verbal report in the detection of masked stimuli. *Journal of Experimental Psychology*, *64*, 126–130.
Hartcher-O'Brien, J., Gallace, A., Krings, B., Koppen, C., & Spence, C. (2008). When vision 'extinguishes' touch in neurologically-normal people: Extending the Colavita visual dominance effect. *Experimental Brain Research*, *186*, 643–658.
Hecht, D., & Reiner, M. (2009). Sensory dominance in combinations of audio, visual and haptic stimuli. *Experimental Brain Research*, *193*, 307–314.
Jaśkowski, P. (1996). Simple reaction time and perception of temporal order: Dissociations and hypotheses. *Perceptual and Motor Skills*, *82*, 707–730.
Johnson, H., & Haggard, P. (2003). The effect of attentional cuing on conscious awareness of stimulus and response. *Experimental Brain Research*, *150*, 490–496.
Johnson, T. L., & Shapiro, K. L. (1989). Attention to auditory and peripheral visual stimuli: Effects of arousal and predictability. *Acta Psychologica*, *72*, 233–245.
Koppen, C., Alsius, A., & Spence, C. (2008). Semantic congruency and the Colavita visual dominance effect. *Experimental Brain Research*, *184*, 533–546.
Koppen, C., Levitan, C., & Spence, C. (2009). A signal detection study of the Colavita effect. *Experimental Brain Research*, *196*, 353–360.
Koppen, C., & Spence, C. (2007a). Seeing the light: Exploring the Colavita visual dominance effect. *Experimental Brain Research*, *180*, 737–754.
Koppen, C., & Spence, C. (2007b). Assessing the role of stimulus probability on the Colavita visual dominance effect. *Neuroscience Letters*, *418*, 266–271.
Koppen, C., & Spence, C. (2007c). Spatial coincidence modulates the Colavita visual dominance effect. *Neuroscience Letters*, *417*, 107–111.
Koppen, C., & Spence, C. (2007d). Audiovisual asynchrony modulates the Colavita visual dominance effect. *Brain Research*, *1186*, 224–232.
Libet, B., Gleason, C. A., Wright, E. W., & Pearl, D. K. (1983). Time of conscious intention to act in relation to onset of cerebral activities (readiness-potential): The unconscious initiation of a freely voluntary act. *Brain*, *106*, 623–642.
Miller, J. O. (1982). Divided attention: Evidence for coactivation with redundant signals. *Cognitive Psychology*, *14*, 247–279.
Miller, J. O. (1991). Channel interaction and the redundant targets effect in bimodal divided attention. *Journal of Experimental Psychology: Human Perception and Performance*, *17*, 160–169.
Müssler, J., & Hommel, B. (1997). Blindness to response-compatible stimuli. *Journal of Experimental Psychology: Human Perception and Performance*, *23*, 861–872.
Neumann, O. (1990). Direct parameter specification and the concept of perception. *Psychiatric Research*, *52*, 207–215.
Occelli, V., Hartcher O'Brien, J., Spence, C., & Zampini, M. (submitted). Visual dominance: Is the Colavita effect an exclusively visual phenomenon? *Experimental Brain Research*.
O'Connor, N., & Hermelin, B. (1963). Sensory dominance in autistic children and subnormal controls. *Perceptual and Motor Skills*, *16*, 920.
Osborn, W. C., Sheldon, R. W., & Baker, R. A. (1963). Vigilance performance under conditions of redundant and nonredundant signal presentation. *Journal of Applied Psychology*, *47*, 130–134.
Pashler, H. (1994). Dual-task interference in simple tasks: Data and theory. *Psychological Bulletin*, *116*, 220–244.
Posner, M. I., Nissen, M. J., & Klein, R. M. (1976). Visual dominance: An information-processing account of its origins and significance. *Psychological Review*, *83*, 157–171.
Quinlan, P. (2000). The "late" locus of visual dominance. *Abstracts of the Psychonomic Society*, *5*, 64.
Randich, A., Klein, R. M., & LoLordo, V. M. (1978). Visual dominance in the pigeon. *Journal of the Experimental Analysis of Behavior*, *30*, 129–137.
Rapp, B., & Hendel, S. K. (2003). Principles of cross-modal competition: Evidence from deficits of attention. *Psychonomic Bulletin and Review*, *10*, 210–219.
Rodway, P. (2005). The modality shift effect and the effectiveness of warning signals in different modalities. *Acta Psychologica*, *120*, 199–226.
Rutschmann, J., & Link, R. (1964). Perception of temporal order of stimuli differing in sense mode and simple reaction time. *Perceptual and Motor Skills*, *18*, 345–352.

Sarri, M., Blankenburg, F., & Driver, J. (2006). Neural correlates of crossmodal visual-tactile extinction and of tactile awareness revealed by fMRI in a right-hemisphere stroke patient. *Neuropsychologia*, *44*, 2398–2410.

Schumacher, E. H., Seymour, T. L., Glass, J. M., Fencsik, D. E., Lauber, E. J., Kieras, D. E., et al. (2001). Virtually perfect time sharing in dual-task performance: Uncorking the central cognitive bottleneck. *Psychological Science*, *12*, 101–108.

Shapiro, K. L., Egerman, B., & Klein, R. M. (1984). Effects of arousal on human visual dominance. *Perception & Psychophysics*, *35*, 547–552.

Shapiro, K. L., & Johnson, T. L. (1987). Effects of arousal on attention to central and peripheral visual stimuli. *Acta Psychologica*, *66*, 157–172.

Sinnett, S., Spence, C., & Soto-Faraco, S. (2007). Visual dominance and attention: The Colavita effect revisited. *Perception & Psychophysics*, *69*, 673–686.

Sinnett, S., Soto-Faraco, S., & Spence, C. (2008). The co-occurrence of multisensory competition and facilitation. *Acta Psychologica*, *128*, 153–161.

Snodgrass, J. G., & Vanderwart, M. (1980). A standardized set of 260 pictures: Norms for name agreement, image agreement, familiarity, and visual complexity. *Journal of Experimental Psychology: Human Learning and Memory*, *6*, 174–215.

Spence, C. (2007). Audiovisual multisensory integration. *Acoustical Science and Technology*, *28*, 61–70.

Spence, C. (2008). Cognitive neuroscience: Searching for the bottleneck in the brain. *Current Biology*, *18*, R965–R968.

Spence, C. (in press). Prior entry: Attention and temporal perception. In: A. C. Nobre & J. T. Coull (Eds.), *Attention and time*. Oxford: Oxford University Press.

Spence, C., Nicholls, M. E. R., & Driver, J. (2001). The cost of expecting events in the wrong sensory modality. *Perception & Psychophysics*, *63*, 330–336.

Taylor, J. L., & McCloskey, D. I. (1990). Triggering of preprogrammed movements as reactions to masked stimuli. *Journal of Neurophysiology*, *63*, 439–446.

Taylor, J. L., & McCloskey, D. I. (1996). Selection of motor responses on the basis of unperceived stimuli. *Experimental Brain Research*, *110*, 62–66.

Terbeck, S., Chesterman, P., Fischmeister, F. Ph.S., Leodolter, U., & Bauer, H. (2008). Attribution and social cognitive neuroscience: A new approach for the "online-assessment" of causality ascriptions and their emotional consequences. *Journal of Neuroscience Methods*, *173*, 13–19.

Thompson, R. F., Voss, J. F., & Brogden, W. J. (1958). Effect of brightness of simultaneous visual stimulation on absolute auditory sensitivity. *Journal of Experimental Psychology*, *55*, 45–50.

Titchener, E. B. (1908). *Lectures on the elementary psychology of feeling and attention*. New York: Macmillan.

Turatto, M., Benso, F., Galfano, G., Gamberini, L., & Umilta, C. (2002). Non-spatial attentional shifts between audition and vision. *Journal of Experimental Psychology: Human Perception and Performance*, *28*, 628–639.

Van Damme, S., Crombez, G., & Spence, C. (2009). Is the visual dominance effect modulated by the threat value of visual and auditory stimuli? *Experimental Brain Research*, *193*, 197–204.

Zahn, T. P., Pickar, D., & Haier, R. J. (1994). Effects of clozapine, fluphenazine, and placebo on reaction time measures of attention and sensory dominance in schizophrenia. *Schizophrenia Research*, *13*, 133–144.

N. Srinivasan (Ed.)
Progress in Brain Research, Vol. 176
ISSN 0079-6123

CHAPTER 17

Development of attentional processes in ADHD and normal children

Rashmi Gupta* and Bhoomika R. Kar

Centre of Behavioural and Cognitive Sciences, University of Allahabad, Allahabad, India

Abstract: Attention Deficit Hyperactivity Disorder (ADHD) is a developmental disorder. Typical development of attentional processes is rapid during early childhood. ADHD results in impairment in response inhibition, error monitoring, attentional disengagement, executive attention, and delay aversion and may effect the ongoing development of these processes during childhood. We examined the development of attentional processes in children with ADHD and normal children. Two hundred forty children (120 in each group) in the age range of 6–9 years participated in the study. Four tasks: Stop-Signal, attentional disengagement, attention network, and choice delay task were administered. Stop signal reaction time, switch costs, conflict effect, and percentage choice of short delay reward was higher in ADHD group compared to normal group. Post error of slowing was less in ADHD children. Endogenous orienting effect was more in normal children compared to ADHD children. Different developmental trajectories were observed for control functions in normal children. Major development in response inhibition occurred in 7–8 years, error monitoring in 6–9 years, and attentional disengagement in 7–9 years. Late development in alerting network was observed in normal children at age 9 years. No developmental changes occurred on these control functions in ADHD children aged 6–9 years. Age related changes were observed on delay aversion between 6 and 9 years in normal children, while it changed between 6 and 7 years in ADHD children. Performance was not changed on orienting and conflict attentional networks in both the children except conflict effect reduced between 7 and 9 years in ADHD children under double cue condition. Conflict network was interacted with the alerting and orienting network in normal children; specifically conflict network interacted with the orienting network in younger children (age 6 years) and with alerting network in older children (age 9 years). In ADHD group interaction between alerting and conflict network was observed only in the double cue condition. Together these results indicated that the deficits in control processes accumulate with age in ADHD children Present study favors the conceptual view of ADHD as a stable deficit in cognitive control functions, which are implicated in the pathology of ADHD. These results have theoretical implication for the theories of executive control and ADHD.

Keywords: ADHD; development; response inhibition; error monitoring; attentional disengagement; attentional networks; delay aversion; stable deficit

Introduction

Age-related changes have been identified with control processes such as response inhibition,

*Corresponding author.
Tel./Fax: +91 5322460738;
E-mail: rash_cogsci@yahoo.com

DOI: 10.1016/S0079-6123(09)17614-8

working memory, task switching, and error monitoring that are critical for perception and action. These processes have also been found to be deficient in certain developmental disorders such as Attention Deficit Hyperactivity Disorder (ADHD).

Development of attention-executive processes in normal children

A number of studies have examined the developmental trajectories of attention-executive processes in normal children (Bunge et al., 2002; Gupta et al., submitted a). For example, it has been reported that inhibitory control develops over childhood and does not reach full maturity until 12 years of age or later (Bunge et al., 2002). Substantial improvement in inhibitory control occurs during childhood and it declines during late adulthood (Williams et al., 1999), suggesting an inverted U-shaped relationship between inhibitory control and age. Paradigms such as go–no-go task, stop-signal task, and Stroop task have been used to study inhibitory control. Performance on Stroop-like tasks improves through 3–7 years of age (Gerstadt et al., 1994) and declines during late adulthood (Spieler et al., 1996). Christ et al. (2001) investigated the ability to inhibit a prepotent response and generate an incompatible response in individuals ranging from 6 to 82 years of age. They found that inhibitory control effect was larger in children and older adults than in young adults, and larger in older adults than children. They further argue that childhood is a critical period in terms of frontal lobe and cognitive development. Thus, changes in inhibitory control could have occurred within 6–15 years old age group. They divided the children group into two age groups: 6–9 and 10–15 years. Raw reaction time data suggested that younger children respond more slowly than older children. To determine if the discrepancy in the magnitude of the effect was attributable to differences in processing speed rather than inhibitory control, the data were reanalyzed following proportional and z-score transformation. Results of the proportional score supported the raw reaction time data. However, using the more rigorous z-score procedure, group effect as well as interaction between group and condition (congruent vs. incongruent) was not found to be significant. These findings indicated that early age-related differences in inhibitory control were due to the difference in processing speed rather than true differences in inhibitory control.

It has been argued that inhibition of task set is one of the contributors to attentional disengagement measured by switch costs (SCs) (Monsell, 2003). Cepeda et al. (2001) examined age-related differences in task switching with respect to the processes responsible for preparation and interference control that underlie the ability to flexibly alternate between two different tasks. They observed larger SCs among young children (7–9 years) and it decreased with age. Crone et al. (2006) found greater SCs with young children (7–8 years) compared to adults for task switching with repeating responses. This age-related difference decreased with an increase in the interval between the previous response and the upcoming stimulus. Young children experienced more interference from the previous stimulus–response (S–R) association, suggesting larger carryover effects from the previous trial.

Errors may occur while switching between actions. Monitoring for such errors online and making subsequent adjustments in processing speed is important for cognitive control. Error monitoring is evident in the slowing of responses following errors (post-error slowing, PES) (Rabbit, 1966), which varies with age. PES varied with age within the range of 7–16 years, older children slowed less than younger ones (Schachar et al., 2004). Kramer et al. (1994) found that elderly participants showed larger PES than younger adults following trials where an error had been made (50 ms vs. 21 ms). Together, these studies indicate that a curvilinear pattern of development in PES is indicated across the life span, showing a decrease as children get older followed by an increase among older adults. Posner and Rothbart (1998) found that children were able to detect an error as early as 48 months of age. Error monitoring was studied with event-related potentials between 7 and 25 years of age and it has been found that error-related negativity

(ERN) (reflects unconscious detection of an error) amplitude increased with age. However, the error-positivity (*Pe*) amplitude (reflects conscious error recognition and performance adjustment after an error) did not change with age (Davies et al., 2004).

In addition to the specific higher order executive processes, development of attentional networks (alerting, orienting, and executive) from 4 years of age to adulthood has been reported (Rueda et al., 2004). These studies indicate that there was a steady decline in overall reaction time from 4 years of age to adulthood. Improvement in conflict resolution was found until age 7 years. Alerting scores showed some improvement in late childhood and continued development between 10-year olds and adults. The orienting score was similar to adult levels at the youngest ages.

Attention-executive deficits in ADHD children

Attention-executive dysfunction is characterized by deficits in response inhibition, error monitoring, attentional disengagement, attentional networks, and motivational style (delay aversion) in children with ADHD (see Pennington and Ozonoff, 1996; Gupta et al., 2006, for reviews). Barkley (1997) suggested that response inhibition is the primary deficit in ADHD, which in turn affects the other executive functions. The evidence supporting a deficiency in behavioral inhibition in ADHD comes from studies that used motor inhibition tasks, such as go–no-go task (Iaboni et al., 1995), the stop-signal task (Oosterlaan and Sergeant, 1998), and delayed response tasks (Sonuga-Barke et al., 1992). Further support for deficient inhibitory control in ADHD is suggested by neuroimaging research indicating both structural and functional deficits in the right inferior frontal cortex (Aron and Poldrack, 2005). Schachar et al. (2004) studied error monitoring by looking at the slowing of responses after inhibition error (PES) in a stop-signal task. ADHD children slowed to a lesser extent after fewer inhibition failures suggesting deficits in error detection as well as in behavioral adjustment to errors. Cepeda et al. (2000) suggested that ADHD children show deficient control processes necessary for disengagement from one task and preparation for a subsequent task.

ADHD children also show impairments in an executive and alerting network due to the inability to maintain the alert state when no warning signal was used (Blane and Marrocco, 2004). In addition to executive function deficits children with ADHD are also characterized by a specific motivational style called delay aversion, which is the motivation to avoid delay and results in preference for small, immediate over large, delayed rewards (Sonuga-Barke, 2002).

Most of the studies have examined the development of attentional processes in normal children. Very few studies have investigated the development of attentional processes in ADHD children. To our knowledge, only one study has examined the development of attentional processes particularly selective attention in ADHD children compared to normal children aged between 6 and 11 years (Brodeur and Pond, 2001). In this study, two age groups of children (6–8 and 9–11 years) with ADHD and normal children were tested using a timed computer task. The task consisted of identifying visual target stimuli under various distracter conditions. Children with ADHD were less efficient on the selective attention task than were children without ADHD, and older children were more efficient than younger children in both groups. Children without ADHD were influenced more by the nature of distracters than were children with ADHD. This study talked about only one component of attention such as selective attention. Yet another study longitudinally examined the brain development especially the cortical maturation of ADHD children from 10 to 17 years of age. They estimated the cortical thickness at various cerebral points. They found that sequence of brain maturation in children with ADHD follows the normal pattern but is delayed by 2–3 years in prefrontal regions that are important for control of cognitive processes including attention and motor planning (Shaw et al., 2007). This study reported delay in the structural development of the brain in ADHD. There is no evidence on delay in functional maturation of the brain and cognitive processes in ADHD. In other words, it is still an open question whether ADHD is characterized by a delay in normal ongoing development of, or a stable deficit in, control processes (Brocki and Bohlin, 2006). Brocki and Bohlin (2006) support the conceptual view of ADHD as a developmental delay, as ADHD symptoms changed with maturation. However, developmental delay in control processes in ADHD was not studied. In addition, they only included the nonclinical sample in their study. The key to understanding ADHD as either a developmental or a categorical disorder lies in comparing developmental trends in a normal group with a clinical ADHD group. Hence, a study on development of attentional processes in children with a clinical diagnosis of ADHD was needed to understand if attentional processes in ADHD represent a delay or a complete deviation from typical development. In the present study we focused on the development of various attentional control functions such as response inhibition, attentional disengagement, error monitoring, attentional networks, and motivational style in ADHD compared to normal children aged 6–9 years. These control functions are important in developing a theory about ADHD (see Gupta et al., 2006, for review) and were found to be sensitive in the diagnosis of ADHD from normals and also helpful in differential diagnosis of ADHD with other developmental disorders such as Oppositional Defiant Disorder (ODD) (Gupta et al., submitted b).

Previous studies have examined the development of control processes by employing a relatively small number of different of age groups among normal children (Johnstone et al., 2005; Cepeda et al., 2001). For example, Johnstone et al. (2005) examined 7–47 years old participants to study the development of response inhibition. Studies on task switching and error monitoring have examined fewer cases or coarse groupings of ages. For example, Cepeda et al. (2001) studied task switching with two groups: 7–9 and 10–12 years old children. Combining children with different ages makes it difficult to track developmental changes in these executive processes. We have examined four age levels (6–9 years) to closely evaluate developmental patterns of control processes among children with and without ADHD. Six to nine years of age is the period of school age, hence, most of the ADHD cases are reported in this age range. DSM-IV also suggests that ADHD should be diagnosed before the age of 7 years. Examination of different control functions in the same group of children enabled us to control for difference in experimental procedures used in different studies and participant demographics. It is also important to determine whether development of control functions occurs in parallel in a particular age group. If so, then a possible interaction among these control functions may be implicated in the pathology underlying ADHD.

Method

Participants

One hundred and twenty children with ADHD ($N = 120$; 117 boys) and 120 normal children (114 boys) in the age range of 6–9 years (30 children in each age level) participated in the study. Participants were recruited from two districts of the state of Uttar Pradesh like Allahabad and Lucknow. The participants with ADHD were referred by consultant psychiatrists. Both the groups were matched on socioeconomic status. Practicing psychiatrists referred children with ADHD. All the participants with ADHD fulfilled the DSM-IV (APA, 1994) criteria for combined type ADHD. They scored the clinical cutoff ($T > 60$) on Conners Parents Rating Scale – Revised Long form (CPRS-R:L; Conners, 2002). CPRS-R:L was also used to rule out any behavioral and attentional problems in normal children. Twenty percent participants with ADHD had a comorbid diagnosis of ODD. Numbers of children with comorbid ODD were equally distributed across the different age groups and this subgroup did not differ in performance profile from the pure ADHD group. Therefore, data from children with pure ADHD as well as those with ADHD with comorbid ODD were analyzed together as one group. All the children were average or above average in intellectual functions with scores in the range of 50–95 percentile on Colored Progressive

Table 1. Demographic characteristics of the children with ADHD and normal controls

Characteristics	ADHD ($N = 120$), 6–9 years		Normal controls ($N = 120$), 6–9 years		F
	M	SD	M	SD	
Age (years)	7.78	1.13	7.80	1.11	0.62
Intellectual Ability Raw Score (percentile)	27.2 (85.2)	3.5 (18.6)	28.2 (88.6)	3.88 (15.7)	2.24
CPRS-R:L DSM-IV total score for ADHD (T score)	43.4 (80.3)	3.7 (3.7)	3.8 (43.2)	1.0 (1.5)	11191.75*

Note: $^{*}p < 0.001$.

Matrices (CPM) (Raven et al., 1998). Table 1 shows the demographic details of two groups of participants.

Procedure

Participants were tested individually on four cognitive tests. Order of the tests was randomized across participants. Each participant was seated in front of the computer at a distance of 60.0 cm and received instructions for each test. CPRS-R:L was administered to the parents. Whole assessment took approximately 1 h per child. All the participants had normal or corrected to normal vision. A written informed consent was taken from the parents. The ethics committee of the Centre of Behavioural and Cognitive Sciences, University of Allahabad, had approved the study.

Measures of cognitive functions

Attentional network test: attention networks

The child version of attentional network test (ANT) (Rueda et al., 2004) was employed to examine the attentional networks: alerting, orienting, and executive networks. The target array was a yellow-colored line drawing of either a single yellow fish or a horizontal row of five yellow fish, presented above or below fixation, over a blue–green background. The participant had to respond whether the central fish was pointing to the left or right by pressing the corresponding left or right key on the keyboard. The ANT consisted of a total of 24 practice trials and three experimental blocks of 48 trials each. Each trial represented 1 of 12 conditions in equal proportions: three target types (congruent, incongruent, and neutral) and four cues (no cue, central cue, double cue, and spatial cue). ANT test does not have invalid trials, hence all the cues are endogenous cues. However, unlike the spatial cue, the double cue did not predict the target exactly because double cues appeared both above and below the fixation and target appeared only on one location either above or below the fixation. Hence, we argue that double cue may function as an exogenous cue. We expected a difference in performance in normal and ADHD children with respect to cue conditions, specifically with double and spatial cues. Therefore, in addition to orienting effect score (difference between center and spatial cues) of ANT, we also computed another orienting score by subtracting the median RT for double cue from the RT for central cue. Former orienting score was called as endogenous orienting and later as exogenous orienting. Alerting, orienting, and conflict effects were calculated to measure alerting, orienting, and conflict networks, respectively. To find out the orienting and alerting scores per subject we computed the median RT per cue condition (across the flanker conditions). The alerting score was obtained by subtracting the median RT for the double cue from median RT for the no cue condition; the endogenous orienting score by subtracting the median RT for spatial cue from the RT for central cue; exogenous orienting score by subtracting the median RT for double cue from the RT for central cue. To obtain the conflict score, we computed the participant's median RT for each flanker condition (congruent vs. incongruent) (across cue conditions) and subtracted the congruent from the incongruent

RTs. The mean score, across subjects, was then computed for each network.

Stop-signal test: response inhibition

The stop-signal test (SST) involved two concurrent tasks. The primary or go task involved discrimination between an X or an O presented in the center of a computer screen for 1000 ms following a 500-ms fixation point. The go stimulus was followed by a blank screen for 2000 ms allowing 3000 ms for key press and total trial duration of 3500 ms. The secondary or stop task involved the presentation of a green visual circle indicating that participants should not respond to the primary task. The green circle occurred randomly and with equal frequency across blocks on 25% of trials. The session consisted of four blocks of 40 trials. We used a dynamic tracking procedure to set the timing of the circle (stop-signal delay) (Logan, 1994). At the beginning of the task, stop delay was set at 250 ms. If a participant was able to stop successfully, the delay was lengthened (by 50 ms) on the succeeding trial (see Schachar et al., 2004, for more details). Stop-signal reaction time (SSRT) and PES were calculated to measure response inhibition and error monitoring, respectively.

Attentional disengagement test: attentional disengagement

Stimuli were presented at fixation. The four possible stimuli were either a single digit (1 or 3) or three digits (1 1 1 or 3 3 3). In other words, either one or three numeric 1s or 3s were presented. On each trial, either the cue "what number?" or the cue "how many?" appeared above the target stimulus. Participants were required to switch their attention between two different tasks with respect to the cue that appeared above the target stimulus: discriminating the value of a number presented on a computer screen or deciding how many numbers were present on the screen. Stimuli stayed on the screen until response was made. Feedback (100 Hz tone) was given whenever participants made an error. A practice session with 75 trials preceded the experimental session consisting of 200 trials. SC was calculated to measure attentional disengagement.

Choice delay test: delay aversion

In the choice delay test (CDT) participants were presented with a series of trials and asked to choose between a small reward (1 point) to be delivered after a short delay (1 s) or a large reward (2 points) to be delivered after a long delay (20 s). The child could choose between a small, immediate reward and a large, delayed reward, and the total length of the trial depended on the percentage of choices for the large, delayed reward. A practice session with 5 trials proceeded the experimental session consisting of 30 trials. Percentage choice of long delay reward (%LDR) and short delay reward (%SDR) was calculated to measure delay aversion.

Data analysis

For each score, data were submitted to 2 (Group: Normal and ADHD) $\times$ 4 (Age: 6, 7, 8, and 9) between factor design.

Results

Age effects

Response inhibition: stop-signal reaction time

The analysis yielded a significant effect of group, $F(1, 232) = 307.1$, $p < 0.001$, with higher SSRT for ADHD ($M = 646.0$ ms) as compared to normals ($M = 310.8$ ms). The effect of age was also significant, $F(3, 232) = 4.64$, $p < 0.01$. SSRT was more for 6-year-old children and it decreased as age increased. To ensure whether the age effect was significant in both the groups, we performed one-way ANOVA with age as a between factor for both the groups separately. Age effect was found only in the normal group, $F(3, 116) = 8.05$, $p < 0.0001$. Tukey's HSD for post-hoc comparisons was performed. Performance improved in children between 7 and 8 years old, $F(1, 116) = 4.50$, $p < 0.01$ (Fig. 1).

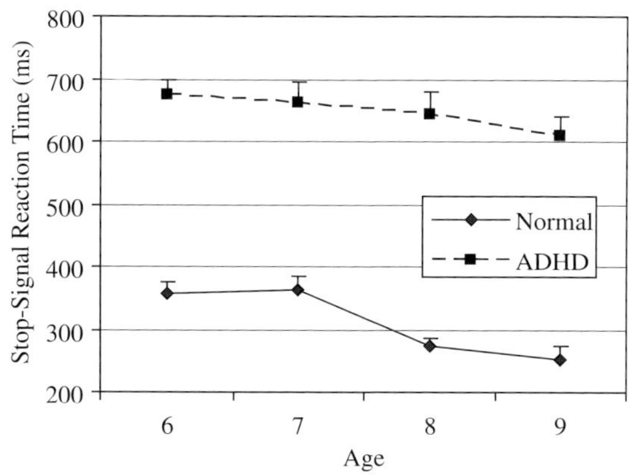

Fig. 1. Mean reaction time for stop-signal reaction time for all the four age groups of normal and ADHD children.

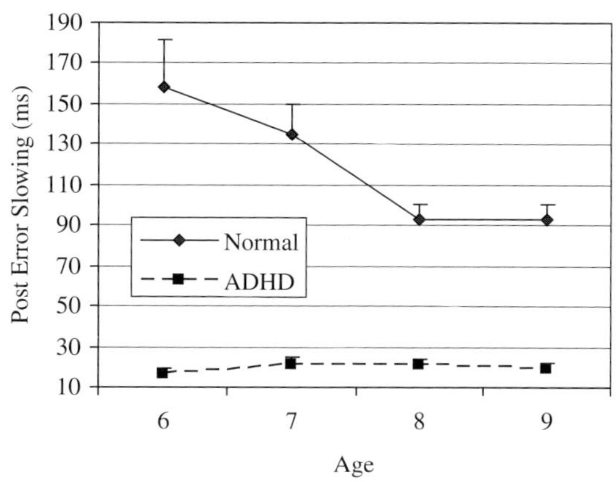

Fig. 2. Post-error slowing of all the four age groups of normal and ADHD children.

Error monitoring: post-error slowing

The analysis yielded a significant effect of group, $F(1, 232) = 174.7$, $p<0.001$, with lesser extent of slowing for ADHD children ($M = 19.7$ ms) compared to normals ($M = 119.7$ ms). The effect of age was also significant, $F(3, 232) = 4.18$, $p<0.01$, with children of 6 years being significantly slower. Performance speeded with age. Performance of 6-year-old children was significantly different from the performance of 8-year-old children, $F(1, 232) = 3.97$, $p<0.05$. Group $\times$ age interaction effect was found significant, $F(3, 232) = 4.99$, $p<0.01$. Post-hoc comparisons indicated that performance of normal children changed from 6 to 8 years of age, $F(1, 232) = 6.07$, $p<0.001$. There was no change in performance of ADHD children between 6 and 9 years of age (Fig. 2).

Attentional disengagement: switch costs

Group effect was significant, $F(1, 232) = 391$, $p<0.001$. SC was higher for ADHD ($M = 633.4$ ms) as compared to normals ($M = 224.0$ ms). Age effect was also significant, $F(3, 232) = 11.8$, $p<0.01$. SC was more for 6-year-old children and it decreased with increase in age. To ensure the age effect in both the groups, we performed one-way ANOVA with age as a between factor for both the groups separately. Age effect was found only in the normal group, $F(3, 116) = 21.0$,

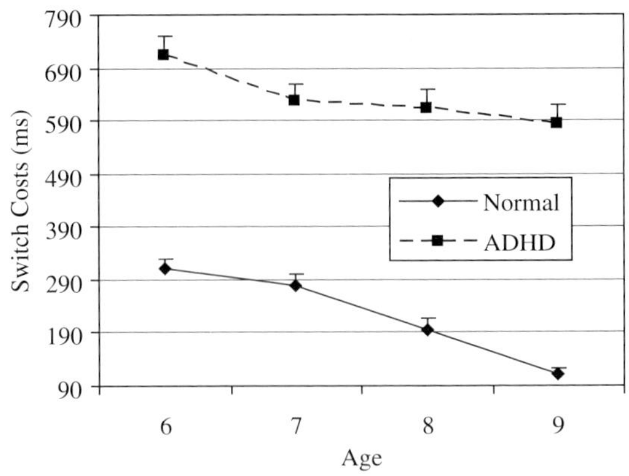

Fig. 3. Switch costs of all the four age groups of normal and ADHD children.

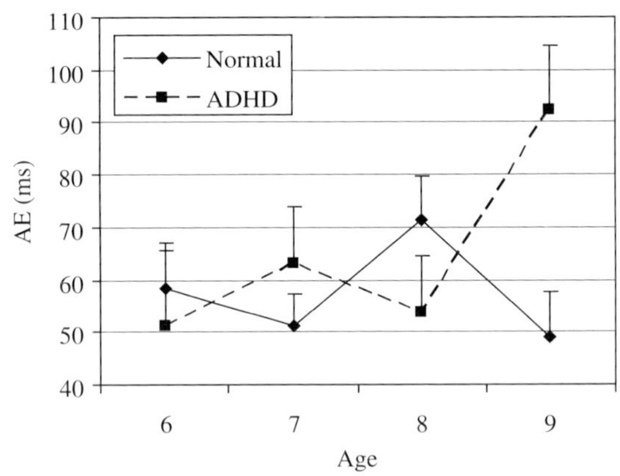

Fig. 4. Alerting effect of all the four age groups of normal and ADHD children.

$p<0.001$. Task switching improved from 7 to 8 years, $F(1, 116) = 4.15$, $p<0.05$, and 8 to 9 years of age, $F(1, 116) = 4.37$, $p<0.01$, in normal children (Fig. 3).

Attentional networks: alerting effect, endogenous and exogenous orienting effects, and conflict effect

A two-way ANOVA was computed with group and age as between factors in order to assay the developmental trend of each attentional network. Group effect was significant for endogenous orienting, $F(1, 232) = 9.44$, $p<0.01$, and conflict score, $F(1, 232) = 8.36$, $p<0.01$. Orienting effect was more for normal ($M = 43.15$ ms) compared to ADHD children ($M = 19.91$ ms). In addition, conflict effect was more for ADHD ($M = 100.8$ ms) compared to normal children ($M = 73.2$ ms). Age effect was not significant for any of the networks. Group × age interaction was significant only for alerting network, $F(3, 232) = 3.27$, $p<0.05$. Post-hoc comparison indicated that performance of 9-year-old normal children was marginally different from the performance of 9-year-old ADHD children, $F(1, 232) = 4.13$, $p = 0.07$ (Fig. 4).

An ANOVA was also performed to compare the two groups across age, flanker type, and cue conditions in a 2 (Group: Normal vs. ADHD) × 4 (Age: 6, 7, 8, and 9) × 4 (Cue type: no cue, center cue, spatial cue, and double cue) × 3 (Flanker type: congruent, incongruent, and neutral) design. This analysis showed significant main effects of group,

$F(1, 2332) = 404.5, p < 0.001$, age, $F(3, 232) = 8.22$, $p < 0.001$, cue type, $F(3, 696) = 143.9$, $p < 0.001$, and flanker type, $F(2, 464) = 404.6$, $p < 0.001$. Children with ADHD were slower in responding to the target ($M = 1066.7$ ms) compared to normal children ($M = 730.1$ ms). The performance of 6-year-old children was significantly different from 8-year-old children, $F(1, 232) = 5.15$, $p < 0.001$, and 7-year-old children was different from 9-year-old children, $F(1, 232) = 3.59$, $p < 0.05$. Performance was significantly different among all the cue and flanker conditions. Reaction time was faster for the spatial cue ($M = 867.6$ ms), followed by the double cue ($M = 882.5$ ms), central cue ($M = 899.3$ ms), and no cue ($M = 944.1$ ms) conditions. Reaction time was faster for the neutral condition ($M = 846.4$ ms), followed by congruent ($M = 880.7$ ms) and incongruent ($M = 968.1$ ms) conditions.

Interaction between group and cue conditions was marginally significant, $F(3, 696) = 3.60$, $p = 0.093$. A difference in performance between ADHD and normal children was expected with respect to processing the cues, specifically spatial cue and double cue conditions. Therefore, a planned comparison between double and spatial cues was performed in both the groups. Significant difference in performance between double cue and spatial cue conditions, $F(1, 696) = 6.48$, $p < 0.00001$, in normal group were found. However, no difference was found between these cue conditions in ADHD group. To further validate these results, an ANOVA was performed for 2 (Group: Normal vs. ADHD) × 4 (Age: 6, 7, 8, and 9) × 2 (Orienting: exogenous orienting vs. endogenous orienting). Group effect, $F(1, 232) = 4.33$, $p < 0.05$, and orienting effect, $F(1, 232) = 14.12$, $p < 0.001$, were significant. Orienting effect was more for endogenous orienting ($M = 31.5$ ms) compared to exogenous orienting ($M = 16.7$ ms). There was significant difference in endogenous orienting between normal and ADHD children, $F(1, 232) = 5.89$, $p < 0.001$; normal children were oriented better to spatial cue ($M = 43.1$ ms) compared to ADHD children ($M = 19.9$ ms). There was no difference in exogenous orienting between the two groups (Fig. 5(a and b)).

Given the difference in performance between the cue conditions, we also analyzed conflict effect for each cue condition separately. Data were submitted to 2 (Group: Normal and ADHD) × 4 (Age: 6, 7, 8, and 9) between factor ANOVA. A group effect was found for all the cues except double cue; conflict effect was more for ADHD children compared to normal children for center cue, $F(1, 232) = 6.95$, $p < 0.01$, spatial cue,

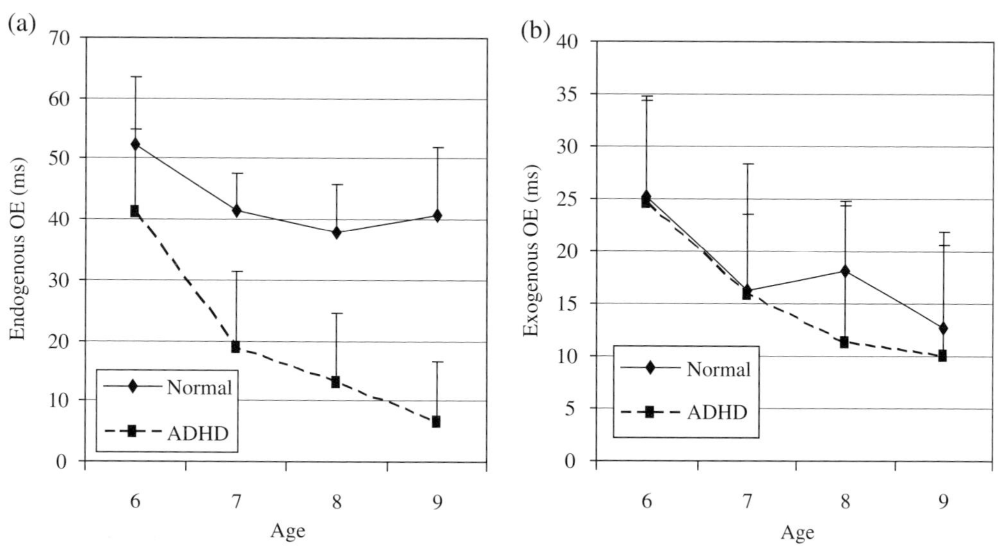

Fig. 5. Endogenous (a) and exogenous (b) orienting effects of all the four age groups of normal and ADHD children.

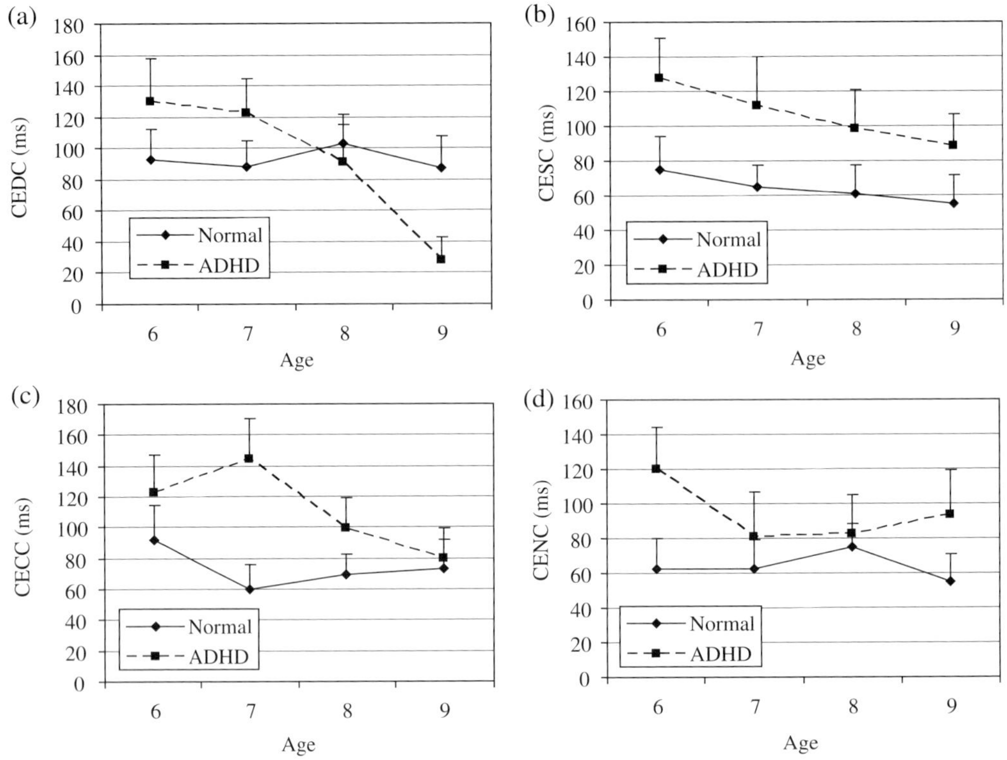

Fig. 6. Conflict effect for double cue (a), spatial cue (b), center cue (c), and no cue (d) conditions for all the four age groups of normal and ADHD children.

$F(1, 232) = 8.90$, $p<0.01$, and no cue, $F(1, 232) = 4.31$, $p<0.05$, conditions. An age effect was found only for the double cue, $F(3, 232) = 2.71$, $p<0.05$. The performance of 6-year-old children was significantly different from 9-year-old children, $F(1, 232) = 3.65$, $p<0.05$. An interaction between group and age was marginally significant, $F(3, 232) = 2.39$, $p = 0.06$. Planned comparisons indicated that 7-year-old ADHD children were significantly different from 9-year-old ADHD children, $F(1, 232) = 4.54$, $p<0.05$. There was no change in performance between 6- and 9-year-old normal children (Fig. 6(a–d)).

To examine whether the attention networks were independent or not, a correlation among the three network scores in both the groups was performed once the effect of age was adjusted. Correlation was also performed for each age of both the groups separately. In normal children, overall conflict score was significantly correlated with the alerting, $r = -0.223$, $p<0.01$, and endogenous orienting, $r = 0.218$, $p<0.01$, scores. There was a trend of an association between conflict and exogenous orienting scores, $r = 0.160$, $p = 0.08$. This effect was not found for all the age groups. Only at age 6 years, conflict score was significantly correlated with the endogenous orienting score, $r = 0.423$, $p<0.05$, and at age 9 years, conflict score was significantly correlated with the alerting score, $r = -0.371$, $p<0.05$. In children with ADHD, significant correlation was only observed in alerting and exogenous orienting scores, $r = 0.333$, $p<0.001$, which was found for ages 7 years, $r = 0.385$, $p<0.05$, 8 years, $r = 0.507$, $p<0.01$, and 9 years, $r = 0.532$, $p<0.01$.

Delay aversion: %SDR and %LDR

The analysis yielded a significant main effect of group, $F(1, 232) = 785.1$, $p<0.001$. ADHD children chose small, immediate rewards ($M = 76.2\%$)

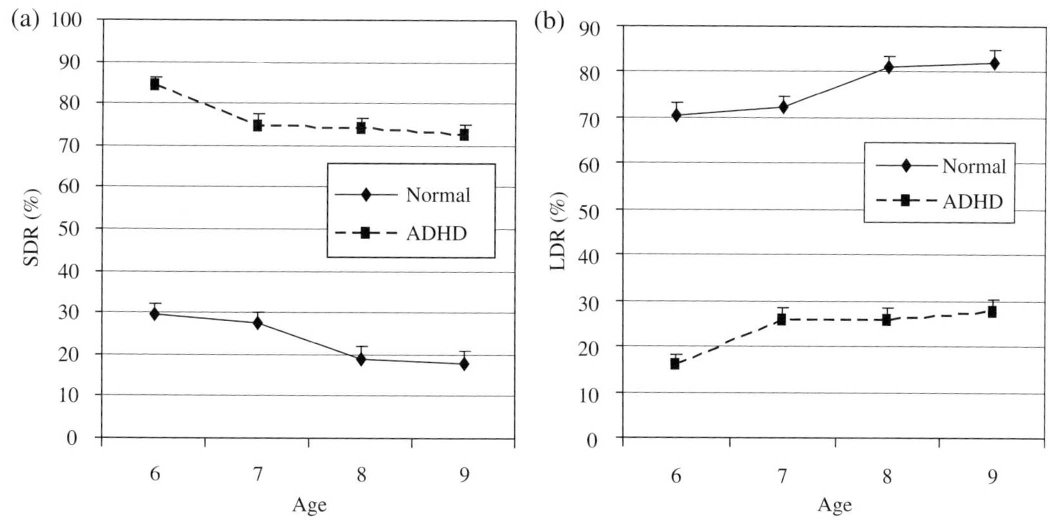

Fig. 7. %SDR (a) and %LDR (b) of all the four age groups of normal and ADHD children.

over large, delayed rewards more than controls ($M = 23.4\%$). The age effect was also found to be significant, $F(3, 232) = 7.79$, $p < 0.001$. The performance of 6-year-old children was significantly different from 8-year-old children, $F(1, 232) = 5.47$, $p < 0.001$. One-way ANOVA with age as a between factor was performed separately for both the groups. Age effect was found in both normal, $F(3, 116) = 4.87$, $p < 0.01$, and ADHD children, $F(3, 116) = 3.96$, $p < 0.01$. Performance was significantly different between 6- and 8-year-old normal children, $F(1, 116) = 3.97$, $p < 0.05$, and that of 7 years old was significantly different from 9 years old, $F(1, 116) = 3.60$, $p < 0.05$. In the ADHD group, performance of 6 years old was significantly different from 7 years old, $F(1, 116) = 3.61$, $p < 0.05$, and there was no difference in performance between 7 and 9 years of age. Similar results were observed with %LDR. ADHD children chose large, delayed rewards ($M = 23.7\%$) over small, short delay rewards much less than controls ($M = 76.4\%$), $F(1, 232) = 784.0$, $p < 0.001$ (Fig. 7(a and b)).

Discussion

We examined the developmental trajectories of control processes such as response inhibition, error monitoring, attentional disengagement, attentional networks, and motivational style in ADHD children as compared to normal children aged 6–9 years. We found that response inhibition, error monitoring, and attentional disengagement develops between 6 and 9 years in normal children. Age-related differences in performance with respect to the control functions were not observed in ADHD children aged 6–9 years. Age-related changes were also observed in the motivational style in normal children between 6 and 9 years of age, while it improved only between 6 and 7 years in ADHD children. Age-related changes were also observed on conflict score under double cue conditions in ADHD children aged 7–9 years.

Development of response inhibition in normal and ADHD children

Major developments in response inhibition seem to occur between 7 and 8 years of age in normal children. Other developmental studies on inhibitory control have also shown significant development between 7.5 and 9.5 years of age followed by 9.6–11.5 years (Brocki and Bohlin, 2004). Becker et al. (1987) reported a developmental transition in inhibitory control between 6 and 8 years of age. Active development of response inhibition between 7 and 8 years of age is consistent with the maturational patterns of the frontal cortex, which mediates inhibitory control (Hudspeth and Pribram, 1992).

Age-related improvement in inhibitory control was not observed in children with ADHD between 6 and 9 years of age. The immature inhibitory control in ADHD children could be related to the behavioral findings that SSRT is larger in patients with ADHD (Oosterlaan et al., 1998). It has been argued that inferior frontal cortex (IFC) is a critical brain region for the process of inhibiting an already initiated response (Aron and Poldrack, 2005), in which structural and functional deficits were observed in ADHD (Rubia et al., 2008). Rubia et al. (2008) found that during successful inhibition ADHD children between 9 and 16 years of age showed reduced activation in the left dorsolateral/inferior prefrontal cortex. Immature inhibitory control in young ADHD children is also supported by EEG studies. Spronk et al. (2008) investigated ERP measures of conflict monitoring and inhibition (Nogo-N2 and Nogo-P3), cue orientation and pre-stimulus target expectation (Cue-P2 and Cue-P3) for 5–7 years old children with and without ADHD. They found that ADHD children detected fewer targets and had higher inattention scores accompanied by reduced centro-parietal Cue- and Go-P3 activity and reduced Nogo-P3 at fronto-central leads, which indicates early signs of delayed attention development and immature inhibitory processing between 5 and 7 years of age in ADHD children.

Development of error monitoring in normal and ADHD children

Age-related improvements in error monitoring were observed between 6 and 8 years of age in normal children. This is consistent with another study in which we examined development of error monitoring in normal children with larger age group of 6–11 years (Gupta et al., submitted a). We found that a major development in error monitoring as measured by PES takes place between 6 and 10 years of age with an initial increase in PES followed by subsequent decrease indicating a curvilinear relationship between PES and age. PES was not found to uniformly decrease across the age range of 7–10 years as the decrease in PES was more substantial between 9 and 10 years as compared to 7–8 years of age. Together, these findings suggested that children are able to recognize the errors and are able to adjust their performance after an error. EEG/ERP studies have also reported that ERN amplitude increased with age, and children with 7–12 years of age are able to consciously recognize the errors and are able to adjust their performance after an error (Davies et al., 2004).

So far, no study has closely examined the development of error monitoring using PES in children between 6 and 9 years, which is a period of major development of executive functions in children. A few ERP studies have briefly discussed the behavioral component (PES) of error monitoring in their studies (Wiersema et al., 2007; Davies et al., 2004; Hogan et al., 2005). For example, Wiersema et al. (2007) examined the development trajectory of error monitoring in children (aged 7–8 years), young adolescent (13–14 years), and adults (23–24 years). They found no difference between age groups with respect to PES. Davies et al. (2004) also reported no difference between age groups for PES. In contrast, Hogan et al. (2005) observed an increase in the amount of PES from adolescence (aged 12–19 years) to adulthood (19–22 years). The diverging results could be due to combining wide age groups and difference in task requirements. The complexity of the task may also play a critical role in highlighting developmental effects in error monitoring.

Our results extend the findings of Schachar et al. (2004) showing that in addition to lesser extent of slowing, error monitoring also does not show age-related improvement in ADHD children between 6 and 9 years of age. An immature control of error monitoring in ADHD children aged 6–9 years is consistent with the ERP correlates of error monitoring. Reduced PES was accompanied by smaller ERN (Liotti et al., 2005) and *Pe* (Wiersema et al., 2005) in ADHD children between 9 and 11 years. Thus, ADHD children might show deficiencies in both unconscious and conscious detections of errors between 9 and 11 years of age. Rubia et al. (2008) reported that children with ADHD showed reduced activation in the posterior cingulate gyrus (error monitoring network), which is implicated in error detection and subsequent enhanced functions of arousal, attention allocation, and performance monitoring that are necessary to avoid future mistakes.

Development of attentional disengagement in normal and ADHD children

Attentional disengagement/task switching also follows different developmental patterns for normal and ADHD children. We found that the overall SC reduced from 7 to 9 years of age in normal children, indicating significant development in executive control processes during this period. This is consistent with another study in which we explored the possibility of shared mechanisms underlying task switching and error monitoring through a developmental framework in the age group of 6–11 years (Gupta et al., submitted a). We found that overall SCs reduced from 7 to 10 years of age. There was no difference in task switching between 10 and 11 years of age. Task switching and error monitoring may share a common mechanism such as response inhibition. This is further supported by neuropsychological and neuroimaging studies, which have reported the role of inferior frontal regions in switching between stimuli (Jemel et al., 2002) also found to be involved in inhibitory control (Rubia et al., 2008).

Age-related differences in attentional disengagement were not observed in ADHD children between 6 and 9 years of age. Lack of age-related improvement in attentional disengagement in ADHD children may be related to the finding of reduced dopamine activity during task switching in children with ADHD (Smith et al., 2006). Dopamine is known to be suboptimally active in ADHD patients (Oades, 2006).

Development of attentional networks in normal and ADHD children

Age-related changes were not observed in ADHD children between 6 and 9 years of age in an alerting network. Both the groups were different at 9 years of age relative to earlier ages, which suggests a late development of alerting network in normal children. An alerting network of attention is not yet mature in ADHD children aged 9 years. Previous studies have also suggested a slow development of alerting network in normal children (Ridderinkhof et al., 1997). Rueda et al. (2004) observed stability in middle childhood (6–9 years) and found some improvement in late childhood (10 years) in the alerting network. Delayed development of alerting network in ADHD may be a direct consequence of low levels of noradrenergic neurotransmitter (Beane and Marrocco, 2004), which was found to be critical for the alerting system (Marrocco and Davidson, 1998).

The overall conflict effect did not reduce between 6 and 9 years of age in both the groups. However, in the double cue condition the conflict effect was reduced between 7 and 9 years of age in ADHD children. It supports the previous findings that double cue can rescue the attention deficit of the ADHD children, presumably via a phasic increase in alertness. Children with ADHD made less omission errors in the double cue condition compared to no cue and center cue conditions (Johnson et al., 2008). Posner and Petersen (1990) reported that patients with frontal lesions were slow to initiate responses when a target stimulus was not preceded by a warning cue, relative to when this cue was present. These findings indicate a problem with the "tonic" or internal aspects of alertness but an intact ability to use cues to improve performance. Tonic levels of alertness are thought to be modulated by noradrenaline and difficulties with alertness might arise due to deficient fronto-parietal control over the locus coeruleus (Halperin and Schulz, 2006). These results are consistent with the current theories of ADHD, which emphasize a problem of regulation of arousal in ADHD (Johnson et al., 2007).

Children with ADHD may find it more difficult to regulate their arousal in the absence of an alerting cue than when it is present.

We did not find any age-related improvement in exogenous and endogenous orienting between 6 and 9 years of age in both the groups, which is consistent with previous literature (Rueda et al., 2004). Early developmental changes occurred in the orienting network. ADHD children were impaired in endogenous orienting and had intact exogenous orienting (Jonkman, 2005). Carter et al. (1995) found that anticipation error was higher in ADHD children aged between 9 and 12 years only in an endogenous cue condition, hence, the tendency to make more anticipation (impulsive) errors when stimulus appearance was predictable.

Interaction among the four networks in children

Interaction among the attentional networks was observed in the present study, which was not consistent with a previous study (Rueda et al., 2004). It was reported that children of 7 years of age showed independence among the three networks. This difference could be attributed to difference in sample size, as the number of participants was fewer in their study ($N = 44$) compared to the present study ($N = 120$). We observed a developmental shift in the interaction among the attentional networks, which may be related to the progressive formation of neural circuits. Akhtar and Enns (1989) also found an interaction between orienting and conflict effects that reduced from 5 years of age through adulthood. In children with ADHD, the alerting score was found to be correlated with the exogenous orienting and conflict scores of the double cue condition, which further strengthens the finding that double cue can rescue the attention deficit of the ADHD children, presumably via a phasic increase in alertness (Johnson et al., 2008).

Development of motivational style in normal and ADHD children

We found that the tendency to avoid delay improved between 6 and 9 years of age in normal children while it improved between 6 and 7 years of age in ADHD children and did not improve between 7 and 9 years. Our results extend the findings of delay aversion theory (Sonuga-Barke, 2002) showing that ADHD children usually prefer not to wait, which did not change between 7 and 9 years of age.

Data showed different developmental trajectories depending on the type of control functions. In normal children, major development in response inhibition was observed between 7 and 8 years, error monitoring in 6–9 years, attentional disengagement in 7–9 years, and delay aversion in 6–9 years. Late development in alerting network was observed at 9 years of age in normal children. It appears that 6–9 years is a critical period showing developmental changes in most of the cognitive functions. Therefore, it is possible that development of one function may affect the development of another function between 6 and 9 years of age. For example, it has been reported that the anterior system mediates executive functions such as developing and maintaining expectations (Carr, 1992), which may help in performing the tests of executive attention such as stop-signal task. ADHD children performed poorly on stop-signal task that may be because of the problem in endogenous attentional system in ADHD children. In addition, we also found that the extent of slowing after failed inhibition error was less in ADHD children compared to normal children, which may be because of a motivation to avoid delay found in ADHD children. Comparator theories of error monitoring suggested that slowing after an error may result from a comparison of the representation of recently executed responses with memory of the instruction set for the task (Scheffers and Coles, 2000). This comparison process takes time and delays the response on subsequent trials. Therefore, in order to avoid delay, ADHD children may not be willing to compare the executed responses with memory of the instruction set of the task and hence do not show subsequent behavioral adjustment following an error. Thus, no change with age in delay aversion may hinder the development of error monitoring in ADHD children.

Our results also throw some light on the long-lasting debate of unitary and component views of executive control. Unitary theory of executive control posits a unified mechanism or a common resource underlying various aspects of executive control. The component theories of executive control argue that executive control consists of a number of distinct components such as response inhibition, error monitoring, attentional disengagement, delay aversion, task switching, and error monitoring. The results of the present study support the component theory of executive control. Different developmental trajectories observed for each of the control functions indicate different mechanisms underlying the control functions. These findings also have implications for the theory of ADHD. Barkley (1997) argued that response inhibition is the primary deficit found in ADHD, which in turn affects the other executive functions such as working memory, error monitoring, and so on. If this is the case then response inhibition should develop earlier than error monitoring. However, we found developmental changes in response inhibition between 7 and 8 years of age while for error monitoring between 6 and 8 years of age, which indicates an early development of error monitoring as compared to response inhibition. Posner and Rothbart (1998) also found that the ability to detect an error (indicated by PES) develops at an earlier age than the ability to inhibit. Further studies on the relationship between inhibition and executive functions are needed to determine whether there is a need to modify Barkley's model to account for the component view of executive functions rather than a unified perspective to executive function deficit as the underlying mechanism for ADHD.

In addition, because of smaller age range of children in the present study, we can not comment on whether ADHD could involve a developmental delay or a stable deficit in control processes. However, the developmental pattern obtained in the present study favors the conceptual view of ADHD as a stable deficit in cognitive control functions. For example, performance of 9-year-old ADHD children was poor from the performance of 6-year-old normal children on all the tasks. These results are consistent with studies that report a deficit in control functions as response inhibition, error monitoring, attentional disengagement, attentional networks, and motivational style in adults with ADHD (Aron et al., 2003; O'Connell et al., 2009; Tucha et al., 2005; Oberlin et al., 2005; Plichta et al., 2009). Results of the present study should be further validated with larger age group of children with a longitudinal design on both structural as well as functional brain developments. This would inform whether delay in structural brain development can be mapped onto the delay in functional brain and cognitive development in ADHD.

Summary and conclusions

Mapping of the developmental trajectories of various control processes, such as response inhibition, error monitoring, attentional disengagement, attentional networks, and motivational style, from 6 to 9 years of age is important in understanding the pathology of ADHD. We demonstrate the period of 6–9 years to be developmentally active with respect to the control processes in normal children but not in children with ADHD. Since the difference across the age range of 6–9 years in ADHD children was not significant, it appears that the deficits in control processes accumulate with age. The developmental patterns of the control processes in normal children supports the component view of executive control. Present study favors the conceptual view of ADHD as a stable deficit in cognitive control functions rather than developmental delay in control functions, which are implicated in the pathology of ADHD. Such developmental trends of control functions in ADHD are consistent with a multifactorial cognitive etiology model of ADHD. In this model, an impulsive cognitive style is attributed to an additive or interactive dysfunction in multiple cognitive systems and their closely related mediating neural networks (Willcutt et al., 2005). Developmental trajectories of these processes in late childhood could be further examined, which is also marked by the relative improvement in symptoms of ADHD.

References

Akhtar, N., & Enns, J. T. (1989). Relations between covert orienting and filtering in the development of visual attention. *Journal of Experimental Child Psychology*, *48*, 315–344.

American Psychiatric Association. (1994). *Diagnostic and statistical manual of mental disorders* (4th ed.). Washington, DC: Author.

Aron, A. R., Dowson, J. H., Sahakian, B. J., & Robbins, T. W. (2003). Methylphenidate improves response inhibition in adults with attention-deficit/hyperactivity disorder. *Biological Psychiatry*, *54*, 1465–1468.

Aron, A. R., & Poldrack, R. A. (2005). The cognitive neuroscience of response inhibition: Relevance for genetic research in Attention-Deficit/Hyperactivity Disorder. *Biological Psychiatry*, *57*, 1285–1292.

Barkley, R. A. (1997). Behavioral inhibition, sustained attention, and executive functions: Constructing a unifying theory of AD/HD. *Psychological Bulletin*, *121*, 65–94.

Beane, M., & Marrocco, R. T. (2004). Norepinephrine and acetylcholine mediation of the components of reflexive attention: Implications for attention deficit disorders. *Progress in Neurobiology*, *74*, 167–181.

Becker, M. G., Isaac, W., & Hynd, G. W. (1987). Neuropsychological development of nonverbal behaviors attributed to 'frontal lobe' functioning. *Developmental Neuropsychology*, *3*, 275–298.

Blane, M., & Marrocco, R. (2004). Cholinergic and noradrenergic inputs to the posterior parietal cortex modulate the components of exogenous attention. In M. I. Posner (Ed.), *Cognitive neuroscience of attention* (pp. 313–325). New York, NY: Guilford.

Brocki, K. C., & Bohlin, G. (2004). Executive functions in children aged 6 to 13: A dimensional and developmental study. *Developmental Neuropsychology*, *26*, 571–593.

Brocki, K. C., & Bohlin, G. (2006). Developmental changes in the relation between executive functions and symptoms of ADHD and co-occurring behaviour problems. *Infant and Child Development*, *15*, 19–40.

Brodeur, D. A., & Pond, M. (2001). The development of selective attention in children with attention deficit hyperactivity disorder-statistical data included. *Journal of Abnormal Child Psychology*, *29*, 229–239.

Bunge, S. A., Dudukovic, N. M., Thomason, M. E., Vaidya, C. J., & Gabrieli, J. D. E. (2002). Immature frontal lobe contributions to cognitive control in children: Evidence from fMRI. *Neuron*, *33*, 301–311.

Carr, T. H. (1992). Automaticity and cognitive anatomy: Is word recognition automatic? *American Journal of Psychology*, *105*, 201–237.

Carter, C. S., Krener, P., Chaderjian, M., Northcutt, C., & Wolfe, V. (1995). Asymmetrical visual-spatial attentional performance in ADHD: Evidence for a right hemisphere deficit. *Biological Psychiatry*, *37*, 789–797.

Cepeda, N. J., Cepeda, M. L., & Kramer, A. F. (2000). Task switching and attention deficit hyperactivity disorder. *Journal of Abnormal Child Psychology*, *28*, 213–226.

Cepeda, N. J., Kramer, A. F., & Gonzalez de Sather, J. C. M. G. (2001). Changes in executive control across the life span: Examination of task-switching performance. *Developmental Psychology*, *37*, 715–730.

Christ, S. E., White, D. A., Mandernach, T., & Keys, B. A. (2001). Inhibitory control across the life span. *Developmental Neuropsychology*, *20*, 653–669.

Conners, C. K. (2002). *Manual for Conners' rating scales*. Revised ed. N. Tonoawanda, NY: Multi-Health Systems Inc.

Crone, E. A., Bunge, S. A., van der Molen, M. W., & Ridderinkhof, K. R. (2006). Switching between tasks and responses: A developmental study. *Developmental Science*, *9*, 278–287.

Davies, P. L., Segalowitz, S. L., & Gavin, W. J. (2004). Development of response-monitoring ERPs in 7-to 25-year-olds. *Developmental Neuropsychology*, *25*, 355–376.

Gerstadt, C. L., Hong, Y. J., & Diamond, A. (1994). The relationship between cognition and action: Performance of 3½ years old on a Stroop-like day-night test. *Cognition*, *53*, 129–153.

Gupta, R., Kar, B. R., & Srinivasan, N. (submitted a). Development of task switching and error monitoring in children.

Gupta, R., Kar, B. R., & Srinivasan, N. (submitted b). Cognitive markers of ADHD: Development of a diagnostic system.

Gupta, R., Kar, B. R., & Thapa, K. (2006). Specific cognitive dysfunction in ADHD: An overview. In: J. Mukherjee & V. Prakash (Eds.), *Recent developments in psychology* (pp. 153–170). Delhi: Defence Institute of Psychological Research.

Halperin, J. M., & Schulz, K. P. (2006). Revisiting the role of the prefrontal cortex in the pathophysiology of attention-deficit/hyperactivity disorder. *Psychological Bulletin*, *132*, 560–581.

Hogan, A. M., Vargha-Khadem, F., Kirkham, F. J., & Baldeweg, T. (2005). Maturation of action monitoring from adolescence to adulthood: An ERP study. *Developmental Science*, *8*, 525–534.

Hudspeth, W. J., & Pribram, K. H. (1992). Psychophysiological indices of cerebral maturation. *International Journal of Psychophysiology*, *12*, 19–29.

Iaboni, F., Douglas, V. I., & Baker, A. G. (1995). Effects of reward and response cost on inhibition in ADHD children. *Journal of Abnormal Psychology*, *104*, 232–240.

Jemel, B., Achenbach, C., Muiller, B., Ropcke, B., & Oades, R. D. (2002). Mismatch negativity results from bilateral asymmetric dipole sources in the frontal and temporal lobes. *Brain Topography*, *15*, 13–27.

Johnson, K. A., Kelly, S. P., Bellgrove, M. A., Barry, E., Cox, E., Gill, M., et al. (2007). Response variability in attention deficit hyperactivity disorder: Evidence for neuropsychological heterogeneity. *Neuropsychologia*, *45*, 630–638.

Johnson, K. A., Robertson, I. H., Barry, E., Mulligan, A., Daibhis, A., Daly, M., et al. (2008). Impaired conflict resolution and alerting in children with ADHD: Evidence from the Attention Network Task (ANT). *The Journal of Child Psychology and Psychiatry*, *49*, 1339–1347.

Johnstone, S. J., Carly, B. P., Robert, J. B., Adam, R. C., & Janette, L. S. (2005). Development of inhibitory processing during the Go/NoGo task. *Journal of Psychophysiology*, *19*, 11–23.

Jonkman, L. M. (2005). Selective attention deficits in children with attention deficit hyperactivity disorder: A review of behavioral and electrophysiological studies. In: D. Gozal & D. L. Molfese (Eds.), *Attention deficit hyperactivity disorder: From genes to patients* (pp. 255–274). Totowa, NJ: Human Press.

Kramer, A. F., Humphrey, D. G., Larish, J. F., Logan, G. D., & Strayer, D. L. (1994). Aging and inhibition: Beyond a unitary view of inhibition processing in attention. *Psychology of Aging*, *9*, 491–512.

Liotti, M., Pliszka, S. R., Perez, R., Kothmann, D., & Woldorff, M. G. (2005). Abnormal brain activity related to performance monitoring and error detection in children with ADHD. *Cortex*, *41*, 377–388.

Logan, G. D. (1994). On the ability to inhibit thought and action. A users' guide to the stop signal paradigm. In D. Dagenbach & T. H. Carr (Eds.), *Inhibitory processes in attention, memory and language* (pp. 189–236). San Diego, CA: Academic Press.

Marrocco, R. T., & Davidson, M. C. (1998). Neurochemistry of attention. In R. Parasuraman (Ed.), *The attentive brain*. Cambridge: Cambridge University Press.

Monsell, S. (2003). Task switching. *Trends in Cognitive Science*, *7*, 134–140.

Oades, R. D. (2006). Function and dysfunction of monoamine interactions in children and adolescents with AD/HD. In E. D. Levin (Ed.), *Neurotransmitter interactions and cognitive function* (pp. 207–244). Basel: Birkhauser Verlag.

Oberlin, B. G., Alford, J. L., & Marrocco, R. T. (2005). Normal attention orienting but abnormal stimulus alerting and conflict effect in combined subtype of ADHD. *Behavioural Brain Research*, *165*, 1–11.

O'Connell, R. G., Bellqrove, M. A., Dockree, P. M., Lau, A., Hester, R., Garavan, H., et al. (2009). The neural correlates of deficient error awareness in attention-deficit hyperactivity disorder (ADHD). *Neuropsychologia*, *47*, 1149–1159.

Oosterlaan, J., Logan, G. D., & Sergeant, J. A. (1998). Response inhibition in AD/HD, CD, comorbid AD/HD+CD, anxious, and control children: A meta-analysis of studies with the stop task. *Journal of Child Psychology and Psychiatry*, *39*, 411–425.

Oosterlaan, J., & Sergeant, J. A. (1998). Response inhibition and response re-engagement in attention deficit/hyperactivity disorder, disruptive, anxious and normal children. *Behavioral Brain Research*, *94*, 33–43.

Pennington, B. F., & Ozonoff, S. (1996). Executive functions and development of psychopathology. *Journal of Child Psychology and Psychiatry and Allied Disciplines*, *37*, 51–87.

Plichta, M. M., Vasic, N., Wolf, R. C., Lesch, K. P., Brummer, D., Jacob, C., et al. (2009). Neural hyporesponsiveness and hyperresponsiveness during immediate and delayed reward processing in adult attention-deficit/hyperactivity disorder. *Biological Psychiatry*, *65*, 5–6.

Posner, M. I., & Petersen, S. E. (1990). The attention system of the human brain. *Annual Review of Neuroscience*, *13*, 25–42.

Posner, M. I., & Rothbart, M. K. (1998). Attention, self-regulation, and consciousness. *Philosophical Transactions of the Royal Society of London B*, *353*, 1915–1927.

Rabbit, P. M. A. (1966). Error correction time without external error signals. *Nature*, *212*, 438.

Raven, J., Raven, J. C., & Court, J. H. (1998). *Colored progressive matrices*. Oxford: Oxford Psychologists Press.

Ridderinkhof, K. R., van der Molen, M. W., Band, P. H., & Bashore, T. R. (1997). Source of interference from irrelevant information: A developmental study. *Journal of Experimental Child Psychology*, *65*, 315–341.

Rubia, K., Halari, R., Smith, A. B., Mohammed, M., Scott, S., Giampietro, V., et al. (2008). Dissociated functional brain abnormalities of inhibition in boys with pure conduct disorder and in boys with pure attention deficit hyperactivity disorder. *American Journal of Psychiatry*, *165*, 889–897.

Rueda, M. R., Fan, J., McCandliss, B. D., Halparin, J. D., Gruber, D. B., Lercari, L. P., et al. (2004). Development of attention during childhood. *Neuropsychologia*, *42*, 1029–1040.

Schachar, R. J., Chen, S., Logan, G. D., Ornstein, T. J., Crosbie, J., Ickowicz, A., et al. (2004). Evidence for an error monitoring deficit in attention deficit hyperactivity disorder. *Journal of Abnormal Child Psychology*, *32*, 285–293.

Scheffers, M. K., & Coles, M. G. (2000). Performance monitoring in a confusing world: Error related brain activity, judgments of response accuracy, and types of errors. *Journal of Experimental Psychology: Human Perception and Performance*, *26*, 141–151.

Shaw, P., Eckstrand, K., Sharp, W., Blumenthal, J., Lerch, J. P., Greenstein, D., et al. (2007). Attention-deficit/hyperactivity disorder is characterized by a delay in cortical maturation. *Proceedings of the National Academy of the Sciences of the United States of America*, *104*, 19649–19654.

Smith, A. B., Taylor, E., Brammer, M., Toone, B., & Rubia, K. (2006). Task-specific hypoactivation in prefrontal and temporoparietal brain regions during motor inhibition and task switching in medication-naïve children and adolescents with attention deficit hyperactivity disorder. *American Journal of Psychiatry*, *163*, 957–960.

Sonuga-Barke, E. J. (2002). Psychological heterogeneity in AD/HD — a dual pathway model of behavior and cognition. *Behavioral Brain Research*, *130*, 29–36.

Sonuga-Barke, E. J., Taylor, E., Sembi, S., & Smith, J. (1992). Hyperactivity and delay aversion — 1. The effect of delay on choice. *Journal of Child Psychology and Psychiatry*, *33*, 387–398.

Spieler, D. H., Balota, D. A., & Faust, M. E. (1996). Stroop performance in healthy younger and older adults and in individual with dementia of the Alzheimer's type. *Journal of*

Experimental Psychology: Human Perception and Performance, *22*, 461–479.

Spronk, M., Jonkman, L. M., & Kemner, C. (2008). Response inhibition and attention processing in 5-to7-year old children with and without symptoms of ADHD: An ERP study. *Clinical Neurophysiology*, *119*, 2738–2752.

Tucha, O., Mecklinger, L., Laufkotter, R., Kaunzinger, I., Paul, G. M., Klein, H. E., et al. (2005). Clustering and switching on verbal and figural fluency functions in adults with attention deficit hyperactivity disorder. *Cognitive Neuropsychiatry*, *10*, 231–248.

Wiersema, J. R., van der Meere, J. J., & Roeyers, H. (2005). ERP correlates of impaired error monitoring in children with ADHD. *Journal of Neural Transmission*, *112*, 1417–1430.

Wiersema, J. R., van der Meere, J. J., & Roeyers, H. (2007). Developmental change in error monitoring: An event-related potential study. *Neuropsychologia*, *45*, 1649–1657.

Willcutt, E. G., Doyle, A. E., Nigg, J. T., Faraone, S. V., & Pennington, B. F. (2005). Validity of the executive function theory of attention deficit/hyperactivity disorder: A meta-analytic review. *Biological Psychiatry*, *57*, 1336–1346.

Williams, B. R., Ponesse, J. S., Schachar, R. J., Logan, G. D., & Tannock, R. (1999). Development of inhibitory control across the life span. *Developmental Psychology*, *35*, 205–213.

N. Srinivasan (Ed.)
Progress in Brain Research, Vol. 176
ISSN 0079-6123

CHAPTER 18

Interaction of language and visual attention: evidence from production and comprehension

Ramesh Kumar Mishra*

Centre for Behavioral and Cognitive Sciences, Allahabad University, Allahabad, UP, India

Abstract: Recent developments in multimodal interactions in language processing have revealed important facts about the important controlling influence of the attentional system on language behavior. In this chapter, we cover the attentional system that is important for language processing and its interaction with the microstructure of the linguistic system. We argue that to understand the origin of language-mediated eye movements; we must integrate neuronal mechanisms of saccade generation with language processing. We discuss the various facets of interaction of language processing with visual attentional mechanism using eye-tracking data from Hindi in the production and comprehension of language.

Keywords: eye movements; grammar; language comprehension; visual attention; saccades; Hindi

Language and attention interface

The current theoretical perspective on attentional mechanism in language stems largely from the work of Talmy (2000). It is assumed that during language processing, speakers and hearers assign different amount of attention to different parts of an utterance. Often salient and grammatical aspects of language attract more attention. This salience of linguistic structures (Talmy, 1978) can be comparable to the salience of an image (Itti and Koch, 2001) that has been shown to attract visual attention in a bottom-up manner. The other important concept in Talmy's framework is the *attentional windowing* mechanism that languages utilise for facilitating perception and quantification of the external world. Different languages choose different attentional windowing mechanisms to bring constituents to the foreground for processing. Attention is said to be the highest on the semantic content of the material rather than on its form.

Attention to different elements of a sentence, for example, makes us perceive their individual meanings. But understanding a sentence is more than understanding its components. In this context, it is important to understand how attentional resources are distributed in real time when processing sentences. Talmy (2000) makes the figure-ground distinction in a sentence, where some parts are the figure, for example the subject of the sentence, and the remaining is the ground. Therefore, from a processing point of view, sentences are figure-ground representations that require differential utilisation of attention. Therefore, attentional mechanisms that play an important role in binding constituents have become a major area of interest in the multidimensional aspects of language processing. Recent methodological developments in eye movement analysis, which provides a direct

*Corresponding author.
Tel.: 91-0532-2460738; Fax: 91-0532-2460738;
E-mail: rkmishra@cbcs.ac.in

DOI: 10.1016/S0079-6123(09)17616-1

quantifiable measurement of overt visual attention, offer a possibility to explore language and its interaction with attentional and memory systems in a more direct manner. Spatial structuring of the space, and its expression using an abstract symbolic representation, is a major function of language. For example, the availability of extensive prepositional systems in languages can map space and temporality precisely and bring into focused attention for cognition. Language is capable of mapping space (Fauconnier, 1994; Levinson, 2003) using its very precise deictic system. The interaction of language with other cognitive systems (Jackendoff, 2007) in various modes allows one to study empirically the role attentional mechanisms play in spoken language comprehension and also how linguistic systems help in the selective deployment of attention. It should be noted that linguistic elements can also take part in deployment of covert as well as overt attention, and the same dualism applies here as in the case of other psychophysical studies of visual attention.

Language comprehension, as well as language production, uses attentional mechanisms differentially. Recent integration of behavioral and neuroimaging data in explaining language production mechanisms clearly demonstrates this (Roelofs, 2008). Many neuroimaging studies of spoken word planning and production have consistently found activation in anterior cingulate cortex and left prefrontal cortex (Roelofs, 2006; Indefrey and Levelt, 2004). Apart from the traditional role of the anterior ciangulate cortex in attention, its role in supervising linguistic processing in naming is becoming clear. Experimental evidence concerning the modulation of attentional networks in language processing has come from studies on spoken word planning and picture naming. In such studies when two pictures appear and the subject has to name one, it has been observed that attention to the second object shifts after the phonological planning and articulatory gesture of the first word is finished (Meyer et al., 2003). It seems that attentional deployment in such linguistic tasks is graded and time bound and hence can be tracked with high-resolution tools.

Different morpho-syntactic elements of language, such as classifiers, deixis, and case markers, actually perform this important function of windowing attention to specific objects in the outer or inner world. Basic perceptual mechanisms that allow us to recognize and process objects through our various sense organs are often modulated by such elements of grammar, as they are symbolic as well as perceptually grounded (Barsalou, 2008). Hence processing language includes shifting attention toward objects that are important for the task in hand. Drawing parallelism to figure-ground segregation in visual perception, Talmy (2000) considers sentential structures of various types basically as figure-ground representations. It seems that there is a general-purpose attentional modulation mechanism where things that are relevant and important are brought under conscious focus and led to goal-directed action. This emerging cognitivist and embodied approach to language processing brings theories of action and perception closer as never before in the history of purely symbolic and amodal psycholinguistics. Recent work in psycholinguistics considers language processing as action (Trueswell and Tanenhaus, 2005) and provides a framework for further exploration of the temporal coupling of attentional mechanism and linguistic processing.

As mentioned earlier, language is a symbolic representation that allows windowing of attention to specific objects and referents for perception and cognition. There are many elements in the grammar of a language that trigger attentional shifts during sentence processing, and those also relate conceptually distinct objects because of their grammatical link. This windowing of attention to particular objects and referents at particular time points during the course of language processing is relevant for understanding the binding problem of linguistic perception. The relationship of language with thought and worldview has classically rested on intuition of the native speakers and without much detailed experimentation. Linguistic research since long has hypothesized the constraining nature of language on a speaker's basic conceptual processes. For example, recently Huettig et al. (2008) have found that Chinese speakers trigger attentional shifts to objects that share classifiers. Classifiers are unique to particular languages and have a wide range of typological variations that quantify objects and states. Such language-triggered eye movements and attention shifts have been taken as evidence for language affecting thinking. Visual world studies on comprehension of gender markers and lexical access have shown that in French, subjects launch eye movements to objects that share the gender marker (Dahan et al., 2000).

These evidences indicate a close relationship between elements of grammar and orientation of attention. Recent developments in eye-tracking technology have enabled researchers to study online dynamics of language and vision interaction, providing fine-grained temporal data that suggest that selective visual attention is often triggered by phonological, semantic, and morphosyntactic elements of languages.

In this chapter we first review evidence from eye-tracking paradigm that demonstrates the nature of language-mediated anticipatory eye movements. These eye movements offer a very clear illustration of the close coupling between visual attention and language processing mechanism. We review important studies in English language where eye movements have been used to explore online interaction of language and vision in a range of issues. Then we describe three different eye-tracking experiments in Hindi, where we show the close interaction of visual attention and linguistic structure. We show the effect of language on simulation of motion in a visual world study, supporting earlier findings in English. In a sentence production task, while looking at natural full color pictures, we show the difference between children and adults in their eye movement control. Again in another study on spoken word comprehension, we demonstrate that Hindi listeners activate conceptual structures as soon as they hear the gender marker on adjectives. With these results and others reviewed, we argue that language processing in a visual world involves coordinated interplay of visual attention and linguistic structure and they are evident in both comprehension and production. Most importantly we discuss the cross-linguistic aspects of our results in understanding the broader research question involving language and vision as measured by eye movements.

Language-guided visual attention and eye movements

The most important aspect of visual attention is to bring objects under foveal inspection and process them for further information. Visuospatial attention particularly plays a crucial role in spatial quantification of the external world, mapping knowledge from the episodic memory about objects and their referents. Visual attention to scenes is reflected by specific patterns of eye movements and the most crucial information is often acquired thorough foveal fixations. We have learnt from several recent eye movement studies that visual attention to objects in scenes are constrained by several factors, such as task demand, image properties, semantics of scene representation, as well as cultural background of the viewer (Chua et al., 2005). Eye movements reveal a great deal about automatized as well as unconscious nature of visual cognition and oculomotor behavior in general and hence their importance in studies of language and cognition (Rayner, 1998). Although there is still debate on the distinction between overt and covert attention, it is agreed that eye movements, particularly fixational as well saccadic, are very fine indicators of moment-by-moment cognitive processing that measure overt orienting of visual attention as well as anticipatory behavior.

What is the role of attention in language processing and cognitive processing in general? It is now well known from several studies on infant language development as well as autism that joint attention plays a very crucial role in modulating acquisition of syntactic competence and overall language behavior. Whether one believes in the purely symbolic and amodal theories of language learning and representation (Chomsky, 1965, Jackendoff, 2002) or connectionist and learning-based approaches (Bybee and Mcclelland, 2005; Altmann, 2009), it is clear that the computational structure that is biologically endowed and is capable of making categorical sense of signals must depend also on the system's ability to attend and maintain attention for accurate as well as rapid information processing. Visual attention to specific patterns in a range of stimulus dynamically modulates language production (Griffin, 2001) as well as comprehension processes and other higher-level processing concerning knowledge representation. However, in spite of a great deal of work with fine measurement of eye movements on different types of stimuli, it is still unclear what causal role visual attention plays in linguistic and conceptual representation in general, except that attention is a necessary cognitive requirement.

Recent findings in the domain of multimodal interaction in cognitive processing have revealed the fine-tuned coupling of visual attention shifts triggered by linguistic as well as conceptual activation. The interaction of visual cognitive processes with linguistic processes has been studied in a range of domains, e.g., reading, scene perception, and sentence comprehension. We focus next on the basic mechanisms of vision language interaction and their temporal patterns as revealed in several eye-tracking studies on spoken sentence comprehension after a brief discussion of neuronal mechanisms in visual attention and eye movements. This is relevant for understanding neuronal as well as behavioral dynamics of language-meditated eye movements that are at the core of such visual world studies. Below we review some important neurophysiological evidence that reveals the time course of shift of visual attention because this is crucial in understanding temporal events in language-mediated eye movements.

Neuronal mechanisms of visual attention and eye movements

Visual attention solves the binding problem by integrating feature information from several neurons and allowing perceptual processes to focus and make sense. The shift of visual attention is extremely rapid to new locations (Desimone and Duncan, 1995). It takes about 200 msec to program a saccade and thus shift visual attention to spatially isolated locations. This saccadic latency involves precise planning and target selection apart from basic physiological reasons of oculomotor delay. It is at this intermediate time scale that language plays a role in channelizing attention, perhaps as time series data from several eye movement studies would indicate (Altmann and Kamide, 2009).

Saccades to targets in the space over time are guided by information on where things are or where they should be. Visual processing is important for saccade target selection and eye movements (Schall, 1995). Neurons in the frontal eye field have specialized control mechanisms, where they program purposeful saccades to only selected targets and prevent saccades to objects that are not targets (Schall and Hanes, 1993). Launching a saccade to any region after initial identification is a very important goal-directed activity as well as fine-tuned decision-making (Schall, 2001). Comparison of neuronal basis of occulomotor functioning between humans and primates suggests similar pathways and activation patterns for target selection and eye movements, leading to deployment of overt attention (Nakahara et al., 2007).

Eye movements to specific areas of a scene involve selection of a saccade target. It has been shown with single-cell recordings in primates that when the animal is covertly attending possible spatial location, neurons the posterior parietal cortex (Bushness et al., 1981) and interconnected lateral pulvinar nucleus of the thalamus show greater activity (Petersen et al., 1985). This indicates that during the planning of a saccade in language-related tasks, there are parallel processes going on regarding possible saccade target selection and an eventual eye movement. This process of spatial target selection must happen simultaneously during lexical-semantic or conceptual activation, and hence there is a high probability of overlapping brain areas to show coordinated activity. In contrast to deployment of covert attention without the promise of an eye movement, when the animal prepares for a saccade, neurons in the frontal eye field (Bruce and Goldberg, 1985) and superior colliculus fire (Wurtz and Goldberg, 1972).

Therefore it is possible that this activity is highly sequential in nature and is important for a complete understanding of the timing issues involved in language-mediated eye movements under various task requirements in humans. Based on the above-mentioned evidence, it can be argued that eye movements during language processing cannot be understood without exploring the neuronal dynamics of both attentional and oculomotor systems. What we know so far from eye-tracking studies of language processing is that language modulates attentional mechanism and guides action, but what remains unknown is its precise neuronal basis.

Visual world studies: language-mediated eye movements

Accurate measurements of important eye movements such as saccades and fixations have been consistently used in several disciplines, providing valuable insights into action, perception, and cognition (Liversedge and Findlay, 2000; Rayner, 1998). Studying eye movements in a range of language-processing tasks has revealed the close link between language, perception, and action and also about the embodied nature of visuolinguistic cognition in humans (see Henderson and Ferreira, 2004 for detailed discussions). It has also bridged the so-called language as action tradition with "language as product" tradition (Trueswell and Tanenhaus, 2005).

It has been observed that visual attention to objects in a display shifts synchronically with spoken language input, and this shift in attention shows a time-coupled behavior. Most visual world studies have consistently followed a paradigm where a visual display of few objects is presented roughly one second before an auditory sentence of a particular type is presented and the participant's eye movements to different objects in the display are recorded. The time lag allows subjects to activate lexical, conceptual, and spatial information of the objects for later integration during comprehension. There exist controversies in the literature on whether subjects use covert naming in launching eye movements to the respective pictures later or they merely use a template-matching technique.

The orientation and shifts in overt visual attention, as evident in fixations and saccades to either named or unnamed objects, have been found to be influenced by particular syntactic, semantic, and lexical inputs. This has been dubbed in recent writings (Altmann, 2009) as language-mediated eye movements to visual scenes that show probabilistic as well as anticipatory nature of spoken language processing. Studies using this technique often measure eye movements to particular objects as compared to other unrelated distracters as a function of acoustic onsets and offsets of particular words in the auditory sentence. Often critical words appear after 4 or 5 s of sentence onset and eye movements triggered by the acoustic unfolding of the word to specific objects in the display, indicating orientation of overt visual attention in the form of eye movements, which is important for the language parser. Studies in spoken word processing have shown how cohort words are activated based on phonological similarity, lending support to probabilistic models, i.e., COHORT and TRACE model in auditory word recognition. For example, in the study of spoken word recognition by Dahan et al. (2000) when *candle* unfolded acoustically, participants' eye movements rose to both *candy* and *candle*, and then after the acoustic offset of the word *candle*, looks kept rising to *candle* but fell on *candy*. This is fairly predictable where the target is present in the display. But most important results in the visual world paradigm have shown that listeners covertly activate the lexical or conceptual meaning of a semantic or a phonological competitor even in the absence of the target or during processing it. This clearly negates the template-matching or covert-naming of the objects theory (see Huettig and Altmann, 2007 for a thorough discussion of this point).

Anticipatory eye movements to phonologically or semantically related words suggest that the language parser activates many competitors based on the lexical input and there is underlying competition in real time, which in turn is reflected in language-guided shifts in overt visual attention. But it must be noted that such activations are constrained by the types and number of objects present in the display, and this could be one of the major methodological weaknesses of this paradigm.

Huettig et al. (2006) found that participants looked at the trumpet while they listened to the piano compared to semantically unrelated distracters, and this has been taken as evidence of the fact that words that are conceptually closer reside in a *high-dimensional semantic space* and activation of this space directly modulates the attentional system. Looks to specific objects at particular time points of the word, even during the life time of the word itself, are closely time locked to images that match visually to the emerging linguistic–conceptual representation. This was first discovered by Cooper (1974) who found that subjects looked more toward the word zebra when they heard something about Africa. This pattern of activation seems to be also constrained by the subjects' general knowledge and proficiency level.

Recently it has also been shown that it is not only phonological or semantic similarity between lexical items that triggers eye movements to specific objects in the display, but also other factors such as shape and color. Huettig and Altmann (2007) showed that subjects looked more at the image of a *cable* when they heard *snake* in a sentence because it is possible that while lexically activating *snake*, they also activated the shape of a snake and many other objects that share visual features with a snake. More importantly, it was also found that in a biasing sentence, where the context clearly biased the parser to orient its attention toward the snake, *cable* still received more fixations compared to other incongruent distracters, indicating the unavoidable and involuntary nature of underlying competition. Such evidence indicates that conceptual representation of words in episodic memory contains many features and it is the lexical input and not the sentential context that affects orientation and goal-directed shifts in overt visual attention in a visual world study.

Studies highlighting the probabilistic nature of syntactic processing and anticipatory behavior show that subjects can anticipate the nature and even changeable state of an event and trigger eye movements to objects in the visual display. Often subjects seem to use world knowledge and background information about events and objects in launching such eye movements. In spite of such important discoveries about the incremental and probabilistic nature of spoken language processing and the time course of mapping between visual and linguistic knowledge, there are questions that remain unexplored. One such important question is why subjects look back again to objects that they have already seen in the display and when they listen to it again in the sentence, and also the question related to the very nature of such eye movements, i.e., what they are for?

Opinion seems to be divided on this issue, as it is argued from the embodied, perceptual symbol processing approach to cognitive processing that eye movements basically index objects in their spatial domain and looks indicate retrieving information again when in demand. The other view suggests that it is similar to visual search in general and template matching (Huettig and Altmann, 2007). The presentation of fixed number of items in a display may also constrain the full range of lexical-semantic activations, and hence eye movements to objects may be just associational and frequency-based rather than reflecting any special event.

Visual world studies can be both action based and nonaction based. In the action-based approach, the subject listens to some spoken instruction and performs some task, mostly clicking with a mouse on the target picture. In the nonaction-based approach, participants listen passively to a spoken sentence and eye movements are recorded to particular objects. This passive listening paradigm departs radically from mainstream psychological or psycholinguistic research as this does not explicitly call for the metalinguistic judgment of the subject; rather it depends on the involuntary aspects of simulations and taps more automatic activations while at the same time being more ecologically valid.

Dynamics of lexical competition and attentional shift: Hindi gender processing

It has already been reviewed that specific morpho-syntactic elements of grammar of particular languages induce anticipatory eye movement behavior, causing overt shifts in visual attention in a visual world study. We explored how gender marking on adjectives in Hindi may trigger lexical access in a probabilistic manner.

Gender is a grammatical element of language that has considerable variability across languages. For example, in German, there are three definite articles, i.e., *der* (masculine), *die* (feminine), and *das* (neutral) and in French there are two, i.e., *le* (masculine) and *la* (feminine). In Hindi, there is no such arrangement of different articles for marking gender on nouns but gender marking is expressed through inflectional attachments on verbs, adjectives, and nouns. In Hindi, adjectives precede nouns in sentences and agree in gender with the noun. For example, *lamba ladka* (tall boy – masculine) and *lambi ladki* (tall girl – feminine). Hence the gender marker is mostly consistent, i.e., /*a*/ indicating the masculine form and /*i*/ indicating the feminine one. These phonological entities should be the triggering agents for attentional shifts, also providing clues about the upcoming nouns for Hindi speakers.

Contemporary psycholinguistic theories assume that when we listen to a spoken word, multiple words that share phonological onsets with the target words get activated covertly and compete against each other (Marslen-Wilson and Welsh, 1978; McQueen et al., 1994). This competition often leads to shifts in visual attention to pictures of competitor objects. Many eye-tracking studies of spoken word recognition have found very robust effects for cohort competition. In an eye-tracking visual world study, Dahan et al. (2000) examined the impact of morphosyntactic context on competitor activation. They tested whether gender marking on definite articles influences the recognition of subsequent nouns. They found that listeners very quickly use the gender information on the article in correctly anticipating the forthcoming nouns. However, when they used gender-marked adjectives preceding nouns, they did not find strong anticipatory effects. Other studies have explored lexical gender effects in word recognition using experimental paradigms other than eye tracking (e.g., Bates et al., 1996; Colé and Segui, 1994; Grosjean et al., 1994). Results from these studies have found that the presence of gender-congruent articles or adjectives enhances the recognition of target nouns whereas gender-incongruent forms slow down recognition.

We used a passive listening paradigm and tracked eye movements as subjects listened to the sentence and saw displays. Hindi gender marking is a unique grammatical property of the language that speakers could use to window attention to a specific object. From a probabilistic and connectionist perspective, it can also be said that Hindi speakers would activate all those objects that share the adjective as well as the gender. Therefore, matching with the adjective will be the first level of competition and matching both gender as well as adjective will be an additional competition in online spoken sentence processing. There have been two different theoretical views concerning the priming of gender (Friederichi and Jacobsen, 1999). One view suggests that gender cues can act immediately and hence are processed prelexically, contributing to both facilitation and inhibition; the other view supports postlexical processing where only inhibition is observed. Different languages, depending on their unique grammatical features, show these different activation patterns. We explored such a possibility in Hindi with gender-marked adjectives preceding nouns.

In this study, we used both the masculine and feminine versions of the same adjective, e.g., *lamba*/*lambi* that only differed in the last syllable from a phonological perspective. We also used a particle *wala* or *wali* that immediately followed the gender-marked adjectives to allow some time for gender-related activations to build up (Fig. 1). We also used a gender-marked particle (e.g., *wala* and *wali*) between the adjective and the noun to allow some time for grammar-mediated eye movements to trigger. This has been a consistently used technique in many such visual world studies, as it is known from oculomotor studies that it takes around 200 ms to program a saccade (Martin et al., 1993). But the conceptual trigger for the saccade must have been activated before 200 ms and much before the actual acoustic offset of the critical word. We calculated probabilities of fixations to different objects 200 ms after the appearance of the gender marker.

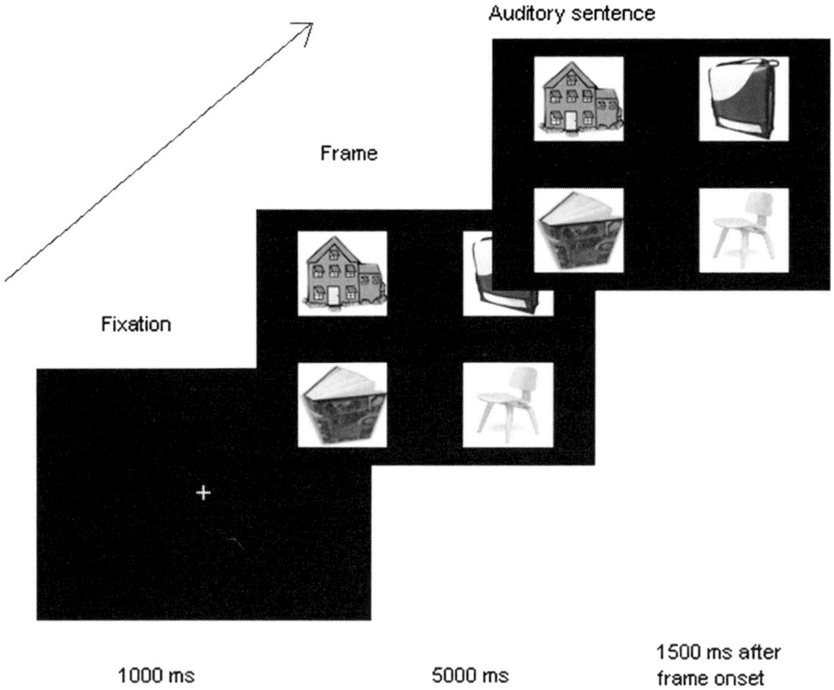

Fig. 1. Time sequence showing the task.

Example of sentences in Hindi.

Yeh jo lambi wali gadi hai wo kafi purani hai.

This long car is very old.

Yeh jo lamba wala ladkaa hai wo mera dost hai.

This tall (long) boy is my friend.

We expected that it is the evolving phonological information from the gender inflection that will help listeners to accurately target a saccade toward the adjective and gender-congruent picture compared to others. It was predicted that participants would launch more fixations toward the gender- and adjective-congruent object in the display compared to the incongruent object and the distracters.

In this exploratory study on the influence of gender information on adjectives, we found mixed results on recognition of subsequent nouns, and the results are not similar to earlier findings on gender-marking effects on spoken word processing using eye movements. We observed that on the point of acoustic onset of the adjective that occurred after 2,000 ms of the sentence onset on an average probability of fixations to distracters target and competitor are not same (Fig. 2). But we see a rise in probability of fixation to the target immediately, since the onset of the adjective till 700 ms. During this period, probability of fixations to the competitor and the distracters decreases. This is the time period when the gender marker of the adjective unfolds (roughly after 280 ms). This pattern of looks indicate immediate conceptual activations leading to more eye movements to the object that shared both the adjective and the gender marker compared to the one that shared only the adjective. So this certainly suggests that Hindi listeners make very rapid use of the gender information on the adjectives in predicting the upcoming nouns or, in other words, constraining the possibility of a word in a display.

After 700 ms post onset of the adjective, we saw a rise in probability of fixations to the competitor and also to the distracters and this trend continues till the actual noun appears in the sentence. Please note that in our experimental conditions, the noun mentioned in the sentence was not in the display. Therefore after the automatic activation of conceptual information, the looks to objects during

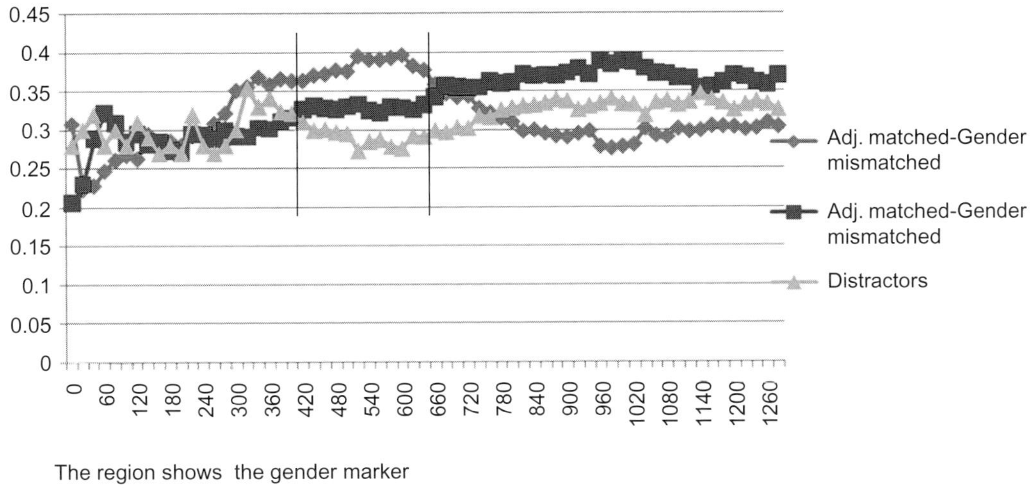

Fig. 2. Time course of probabilities of fixations in 20 m intervals to the adjective- and gender-congruent noun, to the adjective-congruent but gender-incongruent noun, and two averaged distracters from the onset of the adjective till the onset of the noun. Bars indicate the time period where the difference between the target and competitor and distracters is significant.

the noun are immaterial for us. We were interested in whether subjects could activate the lexical information of the noun that shared the adjective and the gender information. The probability of fixations plotted in Fig. 2, since the onset of the adjective till the onset of the noun, clearly indicate so. However, most importantly, we found that this activation period was very brief and subjects diverted their visual attention to the competitor noun that shared only the adjective but not the gender. This pattern of eye movements is not similar to earlier findings where we often see a continued probability of looks to the target, even after the acoustic onset of the noun. This indicates that language-triggered shifts in visual attention can be transient or sustained, and this depends on the level of competition among the display items.

Such results indicate that visual attention is very much dependent on linguistic processing in a dynamic manner and eye movement analysis can offer very valuable clues about the nature of interaction between visual attention and conceptual-linguistic processing. The findings of our experiments raise many interesting questions. For example, when there is no target present in the display other than two competitors, i.e., one adjective and gender marked and one only adjective matched, what will be the sequence of shifts of visual attention? Our results suggest that the most closely matching object will receive immediate visual attention and the second one will follow. Moreover, it is not possible to guarantee the sustenance of overt visual attention over a period of time, when subjects recognize that the object mentioned in the spoken sentence is not there and they indulge in searching behavior.

For example, in the study of Dahan et al. (2000), the effect of gender-marked articles on activations of upcoming nouns was not robust. Most importantly, in that study, the target and the competitor also shared initial phonological features, i.e., onsets. Our stimulus did not share any phonological features and hence were difficult, as the only thing that linked them was their occurrence with that adjective and also their gender. Our results will be more similar to such earlier patterns of eye movements when picture names are controlled for phonology, as this immediately restricts the domain of activation. The only study in Russian where gender-marked color adjectives influence eye movements to congruent nouns was by Sekerina (2003). In this study, anticipatory eye movements to gender-congruent objects were observed when color terms preceded nouns. The other important event is the time it takes to activate lexical and conceptual information. In this regard, our results are similar to earlier studies, for example, Dahan et al. (2000) in which eye movements to the congruent picture rose approximately after 300 ms of the onset of the word. But in their study, it was the object that was spoken and was also present in the display along with a competitor.

The most intriguing part is why subjects looked at the competitor (the objects that differed with the gender but took the adjective) after initial inhibition? This suggests a very dynamic level of competition where the parser selectively allows activation of the most relevant concept and hence more visual attention and then slows down while allowing attention to the other competitor. If this had not been the case, we would have seen a constantly rising probability to the most congruent picture.

Many contemporary studies using the visual world paradigm have found different effects on semantic and conceptual activation: phonological competitor effect (Allopenna et al., 1998), semantic competitor effect (Huettig et al., 2006), and shape and color effect (Huettig and Altmann, 2007). In the Allopenna study, when *candy* acoustically unfolded, looks to both *candle* and *candy* increased, but on the offset of *candy*, looks to *candle* decreased while those to *candy* kept increasing. Our results show a different pattern from this. In our results, the frequency of looks to the gender- and adjective-matched object decreased after the acoustic offset of the adjective and looks to the competitor increased after this point, suggesting a dynamic shift of attention. This could be because stimulus targets were absent and hence the parser, after having given sufficient attention to the most probable object, withdrew attention from it and deployed on the competitor. This is surprising from several perspectives.

Huettig et al. (2004) in a study of the shape competitor effect found that participants kept looking at the target and competitor even after the acoustic offset of the word. It is interesting to ask why overt attention would still be engaged to an object that the parser has already identified as either necessary or completely unnecessary. As in our case, when the object mentioned in the sentence was not in the display and there were two equally strong competitors, one more favored by virtue of having shared the adjective and gender and the other only gender, it seems logical that after parallel activation of both these concepts, there may be a period of selective inhibition of overt visual attention to one object after its suitability for the sentence was assessed. If this had not been the case, we would have observed increased fixations to both the objects throughout and at least a decreased probability of fixation to the object that only shared the gender and not the adjective.

Such results indicate that visual attention is very much dependent on linguistic processing in a dynamic manner and eye movement analysis can offer very valuable clue about the nature of interaction between visual attention and conceptual-linguistic processing. The findings from our experiment raise many interesting questions. For example, when there is no target present in the display but only two competitors instead i.e. one adjective and gender marked and one only adjective matched, what will be the sequence of shifts of visual attention? Our results suggest that the most closely matching object will receive immediate visual attention and the second one will follow. It is not possible to guarantee the sustenance of overt visual attention over a period of time, when subjects have recognized that the object mentioned in the spoken sentence is not there and then they indulge in searching behavior. After examining at length the issue of visual attention in sentence comprehension, it is important to consider how attention is channelized in sentence production. Below we describe results from an experiment, where we observed systematic deployment of visual attention in sentence conceptualization and speech in Hindi in children and adults.

Visual attentional shifts in sentence generation

Recent eye movement studies on picture naming and sentence production have revealed that fixations to particular parts of the image are closely time locked to production units (Griffin, 2001). Visual attention to objects to be named indicates processing related to planning and conceptualizing stages of language production (Levelt et al., 1999). The semantic complexity of the image certainly plays a role in patterns in saccades and fixations that one sees during language production. Recent studies in naming have shown that subjects shift their visual attention to other objects after they have completed phonological planning of the object that is under foveal inspection. Hence such behavior provides clear evidence that attentional mechanisms closely participate in early processes of language generation, and that we can measure these effects by measuring fixations to selected parts of the image as also by analyzing sequences of eye movements prior to actual sentence production.

We explored eye movement patterns among children and adults while they saw complex scenes and generated sentences freely. Scenes included full color photographs of individuals engaged in various activities of daily life. We divided the scenes into three broad types depending on the number of actors it depicted, as this information would have a direct effect on the transitivity of the verb to be generated. In one set of pictures, individual subjects were shown performing intransitive actions like dancing, sleeping, etc., while in the transitive category of images, subjects were shown to be engaged in an action with some objects, i.e., pulling a box. In the third category of pictures, two subjects were shown to be engaged in action with each other, e.g., a shopkeeper selling something to a buyer, or a doctor treating a patient. Psycholinguistic theories would assume that the verb is the most important lexical structure in the sentence that provides detailed knowledge about the other arguments that would appear in the sentence. For example, in the English sentence, "He will put the ball on the table," the subcategorization principle (Chomsky, 1965) makes *put* a transitive verb. This means there is some object that the actor will put somewhere. Therefore, if English speakers say this sentence with *put,* they would already have an idea about the object to be used with this verb. In contrast to this, sentences in Hindi generally have verbs in the final position. Hence it was worth exploring what happens in a language like Hindi when the verb comes at the end of the sentences (SOV canonical structures) and one tries to generate sentences to pictures and how fixations to different entities of the picture and their sequences match sequences of words spoken in a verb final sentence. We also explored how children and adults differed in their channelization of visual attention during sentence generation. This was important in understanding the development of the interaction between language and vision.

Visual attention on objects or entities to be named is important for retrieving conceptual and linguistic information, as this in turn feeds the emerging phonological representations. As shown in the following figure, for children and adults, there is a very striking difference between the areas that have received maximum visual attention while producing a sentence (Fig. 3a and b). This was an action picture where one person is giving milk to somebody else. Hence to conceptualize and say the sentence in Hindi, it is important that attention be focused on that part of the picture that will provide maximum information for the lexical retrieval of the verb in formation and then this verbal information will guide the framing of the sentence with arguments. We see that for children this is quite dispersed throughout the picture. This reflects a developing ability to narrow down the spatial area of visual attention, depending on the linguistic task at hand. This may also depend on the cortical maturational states related to linguistic conceptualization and saccadic eye movements. In contrast, for the adults we see a more focused area of visual attention, which largely falls on the hand of the milkman and is not dispersed widely. This pattern of visual attention indicates a much more neuronally matured system in terms of conceptualization and attention channelization.

Comprehension of language in the presence of a visual scene is a dynamic process involving interplay of attention and other cognitive resources. Language can produce simulation of abstract motion when there is actually none. Below we describe findings from an experiment where we observed simulation of fictive emotion when Hindi listeners heard sentences containing fictive motion verbs. These data provide very strong support for the fact that language has the capability to induce changes in physiological measures as evidenced through eye movements.

Fig. 3. (a) Visual attention map for children during sentence generation. (b) Visual attention map for adults during sentence generation.

Language-guided motion simulation

Language comprehension, in the embodied cognition framework (Barsalou, 2007), involves perceptual motor simulation. Talmy in his work on attentional mechanisms of language (Talmy, 2000) classifies fictive motion verbs. There are verbs whose grammatical patterning in sentences would induce simulation of fictive motion. In the first eye-tracking study of fictive motion in English, comprehending sentences like *The road runs through the valley* triggers mental simulation of motion. Richardson and Matlock (2007) found that subjects spent more time looking at a named entity (the subject–noun phrase) in the fictive motion form of the sentences. This is interesting because it engages simulation of real motion as the agent associated with motion verb is itself immobile and static. So nonliteral use of motion verbs induces simulation of motion in a way similar to that when used literally. Eye movement patterns observed directly mirror the mental simulation of motion. Studying language-induced motion simulations is important as it brings the two modalities, i.e., visual and linguistic, much closer in terms of perception.

We replicated Richardson and Matlock's original eye-tracking experiment of simulation of fictive motion in English sentences, with a slightly different stimulus in Hindi. Cross-linguistic evidences of psycholinguistic phenomena are crucial for generalization of results. We employed two versions of sentences where one sentence described the literal representation of the scene (i.e., *Yeh pull nadi ke upar hai*) and the fictive motion counterpart (*Yeh pull nadi ke upar se hokar guzarta hai*) described the same event but with the use of a verb that induced fictive motion. Talmy (2000) makes a distinction between travelable and nontravelable paths, and there can be a difference between these two in specific languages.

In Hindi, the verb comes at the end of the sentences as it is a verb-final language. Travelable and nontravelable entities, i.e., *fence* versus *road*, can be used along with verbs to simulate fictive motion. In a visual-world eye-tracking study, using the passive listening paradigm, we gave Hindi-speaking subjects fictive and nonfictive

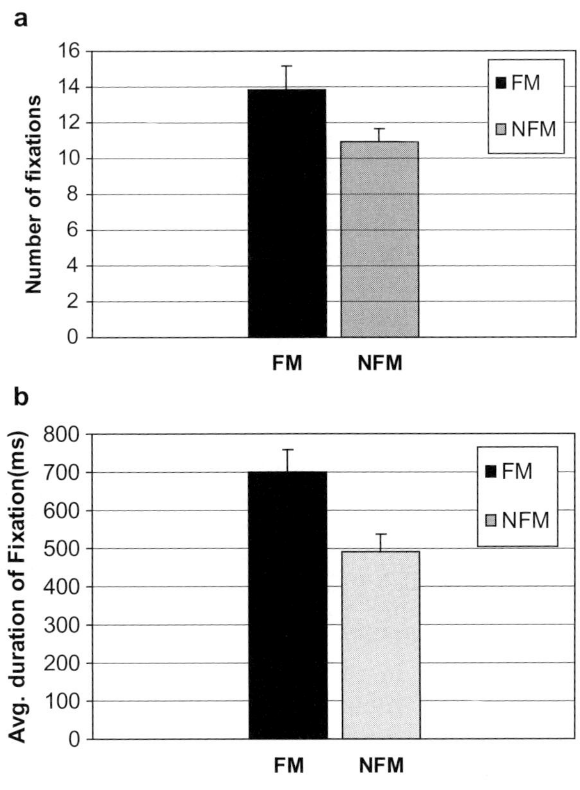

Fig. 4. (a) Mean fixations in fictive motion and nonfictive motion conditions. (b) Average duration of fixations for the fictive motion and nonfictive motion sentences.

versions of the sentences for the same display corresponding to the description. Sentences described spatial positions of entities in various contexts. We measured fixation durations, total gaze duration, and number of fixations for both the conditions and found statistically significant differences for number of fixations and average duration of fixations (Fig. 4a and b). The longer average duration and more number of fixations in the fictive motion sentences indicate simulation of motion as the findings are in agreement with that of Richardson and Matlock (2007). As the display was same for both types of sentences, and only sentences accompanying them differed, it can be concluded that it was simulation of fictive motion in figurative contexts that gave rise to a different pattern of eye movements. Therefore language comprehension seems to affect motor resonance in an embodied manner.

Conclusion

Evidence reviewed in this chapter suggests that language affects the dynamics of visual attention in a very remarkable way. However, in spite of such robust online evidence on the interaction of visual processing with linguistic processing, it is still not clear how *attention* per se plays a causal role in language comprehension or what is the precise nature of mental representation of conceptual knowledge. We have discussed evidence from both production and comprehension of language where it was observed that language input affects visual cognition as reflected in eye movements. As in the comprehension of fictive motion sentences, subjects simulate illusory motion in an embodied fashion and spend more time viewing the scenes. During comprehension of sentences, subjects direct their visual attention immediately and in a probabilistic manner to those objects in the scene that are congruent with the evolving linguistic representation as in the gender study. These evidences provide robust support for the multimodal interactionist theories of language and vision interaction. These findings from Hindi, a lesser-studied language, provide support for findings in other language that have used eye-tracking as a paradigm. However, there remain questions and issues that future research must take up to explore completely the representational issues that arise when linguistic and visual information interact in real time (Mishra and Marmolejo-Ramos, submitted).

To address such questions one obviously needs to look at the real-time neural mechanisms of both language processing and attentional orientation that happens in parallel and in distributed cortical networks. It would be interesting to explore how evolving linguistic representations at the neuronal level may trigger and modulate neural mechanisms of saccade generation and attentional shifts, as discussed earlier. For example, it is now well known that generation of saccades is programmed by specialized neurons in the superior colliculus and the frontal eye fields. However, most such studies have come from single-cell recordings of primates. Moreover, it is also known that important cortical areas of the perisylvian language network participate in a range of language-related tasks in a distributed fashion. Functional and effective connectivity studies of brain imaging focusing on cortical interaction patterns in both their temporal and spatial domain between the language and attentional areas would certainly yield fruitful results. Most importantly, the very origin of anticipatory behavior, reflected in eye movements triggered by activation of linguistic and conceptual knowledge, needs to be understood in its neurobiological context.

Acknowledgments

I would like to thank Falk Huettig for his advice. Thanks to Aparna Pandey for helping in computer programming. Thanks also go to Niharika Singh and Dhruv Raj Sharma for helping in data collection and analysis.

References

Allopenna, P., Magnuson, J., & Tanenhaus, M. (1998). Tracking the time course of spoken-word recognition using eye movements: Evidence for continuous mapping models. *Journal of Memory and Language*, *38*, 419–439.

Altmann, G. T. M., & Kamide, Y. (2009). Discourse-mediation of the mapping between language and the visual world. *Cognition, 111*, 55–71.

Altmann, G. T. M., & Mirkovic, J. (2009). Incrementality and prediction in human sentence processing. *Cognitive Science, 33*(4), 583–609.

Barsalou, L. W. (2007). Grounded cognition. *Annual Review of Psychology, 59*, 617–645.

Barsalou, R. (2008). Grounded cognition. *Annual Review of Psychology, 59*, 617–645.

Bates, E., Devescovi, A., Hernandez, A., & Pizzamiglio, L. (1996). Gender priming in Italian. *Perception & Psychophysics, 58*, 992–1004.

Bruce, C. J., & Goldberg, M. E. (1985). Primate frontal eye fields. I. Single neurons discharging before saccades. *Journal of Neurophysiology, 3*, 603–635.

Bushness, M. C., Goldberg, M. E., & Robinson, D. L. (1981). Behavioural enhancement of visual responses in monkey cerebral cortex. I. Modulation in posterior parietal cortex related to selective visual attention. *Journal of Neurophysiology, 46*, 755–772.

Bybee, J., & Mcclelland, (2005). Alternative to the combinatorial paradigm of linguistic theory based on domain general principles of human cognition. *The Linguistic Review, 22*, 381–410.

Chomsky, N. (1965). *Aspects of the theory of syntax*. Cambridge, MA: MIT Press.

Chua, H. F., Boland, J. E., & Nisbett, R. E. (2005). Cultural variation in eye movements during scene perception. *Proceedings of the National Academy of Science, 102*, 12629–12633.

Colé, P., & Segui, J. (1994). Grammatical incongruency and vocabulary types. *Memory & Cognition, 22*, 387–394.

Cooper, R. M. (1974). The control of eye fixation by the meaning of spoken language: A new methodology for the real-time investigation of speech perception, memory, and language processing. *Cognitive Psychology, 6*, 84–107.

Dahan, D., Swingley, D., Tanenhaus, M., & Magnuson, J. S. (2000). Linguistic gender and spoken word recognition in French. *Journal of Memory and Language, 42*, 465–480.

Desimone, R., & Duncan, J. (1995). Neural mechanisms of selective visual attention. *Annual Reviews of Neuroscience, 18*, 193–222.

Fauconnier, G. (1994). *Mental spaces*. Cambridge: Cambridge University Press.

Friederichi, A. D., & Jacobsen, T. (1999). Processing grammatical gender during language comprehension. *Journal of Psycholinguistic Research, 28*, 467–484.

Griffin, Z. M. (2001). Gaze durations during speech reflect word selection and phonological encoding. *Cognition, 82*, B1–B14.

Grosjean, F., Dommergues, J., Cornu, E., Guillelmon, D., & Besson, C. (1994). The gender-marking effect in spoken word recognition. *Perception & Psychophysics, 56*, 590–598.

Henderson, J. M., & Ferreira, F. (2004). *The interface of language, vision, and action: Eye movements and the visual world*. New York: Psychology Press.

Huettig, F., & Altmann, G. T. M. (2007). Visual-shape competition during language mediated attention is based on lexical input and not modulated by contextual appropriateness. *Visual Cognition, 15*, 985–1018.

Huettig, F., Chen, J., Bowerman, M., & Majid, A. (submitted). Linguistic relativity: Evidence from Mandarin speaker's eye movements.

Huettig, F., Gaskell, M. G. & Quinlan, P. T. (2004). How speech processing affects our attention to visually similar objects. *Proceedings of the Twenty-Sixth Annual Conference of the Cognitive Science Society*. Mahwah, NJ: Lawrence Erlbaum Associates.

Huettig, F., Quinlan, P. T., McDonald, S. A., & Altmann, G. T. M. (2006). Models of high dimensional semantic space predicts language–mediated eye movements in the visual world. *Acta Psychologia, 121*, 65–80.

Indefrey, P., & Levelt, W. J. M. (2004). The spatial and temporal signatures of word production components. *Cognition, 92*, 101–144.

Itti, L., & Koch, C. (2001). Computational modeling of visual attention. *Nature Review Neuroscience, 2*, 194–203.

Jackendoff, R. (2002). *Foundations of language: Brain, meaning, grammar, evolution*. New York: Oxford University Press.

Jackendoff, R. (2007). A parallel architecture perspective on language processing. *Brain Research, 1146*, 2–22.

Levelt, W. J. M., Roelofs, A., & Meyer, A. (1999). A theory of lexical access in speech production. *Behavioral and Brain Sciences, 22*, 1–38.

Levinson, S. (2003). *Space in language and cognition*. Cambridge: Cambridge University Press.

Liversedge, S., & Findlay, J. M. (2000). Saccadic eye movements and cognition. *Trends in Cognitive Science, 4*, 6–14.

Marslen-Wilson, W., & Welsh, A. (1978). Processing interactions and lexical access during word-recognition in continuous speech. *Cognitive Psychology, 10*, 29–63.

Martin, E., Shao, K. C., & Boff, K. R. (1993). Saccadic overhead: Information processing time with and without saccades. *Perception & Psychophysics, 53*, 372–380.

McQueen, J., Norris, D., & Cutler, A. (1994). Competition in spoken word recognition: Spotting words in other words. *Journal of Experimental Psychology: Learning, Memory, and Cognition, 20*, 621–638.

Meyer, A. S., Roelofs, A., & Levelt, W. J. M. (2003). Word length effects in object naming: The role of response criterion. *Journal of Memory and Language, 48*, 131–147.

Mishra, R. K., & Marmolejo-Ramos, F. (submitted). How does language comprehension affect shifts in visual attention in complex scene viewing?

Nakahara, K., Adachi, Y., Osada, T., & Miyashita, Y. (2007). Exploring the neural basis of cognition: Multi-modal links between human fMRI and macaque neurophysiology. *Trends in Cognitive Sciences, 11*, 84–92.

Petersen, S. E., Robinson, D. L., & Keys, W. (1985). Pulvinar nuclei of the behaving rhesus monkey: Visual responses and their modulation. *Journal of Neurophysiology, 54*, 867–886.

Rayner, K. (1998). Eye movements in reading and information processing: 20 years of research. *Psychological Bulletin*, *124*(3), 372–422.

Richardson, D., & Matlock, T. (2007). The integration of figurative language and static depictions: An eye movement study of fictive motion. *Cognition*, *102*, 129–138.

Roelofs, A. (2006). Context effects of pictures and words in naming objects, reading words, generating simple phrases. *Quarterly Journal of Experimental Psychology*, *59*, 1764–1784.

Roelofs, A. (2008). Attention to spoken word planning: Chronometric and neuroimaging evidence. *Language and Linguistic Compass*, *2/3*, 389–405.

Schall, J. D. (1995). Neural basis of saccade target selection. *Reviews in Neuroscience*, *6*, 63–85.

Schall, J. D. (2001). Neural basis of deciding, choosing and acting. *Nature Reviews Neuroscience*, *2*, 33–42.

Schall, J. D., & Hanes, D. P. (1993). Neural basis of saccade target selection in frontal eye field during visual search. *Nature*, *366*, 467–469.

Sekerina, I. (2003). Grammatical gender and mapping of referential expressions in Russian. Talk presented at the *9th Annual Conference on Architectures and Mechanisms for Language Processing*, Glasgow, Scotland.

Talmy, L. (1978). Figure and ground in complex sentences. In J. H. Greenberg (Ed.), *Universals of human language, vol. 4: Syntax* (pp. 625–649). Stanford, CA: Stanford University Press.

Talmy, L. (2000). *Toward a cognitive semantics*. Cambridge: MIT Press.

Trueswell, J. C., & Tanenhaus, M. K. (Eds.). (2005). *Processing world-situated language: Bridging the language-as-action and language-as-product traditions*. Cambridge, MA: MIT Press.

Wurtz, R. H., & Goldberg, M. E. (1972). Activity of superior colliculus in behaving monkey. 3. Cells discharging before eye movements. *Journal of Neurophysiology*, *35*, 575–586.

N. Srinivasan (Ed.)
Progress in Brain Research, Vol. 176
ISSN 0079-6123

CHAPTER 19

Interactions of attention, emotion and motivation

Jane Raymond*

School of Psychology, Bangor University, Bangor, Gwynedd, UK

Abstract: Although successful visually guided action begins with sensory processes and ends with motor control, the intervening processes related to the appropriate selection of information for processing are especially critical because of the brain's limited capacity to handle information. Three important mechanisms — attention, emotion and motivation — contribute to the prioritization and selection of information. In this chapter, the interplay between these systems is discussed with emphasis placed on interactions between attention (or immediate task relevance of stimuli) and emotion (or affective evaluation of stimuli), and between attention and motivation (or the predicted value of stimuli). Although numerous studies have shown that emotional stimuli modulate mechanisms of selective attention in humans, little work has been directed at exploring whether such interactions can be reciprocal, that is, whether attention can influence emotional response. Recent work on this question (showing that distracting information is typically devalued upon later encounters) is reviewed in the first half of the chapter. In the second half, some recent experiments exploring how prior value-prediction learning (i.e., learning to associate potential outcomes, good or bad, with specific stimuli) plays a role in visual selection and conscious perception. The results indicate that some aspects of motivation act on selection independently of traditionally defined attention and other aspects interact with it.

Keywords: attention; emotion; affective evaluation; faces; distractor devaluation; motivation; attentional blink

The visual environment consists of a changing array of objects, each inviting a potential interaction and promising an outcome. The visual mechanisms of the brain, from low-level sensory systems to higher level visual representation mechanisms, operate so that we can control and guide our actions towards and away from potent objects in our physical world. This enables each of us to make hundreds of rapid, rewarding choices everyday, such as what seat to take on the bus, which piece of food to eat first at a meal, or who to talk to at a party. Such choices depend on visually guided actions that make use of high-level, limited capacity mechanisms, including consciousness and working memory (WM), and are informed by prior learning, emotional states, and current and long-range motivational agendas.

Consider, for example, your possible actions in a summer garden with fruiting raspberry vines. Upon spying some berries amongst the leaves, you might reach out to eat one; but prior to doing so, you would rapidly judge whether it was ripe (good to eat) or rotten (bad to eat) and this might be accompanied by a conscious evaluative feeling of attraction or disgust for the berry. Berries, like

*Corresponding author.
Tel.: +44(0)1248 383787; Fax: +44(0)1248 382599;
E-mail: j.raymond@bangor.ac.uk

DOI: 10.1016/S0079-6123(09)17617-3

other objects in the garden (e.g., a rake left across the garden path, a spider dangling from an overhead branch, a scented rose or a stinging nettle), each afford an action that also affords an affective evaluation and a prediction of the probable outcome, either positive or negative, that would result should a particular action be performed. Each offers an outcome that might meet a current or future motivational state. In other words, for each object in a scene, visual selection processes are likely to be accompanied by both affective evaluation and value prediction in light of current and future goals. This raises the interesting question of how the processes that mediate visual selection interact with the processes that yield the conscious experiences leading to affective (or emotional) evaluation and the not necessarily conscious processes that predict value in terms of short- and long-term motivational agendas.

In this chapter, I focus on the possible interactions among selective visual attention, emotional evaluation and the motivational mechanisms that make use of value prediction. I begin with a brief theoretical outline and a discussion of the potential neural mechanisms that might underpin such interactions. This is followed by a brief discussion of research on how emotional stimuli affect attention. Then, I summarize a recent body of research conducted by my colleagues and I that asks how attentional processes affect emotional evaluation. In the latter half of the chapter, I describe some recent studies that explore the question of how motivational value interacts with attention.

Theoretical outline

A possible theoretical outline of the interplay between attention, emotion and motivation is sketched out in Fig. 1. The black lines indicate the main route from visual sensation to visual awareness and then to action and long-term memory. The solid grey lines indicate the main routes of motivation, that is, a goal setting mechanism, and emotion. The dashed lines represent the more 'automatic' links between sensory input and

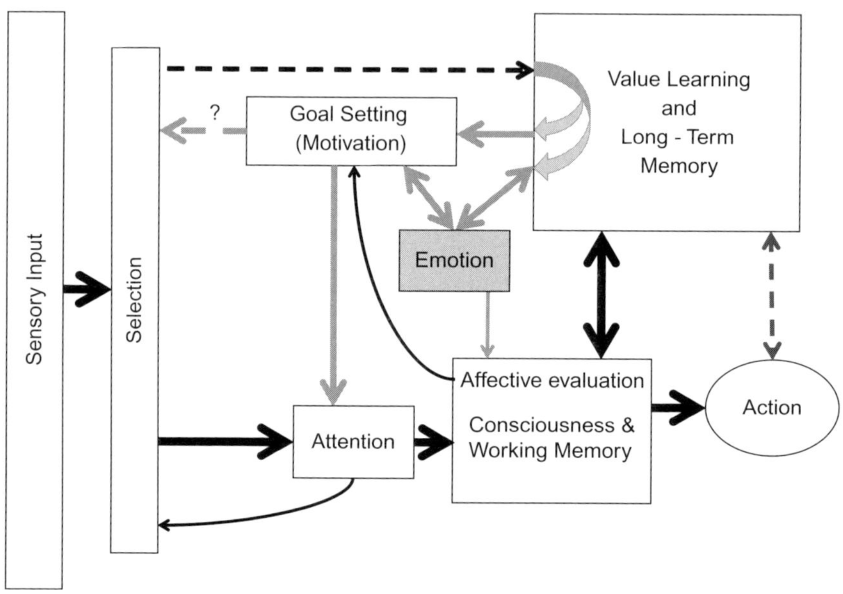

Fig. 1. An outline model of interconnections between attention, motivation and emotion. Conscious perception of sensory information (solid black lines) is achieved by filtering input by attention before making it made available to high-level processes such as working memory and consciousness. Automatic processes linking sensory input and action are shown by the dashed lines. Goal setting mechanisms (motivation), with the aid of emotion (grey lines), determine the attentional filter and may act directly on low-level selection. Learning and long-term memory allow outcome predictions to inform goal setting.

action via well-learned associations. These links also allow relatively automatic access of sensory data to emotional and goal setting mechanisms. (The figure is not meant to indicate all possible connections but rather to highlight those that are especially relevant to the current chapter.) It can be seen in the figure that to prioritize what visual information from the sensory array should gain access to the limited capacity mechanisms of consciousness and WM, a carefully controlled visual selection process that is specifically sensitive to the current goal is needed. The system that performs this type of selection is often called *attention* and can be defined as a set of neural mechanisms that facilitate perceptual processing of stimuli that are task relevant over those that are not and inhibit processing of stimuli that potentially interfere with action needed to achieve an immediate goal. It can be viewed as the main gatekeeper for visual awareness, non-automatic visually controlled action and visual WM.

Another system, sometimes called *motivation*, or goal setting, specifies the current goal (or sequence of goals needed to perform complex actions) and directs attention. Motivation also prioritizes a range of concurrent short- and long-term goals (and thus may defer action for one goal while initiating action for another). This high-level, central, executive system sets the agenda for achieving goals by monitoring internal emotional and biological states as well as external perceptually assessed conditions. It then predicts outcomes of possible actions (if taken now, as opposed to later) in terms of these goals. This mechanism thus relies heavily on prior learning of the likely utility or value (reward or punishment) of an action in response to specific stimuli and is intimately linked to the brain's emotional system (a part of a set of mechanisms that monitors and adjusts external and internal states). An open theoretical question is whether motivation can act directly on visual selection or only do so via attention. The emotional mechanisms of the brain provide direct input to WM and consciousness (in the form of affective evaluations and feelings) as well as informing goal setting agendas.

From this hypothetical picture it is clear that an important question for the study of visual information processing is how attentional selection mechanisms are affected by perceptual stimuli that activate motivation to achieve *other* goals that may or may not be congruent with the current goal. 'Other' goals may be currently deferred but otherwise active in the longer term, or they may be goals that were just previously relevant but are no longer needed. Stimuli that activate such goals are typically called distractors in attention research but in the natural world we tend to think of them as non-neutral objects with emotional (e.g., an angry face) or learned value (e.g., a 'lucky' number) or both. Although motivation sets up the attentional state in the organism so that that the goal designated as foremost and immediate can prioritize current perceptual processing (via attention), the goal setting mechanisms of motivation can allow a sudden change (task switch) in the goal in the face of change in the perceptual or internal (e.g., emotional) environment and can allow latent learning (i.e., processing information about stimuli that are currently irrelevant but nevertheless may be relevant to other, future goals).

Neural mechanisms

At this point it is useful to ask whether there is any evidence that the neural machinery thought to mediate attention, emotion (or, more specifically, affective evaluation) and value prediction could interact. Indeed, there are at least three primary candidate structures that appear to be involved in some combination of these functions and could therefore play a role in their interaction. The first is the *anterior cingulate*, a structure known to be involved in selective attention, emotional evaluation and error monitoring (a critical step in value learning) (Bush et al., 2000; Yamasaki et al., 2002; Kawabata and Zeki, 2004). This structure with its numerous connections to visual processing regions of the brain is ideally suited to coordinate affective, motivational and attentional processes. A second structure that probably plays a pivotal important role in integrating attention, affective evaluation and motivation is the *orbitofrontal cortex (OFC)*. Numerous studies have shown that this structure plays a vital role in affective evaluation (Aharon et al., 2001; O'Doherty et al., 2003; Kawabata and Zeki, 2004) and value learning (e.g., Knutson et al., 2001; Gottfried et al., 2003). It sends and receives input from the primary visual areas of the brain (Rolls, 2000), thus supplying an infrastructure for the reciprocal modulation between visual sensory, evaluative and motivation systems. A third critical area is the *amygdala*. This limbic system structure sends a large efferent pathway to primary visual cortex (Freese and Amaral, 2005) and receives signals from numerous brain areas including OFC. It receives visual input via the secondary visual pathway (i.e., via the superior colliculus and pulvinar, an important attentional structure). Importantly, responses to emotional stimuli by this complex structure are modulated by attention (Vuilleumier et al., 2001; Pessoa et al., 2002; Silvert et al., 2007) and by the extent to which rewards or punishment outcomes differ from those expected (Paton et al., 2006). Taken together, these findings suggest that the amygdala plays a role in linking attention, emotion and motivation.

Although very brief, this overview of the important neural structures involved in linking attention, emotional evaluation and value prediction is sufficient to make clear that the neural infrastructure to support extensive communications among these large neural systems of the brain probably exists. It is then up to behavioural and functional studies to determine how these interactions might work and to quantify how each process influences the others.

The effect of emotional stimuli on attention

The most widely studied connection between attention and emotion concerns the effect of emotionally valenced images on selective visual attention. This is a large, burgeoning area of research and the current chapter does not offer sufficient scope to review this literature in any detail. However, to summarize, numerous empirical studies indicate that emotional or arousing images compared to neutral images, when either task relevant (targets) or irrelevant (distractors), alter performance on simple tasks (such as response time to detect or discriminate pre-specified targets), thus indicating that emotional content of stimuli influences mechanisms of selective attention. These studies have used many of the traditional paradigms for studying spatial attention including spatial cueing effects (Fox et al., 2001): dot probe (Armony and Dolan, 2002), Eriksen flanker tasks (Fenske and Eastwood, 2003), inhibition of return (Rutherford and Raymond, in press) and spatial visual search (e.g., Eastwood et al., 2001; Fox et al., 2001, 2002; Öhman et al., 2001). In general, these studies report that emotional stimuli (e.g., angry faces) attract and hold attention more than neutral or novel stimuli; but note that some studies have failed to find such effects (e.g., Lipp et al., 2004). Other studies have explored the effects of emotional stimuli on temporal attention using the attentional blink (AB) paradigm. (This involves presenting two targets within in a rapid serial visual presentation, RSVP, of filler stimuli. The basic finding is that if the two targets are presented within about a half second of each other, perception of the second target is impaired. This effect is called the attentional blink and is described in more detail later; Raymond et al., 1992.) A general finding has been that if the second of the two targets presented in the AB has emotional or arousing content, then it will be less likely to go undetected during the critical AB interval (Anderson, 2005). However, Fox et al. (2005) showed that this was true for only high anxious individuals, and that emotionally expressive faces do not 'survive' the AB any more effectively than non-emotional faces for normal participants. If the emotional stimulus is presented as the first target, it only affects subsequent (neutral) target perception (making the AB effect larger), if the emotional content is task relevant (Huang et al., 2008). Other temporal attention studies have presented emotional images as distractors in single target RSVP streams (Most et al., 2005). As in AB studies, they report that emotional or arousing stimuli erroneously capture attention, leading to poorer than expected target performance if the critical emotional distractor is presented too close in advance of the target.

The challenge for studies exploring the effect of emotional stimulus content on attention lies in their stimuli. Always, one must ask the question — are the effects on performance really due to the emotional content in the stimuli or can other non-emotional characteristics (e.g., teeth showing in face images, word frequency for word stimuli or average luminance in scene stimuli) account for the outcomes? Indeed, such low-level stimulus issues have probably contributed significantly to conflicting findings in this literature. An additional problem is the assumption that the photographs or words are emotional to the participants in the study. Typically, stimuli are affectively rated by different participants than those in the study so that the affective impact of each image for each participant is not known. Later in the chapter I return to this issue by raising the suggestion that preconditioning or learning of stimuli might be a better strategy when addressing research questions about the impact of emotional stimuli on attention and perception. This has already been adopted in some studies, especially those investigating fear responses (Armony and Dolan, 2002; Smith et al., 2006).

Attentional effects on affective evaluation

Although it makes sense that an emotionally charged stimulus (e.g., a spider) should draw attention or gain priority in processing, it also makes sense that this interaction might work in reverse, that is, attending or ignoring stimuli might modulate emotional responses to them later. Indeed, the effects of emotional stimuli on attention tell us that prior experience (perhaps combined with built-in predispositions, in some cases) plays a role in guiding selection. So, perhaps the simple experience of attending to one thing and ignoring another could lead to the genesis or development of an affective response; the affective response could then eventually be used to prioritize selection on subsequent encounters. The possibility that attentional states could play a role in the genesis of affective evaluation was explored in a recent series of studies from my laboratory (see Fenske and Raymond, 2006, for a review). The general strategy used in all of these experiments was to give participants an initial exposure to affectively neutral stimuli under conditions that controlled their attentional state. Then, a short time later, these and other novel stimuli were re-presented and explicit affective evaluations of each were obtained. The question was would the attentional state in place at the time of initial exposure determine the subsequent affective evaluation.

In the first of these studies, Raymond et al. (2003) presented participants with two different abstract 'Mondrian' patterns (200 ms) on either side of a fixation cross and asked them to locate one and ignore the other (see Fig. 1(A)). About 1500 ms later (or 1000 ms after response), observers rated a previously attended, previously ignored, or a novel pattern on a positive (e.g., how cheerful?) or negative (e.g., how dreary?) emotional dimension. By using both positively and negatively valenced response scales, emotional tone rather than simple response bias could be assessed. The main finding illustrated in Fig. 1(B) was that just previously ignored stimuli (prior distractors), compared to just previously attended stimuli (prior targets), were evaluated as more affectively negative (regardless of the response scale used). Moreover, prior distractors were rated as more negative than similar images that had never been seen before. Evaluations given to items just seen previously as targets and to novel items did not differ. This pattern of results shows that distractors were devalued as a consequence of ignoring. Just previously attended items (targets) were not up-valued. With this finding, we reported the first demonstration that the attentional state active when a novel, emotionally neutral stimulus is viewed determines affective response for the image a short time later. This *distractor devaluation effect* makes the fundamental and important point that attention can modulate affective response. This basic finding has since been reported by other laboratories (Veling et al., 2007; Griffiths and Mitchell, 2008).

To account for their findings, Raymond et al. (2003) first considered and then rejected accounts used to explain the mere exposure effect (Zajonc, 2001). Mere exposure effect is the increase in affective appraisal of stimuli that have been passively viewed several times previously. One of the most widely accepted explanations for this effect is called perceptual fluency (Reber et al., 1998). According to this theory, repeated exposure to stimuli makes them easier to perceptual process and this ease of processing, or fluency, is misinterpreted as liking. However, such ideas cannot explain how prior exposure to a stimulus could make it less fluent to process than a novel stimulus. Instead of fluency theory, we proposed a devaluation-by-inhibition account of the influence of attention on emotion. We proposed that when an inappropriate stimulus (distractor) competes for responding, attentional inhibition is applied and encoded with the distractor's representation. When the previously ignored stimulus is again encountered, this inhibition is re-instantiated and, when applied to the current evaluative task, leads to emotional devaluation.

This interesting effect and our working hypothesis to explain it raised several questions. First, would the effect be evident with other, more representational stimuli, such as faces? Second, was the inhibition applied to the ignored object or to the to-be-ignored feature that had to be ignored? Third, if the effect was based on associating inhibition with the representation of the distractor, would it depend on the availability of visual WM, assuming association requires WM resources? However, before addressing these questions it was first necessary to more fully establish and test the notion that devaluation occurred via attentional inhibition.

This point was established in three main ways. First, using that basic strategy of Raymond et al. (2003), my colleagues and I exploited other visual attention paradigms that posit the application of attentional inhibition. If items in these specific situations are subjected to inhibition and inhibition leads to devaluation, then such items should later be devalued compared to items not subjected to inhibition. We conducted two studies using spatial visual search in this endeavour. The first exploited a phenomenon called the preview effect, or visual marking (Watson and Humphreys, 1997). In this task the participant is required to find a target defined by the conjunction of two different features (e.g., red and square) as quickly as possible. Distractors can contain one of these features but not both (e.g., they could be red circles or green squares). On some trials, the target and all the distractors are presented simultaneously (*non-preview* trials) and on other, *preview* trials, half the distractors of one type (e.g., green squares) are presented for one full second prior to the presentation of the rest of the display (all the red circles and the target). On non-preview trials, search is slow and heavily dependent on the number of distractors in the total array (i.e., the slope of the function relating search time and array set size is quite steep). On preview trials, however, search times are much faster and search functions are flatter. This difference in response time is called the preview effect and is thought to occur because the previewed items are 'marked' with inhibition, so that when the rest of the array is presented, only the new subset of distractors need be searched through, making the task an easy single-feature search task instead of a difficult conjunction search task. Fenske et al. (2004) reasoned that if this was the case, then previewed distractors, having been subjected to attentional inhibition, should be affectively devalued relative to non-previewed (and non-inhibited) distractors. This paradigm also allows a direct test of the fluency theory because previewed distractors are seen for one full second longer than non-previewed items and, according to this theory, should be evaluated more positively than non-previewed items. Contrary to the predictions of fluency theory and consistent with devaluation-by-inhibition theory, Fenske et al. found that items presented as preview distractors on preview trials were later evaluated more negatively than similar items seen as distractors on non-preview trials. This result thus strongly indicates that task irrelevance, not simple fluency, determines subsequent affective evaluation of stimuli.

Using a similar line of reasoning, Raymond et al. (2005) examined affective evaluations given to abstract Mondrian patterns seen as distractors in a conjunction visual search task. Previous studies showed that facilitated processing of the target in a visual search task is accompanied by a ring of inhibition that is strongest near the target and reduces as the distance from the target increases (Mounts, 2000; Slotnick et al., 2002). According to the devaluation-by-inhibition theory, distractors seen near the target should be evaluated more negatively than distractors presented far from the target. Indeed, this was exactly what Raymond et al. found, a result that again supports the devaluation-by-inhibition theory.

To search for physiological evidence for devaluation-by-inhibition theory, I turned to electrophysiological measures related to attention. Specifically, Kiss et al. (2007) used an experimental procedure that mimicked the simple search-then-evaluate paradigm of Raymond et al. (2003), this time replacing Mondrian patterns with tinted greyscale face images. The basic task is shown in Fig. 2(A). A search display containing two faces,

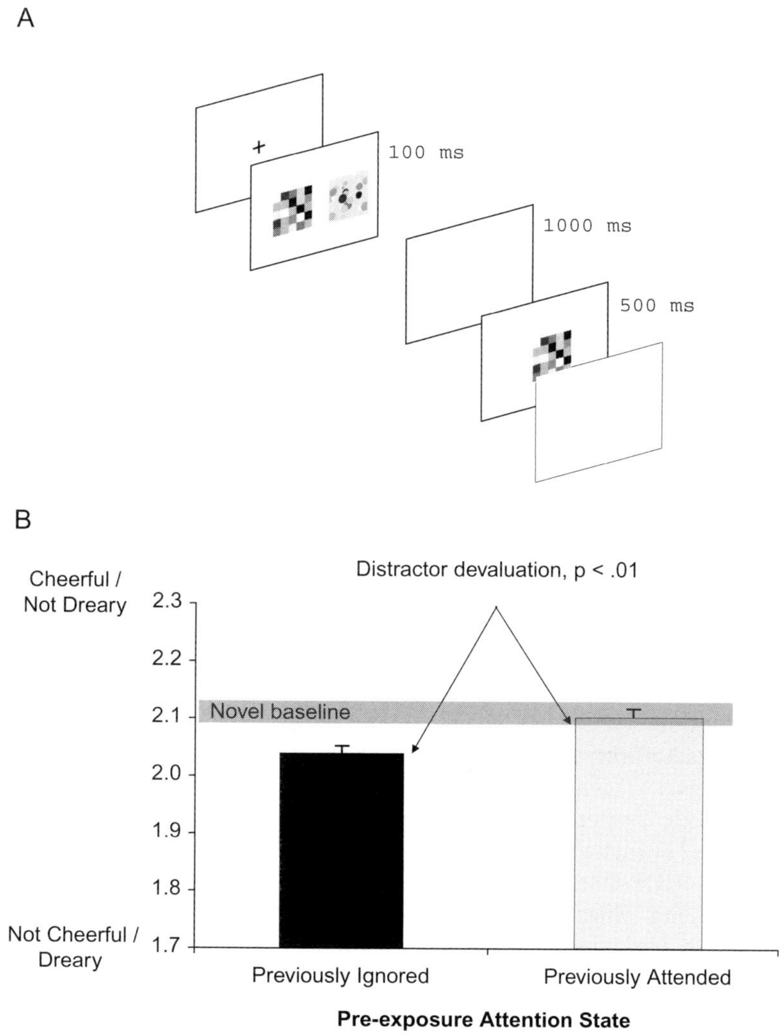

Fig. 2. (A) A schematic of the essential elements of a trial in the initial distractor devaluation experiment of Raymond et al. (2003). A pair of abstract patterns appeared; the task was to locate (left/right) the target item as pre-defined by its textural elements (e.g., squares) as quickly as possible. After responding, a 1000-ms blank interval occurred and then a single abstract pattern was presented. Participants rated the pattern on a three-point 'cheerfulness' or 'dreariness' scale. (B) The average rating of the to-be-evaluated image (recoded for emotional tone) plotted for conditions where it had been a previous target (black bar) or a previous distractor (grey bar).

one male and one female, one tinted transparently with blue and the other yellow, was presented for 200 ms. The task was to find the male (or female in different blocks) and report its tint colour (yellow or blue) by pressing the appropriate computer key. A few seconds later, one of these faces was presented (without a tint) for evaluation of trustworthiness. The behavioural result mimicked that of the earlier study (Raymond et al., 2003). Faces seen previously as distractors were rated as less trustworthy than faces seen previously as targets. Thus this study showed that the distractor devaluation effect could be obtained with concrete images and was not limited to abstract images. In the Kiss et al. study, we recorded the EEG during the attention search task with the intention of calculating an event-related potential (ERP) time-locked to the onset of the search display. Specifically, we were interested in measuring the N2pc difference waveform. This is derived by subtracting activity (specifically a negative potential occurring about 200 ms post stimulus) over posterior sites ipsilateral to the target's visual field from those at the corresponding contralateral site. The magnitude and latency of this difference is highly correlated with selective attention (Eimer, 1996) in extrastriate visual cortex triggered by re-entrant feedback signals from higher order attentional control regions in posterior parietal cortex (Woodman and Luck, 1999). We reasoned that if the distractor devaluation effect was truly attentional in nature and if the effect was related to ignoring rather than attending, then the N2pc measured during the attention search task should predict the subsequent evaluation of distractors but not targets. To test this we binned the EEG data according to the rating response given on each trial for each trial type (evaluate prior targets; evaluate prior distractors). ERPs were computed separately for trials yielding low versus high trustworthiness evaluations, for target trials and distractor trials. We found that for target trials, the N2pc did not differ for high-rated versus low-rated faces. However, the N2pc for distractor trials was significantly larger and occurred earlier for trials resulting in low versus high trustworthiness evaluations. This finding provides important evidence that distractor devaluation is linked to attentional states active during exposure to stimuli and, moreover, that the effect is specifically inhibitory, acting on distractors.

A third line of evidence to support the idea that the distractor devaluation effect is inhibitory in nature comes from a series of studies using response inhibition paradigms. In a study by Fenske et al. (2005), participants were shown a series of face pairs for 1000 ms each. On different trials, a large red or green translucent circle (cue) was then superimposed over the left or right face. Participants were instructed to depress a response key as quickly as possible when the cue was green and to inhibit responding when it was red. A few minutes later the same face pair was seen again (without a cue), this time preceded by a question asking the participant to choose the more trustworthy or less trustworthy of the two faces; or to choose the face on the lighter or darker background. The important finding was that faces previously associated with 'no-go' cues were chosen as the less trustworthy face more frequently and as the more trustworthy face less frequently than their uncued mates. A similar but much smaller bias against cued faces from no-go trials was seen when perceptual judgements were required. Importantly, unlike the affective questions, this bias did not depend on the valence (bright/darker) of the questions. This study made the important point that inhibition needed to stop a pre-potent action can become associated with stimuli in such a way that specifically affective (not perceptual) evaluations are modulated.

Using a somewhat different go/no-go paradigm, Kiss et al. (2008) presented observers with a sequence of faces of two different races (Asian and Caucasian). In different blocks of trials, participants were instructed to 'go' (i.e., press a response key) to Asian faces and withhold responding to Caucasian faces. In other blocks, the reverse instruction was given. During this task, ERPs were recorded time-locked to the presentation of each face so that the N2 over frontal sites could be used to index the level of inhibition applied on no-go trials (for a review see Folstein and Van Petten, 2008). After each go/no-go block of trials, the same sequence of faces was presented but this time the participant was asked to rate each face for trustworthiness. We found that faces presented as no-go stimuli were rated as significantly less trustworthy than faces presented as go stimuli, once again demonstrating that response inhibition associated with a particular stimulus leads to affective devaluation at a later time. As in the previous ERP study, ERPs were averaged separately for trials on which no-go faces were subsequently rated as low versus high in trustworthiness. A larger N2 component at frontal sites was found for faces that were later rated low versus high, showing directly that attentional inhibition causes subsequent affective devaluation.

An important question for the understanding of distractor devaluation is whether this effect is caused by inhibition being associated with the ignored object or with the feature needing to be ignored during the attentional task. To address this question, we conducted a behavioural study using the procedure illustrated in Fig. 2 (and identical to that used by Kiss et al., 2007). In one experiment participants were required to find a face based on its gender (e.g., select male) and report its tint colour. In a second experiment, using identical stimuli, participants were required to find a face based on its tint colour and report its gender. In both cases, the to-be-evaluated face was seen in greyscale. If distractor devaluation was object-based, then we would expect to find devaluation in both experiments because the to-be-evaluated object was present in the search array in both cases. However, if the effect was feature-based, then we would expect to see distractor devaluation in the first experiment but not in the second because in first experiment the to-be-ignored feature, that is, gender, was also present in the to-be-evaluated face. In contrast, in the second experiment the to-be-ignored feature (colour) was absent in a greyscale face during evaluation. If distractor devaluation is feature-based, then no devaluation of distractors should be evident in that experiment. Indeed, this was the pattern of results found. Robust distractor devaluation was found in the first experiment but the effect was absent in the second. To further test the conclusion that distractor devaluation is feature-based, we ran a third experiment in which items in the search array were images of houses or buildings each tinted either blue or yellow. The participant's task was to select an item based on colour and report the type of image (house or building). A few seconds later, a novel face was presented tinted either in the colour of the prior target, in the colour of the prior distractor or in a different colour entirely. Robust devaluation of faces presented in the distractor colour was found, relative to faces presented in the target colour or in a different colour. This result confirms that distractor devaluation is feature-based, that is, inhibition is applied to task-irrelevant features (not necessarily object) that directly compete to control responding, and that this inhibition persists and can be applied to other subsequently presented objects if they possess that feature.

A key element of the devaluation-by-inhibition theory is that a neural representation of the to-be-ignored feature becomes associated with inhibition during the attentional task and that this association persists in memory so as to affect subsequent evaluations. This reliance on association and memory suggests that resources in visual WM must be engaged if distractor devaluation effects are to be seen. This predicts that given the limited capacity of visual WM (Luck and Vogel, 1997; Cowan, 2001), distractor devaluation effects should be absent if visual WM resources are fully occupied with another task. To test this idea, my colleagues and I (Goolsby et al., 2009) took the basic two-stage task depicted in Fig. 2 and sandwiched it in between the encoding and the test phases of a standard visual WM task (Luck and Vogel, 1997). The sequence of events was as follows. First, a WM study array of zero, one or two faces (or houses) was presented for 2000 ms; then, the bilateral two-face visual search array was presented, as in Fig. 2, followed by the presentation of a greyscale face (either prior target or prior distractor) for evaluation. After this, the same or a different WM array was presented and the participant's task was to say whether it matched that seen previously during encoding. Throughout the trial, a verbal suppression task was used to suppress recruitment of verbal WM. As expected from previous studies (Jackson and Raymond, 2008), performance on the visual WM task declined significantly when the study array was increased from zero to two items. What was of interest to us in this experiment was whether the magnitude of the distractor devaluation effect varied as a function of visual WM load. We found that whereas WM load had no effect on performance on the visual search task, it had a significant effect on distractor devaluation (see Fig. 2(B)). The distractor devaluation effect was large without a WM load, replicating previous experiments; was present but smaller with a WM load of one item; and was entirely absent when the WM load was increased to two items. These data provide strong evidence that associative mechanisms underlie the distractor devaluation effect. Without any WM resources available, an association between the to-be-ignored feature and the attentional state (inhibition) could not be made and could not persist to modulate subsequent affective appraisal.

Other findings obtained thus far on the distractor devaluation effect can be summarized into four main points. First, there is considerable evidence that the effect acts on distractor, not target information (Raymond et al., 2003; Kiss et al., 2007, 2008; Goolsby et al., 2008). Second, the effect is linked to attentional inhibition, whether stimulus-based (Fenske et al., 2004; Raymond et al., 2005; Kiss et al., 2007) or response-based in nature (Fenske et al., 2005; Kiss et al., 2008). Third, the effect acts on recently ignored features (not necessarily objects). Fourth, it requires visual WM resources and therefore must be associative in nature.

Now I address the question of why inhibition might become associated in memory with features of stimuli that compete with targets for control over selection. During a selection task, inhibition of distractor information reduces the likelihood that it will control current action. Perhaps, associating this inhibitory control state with critical distracting features (or objects) leads to their subsequent devaluation so as to aid subsequent selection prioritization. If affectively devalued information was less likely to attract attention on future encounters, then this form of implicit memory could assist selection of relevant information at a later time. Thus attention and affective evaluation mechanisms may be seen as reciprocal in their interactions.

This framework raises an important question — what evidence do we have that the affective value of information in any way determines the capture of attention? This very question suggests another way of looking at the mechanisms that could underlie distractor devaluation. As I mentioned above, during any selection task, distractor information represents a potential error. Making errors provides implicit punishment whereas making correct responses provides implicit reward. Suppose that such punishments and rewards could serve as a currency for simple value learning in the selection tasks described above. If so, then distractor devaluation could be seen as a value-learning effect. This then offers a more tractable approach for addressing the question of how or whether acquired value of stimuli contributes to visual selection.

Value learning and attention

Recently, numerous single unit studies of animals, lesions studies of humans and functional neuroimaging studies of healthy humans have begun to detail how the brain codes and stores information about the value of visual stimuli, acquired through association with rewards and punishers (O'Doherty, 2004). Value learning involves matching the predicted value of interacting with a stimulus before an action is initiated with the actual outcome of the action. Any differences constitute a prediction error signal that can then be used by the brain to update value predictions cumulatively with experience. Understanding the neurobiology of value learning requires distinguishing between the various neural responses that could code value prediction, prediction error, sensory events related to the stimuli predicting rewards or punishers, action events or intentions, and the sensory signals related to the rewards and punishers themselves (Schultz, 2000). Effective separation of these components experimentally is often difficult and has created controversy in this field regarding the specific roles of different brain areas. However, evidence from both humans and animals suggests that value prediction is coded using a dopaminergic circuit involving the OFC (Knutson et al., 2001; O'Doherty et al., 2002; O'Doherty, 2004) and the ventral striatum, including the nucleus accumbens (NAcc). In addition to these areas, the amygdala also contributes to coding and updating value prediction and sends efferents to both the NAcc and the OFC (Gottfried et al., 2003; Paton et al., 2006), and as described previously has extensive connections with visual cortex (Freese and Amaral, 2005). This suggests that the value-coding system could have important modulatory effects on visual cognitive processes, such as selective attention.

Although most studies have investigated the neural system coding predicted value generated by reward, some also address how predicted value generated by punishment or loss is coded. A picture of two somewhat separate systems seems to be emerging. For example, a recent study showed that amygdala appears to use distinct populations to link visual object representations with gains versus losses in a plastic, updatable way (Paton et al., 2006). Yacubian et al. (2006) contrasted predictive brain responses in humans for monetary gain versus loss and found gain-specific modulation of the ventral striatum and right OFC, and loss-specific modulation of the amygdala. If distinct neural networks subserve value coding associated with gains versus losses, then this suggests that positively versus negatively valenced value-prediction codes might differently modulate visual cognitive processes.

Little is currently known about how value-prediction codes might contribute to visual selection (conjointly with or independently of attention) because most studies of selective attention (both animal and human) have fully confounded immediate task relevance with availability of reward (Maunsell, 2004). Nevertheless, indirect evidence supports the possibility that motivation (perhaps in the form of value-prediction codes) contributes to visual selection. For example, studies of addicts indicate that words and images related to the subject of their addiction are more likely to intrude on conscious experience than unrelated items (Hogarth et al., 2003). Similarly, as described earlier in the chapter, considerable evidence shows that emotional content, perhaps itself rewarding or punishing, also enhances visual selection. Although these effects are consistent with the notion that motivation codes may contribute to visual selection, they do not provide any indication that motivation's role in selection is distinct from attention; rather, they support the more parsimonious conclusion that motivation simply enhances selective attention (Della Libera and Chelazzi, 2006).

To study the role of value learning and selective attention my colleague and I used the following strategy (Raymond and O'Brien, 2009). First, we established stable positive (motivating) and negative (aversive) predicted value codes for different stimuli through a simple value-learning procedure (Pessiglione et al., 2006) that used modest monetary wins and losses. This technique also afforded measurement of value prediction for each stimulus for each participant via their choice behaviour in the task. Then, we measured visual selection (recognition) for these stimuli in a subsequent, different task in which all stimuli were *equally* task relevant (thus holding selective attention constant, so as to reveal the effects of motivation). Critically, in the visual task, attention was manipulated across conditions so that the effect of attention on selection of stimuli with the *same* previously established predicted value codes could be observed. Such a two-phase experiment (value learning then visual task) allowed the independent effects of motivation and attention on selection, as well as their interactions to be observed.

The value-learning phase of the experiment involved a simple choice game with modest monetary outcomes. On each trial, participants viewed a pair of faces and chose one. Depending on the face chosen, they could win or lose a small amount of money or have no change to their current total winnings. A total of 12 faces were viewed, always paired with the same mate and assigned, unbeknownst to the participant, with a fixed outcome contingency. For some face pairs, choice sometimes resulted in a gain or no outcome; for other pairs, choice sometimes led to a loss or no outcome; and for a third pair type (controls), monetary outcomes were never forthcoming, regardless of choice (controlling for effects of exposure). For gain and loss pairs, one face produced a monetary outcome with a probability of .80 and its mate with a probability of .20. These response contingencies produced value predictions for each face that varied in valence (predicting gains vs. losses) and probability (.8, .2 or 0), yielding stimuli with five different expected values (EV, an economic term calculated as $v \times p$, where v is the value of the outcome and p the probability of its occurring): $-.8x$, $-.2x$, 0, $.2x$ and $.8x$, where x

is the cash value involved (5*p*). By the end of the learning session (i.e., after 100 randomly ordered trials for each face pair), learning (choosing optimally) approximated the outcome contingencies similarly for gain and loss pairs. In other words, for these pairs, participants selected the faces most likely to yield wins (EV $= .8x$) and faces the least likely to produce losses (EV $= -.2x$) on about 80% of trials.

In the second phase of the experiment, we used a modified AB procedure (Raymond et al., 1992). See Fig. 3(A) for a schematic of the trial sequence and stimuli. In each trial, participants viewed a rapid sequence of briefly presented images containing two targets, an abstract object (T1) and a face (T2). The T2 image was either a face seen in the value-learning task (hereafter referred to as a value face) or a novel face. They were required to discriminate the texture of T1, and then to decide whether T2 was a value face (old) or not (new). No monetary outcomes or other feedback were ever provided in this phase. Critically, the lag between successive target presentations was either short (200 ms), creating a divided-attention condition, or long (800 ms), providing a full-attention condition. Short lags (less than 500 ms) between successive targets typically cause a large impairment in perceptual awareness of T2 (an AB effect) that can be completely eradicated by extending the lag to longer than 500 ms. The AB is thought to index temporal changes in the limited availability of attentional resources initiated by processing T1. Thus, varying the T1–T2 lag makes a tidy manipulation of available

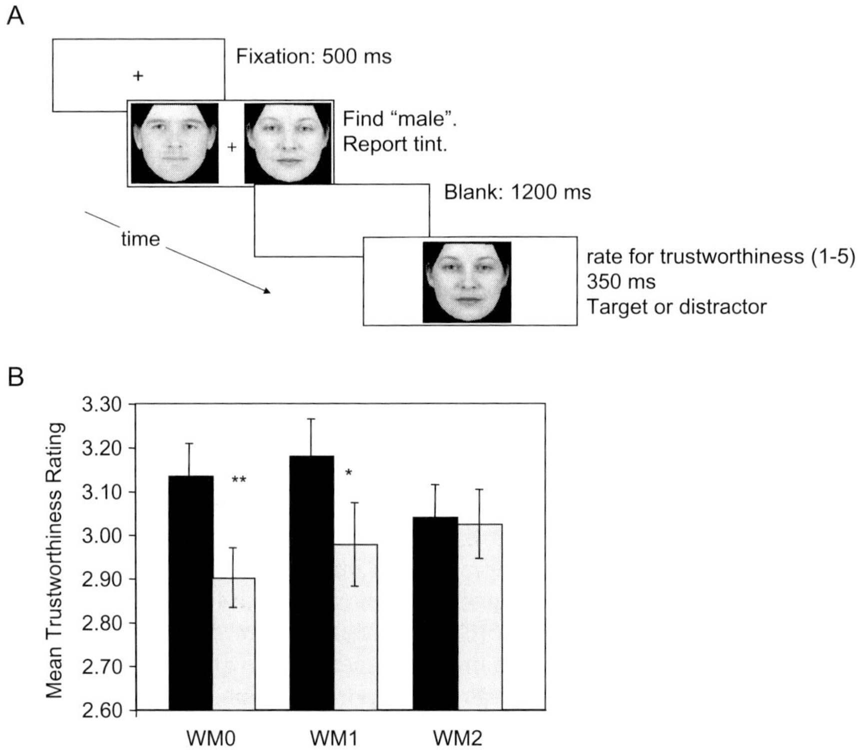

Fig. 3. (A) A schematic of the essential elements of a trial in the face devaluation experiment of Raymond et al. (2003). A pair of faces was presented for 200 ms. The task in different experiments was to select on the basis of face gender (or tint) and to report the face's tint colour (or gender). After responding, one of the faces (either the prior target or the distractor) was presented and then judged for trustworthiness using a five-point scale. (B) Group mean trustworthiness ratings in the working memory (WM) experiment (adapted from Goolsby et al., in press) given for prior targets (black bars) or prior distractors (grey bars) for each WM load condition. See text for details.

cognitive resource for T2, without concurrently changing demands on sensory or response systems. Note that in this phase, all T2 faces, regardless of their previously learned value, were equally relevant in this task and thus should have attracted an equal amount of attention.

Two surprising and interesting results were found (Fig. 3(B)). First, in the full-attention condition (long lag), T2 recognition strongly depended on EV. Recognition of 'old' faces was more accurate for high-probability gain and loss faces, regardless of valence, than for low-probability faces or for faces never associated with a monetary outcome, even though all had been seen previously the same number of times. This effect of learned outcome probability is especially interesting considering that all value faces, regardless of their prior history, were equally relevant in the recognition task and should have engaged top-down attention similarly. The second, surprising result was that the learned valence of value faces determined the AB. For faces associated with loss or no outcome, large AB effects were evident; in stark contrast, recognition of gain-associated faces, regardless of probability, showed no cost of divided attention (Fig. 4).

Finding that T2 performance varied with the outcome probability associated with the stimuli during the learning task, in both long and short SOA conditions, is important because it shows that visual recognition is not determined solely by current attentional demands directed at available sensory data. Value-prediction information regarding the probability of an outcome seems to provide an additional 'top-down' signal that can facilitate the processes underpinning recognition, such as perception or long-term memory. Recent neurobiological studies indicating that outcome prediction can greatly determine activity in visual cortex in rats (Shuler and Bear, 2006) and in lateral intraparietal cortex in monkeys (Bendiksby and Platt, 2006) support this finding. The effect of the valence of the predicted outcome associated with different face stimuli on the AB effect is also highly interesting because it is not readily predicted from the effect of outcome probability seen when attention is unconstrained (long SOA condition). It is, however, consistent with evidence that value coding for reward and punishment may be mediated by different neural networks (Yacubian et al., 2006) and suggests that these may differentially interact with attentional networks. Results from the full-attention condition (long lag) could be interpreted as indicating that outcome probability, in addition to current task relevance, determines how successfully a stimulus can compete for attentional resources. However, if this was the case, then in the harder, reduced attention condition (short SOA), stimuli associated with a high probability of producing an outcome (wins or losses) in the learning task should have retained their competitive advantage and 'escaped' the AB. Instead, large AB effects were observed for high-probability loss stimuli and none for high-probability win stimuli. Clearly, association with gains but not losses enhances attentional competitiveness of stimuli, whereas outcome probability modulates other processes important for recognition, independently of attention. Thus the pattern of results found in this experiment supplies important evidence that attention and motivation provide separable, independent top-down signals for controlling perceptual awareness.

To summarize, recent studies on humans have shown that the attention, emotional evaluation and value learning interact to determine visual selection. Importantly, these interactions allow immediate task-related concerns to become combined with prior learning history to determine prioritization for processing information. There are now a substantial number of reports that emotional stimuli can modulate attentional processes, providing evidence that prior history can override or augment (in different situations) attentional selection, that is, selection based on immediate task relevance and sensory salience. Work in my laboratory and others, reviewed here, have shown that attentional processes can have a subsequent effect on the explicit emotional (or affective) evaluation of stimuli, revealing a distractor devaluation effect. One way to interpret this effect is that the very process of selecting a visual target and ignoring other stimuli that compete for control over responding provides a simple value-learning opportunity that could imbue targets and distractors with positive and negative values, respectively. Such learning could give rise to

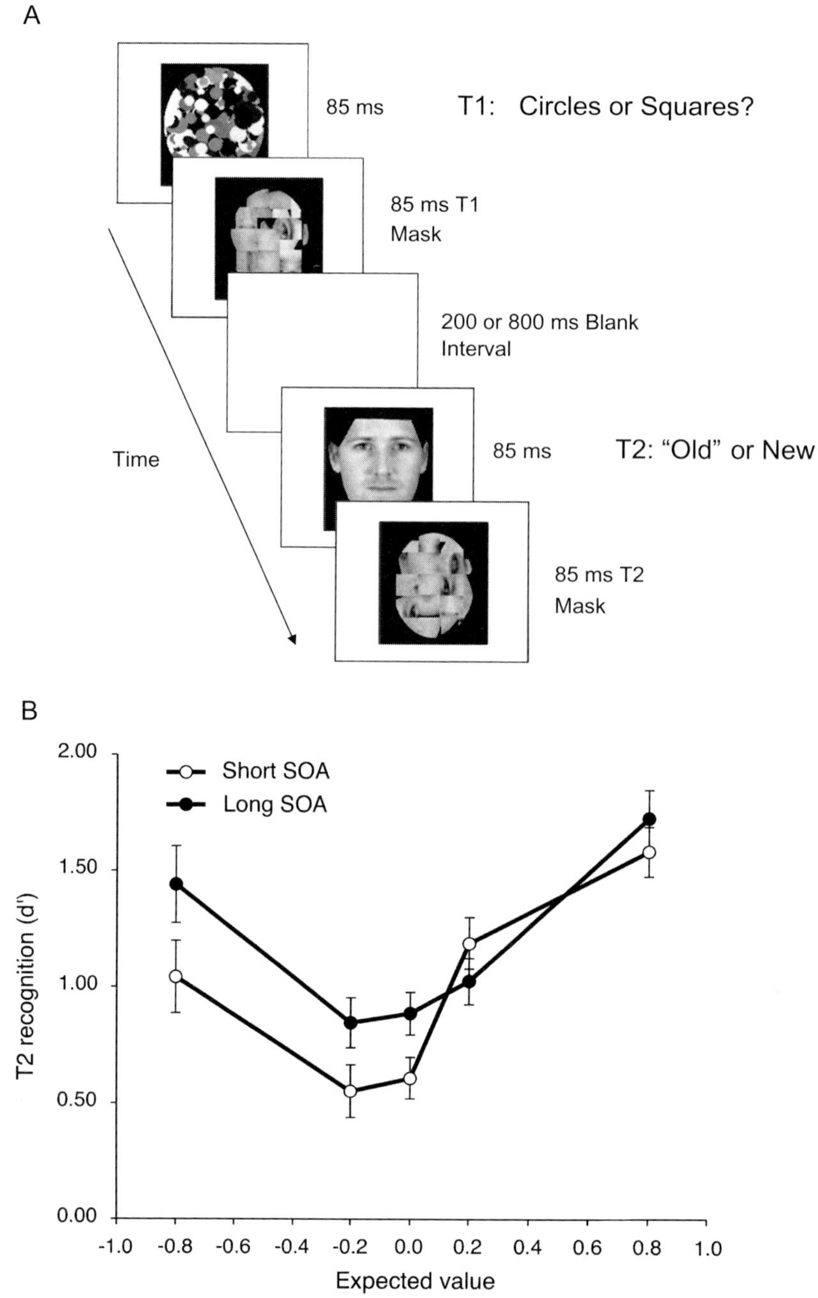

Fig. 4. (A) Sequence of events in the attentional blink task. (B) Recognition (d') of T2 faces (conditional on correct T1 responses) as a function of EV of T2 faces: filled circles for long lag and open circles for short lags; s.e.m. shown.

the consciously accessible feelings of affective evaluation that we tapped in our distractor devaluation effect studies. In support of such speculation, we have recently acquired data using a bi-valenced value-learning paradigm combined with an AB procedure that suggests that value-prediction codes can become associated with specific stimuli and that such codes can exert an effect on selection even when they themselves are irrelevant. These data resonate with studies showing that emotional stimuli can modulate performance on simple 'attention' tasks and suggest that perhaps some of those effects are due to value learning and may be motivational in basis, rather than purely attentional. This opens a new set of possibilities to explore the factors affecting motivational effects on visual selection in healthy and abnormal adults.

Acknowledgements

Some of this work was supported by a grant to JR from ESRC (UK) and other aspects by a grant to JR and colleagues from BBSRC (UK).

References

Aharon, I., Etcoff, N., Ariely, D., Chabris, C. F., O'Connor, E., & Brieter, H. C. (2001). Beautiful faces have variable reward value: fMRI and behavioural evidence. *Neuron*, *32*, 537–551.

Anderson, A. K. (2005). Affective influences on the attentional dynamics supporting awareness. *Journal of Experimental Psychology: General*, *134*(2), 258–281.

Armony, J. L., & Dolan, R. J. (2002). Modulation of spatial attention by fear conditioned stimuli: An event-related fMRI study. *Neuropsychologia*, *40*(7), 817–826.

Bendiksby, M. S., & Platt, M. L. (2006). Neural correlates of reward and attention in macaque area LIP. *Neuropsychologia*, *44*, 2411–2420.

Bush, G., Luu, P., & Posner, M. I. (2000). Cognitive and emotional influences in anterior cingulate cortex. *Trends in Cognitive Sciences*, *4*, 215–222.

Cowan, N. (2001). The magical number 4 in short-term memory: A reconsideration of mental storage capacity. *The Behavioral and Brain Sciences*, *24*, 87–185.

Della Libera, C., & Chelazzi, L. (2006). Visual selective attention and the effects of monetary rewards. *Psychological Science*, *17*, 222–227.

Eastwood, J. D., Smilek, D., & Merikle, P. M. (2001). Differential attentional guidance by unattended faces expressing positive and negative emotion. *Perception & Psychophysics*, *63*, 1004–1013.

Eimer, M. (1996). The N2pc component as an indicator of attentional selectivity. *Electroencephalography and Clinical Neurophysiology*, *99*, 225–234.

Fenske, M. J., & Eastwood, J. D. (2003). Modulation of focused attention by faces expressing emotion: Evidence from flanker tasks. *Emotion*, *3*, 327–343.

Fenske, M. J., & Raymond, J. E. (2006). Emotional influences of selective attention. *Current Directions in Psychology*, *15*(6), 312–316.

Fenske, M. J., Raymond, J. E., Kessler, K., Westoby, W., & Tipper, S. (2005). Attentional inhibition has social-emotional consequences for unfamiliar faces. *Psychological Science*, *16*(10), 753–758.

Fenske, M. J., Raymond, J. E., & Kunar, M. (2004). The affective consequences of visual attention in preview search. *Psychonomic Bulletin & Review*, *11*(6), 1034–1040.

Folstein, J. R., & Van Petten, C. (2008). Influence of cognitive control and mismatch on the N2 component of the ERP: A review. *Psychophysiology*, *45*, 152–170.

Fox, E., Russo, R., Bowles, R., & Dutton, K. (2001). Do threatening stimuli draw or hold visual attention in subclinical anxiety? *Journal of Experimental Psychology: General*, *130*, 681–700.

Fox, E., Russo, R., & Dutton, K. (2002). Attentional bias for threat: Evidence for delayed disengagement from emotional faces. *Cognition and Emotion*, *16*(3), 355–379.

Fox, E., Russo, R., & Georgiou, G. A. (2005). Anxiety modulates the degree of attentive resources required to process emotional faces. *Cognitive, Affective & Behavioral Neuroscience*, *5*, 396–404.

Freese, J. L., & Amaral, D. G. (2005). The organisation of projections from the amydgala to visual cortical areas TE and V1 in the macaque monkey. *The Journal of Comparative Neurology*, *486*, 295–317.

Goolsby, B., Shapiro, K. L., & Raymond, J. E. (2009). Distractor devaluation requires visual working memory. *Psychonomic Bulletin & Review*, *16*, 133–138.

Goolsby, B., Shapiro, K. L., Silvert, L., Kiss, M., Fragopanagos, N., Taylor, J. T., et al. (2008). Feature-based inhibition underlies the affective consequences of attention. *Visual Cognition*, doi:10.1080/13506280801904095.

Gottfried, J., O'Doherty, J., & Dolan, R. J. (2003). Encoding predictive reward value in human amygdala and orbitofrontal cortex. *Science*, *301*, 1104–1107.

Griffiths, O., & Mitchell, C. J. (2008). Negative priming reduces affective ratings. *Emotion*, *22*(6), 1119–1129.

Hogarth, L. C., Mogg, K., Bradley, B. P., et al. (2003). Attentional orienting towards smoking-related stimuli. *Behavioural Pharmacology*, *14*(2), 153–160.

Huang, Y.-M., Baddeley, A., & Young, A. W. (2008). Attentional capture by emotional stimuli is modulated by semantic processing. *Journal of Experimental Psychology: Human Perception and Performance*, *34*(2), 328–399.

Jackson, M. C., & Raymond, J. E. (2008). Familiarity enhances visual working memory for faces. *Journal of Experimental Psychology: Human Perception and Performance*, *34*(3), 556–568.

Kawabata, H., & Zeki, S. (2004). Neural correlates of beauty. *Journal of Neurophysiology*, *91*, 699–1705.

Kiss, M., Goolsby, B., Raymond, J. E., Shapiro, K. L., Silvert, L., Fragopanagos, N., et al. (2007). Efficient attentional selection predicts distractor devaluation: ERP evidence for a direct link between attention and emotion. *Journal of Cognitive Neuroscience*, *19*, 1316–1322.

Kiss, M., Raymond, J. E., Westoby, N., Nobre, A. C., & Eimer, M. (2008). Response inhibition is linked to emotional devaluation: Behavioural and electrophysiological evidence. *Frontiers in Human Neuroscience*, *2*, 13, doi: 10.3389/neuro.09.013.2008.

Knutson, B., Fong, G. W., Adams, C. M., Varner, J. L., & Hommer, D. (2001). Dissociation of reward anticipation and outcome with event related fMRI. *Neuroreport*, *12*, 3683–3687.

Lipp, O. V., Derakshan, N., Waters, A. M., & Logies, S. (2004). Snakes and cats in the flower bed: Fast detection is not specific to pictures of fear-relevant animals. *Emotion*, *4*(3), 233–250.

Luck, S. J., & Vogel, E. K. (1997). The capacity of visual working memory for features and conjunctions. *Nature*, *390*, 279–281.

Maunsell, J. H. R. (2004). Neuronal representations of cognitive state: Reward or attention? *Trends in Cognitive Sciences*, *8*(6), 261–265.

Most, S. B., Chun, M. M., Widders, D. M., & Zald, D. H. (2005). Attentional rubbernecking: Cognitive control and personality in emotion-induced blindness. *Psychonomic Bulletin & Review*, *12*, 654–661.

Mounts, J. R. W. (2000). Evidence for suppressive mechanisms in attentional selection: Feature singletons produce inhibitory surrounds. *Perception & Psychophysics*, *62*, 969–983.

O'Doherty, J., Winston, J., Chritchley, H., Perrett, D., Burt, D. M., & Dolan, R. J. (2003). Beauty in a smile: The role of medial orbitofrontal cortex in facial attractiveness. *Neuropsychologia*, *41*, 147–155.

O'Doherty, J. P. (2004). Reward representations and reward-related learning in the human brain: Insights from neuroimaging. *Current Opinion in Neurobiology*, *14*, 769–776.

O'Doherty, J. P., Deichman, R., Critchley, H. D., & Dolan, R. J. (2002). Neural responses during anticipation of a primary taste reward. *Neuron*, *33*, 815–826.

Öhman, A., Flykt, A., & Esteves, F. (2001). Emotion drives attention: Detecting the snake in the grass. *Journal of Experimental Psychology: General*, *130*(3), 466–478.

Paton, J. J., Belova, M. A., Morrison, S. E., & Salzman, D. (2006). The primate amygdala represents the positive and negative value of visual stimuli during learning. *Nature*, *439*, 865–870.

Pessiglione, M., Seymour, B., Flandin, G., Dolan, R., & Frith, C. (2006). Dopamine-dependent prediction errors underpin reward-seeking behaviour in humans. *Nature*, *442*, 1042–1045.

Pessoa, L., McKenna, M., Gutierrez, E., & Ungerleider, L. G. (2002). Neural processing of emotional faces requires attention. *Proceedings of the National Academy of Sciences of the United States of America*, *99*(17), 11458–11463.

Raymond, J. E., Fenske, M. J., & Tavassoli, N. T. (2003). Selective attention determines emotional responses to novel visual stimuli. *Psychological Science*, *14*, 537–542.

Raymond, J. E., Fenske, M. J., & Westoby, N. (2005). Emotional devaluation of distracting patterns and faces: A consequence of attentional inhibition during visual search? *Journal of Experimental Psychology: Human Perception and Performance*, *31*, 1404–1415.

Raymond, J. E., & O' Brien, J. L. (2009). Selective visual attention and motivation: The consequences of value learning in an attentional blink task. *Psychological Science*, DOI: 10.1111/j.1467-9280.2009.02391.x.

Raymond, J. E., Shapiro, K. L., & Arnell, K. M. (1992). Temporary suppression of visual processing in an RSVP task: An attentional blink? *Journal of Experimental Psychology: Human Perception and Performance*, *18*(3), 849–860.

Reber, R., Winkielman, P., & Schwarz, N. (1998). Effects of perceptual fluency on affective judgments. *Psychological Science*, *9*, 45–48.

Rolls, E. T. (2000). The orbitofrontal cortex and reward. *Cerebral Cortex*, *10*, 284–294.

Rutherford, H. J. V., & Raymond, J. E. (in press). Effects of spatial cues on locating emotional targets. *Psychological Bulletin Review*.

Schultz, W. (2000). Multiple reward signals in the brain. *Nature Reviews. Neuroscience*, *1*, 199–207.

Shuler, M. G., & Bear, M. F. (2006). Reward timing in the primary visual cortex. *Science*, *311*, 1606–1609.

Silvert, L., Lepsien, J., Fragopanagos, N., Goolsby, B., Kiss, M., Taylor, J. G., et al. (2007). Influence of attentional demands on the processing of emotional facial expressions in the amygdala. *NeuroImage*, *38*(2), 357–366.

Slotnick, S. D., Hopfinger, J. B., Klein, S. A., & Sutter, E. E. (2002). Darkness beyond the light: Attentional inhibition surrounding the classic spotlight. *Neuroreport*, *13*, 773–778.

Smith, S. D., Most, S. B., Newsome, L. A., & Zald, D. H. (2006). An "emotional blink" of attention elicited by aversively conditioned stimuli. *Emotion*, *6*(3), 523–527.

Veling, H., Holland, R. W., & van Knippenberg, A. (2007). Devaluation of distracting stimuli. *Cognition and Emotion*, *21*(2), 442–448.

Vuilleumier, P., Armony, J. L., Driver, J., & Dolan, R. J. (2001). Effects of attention and emotion on face processing in the human brain: An event-related fMRI study. *Neuron*, *30*(3), 829–841.

Watson, D. G., & Humphreys, G. W. (1997). Visual marking: Prioritizing selection for new objects by top-down attentional inhibition of old objects. *Psychological Review*, *104*, 90–122.

Woodman, G. F., & Luck, S. J. (1999). Electrophysiological measurement of rapid shifts of attention during visual search. *Nature*, *400*, 867–869.

Yacubian, J., Gläscher, J., Schroeder, K., Sommer, T., Braus, D. F., & Büchel, C. (2006). Dissociable systems for gain- and loss-related value predictions and errors of prediction in the human brain. *The Journal of Neuroscience*, *26*, 9530–9537.

Yamasaki, H., LaBar, K. S., & McCarthy, G. (2002). Dissociable prefrontal brain systems for attention and emotion. *Proceedings of the National Academy of Sciences of the United States of America*, *99*(17), 11447–11451.

Zajonc, R. B. (2001). Mere exposure: A gateway to the subliminal. *Current Directions in Psychological Science*, *10*, 224–228.

N. Srinivasan (Ed.)
Progress in Brain Research, Vol. 176
ISSN 0079-6123

CHAPTER 20

Human social attention

Elina Birmingham[1,*] and Alan Kingstone[2]

[1]*Division of Humanities & Social Sciences, California Institute of Technology, Pasadena, CA, USA*
[2]*Department of Psychology, University of British Columbia, Vancouver, BC, Canada*

Abstract: The present chapter suggests that while there is strong evidence that specific brain systems are preferentially biased toward processing gaze information, this specificity is not mirrored by the behavioral data as measured in highly controlled impoverished model tasks. In less controlled tasks, however, such as when observers are left free to look at whatever they want in complex natural scenes, observers focus on people and their eyes. This agrees with one's intuition, and with the neural evidence, that eyes are special. We discuss the implications of these data, including that there is much to be gained by examining brain and behavioral processes to social stimuli as they occur in complex real-world settings.

Keywords: eye movements; social attention; gaze perception; attentional selection; cueing paradigm; scene perception; visual attention; saccades

Imagine you are riding your bicycle down the road, and you notice that there is a person standing on the sidewalk looking upward. Using this person's gaze direction, you turn your eyes to see what is being looked at. As this simple scenario illustrates, folk knowledge suggests that we are very interested in where other people are directing their attention, and that we use their eyes to infer where, and to what, they are attending. The intuition that we care about the attentional states of others has led to the birth of research in *social attention*. We have a strong intuition that eye gaze is a special social attention cue in that it tells us with a reasonable degree of reliability where someone is attending (Emery, 2000). As such, we would expect that (1) the brain is particularly selective for eye gaze, and (2) humans readily use eye gaze to determine where others are directing their attention.

With regard to point (1), considerable research has been conducted on the neural mechanisms that are critical to processing gaze information, with much of this research suggesting that a key role is played by a region of cortex called the superior temporal sulcus (see Birmingham and Kingstone, 2009, for a review). It seems that one of the many functions of the superior temporal sulcus (STS), including biological motion processing, audiovisual integration, theory of mind, and face processing (see Allison et al., 2000; Hein and Knight, 2008, for reviews) is to process gaze direction. Single-cell studies with macaque monkeys have found populations of cells in the anterior STS that are selective for specific gaze directions (Perrett et al., 1985). Neuropsychological and lesion studies have found evidence for deficits in judging gaze direction associated with

*Corresponding author.
Tel.: (626) 395-4868; Fax: (626) 395-2000;
E-mail: elinab@hss.caltech.edu

DOI: 10.1016/S0079-6123(09)17618-5

damage to the STS (e.g., Heywood et al., 1992; Akiyama et al., 2006). Some neuroimaging studies of human posterior STS activity have also found stronger activation for faces with averted gaze than for faces with direct gaze (e.g., Hoffman and Haxby, 2000), although in some cases the opposite effect was found (e.g., Pelphrey et al., 2004) or no significant difference between direct and averted gaze was found at all (e.g., Wicker et al., 1998; George et al., 2001; Calder et al., 2002). Most recently, researchers have looked at neural adaptation to demonstrate the presence of neurons in anterior STS that are finely tuned for processing left and right gaze directions (Calder et al., 2007).

The evidence for point (2), that humans use eye gaze to determine where others are directing their attention, is much more controversial and much of the present chapter focuses on this issue. We note that while there is evidence derived from laboratory-based attention-cuing studies (e.g., Posner, 1980) that people automatically shift their attention to where other people are looking (e.g., Friesen and Kingstone, 1998); recent research also indicates that people shift their attention automatically in response to a number of other familiar directional cues, most notably arrows (Eimer, 1997; Ristic et al., 2002; Tipples, 2002). Indeed, a number of the latest studies have been dedicated to determining whether the effects of gaze cuing are truly unique from the effects of arrow cuing, and whether gaze and arrow cuing arise from the same underlying neural system. We propose that while some differences do occasionally emerge between gaze and arrow cuing, their general convergence suggests that the cuing paradigm may be failing to capture many of the key aspects of eyes that distinguish them from other stimuli, like arrows. In other words, the general intuition that eyes are very special is correct, but the cuing paradigm is measuring eyes and arrows on a dimension that they share a great deal of similarity, i.e., their ability to communicate directional information (Gibson and Kingstone, 2006). The implication is that researchers may benefit by considering alternative approaches for studying the uniqueness of eyes relative to other stimuli.

At the conclusion of our chapter we present a new approach emphasizing the *selection* of gaze information rather than the *orienting* of attention to where gaze is directed. To return to our initial example of a person on the sidewalk looking upward, our final section examines the selection of the person's gaze rather than subsequent orienting of attention to where that gaze is directed. When gaze information is examined in the laboratory in this way, the evidence indicates that people have a fundamental interest in eye information that far exceeds other information in the environment, including arrows. These data dovetail with the evidence that attentional selection is being driven by neural systems that give weight to the unique social information provided by the eyes of others, and hence they suggest a fruitful direction for future investigations.

The effect of gaze direction on spatial attention

To get at this issue, researchers recently modified a model task popularized by Michael Posner (1980) and used it to investigate whether people are preferentially biased to attend to where someone else is looking. In the model cuing paradigm participants are presented with a central fixation dot that is flanked by two squares. The task is to make a key press as quickly as possible when a target item appears inside one of the squares. This target event is preceded by a cue, i.e., the flashing of one of the squares or the appearance of a central arrowhead pointing toward one of the squares. The standard finding is that the target is detected faster when it appears in the cued square than when it appears in the uncued square. Because the brain processes attended items more quickly than unattended items, it is concluded that target detection time is speeded because attention has been committed to the square that was cued.

It is noteworthy that there are two different ways that attention is manipulated in the cuing task. One way is to flash one of the squares. In this case, attention is directed to the cued square that flashed. This attention shift is considered exogenous (automatic) because people are faster to detect a target in the cued square even when the flashing does not predict where the target will occur (i.e., the target appears in the cued location 50% of the time and in the uncued location 50% of the time). The other way to direct attention in this paradigm is to present a central arrowhead pointing left or right. In this case, attention is directed to the cued square that the arrowhead pointed toward. Since the early 1980s (Posner, 1980; Jonides, 1981) it has been assumed that this orienting happens only when the arrowhead predicts where the target will appear (e.g., the target appears in the cued location 80% of the time and in the uncued location 20% of the time). In other words, orienting to a central direction stimulus cue, like an arrow, does not occur when the cue is spatially nonpredictive. Thus, the attentional shift associated with a central directional cue is considered to be endogenous (voluntary).

Friesen and Kingstone (1998) hypothesized that given the intuition that eye gaze is a special social attention stimulus, perceived shifts in eye direction might automatically trigger attention shifts to gazed-at locations. This idea was tested by modifying the model-cuing task in two significant ways (Fig. 1). First, arrows pointing to the left and right were replaced by a schematic face that looked left or right. Second, the predictive value of the central cue was eliminated, i.e., eye direction did not predict where a target item would appear. Note that because the eyes were centrally located and spatially nonpredictive, the traditional line of thinking predicted that gaze would not lead to shifts of attention. In other words, the assumption was that central directional cues should only produce a shift in attention if they reliably predict where a target is likely to appear. Remarkably, and contrary to traditional thought, spatially nonpredictive eye gaze triggered shifts of attention; target detection was faster for items at the gazed-at location than for items at the other location (see also, Langton and Bruce, 1999; Driver et al., 1999). This discovery led to the proposal that the attention shift to eye gaze was automatic because it emerged rapidly and occurred even when gaze direction did not reliably predict where a target would occur. And most importantly, it was thought that this effect was special to eyes, suggesting that the human brain may be specialized to shift attention automatically in response to where other people are attending/looking. The brain mechanisms for this "gaze cuing" effect were hypothesized to involve parietal cortex, which is involved in spatial orienting, and the STS, which is reciprocally connected with the parietal cortex (e.g., Harries and Perrett, 1991).

Thus, this gaze-cuing paradigm appeared to tap into social attention and the fundamental importance that humans place on the eyes of others. Furthermore, it suggested a "meeting of fields," in that mainstream attention research methods could be used to study questions in the field of social cognition, and that social cognition could enrich our understanding of human attention.

This idea was soon tested by research examining whether other familiar directional cues, like

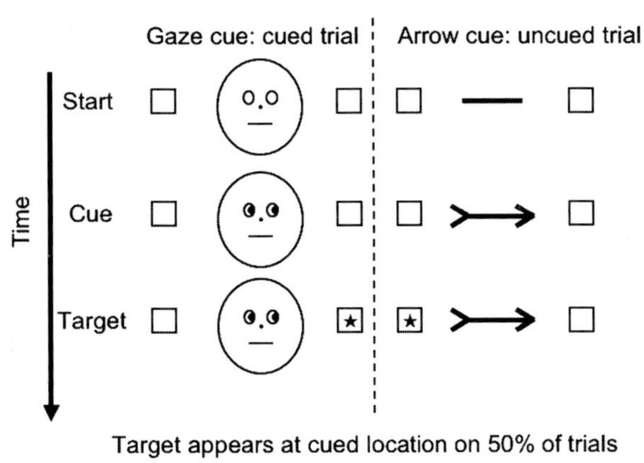

Fig. 1. The gaze-cuing paradigm (left) and the arrow-cuing paradigm (right).

arrows, would produce an automatic shift of attention to the cued location (Ristic et al., 2002; Tipples, 2002). It is very important to recall here that the endogenous (volitional) attention-cuing task that set the standard for all other cuing tasks that followed was founded on the principle that central arrow cues do not produce an orienting effect when they are spatially uninformative (Posner, 1980; Jonides, 1981). After all, if they did, then there would be little credibility to researcher's long-standing claim that informative central arrow cues tap into endogenous mechanisms (e.g., Posner, 1980; Jonides, 1981; Mueller and Rabbitt, 1989; Kingstone, 1992; Berger et al., 2005 to name but a small handful of what are literally hundreds of studies). It was therefore surprising when Ristic et al. (2002) and Tipples (2002) reported in separate investigations that central, spatially nonpredictive arrow cues, produce a robust reflexive orienting effect that is very similar to what is observed for gaze cues.

These findings raise the possibility that gaze cuing is not a unique or special effect. Understandably, this doubt has led to a flurry of research seeking to determine whether gaze cuing is different from arrow cuing, either at a behavioral or neural level. Below we review some of these studies.

One possibility is that while both gaze and arrows can orient attention automatically, and therefore both are important attentional cues, gaze cuing may be more strongly reflexive than arrow cuing, reflecting that eyes are biological stimuli with strong social meaning. In fact, there is some research that suggests a behavioral distinction between gaze and arrow cuing. Friesen et al. (2004) found that counterpredictive gaze cues, but not counterpredictive arrow cues, produced reflexive orienting to the cued location. This suggests that gaze cues are prioritized by the brain because of their social significance, leading to more reflexive shifts of attention than for arrow cues. Certainly, this is consistent with the findings that gaze direction is processed by a specialized neural system. Indeed, Downing et al. (2004) suggested that while almost any nonpredictive cue carrying spatial compatibility with the target will produce reflexive orienting, it is this more complex influence of gaze cues, i.e., the resistance to top-down biases, that sets gaze cues apart from other directional cues. In support of this, Ristic et al. (2007) showed that while arrow cuing is sensitive to arbitrary cue-target color contingencies (i.e., it only occurs when the cue and target share the same color), gaze cuing is not, and therefore can be considered to be more reflexive than arrow cuing.

Further evidence comes from studies testing adults' *overt orienting* (involving eye movements) of attention in response to gaze cues. Ricciardelli et al. (2002) found different overt orienting signatures for central gaze cues and arrow cues. Subjects were asked to make a speeded saccade to the left or right of fixation, as indicated by a central square stimulus. In concert with studies of covert orienting to gaze direction, correct saccade latencies were faster on trials on which a face also gazed at the correct location, relative to when the face gazed at the incorrect location. The same effect occurred for a central arrow stimulus. However, only the *incongruent gaze* stimulus produced unwanted saccades toward the incorrect location; *incongruent arrows* failed in this respect. This is consistent with covert attention studies showing that orienting to gaze cues is more strongly reflexive than to arrow cues and persists despite instructions to orient elsewhere.

Finally, Ristic and Kingstone (2005) demonstrated the uniqueness of gaze cuing relative to a stimulus — not an arrow — that was physically identical to the gaze cue but could be perceived as the wheels on a car. They found that when an ambiguous stimulus was first perceived as eyes, it produced reflexive orienting, even in a later block in which subjects were told the stimulus could be perceived as a car. However, when the stimulus was first perceived as a car, it did not produce reflexive orienting. Reflexive orienting only occurred when subjects were later informed that it could be perceived as containing eyes. This suggests that the stimulus had to be perceived as having eyes before it could trigger orienting, and that once this percept was activated it triggered reflexive orienting even when an alternative percept was suggested. However, later results by the same authors suggest limits to this finding, showing that an enlarged version of the ambiguous stimulus triggered orienting regardless of the percept that was adopted (Kingstone et al., 2004).

Despite the collection of research showing that gaze cuing may yield relatively subtle differences when compared to arrow cuing, other behavioral research has shown that gaze and arrow cues produce nearly identical shifts of attention (Hommel et al., 2001; Tipples, 2002). In contrast to Driver et al. (1999) and Friesen et al. (2004), Hommel et al. (2001) found that arrows *do* produce reflexive shifts in attention despite observers' knowledge that another location was more likely to receive the target. Similarly, Tipples (2008) replicated the conditions of Friesen et al.'s (2004) counterpredictive gaze cue study and reported reflexive orienting to the location cued by arrows and gaze even when a target was far more likely to appear elsewhere (like Friesen et al., this reflexive attention effect occurred concurrently with the volitional attention effect to the predicted target location). Furthermore, in contrast to Ricciardelli et al. (2002), Kuhn and Benson (2007) did not find different reflexive overt orienting signatures for gaze and arrow cues. The authors used a similar voluntary saccade paradigm to Ricciardelli et al., but used more traditional, "arrow-like", cues than did Ricciardelli et al. (who used simple arrowheads, e.g., < >). Using these more effective arrow stimuli, the authors found that the interference effect for arrow cues was of equal magnitude to gaze cues. The only difference Kuhn and Benson found between the two types of cues was in the response latency for erroneous saccades, finding shorter error latencies for gaze cues than for arrow cues. However, a later study found no difference between errors elicited by arrows and gaze stimuli (Kuhn and Kingstone, 2009).

Because the behavioral research has generally failed to reveal robust differences between gaze and arrow cuing, one possibility is that these differences are only detectable by digging into the neural mechanisms underlying each type of cuing. Some neuropsychological studies suggest that there are different neural systems for gaze and arrow cuing (Kingstone et al., 2000; Ristic et al., 2002; Akiyama et al., 2006). For instance, there is evidence from a study with split-brain patients that the reflexive gaze-cuing effect is lateralized to the hemisphere specialized for face processing (Kingstone et al., 2000). In contrast, in a later study this same split-brain patient showed no lateralization of reflexive orienting to nonpredictive arrows, with the cuing effect occurring in both hemispheres (Ristic et al., 2002). In addition, Akiyama et al. (2006) found that a patient with damage to her right superior temporal gyrus (STG) showed no orienting in response to gaze cues but preserved orienting to arrow cues. These findings are consistent with the idea that reflexive orienting to nonbiological cues is underpinned by subcortical brain mechanisms that are shared between the two hemispheres, whereas reflexive orienting to gaze cues is subserved by lateralized cortical mechanisms involved in face/gaze processing (e.g., Kingstone et al., 2004; Friesen and Kingstone, 2003).

However compelling these findings are, they must be interpreted with some caution. In particular, lesion studies which test very few individuals are difficult to interpret because of the natural variation in the gaze cuing effect across individuals. As Frischen et al. (2007) point out, some people do not show gaze cuing. Thus, in studies such as Akiyama et al. (2006), which tested only one participant, it is difficult to know whether the lesion interfered with gaze cuing or whether the patient never showed gaze cuing. Furthermore, it is important to consider the influence of low-level differences between gaze and arrow stimuli when interpreting results with single-patient case studies. For instance, the arrow cues in Akiyama et al.'s (2006) study may have conveyed direction more effectively than the gaze cues did, i.e., whereas the arrow cues had clear directionality, the gaze cues were only partially averted (off-center by 11%). A similar type of concern may be applied to the split-brain studies of Kingstone and colleagues (e.g., Kingstone et al., 2000; Ristic et al., 2002).

Neuroimaging studies with healthy populations have also been conducted in hopes shedding light on whether gaze cuing is unique in some way, with mixed results. For instance, there is evidence that brain activation differences produced for gaze and arrow cuing may be partly due to the recruitment of different brain areas for visually analyzing gaze and arrow cues, and not necessarily for the subsequent shifts of attention (Hietanen et al., 2006). Once these basic visual processing differences are removed, and only the subsequent orienting of attention is examined, there is little evidence that gaze and arrow cues are subserved by distinct attentional mechanisms. Tipper et al. (2008) studied gaze and arrow cuing using an ambiguous stimulus that could be perceived as either an eye or an arrow, thus removing the physical stimulus differences normally present in comparisons of gaze and arrow cuing. Tipper et al. found very few differences between the neural activations underlying gaze and arrow cuing, save for a bigger sensory gain at the cued location for gaze cues than for arrow cues.

However, even these studies must be interpreted with caution, as both suffer from methodological limitations. For instance, the analysis of the Hietanen et al., 2006 study may have underestimated the unique contributions of social information. In particular, the analysis collapsed across valid and invalid trials, which is the critical comparison for the attention effect. And the relatively small number of participants in Tipper et al.'s (2008) fMRI experiment is a potential limitation given that the critical finding was a null difference in STG activity between gaze and arrow cuing.

As a final line of inquiry, Frischen et al. (2007) point out that if gaze and arrow cuing are subserved by separate neural systems, then one might expect that gaze- and arrow-cuing effects may not correlate strongly within an individual. Although little research has been committed to determining whether individuals who show strong gaze cuing also show strong arrow cuing, there is some evidence from studies of gender differences that suggests that gaze and arrow cuing are related. Bayliss et al. (2005) found that males show a weaker orienting effect for gaze cues than do females, consistent with previous findings that male infants make less eye contact than female infants (Lutchmaya et al., 2002) and therefore may be less sensitive to social cues. However, Bayliss et al. found the same gender difference for arrow cuing, suggesting that gaze and arrow cuing are not distinct. That is, Bayliss et al.'s findings run counter to what would have been found if gaze and arrow cuing were unique: "If orienting to the direction of another person's eye gaze is functionally different to the symbolic cuing seen with arrows, for example, then no gender difference would be obtained with arrow cues: Males and females should display attention shifts of equivalent magnitude" (p. 642).

When taken together, the results are rather equivocal with regard to the uniqueness of gaze cuing. On the one hand, there are some studies that find subtle differences between gaze cuing and arrow cuing, but on the other hand, these differences often are not observed. Overall, the evidence that gaze cuing is unique from arrow cuing is weak.

What are the implications of this conclusion? Certainly, the finding that arrow cues produce nearly identical effects to gaze cues runs counter to one's intuition that eyes are unique, special social attention stimuli. However, it could be that arrows are also important social stimuli, which explains why they, too, produce reflexive shifts in attention. This potential status of arrows has not been overlooked. Kingstone et al. (2003) have written that "arrows are obviously very directional in nature, and, like eyes, they have a great deal of social significance. Indeed, it is a challenge to move through one's day without encountering any number of arrows on signs and postings" (p. 178). Thus, perhaps eyes and arrow produce identical effects on attention in the cuing paradigm because they are both important social cues.

An alternative explanation of the data is that the cuing paradigm may be failing to capture key aspects about eyes that distinguish them as special social stimuli that are unlike other stimuli, like arrows. In other words, the general intuition that eyes are special is correct, but the cuing paradigm may not be measuring what makes eyes distinct from arrows. Indeed, the cuing paradigm appears to be measuring eyes and arrows on a dimension that they share a great deal of similarity, i.e., their ability to communicate directional information (Gibson and Kingstone, 2006). This interpretation is supported by growing evidence that cuing effects similar to gaze cuing are found for a variety of biological and nonbiological cues that convey direction. For instance, Downing et al. (2004, Experiment 1) found that a central face with its tongue pointing randomly left or right produced reflexive attention effects that were indistinguishable from gaze cuing effects. Hommel et al. (2001) found reflexive orienting both for non-predictive arrows and for nonpredictive directional words (e.g., "left," "right") presented centrally. Even more striking, Quadflieg et al. (2004) found equivalent cuing effects for drawings of averted eyes within human faces, within animal faces (e.g., tiger, owl), or within an apple or a gloved hand. The same cuing effect was found for a gloved hand containing two arrows instead of eyes. Thus, working from the basic intuition that eyes are very different social stimuli from arrows, one may conclude that the similarity found between eyes and arrows in the cuing paradigm tells us about the limitations of the cuing paradigm.

The results of this growing collection of studies suggest that *any* cue with a directional component, or more specifically, any cue carrying the potential for spatial compatibility with the target (e.g., arrow points left, target appears left) produces reflexive orienting of attention. From this perspective, behavioral differences found between cues could be reattributed to the differences in the cues' ability to convey left/right information. What may be needed in the area of social attention is a different research approach — one that better reflects our intuition that the human attention system cares about eyes in a way that is distinct from other stimuli in the environment. One possible avenue has recently been suggested by Kuhn and Kingstone (2009) "Thus although arrows and eye gaze may be of equal relevance when they are presented to the participant in isolation, key differences between social and non social cues may only become apparent when they are embedded within a richer environment" (p. 41).

A new approach for studying social attention

An alternative approach for studying social attention is provided by considering the different components of attention that can be measured in experiments involving social stimuli. Rather than examining the orienting of attention in response to a cue (i.e., orienting *from* the cue to where the cue is pointing), we propose to study the selection of the cue itself (i.e., orienting *to* the cue). Consider a real world example of social attention: You are walking on campus toward your colleague when you notice that she is looking intensely at something on the ground. Using her gaze direction you orient your attention to see what she is looking at. Now, it is clear from this example that there are at least two distinct stages of social attention: first, you select (orient to) your colleague's eyes as a key social stimulus, and second, you orient your attention from her eyes to select the location/object that she is looking at. Importantly, cuing studies with central symbolic cues are specifically designed to test only one of these attentional components: orienting from the cue. The initial *selection* of the cue is relatively trivial within the context of the cueing paradigm because the cue, that is, a gaze, arrow, word, or number stimulus, is presented at central fixation and typically in advance of the target object (Gibson and Kingstone, 2006). Thus, the experimenter essentially *preselects* the cue and places it at fixation (the current focus of attention). As we found in the preceding section, when there is effectively no selection process on the part of the observer, the prevailing literature indicates that eyes and arrows are generally given equal priority by the attention system. Does this general equivalence hold, however, when the observer's *selection* of social cues is nontrivial and measured? In other words, will eyes and arrows be given equal priority when participants are provided with the opportunity to select them from a complex visual scene?

The fact that no studies have compared the selection of eyes versus arrows is noteworthy, given the strong tradition of research on selective attention (e.g., James, 1890; Broadbent, 1958, 1972; Moray, 1959; Treisman, 1960; Deutsch and Deutsch, 1963; Neisser, 1967). The basic assumption behind all these conceptualizations of selective attention is that humans possess a capacity limitation when it comes to handling information in the world. The implication of this capacity limitation is that we must select some items for processing at the expense of others — hence the term *selective* attention.

Now, before using measures of selection to compare the social relevance of eyes and arrows, one would want to verify that these measures tap into social attention mechanisms. Some of our own work has done just that. We have used the selection approach to demonstrate that observers select — by looking at — the eyes of people within complex scenes because they are interested in the social information provided by the eyes (Birmingham et al., 2008a, b). We presented real-world photographs of scenes with people and a variety objects, and depicting a range of different natural social situations. The results of our work showed that indeed, observers do look mostly at the eyes of other individuals, and they look relatively infrequently at the rest of the scene (e.g., bodies, foreground objects, background objects). Importantly, we also found that this general interest in the eyes of others can be modulated by social factors. For instance, observers in our investigations selected the eyes more frequently in highly social scenes, such as scenes containing multiple people doing something together. They also selected the eyes more frequently when reporting on the social attention within the scenes relative to when completing other less socially focused tasks, such as describing the scenes (Birmingham et al., 2008b; see also Smilek et al., 2006). These findings led us to conclude that our methodology captures some important social attention processes, revealing a preferential selection of gaze information that is enhanced by a social attention task and by the social content of the scene. This agrees also with our finding that the preferential selection of gaze cannot be explained by low-level saliency advantages conferred to eyes relative to the other items in the scenes (Birmingham et al., 2009, under review). Finally, it is noteworthy that our findings agree with the everyday intuition that eyes are unique social stimuli that are prioritized by the human attention system.

With the confidence that our basic approach taps into social attention, we adapted it in a recent study to determine whether observers select eyes and arrows to the same extent (Birmingham et al., 2009). We did this by presenting gaze and arrows within complex scenes and studying what people select to fixate. Scenes were shown for 15 s during which observers simply looked at the images. A representative illustration of the data is shown in Fig. 2. What we found is that observers demonstrate a strong bias to fixate the eyes in the scene with few fixations committed to the arrows (Fig. 2A). Furthermore, while eyes and heads were likely to be prioritized, that is, looked at first, arrows were never fixated first. This general

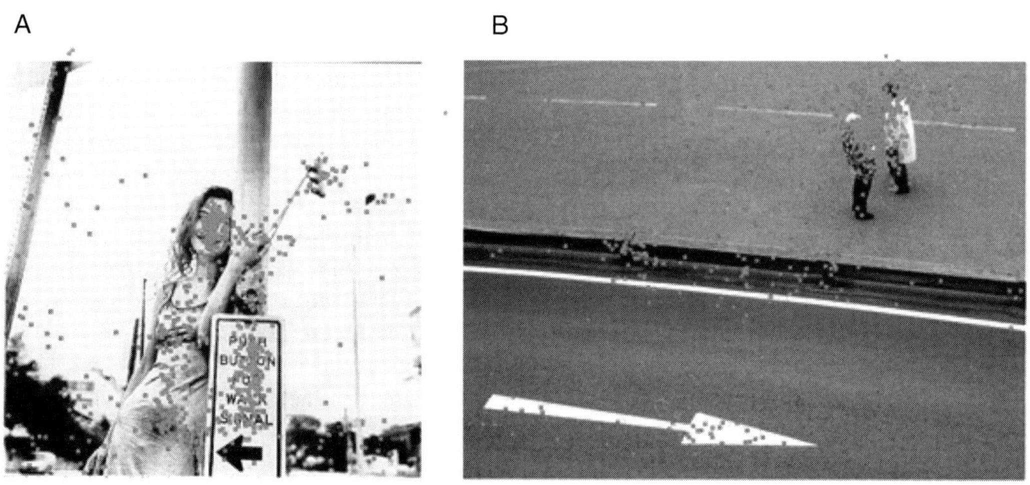

Fig. 2. Fixations (dots) overlaid on an image with eyes and arrows (A) and an image with larger arrows (B).

interest in people, and lack of interest in arrows, persists even when we make the arrow much larger than the people in the scene (Fig. 2B). Note also that these findings cannot be explained by low-level visual characteristics of the eyes or arrows, such as visual saliency (Itti and Koch, 2000), as we computed the saliency at fixated locations and found that it was no higher than what would be expected by chance. Overall, our data show that when one examines attentional selection, rather than orienting in response to a preselected centrally positioned stimulus (as in the cuing paradigm) what one finds is that people care about people, especially their eyes and faces. They rarely look at the arrows when they are small; and they rarely look at them when they are large. These findings indicate that in general, the human attention system does not treat eyes and arrows equivalently. When people are free to look at what they find important, they choose to look at people and their eyes. This profound and clear-cut difference between eyes and arrows has never been observed within the context of the cuing paradigm.

The implications of these preliminary data are both broad and deep. First, they suggest that when one takes a different approach to measuring the impact of eyes and arrows on the spatial attention system — one that moves away from the orienting of attention in response to a directional cue that the observer is forced to select (because it is presented at fixation in an otherwise uncluttered field), and moves toward the selection of items in a complex scene — then one finds that observers tend to select people and eyes rather than arrows.

Second, finding a profound difference between eyes and arrows would appear to lend support to the suggestion raised in the previous section, that the cuing paradigm may not be picking up on basic differences in the social relevance of eyes and arrows, differences that appear to be captured when selection is measured. This dovetails with the finding that when eyes and arrows are inserted as cues into the cuing paradigm, they tend to be treated the same way behaviorally (e.g., Tipples, 2008) and engage the same brain mechanisms (e.g., Tipper et al., 2008). The most reasonable explanation as to why this should be the case is that the factor of interest in the cuing paradigm is cue/target location information, and eyes and arrows are well matched in their ability to deliver this type of information. On the other hand, the many features that make eyes and arrows different types of stimuli are not typically important to the cuing paradigm.

Third, the present data raise the possibility that when researchers place a face stimulus in isolation, as they do in the cuing paradigm, they may be bypassing a critical aspect of attention, i.e., the selection process. As we saw in the study mentioned above, the selection process allows one to assess the importance that observers place on different stimuli. When this selection opportunity is bypassed by the experimenter preselecting and presenting the stimulus to the observer in relative isolation, it is very difficult to gain a sense of the relative importance placed on each stimulus. Additionally, by preselecting and isolating different stimuli one may change the context in which the stimuli are normally embedded, and in doing so, change the meaning that is normally attached to those stimuli.

Fourth, the data and above considerations suggest that it would be wise to move from monitoring the eyes of observers while they view static *images* of people to monitoring the eyes of observers while they view moving images of people, and ultimately to studying how observers look at *real* people. To date virtually all of the research in social attention (including the data presented here) has been confined to situations involving static images of people. By definition, these images of people cannot attend to the observer while the observer is attending to them. This stands in sharp contrast to many situations in real life. Interestingly, while one might be tempted to predict that observers would look even *more* often at the eyes in real social situations than when they are presented with images of eyes, the opposite could just as easily be true. For instance, while eye contact is a functional part of everyday social interactions, social norms indicate that it is often rude to make excessive eye contact or to spend too much time looking at another person. Indeed, in some situations (e.g., being approached by a hostile person) it may be appropriate to avoid eye contact altogether.

Summary and future directions

The present chapter considered a body of behavioral evidence that sought to examine the functional impact of gaze direction on the spatial orienting of attention. Contrary to what had been expected from the neural evidence for a specialized system for processing gaze direction, these studies found that a range of cues — from eyes to arrows — have a similar effect on attentional orienting. In our second section we showed that when observers are left free to select what they want to attend to, they focus on people and their eyes — and not arrows — consistent with one's intuition and the neural evidence that eyes are special. We discussed a range of implications of this discovery, including that when researchers preselect a stimulus and simplify the setting it is normally embedded in, they may profoundly change the way a stimulus is processed relative to more complex real-world settings.

Thus, an exciting direction for future research is to measure social attention in more real-world settings, in which gaze direction is one of several stimuli that make up a rich social context. For instance, Kuhn and Land (2006) showed that the vanishing ball illusion, in which a ball is perceived to have vanished in mid air, relies strongly on social attention cues from the magician performing the trick. That is, when the magician pretends to toss a ball upward but secretly conceals the ball in the palm of his hand, observers are much more likely to perceive the ball traveling upward and vanishing when the magician looks upward with the fake toss than when he looks down at his hand. Furthermore, on real throws on which the ball is physically present, instead of simply tracking the ball with their eyes, observers often make fixations to the magician's face before looking at the ball. This suggests that observers select information about the magician's attention in order to predict the position of the ball. Kuhn and Land's study thus provides an excellent example of how social attention, both with regard to the *selection* and *orienting* components of attention, can be studied successfully using rich, complex stimuli.

Acknowledgment

This chapter is based on a substantially larger article by Birmingham and Kingstone (2009).

References

Akiyama, T., Kato, M., Muramatsu, T., Saito, F., Umeda, S., & Kashima, H. (2006). Gaze but not arrows: A dissociative impairment after right superior temporal gyrus damage. *Neuropsychologia*, *44*, 1804–1810.

Allison, T., Puce, A., & McCarthy, G. (2000). Social perception from visual cues: Role of the STS region. *Trends in Cognitive Sciences*, *4*(7), 267–278.

Bayliss, A. P., di Pellegrino, G., & Tipper, S. P. (2005). Sex differences in eye gaze and symbolic cueing of attention. *Quarterly Journal of Experimental Psychology A*, *58A*, 631–650.

Berger, A., Henik, A., & Rafal, R. (2005). Competition between endogenous and exogenous orienting of visual attention. *Journal of Experimental Psychology General*, *134*, 207–221.

Birmingham, E., Bischof, W. F., & Kingstone, A. (2008a). Social attention and real world scenes: The roles of action, competition, and social content. *Quarterly Journal of Experimental Psychology*, *61*(7), 986–998.

Birmingham, E., Bischof, W. F., & Kingstone, A. (2008b). Gaze selection in complex social scenes. *Visual Cognition*, *16*(2/3), 341–355.

Birmingham, E., Bischof, W. F., & Kingstone, A. (2009). Get real! Resolving the debate about equivalent social stimuli. *Visual Cognition, Special issue, Eye guidance in natural scenes*, 1–21.

Birmingham, E., Bischof, W. F., & Kingstone, A. (under review). Saliency does not account for fixations to eyes within social scenes. *Vision Research*.

Birmingham, E., & Kingstone, A. (2009). Human social attention: A new look at past, present and future investigations. *The Year in Cognitive Neuroscience 2009: Annals of the New York Academy of Sciences*, *1156*, 118–140.

Broadbent, D. E. (1958). *Perception and communication*. London: Pergamon Press.

Broadbent, D. E. (1972). *Decision and stress*. New York: Academic Press.

Calder, A. J., Beaver, J. D., Winston, J. S., Dolan, R. J., Jenkins, R., Eger, E., & Henson, R. N. A. (2007). Separate coding of different gaze directions in the superior temporal sulcus and inferior parietal lobule. *Current Biology*, *17*, 20–25.

Calder, A. J., Lawrence, A. D., Keane, J., Scott, S. K., Owen, A. M., Christoffels, I., & Young, A. W. (2002). Reading the mind from eye gaze. *Neuropsychologia*, *40*, 1129–1138.

Deutsch, J. A., & Deutsch, D. (1963). Attention: Some theoretical considerations. *Psychological Reviews*, *70*, 80–90.

Downing, P. E., Dodds, C. M., & Bray, D. (2004). Why does the gaze of others direct visual attention? *Visual Cognition*, *11*, 71–79.

Driver, J., Davis, G., Ricciardelli, P., Kidd, P., Maxwell, E., & Baron-Cohen, S. (1999). Gaze perception triggers reflexive visuospatial orienting. *Visual Cognition*, *6*, 509–540.

Eimer, M. (1997). Uninformative symbolic cues may bias visual-spatial attention: Behavioral and electrophysiological evidence. *Biological Psychology*, *46*, 67–71.

Emery, N. J. (2000). The eyes have it: The neuroethology, function and evolution of social gaze. *Neuroscience and Biobehavioral Reviews*, *24*, 581–604.

Friesen, C. K., & Kingstone, A. (1998). The eyes have it! Reflexive orienting is triggered by nonpredictive gaze. *Psychonomic Bulletin & Review*, *5*, 490–495.

Friesen, C. K., & Kingstone, A. (2003). Covert and overt orienting to gaze direction and the effects of fixation offset. *NeuroReport*, *14*, 489–493.

Friesen, C. K., Ristic, J., & Kingstone, A. (2004). Attentional effects of counterpredictive gaze and arrow cues. *Journal of Experimental Psychology: Human Perception and Performance*, *30*, 319–329.

Frischen, A., Bayliss, A. P., & Tipper, S. P. (2007). Gaze cueing of attention: Visual attention, social cognition, and individual differences. *Psychological Bulletin*, *133*(4), 694–724.

George, N., Driver, J., & Dolan, R. J. (2001). Seen gaze-direction modulates fusiform activity and its coupling with other brain areas during face processing. *Neuroimage*, *13*, 1102–1112.

Gibson, B. S., & Kingstone, A. (2006). Visual attention and the semantics of space: Beyond central and peripheral cues. *Psychological Science*, *17*, 622–627.

Harries, M. H., & Perrett, D. I. (1991). Visual processing of faces in temporal cortex: Physiological evidence for a modular organization and possible anatomical correlates. *Journal of Cognitive Neuroscience*, *3*, 9–24.

Hein, G., & Knight, R. T. (2008). Superior temporal sulcus — It's my area: Or is it? *Journal of Cognitive Neuroscience*, *20*(12), 2125–2136.

Heywood, C. A., Cowey, A., & Rolls, E. T. (1992). The role of the face cell area in the discrimination and recognition of faces by monkeys. *Philosophical Transactions of the Royal Society of London. Series B, Biological Sciences*, *335*, 31–38.

Hietanen, J. K., Nummenmaa, L., Nyman, M. J., Parkkola, R., & Hämäläinen, H. (2006). Automatic attention orienting by social and symbolic cues activates different neural networks: An fMRI study. *Neuroimage*, *33*, 406–413.

Hoffman, E. A., & Haxby, J. V. (2000). Distinct representations of eye gaze and identity in the distributed human neural system for face perception. *Nature Neuroscience*, *3*, 80–84.

Hommel, B., Pratt, J., Colzato, L., & Godijn, R. (2001). Symbolic control of visual attention. *Psychological Science*, *12*, 360–365.

Itti, L., & Koch, C. (2000). A saliency-based search mechanism for overt and covert shifts of visual attention. *Vision Research*, *40*, 1489–1506.

James, W. (1890). *Principles of psychology*. New York: H. Holt & Co.

Jonides, J. (1981). Voluntary versus automatic control over the mind's eye's movement. In J. B. Long & A. D. Baddeley (Eds.), *Attention and performance* (Vol. IX, pp. 187–203). Hillsdale: Erlbaum.

Kingstone, A. (1992). Combining expectancies. *The Quarterly Journal of Experimental Psychology Section A: Human Experimental Psychology*, *44*(1), 69–104.

Kingstone, A., Friesen, C. K., & Gazzaniga, M. S. (2000). Reflexive joint attention depends on lateralized cortical connections. *Psychological Science*, *11*, 159–166.

Kingstone, A., Smilek, D., Ristic, J., Friesen, C. K., & Eastwood, J. D. (2003). Attention, researchers! It's time to pay attention to the real world. *Current Directory in Psychological Science*, *12*, 176–180.

Kingstone, A., Tipper, C., Ristic, J., & Ngan, E. (2004). The eyes have it! An fMRI investigation. *Brain and Cognition*, *55*, 269–271.

Kuhn, G., & Benson, V. (2007). The influence of eye-gaze and arrow pointing distractor cues on voluntary eye movements. *Attention, Perception & Psychophysics*, *69*(6), 966–971.

Kuhn, G., & Kingstone, A. (2009). Look away! Eyes and arrows engage oculomotor responses automatically. *Attention, Perception & Psychophysics*, *71*, 314–327.

Kuhn, G., & Land, M. F. (2006). There's more to magic than meets the eye. *Current Biology*, *16*(22), R950–R951.

Langton, S. R. H., & Bruce, V. (1999). Reflexive visual orienting in response to the social attention of others. *Visual Cognition*, *6*, 541–568.

Lutchmaya, S., Baron-Cohen, S., & Raggatt, P. (2002). Foetal testosterone and eye contact in 12-month-old human infants. *Infant Behavior & Development*, *25*(3), 327–335.

Moray, N. (1959). Attention in dichotic listening: Affective cues and the influence of instructions. *Quarterly Journal of Experimental Psychology*, *11*, 56–60.

Mueller, H. J., & Rabbitt, P. M. A. (1989). Reflexive and voluntary orienting of visual attention: Time course of activation and resistance to interruption. *Journal of Experimental Psychology: Human Perception and Performance*, *15*(2), 315–330.

Neisser, U. (1967). *Cognitive psychology*. New York: Appleton-Century-Crofts.

Pelphrey, K. A., Viola, R. J., & McCarthy, G. (2004). When strangers pass: Processing of mutual and averted social gaze in the superior temporal sulcus. *Psychological Science*, *15*, 598–603.

Perrett, D. I., Smith, P. A. J., Potter, D. D., Mistlin, A. J., Head, A. S., Milner, A. D., & Jeeves, M. A. (1985). Visual cells in the temporal cortex sensitive to face view and gaze direction. *Proceedings of the Royal SocietyLondon. Series B, Biological Sciences*, *223*, 293–317.

Posner, M. I. (1980). Orienting of attention. *Quarterly Journal of Experimental Psychology*, *32*, 3–25.

Quadflieg, S., Mason, M. F., & Macrae, C. N. (2004). The owl and the pussycat: Gaze cues and visuospatial orienting. *Psychonomic Bulletin & Review*, *11*(5), 826–831.

Ricciardelli, P., Bricolo, E., Aglioti, S. M., & Chelazzi, L. (2002). My eyes want to look where your eyes are looking: Exploring the tendency to imitate another individual's gaze. *Neuroreport*, *13*(17), 2259–2264.

Ristic, J., Friesen, C. K., & Kinsgtone, A. (2002). Are eyes special? It depends on how you look at it. *Psychonomic Bulletin & Review*, *9*, 507–513.

Ristic, J., & Kingstone, A. (2005). Taking control of reflexive social attention. *Cognition*, *94*, B55–B65.

Ristic, J., Wright, A., & Kingstone, A. (2007). Attentional control and reflexive orienting to gaze and arrow cues. *Psychonomic Bulletin & Review*, *14*(5), 964–969.

Smilek, D., Birmingham, E., Cameron, D., Bischof, W. F., & Kingstone, A. (2006). Cognitive ethology and exploring attention in real world scenes. *Brain Research*, *1080*, 101–119.

Tipper, C. M., Handy, T. C., Giesbrecht, B., & Kingstone, A. (2008). Brain responses to biological relevance. *Journal of Cognitive Neuroscience*, *20*(5), 879–891.

Tipples, J. (2002). Eye gaze is not unique: Automatic orienting in response to uninformative arrows. *Psychonomic Bulletin & Review*, *9*, 314–318.

Tipples, J. (2008). Orienting to counterpredictive gaze and arrow cues. *Attention, Perception & Psychophysics*, *70*, 77–87.

Treisman, A. M. (1960). Contexual cues in selective listening. *Quarterly Journal of Experimental Psychology*, *12*, 242–248.

Wicker, B., Michel, F., Henaff, M. A., & Decety, J. (1998). Brain regions involved in the perception of gaze: A PET study. *Neuroimage*, *8*, 221–227.

Subject Index

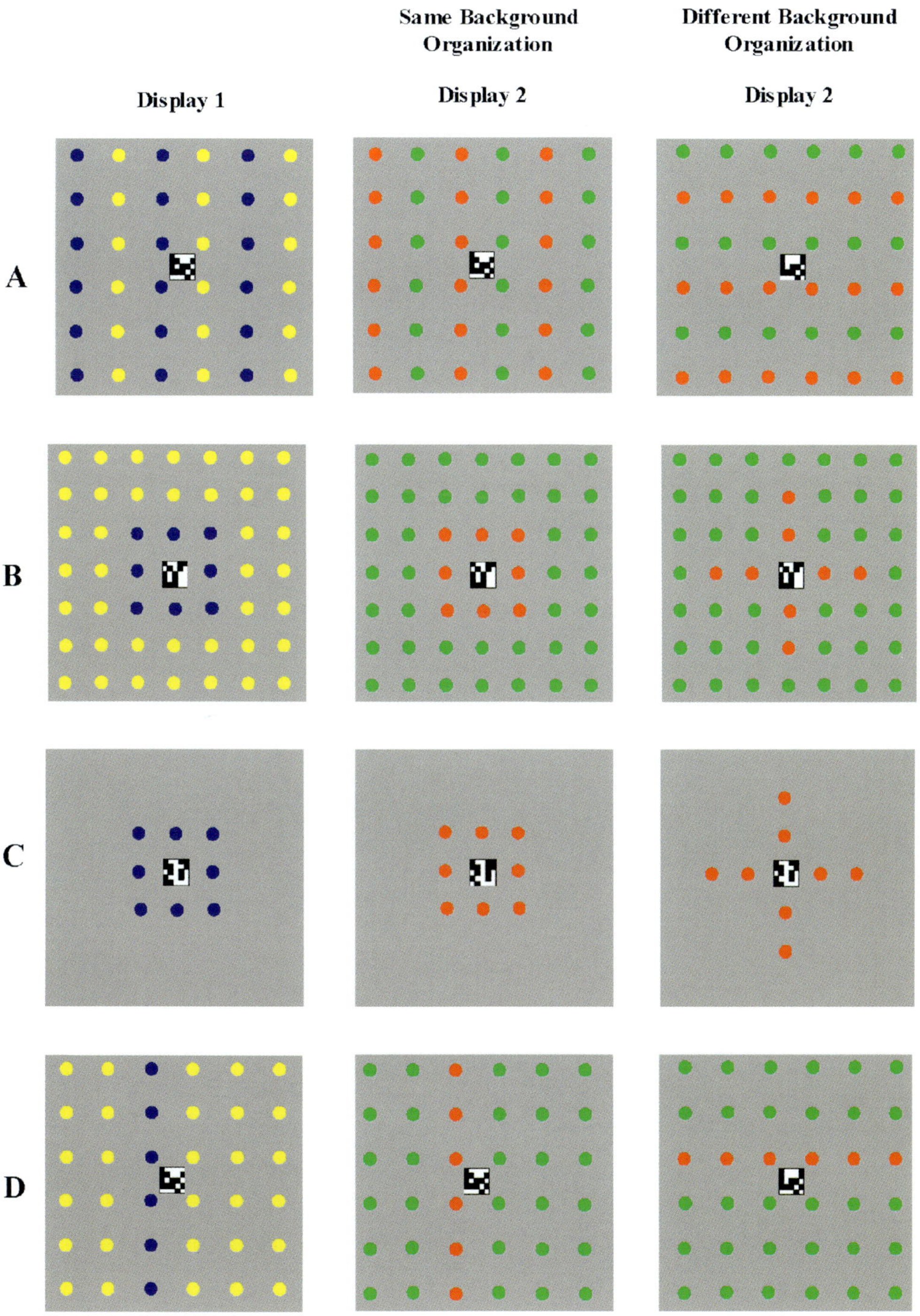

Plate 2.3. Examples of the stimulus displays used by Kimchi and Razpurker Apfeld (2004). Two successive displays were presented on each trial. The central target matrix in Displays 1 and 2 were either the same or different. The surrounding colored elements were grouped into (A) columns/rows by color similarity, (B) a square/cross by color similarity, (C) a square/cross, (D) a vertical/horizontal line by color similarity. This background organization either stayed the same across Displays 1 and 2 or changed, independently of whether the target matrix changed or remained the same. The colors of the background elements always changed between Displays 1 and 2. All colors were equiluminant in the experiment. Adapted with permission from Kimchi and Razpurker Apfeld (2004).

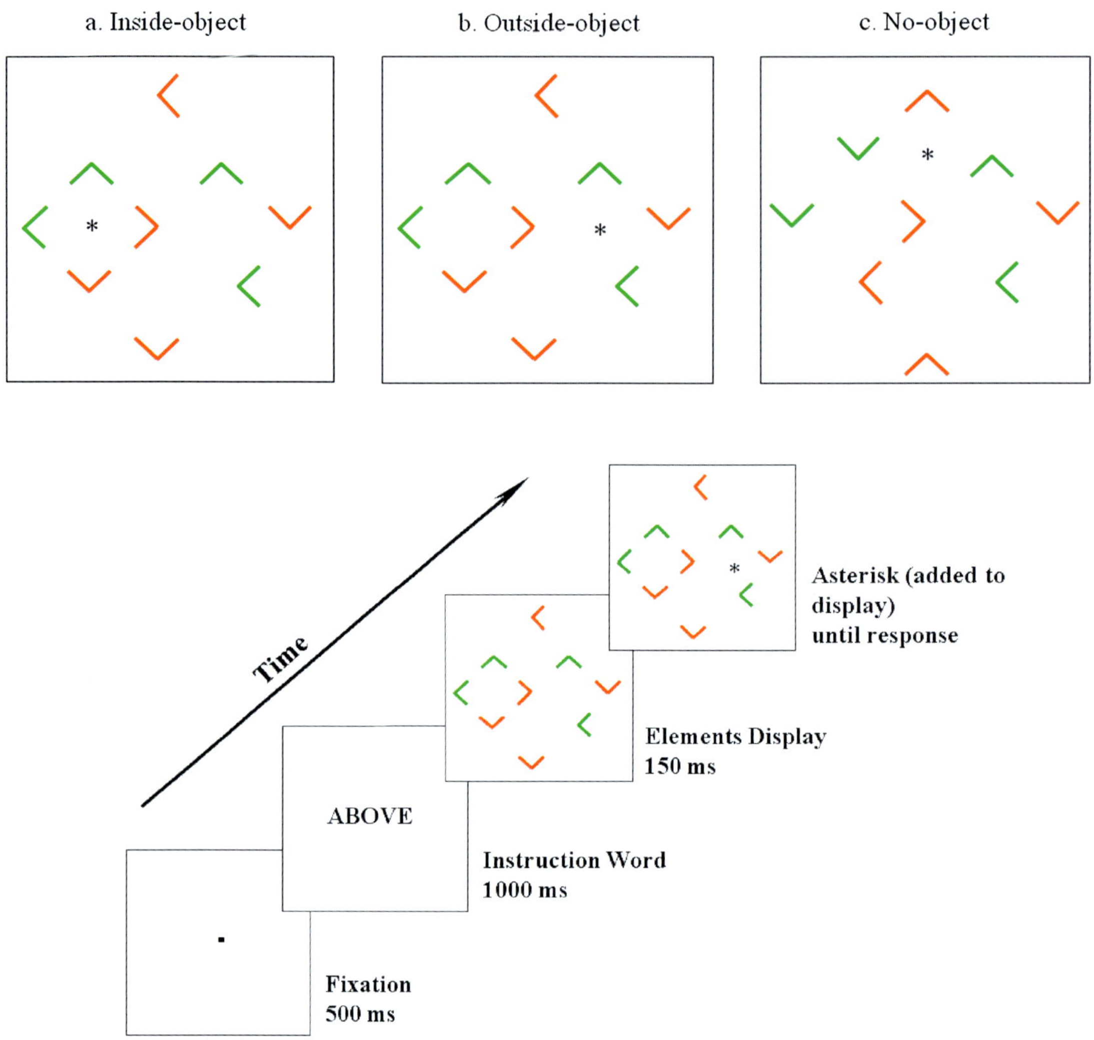

Plate 2.5. Top panel: Examples of the displays used by Kimchi et al. (2007). Each display composed of nine red and green elements. (a) Inside-object condition: object present in display and asterisk appearing in center of object quadrant; (b) Outside-object condition: object present in display and asterisk appearing in center of nonobject quadrant; and (c) No-object condition: no object present in display. Fifty percent of the trials were No-object trials, 12.5% were Inside-object trials, and 37.5% were Outside-object trials. Bottom panel: Sequence of events in a trial. The illustration depicts an Outside-object trial with the instruction word *above*. In this trial, the participants had to identify the color of the element above the asterisk (green). Adapted with permission from Kimchi et al. (2007).

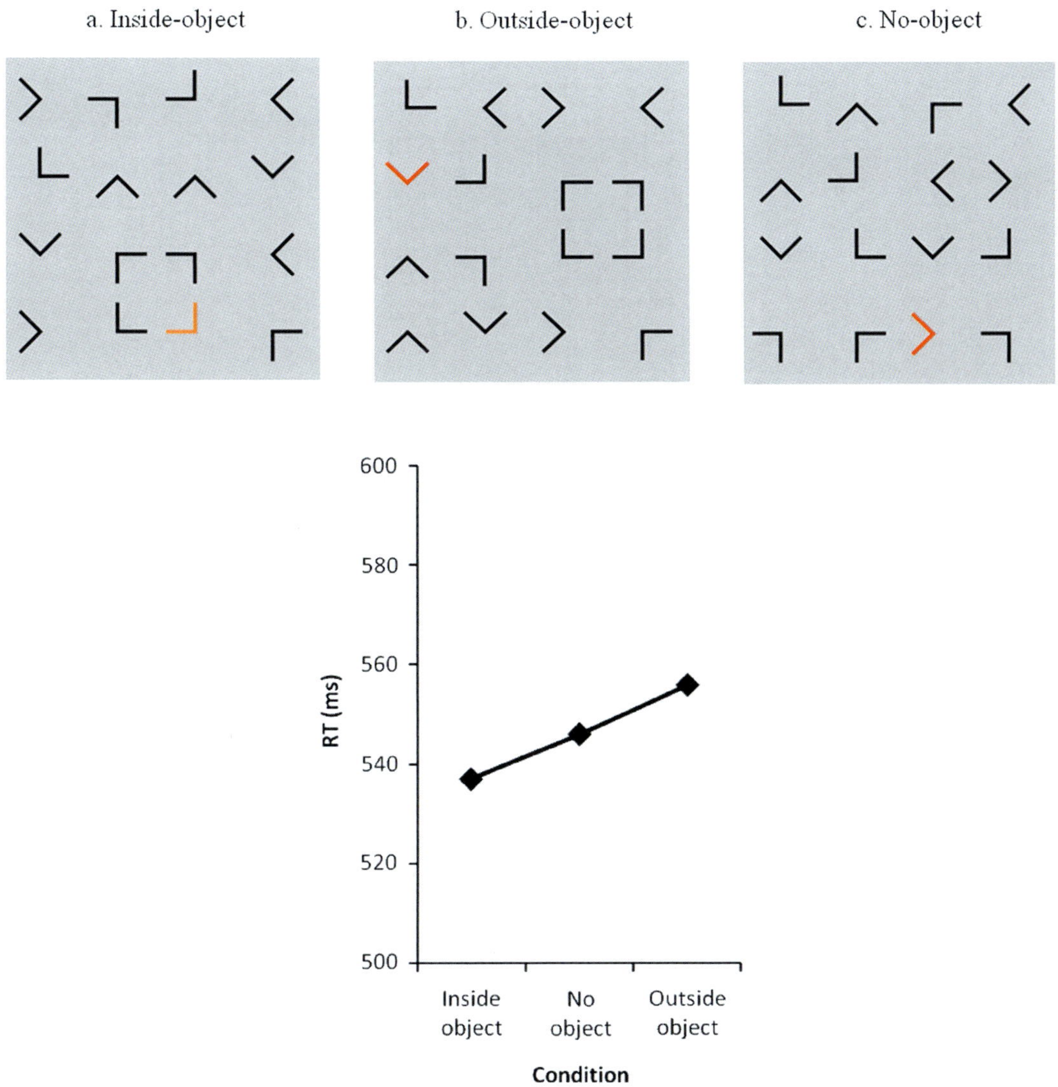

Plate 2.7. Top panel: Examples of the displays in the three conditions. (a) Inside-object condition: object present in display and the target is an object element (16.6% of all trials); (b) Outside-object condition: object present in display and the target is a non-object element (33.4% of all trials); and (c) No-object condition: no object present in display (50% of all trials). Bottom panel: Mean correct reaction times (RTs) as a function of condition.

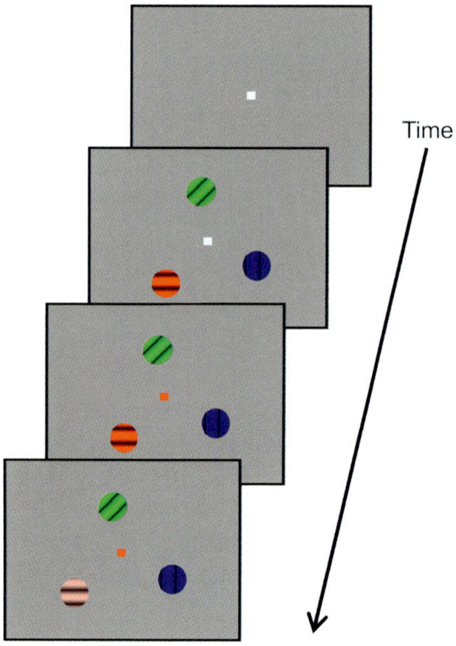

Plate 3.1. Behavioral task. The monkeys had to hold a bar to initiate the trial and subsequently fixate the white fixation spot at the center of the screen. After successful fixation three sinusoidal drifting gratings (red, blue, and green) appeared on the screen, at positions distributed radially around the fixation spot at 120° intervals. The fixation spot was subsequently replaced by a small square cue that matched the color of one of the gratings indicating the color of the stimulus to be attended. The monkeys were required to shift their attention to the target stimulus while maintaining fixation of the cue and monitor the target for a color change. The animals were rewarded with a drop of juice for releasing the bar when the target changed color. On any given trial one or both of the distracter stimuli could also change color before the target but the monkeys were trained to ignore the distracters' change. If the monkeys released the bar to the distracter change, failed to maintain fixation, or did not respond to the target color change within 600 ms, the trial was aborted.

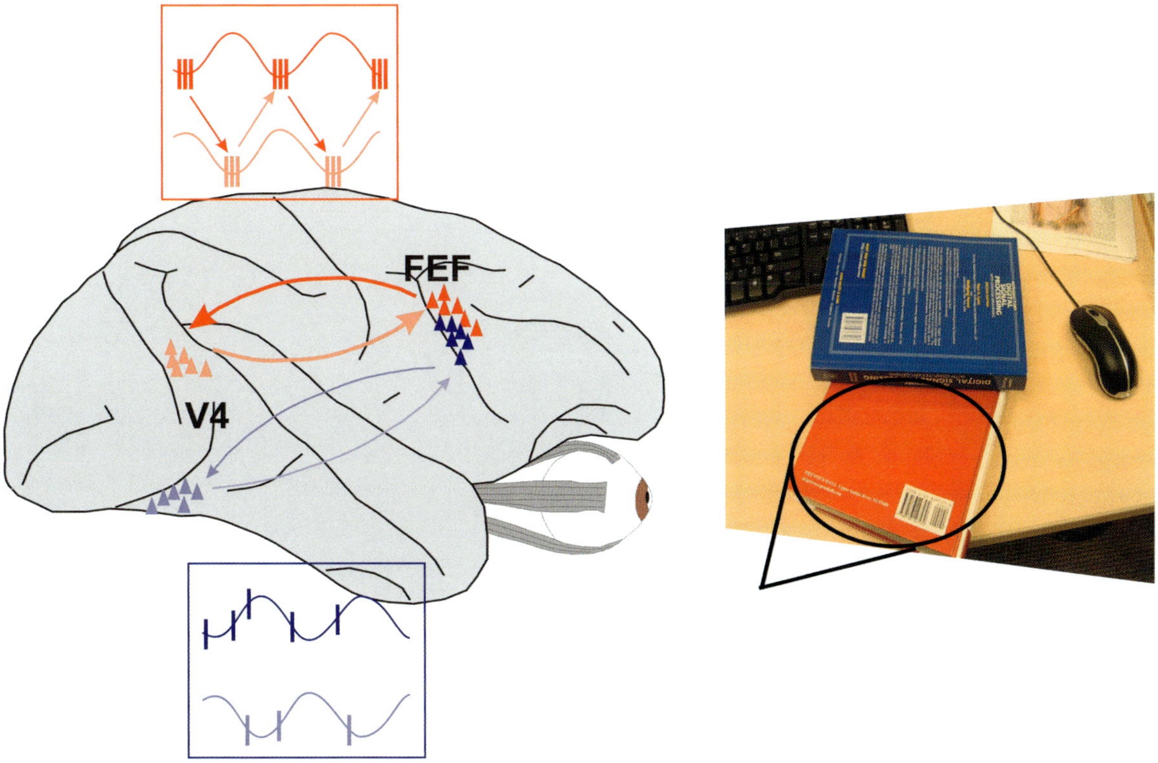

Plate 3.6. Schematic illustration of inter-areal neuronal communication in attention. Dark red and light red triangles illustrate neurons in FEF and V4, respectively, encoding the attended stimulus (red book on the right), whereas dark blue and light blue triangles illustrate FEF and V4 neurons, respectively, encoding the unattended stimulus (blue book). The vertical lines in the boxes above and below the schematic brain illustrate action potentials of neurons in the four groups. Arrows indicate propagation of action potentials between areas along the projecting axons. Coherent spikes which arrive at the phase of maximal excitability increase the probability of generating spikes in the receiving area (red box). Note the phase relationship between excitability fluctuations in the two areas which facilitates neuronal communication. Less coherent spikes corresponding to the unattended stimulus (blue box) are less effective in triggering spikes in the receiving area and result in weak communication between the areas and a weaker representation of the unattended stimulus.

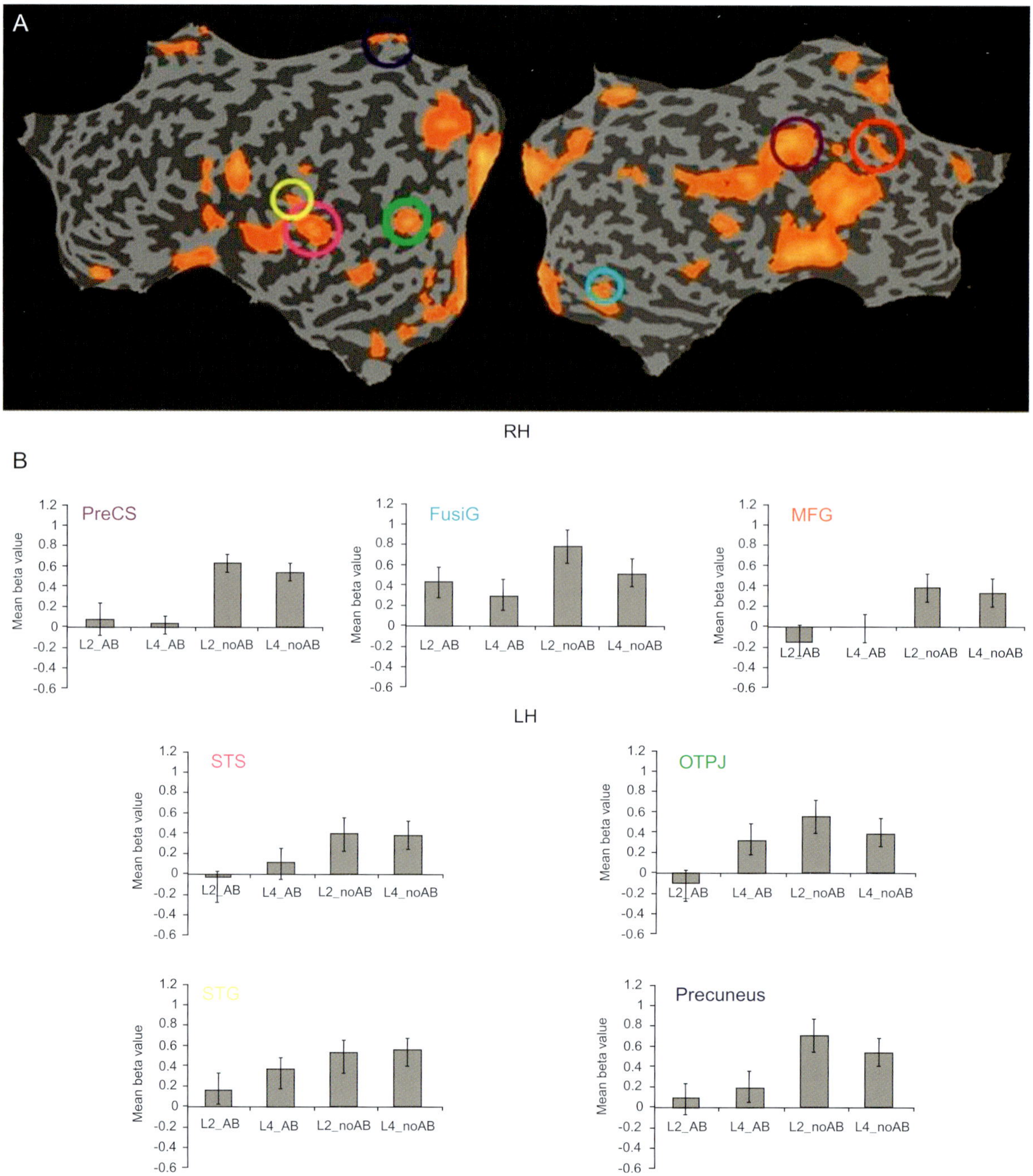

Plate 7.7. Cortex-based group analysis of the experiment. (A) Contrast for the main effect of blink. Contrast maps showing averaged blink and no-blink trial evoked activity during encoding using a contrast threshold value of $P<0.01$, uncorrected. To protect against false positives, a cluster filter correction was implemented. Group-averaged random effect activation maps are superimposed on a flattened MNI template brain. On the flattened template, light and dark grey regions indicate gyri and sulci, respectively. Colour indicates t-value: $t(16)>2.95$ to >8 (red to yellow), positive activity. Marked clusters represent ROIs described in (B). (B) Mean parameter estimates of the peak voxel of selected ROIs for the attentional blink (AB) and no-attentional blink (no-AB) conditions of the main effect of blink during the encoding phase. LH, left hemisphere; RH, right hemisphere; PreCS, precentral sulcus (purple); FusiG, fusiform gyrus (cyan); MFG, middle frontal gyrus (red); STS, superior temporal sulcus (pink); OTPJ, occipitotemporoparietal junction (green); STG, superior temporal gyrus (yellow); Precuneus (dark blue). Error bars $\pm$ SE.

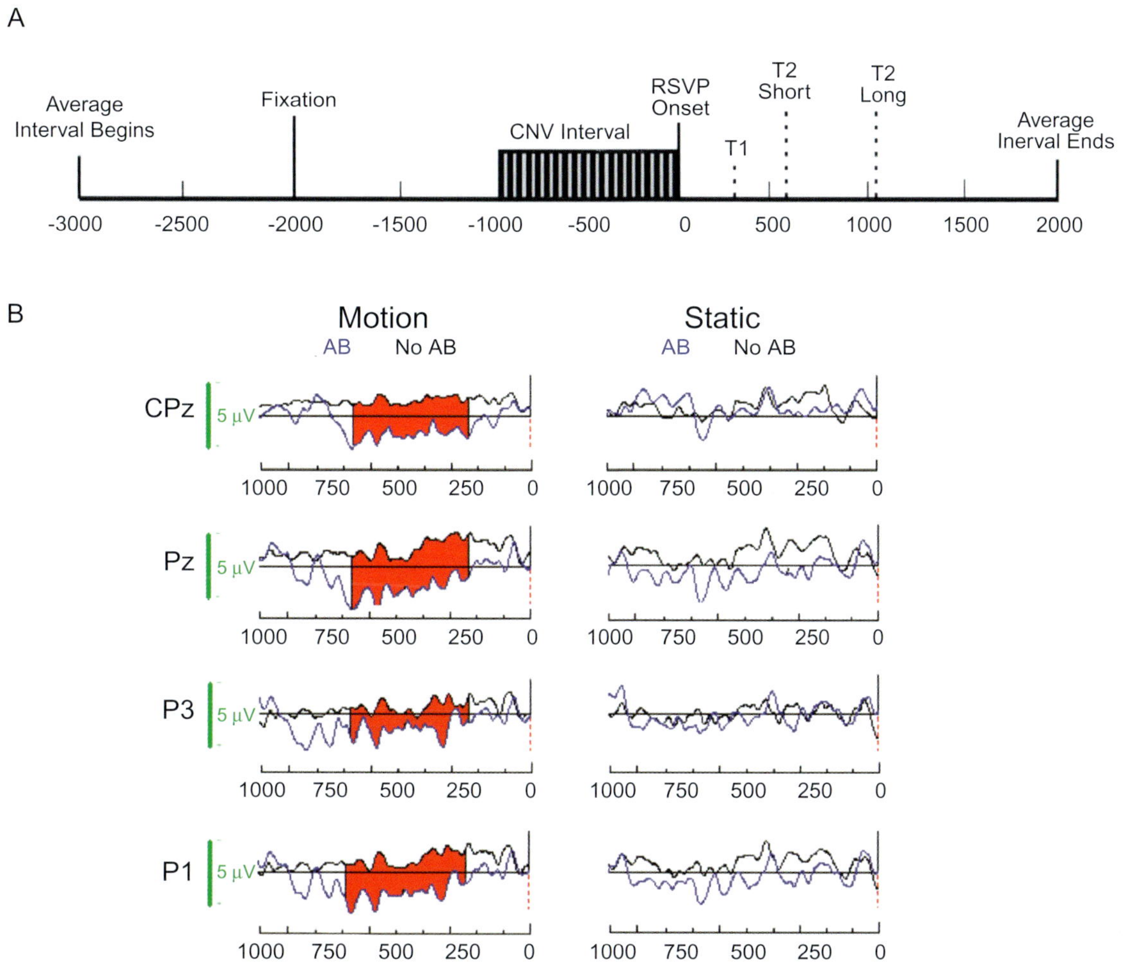

Plate 7.8. Top panel: Critical temporal markers for averaged ERP waveforms. As shown the CNV was measured between fixation and RSVP onset. Bottom panel: Respective ERP waveforms for dual-target trials. Shown are AB versus no-AB trials for the motion and static conditions. Vertical bars mark the temporal interval analysed to reflect CNV amplitude. Red colouring indicates a statistically significant difference between AB and No-AB trials as revealed by post-hoc comparisons.

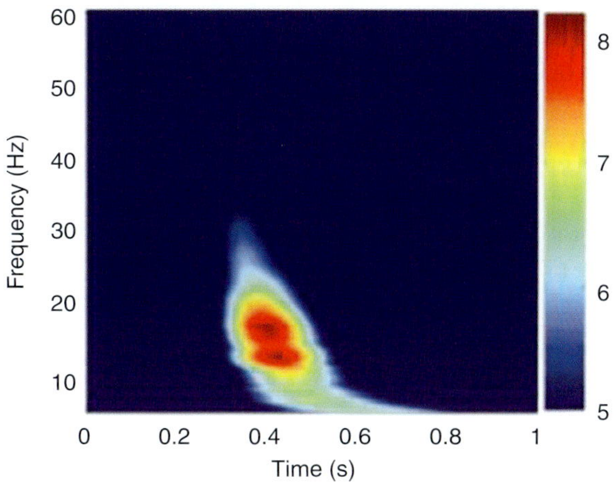

Plate 7.11. TFR for the distractor condition subtracted from the target condition. Time 0 marks the onset of the target. The TFR represents the average across subjects and channels and is displayed in units of standard deviation of the baseline (thresholded at a value of 5). TFRs have been normalised for each frequency before averaging.

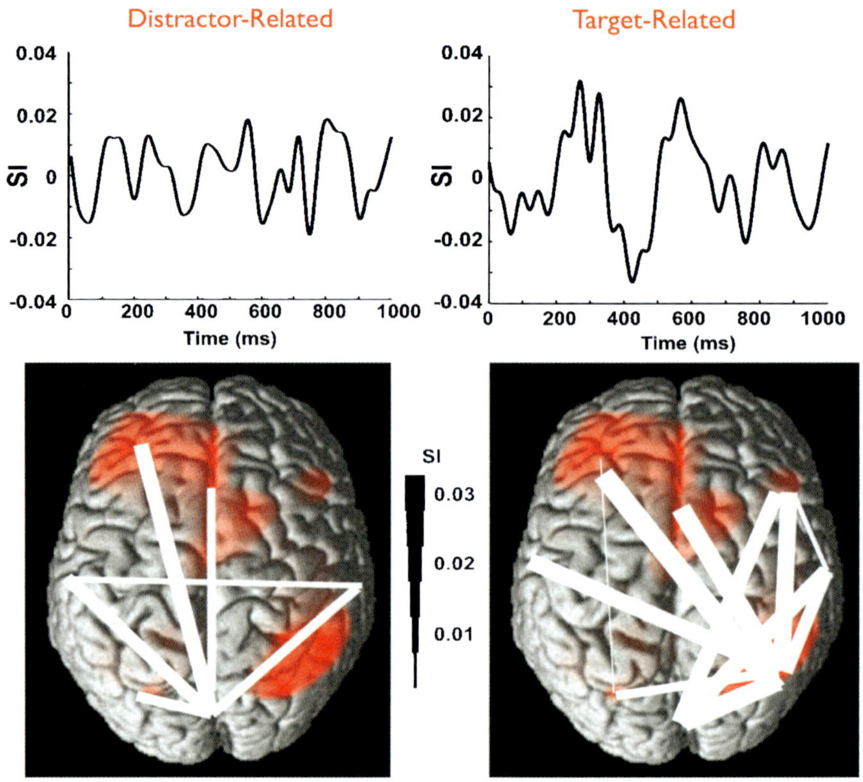

Plate 7.13. Classification of stimulus- and target-related connections. Top panel: SI for one subject for a typical stimulus-related (left, occipital to posterior parietal left) and a typical target-related (right, frontal left to posterior parietal right) connection. SI was computed based on sensor groups that are most sensitive to a given region. Bottom panel: The stimulus-related (left) and target-related (right) networks are shown with linewidth coding for the strength of synchronisation at 260 ms.

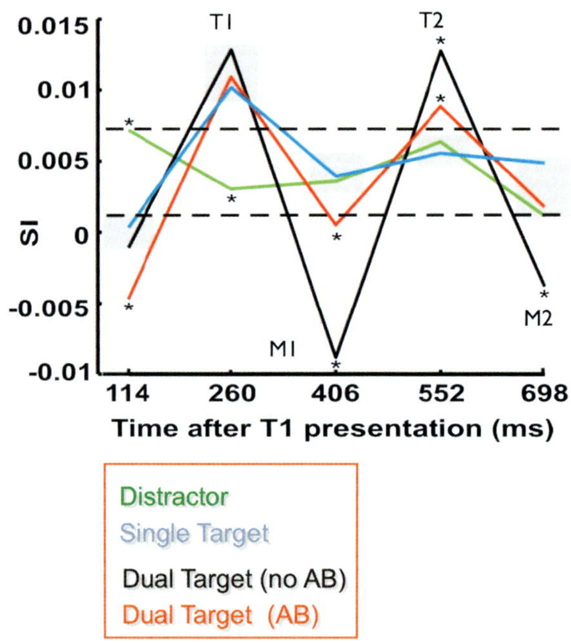

Plate 7.14. SI for the components of five successive stimuli. The x axis specifies time after presentation of the first target. Each point represents the mean SI in a 60-ms window centered at 260 ms after the respective stimulus. Values at 260 ms quantify the network synchronisation to the first target, and values at 114 ms represent the network synchronisation corresponding to the distractor preceding the first target. Conditions are colour-coded (black, no-AB; red, AB; blue, target; green, distractor). The dashed lines mark the extent of SI in trials containing only distractors. Points marked with an asterisk are significantly different from their neighbours at the same position ($P < 0.05$, Kruskal–Wallis test), whereas points within the same shaded area are not significantly different. Negative values arise from the filtering of the SI time courses.

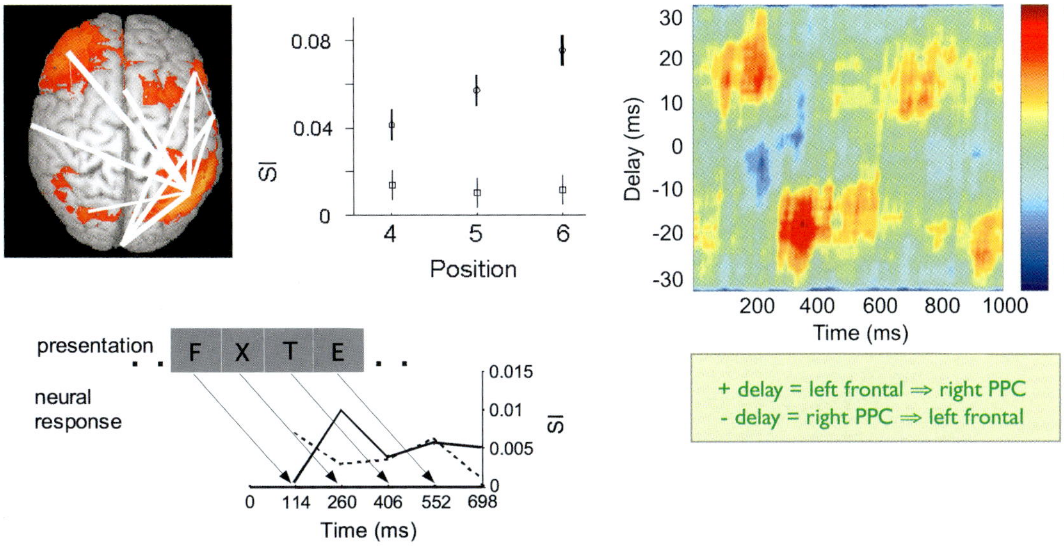

Plate 7.15. Top left panel: Connections in the target-related network. Functional maps of oscillatory power in the beta-band were computed for each subject. The functional maps were spatially normalised using SPM99 and a permutation analysis using SnPM99 was performed. Only areas with a significance below $P<0.05$ (corrected) are shown. Lines mark connections for which the phase synchronisation is significantly modulated by target presentation. The displayed connections form the target-related network. Top middle panel: Modulation of phase synchronisation (SI) by targets at different positions in the presentation stream. The mean of 11 points surrounding the maximum (at about 260 ms) and the minimum (at about 114 ms) was computed for all subjects and connections of the target-related network for targets (circles) and distractors (boxes). Lines extending from the mean indicate the standard error. The modulation (difference of synchronisation and desynchronisation) increases with the position only for target trials. Top right panel: *Y* axis shows delay between left prefrontal and right PPC activation with positive delays indicating left frontal preceding right PPC and negative delays the reverse relative to time since target onset shown on *X* axis. Bottom panel: Modulation of phase synchronisation (SI) by targets as compared with distractors. The solid line shows the SI in trials where a target occurs, whereas the dashed line shows the SI in trials with only distractors. The *X* axis specifies time relative to target onset. Each point represents the mean SI in a 60-ms-long window centered at 260 ms after the respective stimulus. For illustration, part of a possible letter sequence (with target X) is shown in the upper part of the panel. Synchronisation values at 260 ms quantify the network synchronisation to the target. At 114 ms a reduced synchronisation is evident. This may represent the network response to a distractor that is followed by a target. The panel illustrates that target X is already partly processed at 114 ms and obviously affects the processing of the distractor by reducing the synchronisation.

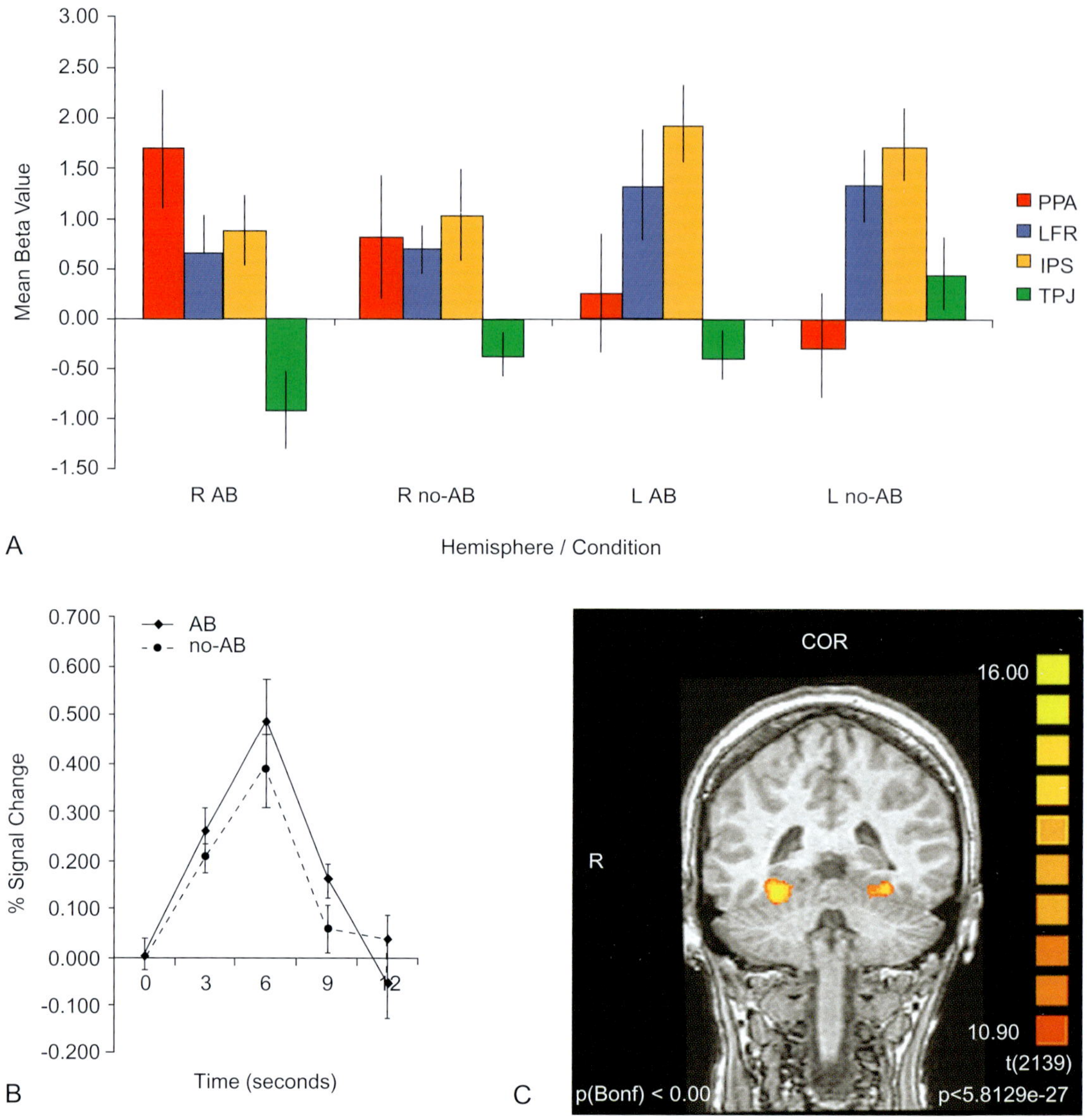

Plate 7.19. (A) Mean parameter estimates for the short SOA, attentional blink and no-attentional blink conditions, separated into right and left hemisphere for each region of interest. Error bars represent the standard error of the mean. (B) Percent signal change versus time (post-stimulus onset) for the short SOA, attentional blink and no-attentional blink conditions extracted from the PPA ROI. Error bars represent the standard error of the mean. (C) A representative coronal slice of one participant showing activations in PPA as a result of the localiser scans. The functional imaging result is shown superimposed on an anatomical MRI data set.

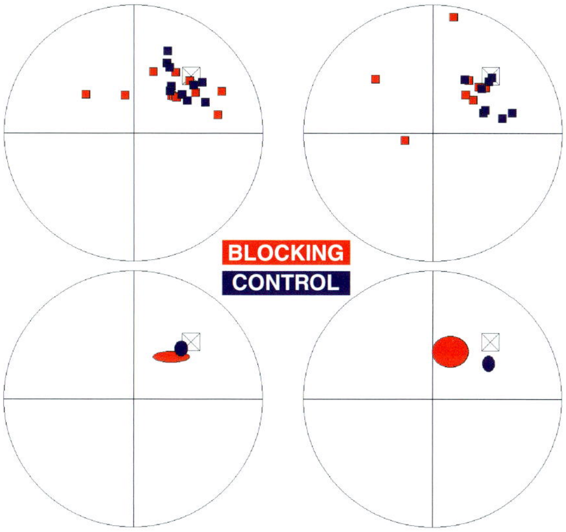

Plate 12.5. Performance on the location accuracy probe trial. The blocking effect for males found in the search persistence probe trial is absent. Both males and females located the target in the actual quadrant, close to its actual location. Although females were more accurate than males in recalling the target location, both males and females knew in which quadrant the target was located and in about what area within this quadrant. Since the two probe trials were administered in succession, and since the actual target position was never revealed during the probe trials nor between them, male subjects must have known the target location during the search persistence probe trial as well, however, for yet unknown reasons they did not continue to exclusively search the target quadrant for it. Upper graphs show location recall for each subject, while lower graphs represent the average of the recalled locations for each group in form of an ellipse, the center of which denotes the arithmetic mean, while major and minor axis represent the standard deviation of the *x* and *y* coordinates of the recalled target positions, respectively.

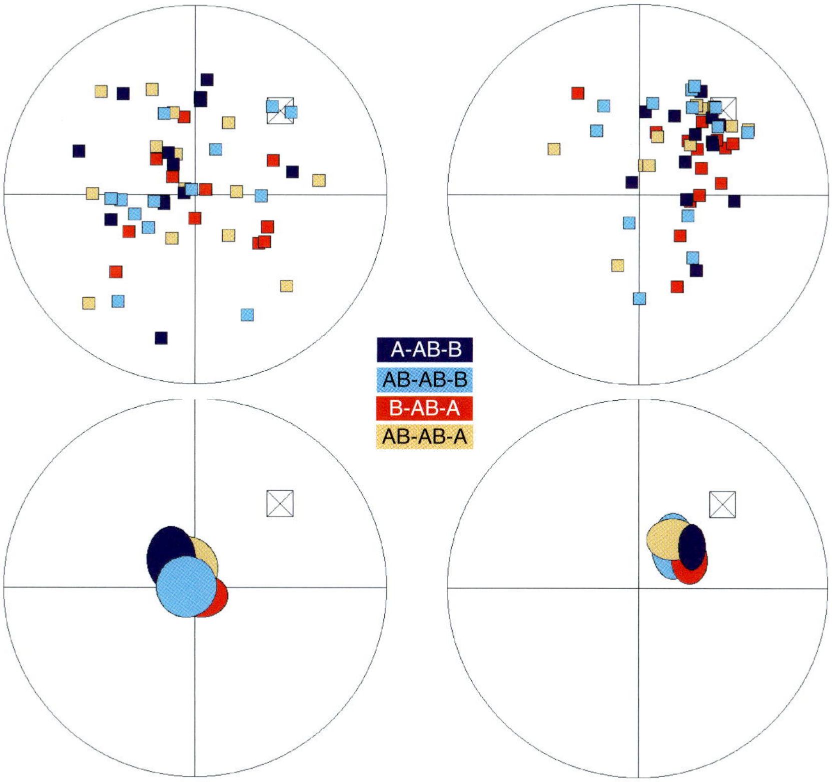

Plate 12.8. Performance during the location accuracy probe trial reveals that location knowledge is the same for all groups (see panels on right hand, showing recall performance of subjects who believed that the target remained in the same location during training). The panels on the left hand show location recall of subjects who believed that the target location changed during training: for them, all quadrants were equally plausible locations for the target. Upper panels show recalled locations for all individuals, lower panels the group averages.

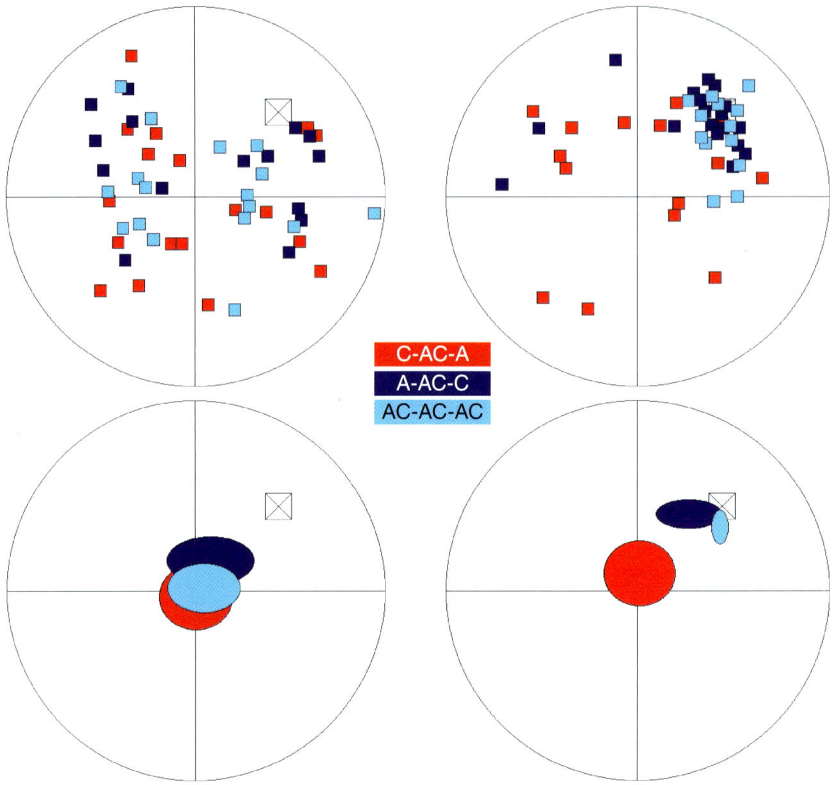

Plate 12.11. Location accuracy probe trial. When only the added abstract cues were available, subjects were not able to accurately recall the target location. This was not the case when concrete cues were present (see right-hand panels). Subjects who believed that the target location was changing recalled the target location equally often in every quadrant (left-hand panels). Upper panels show recall locations of all individuals, lower panels the group averages of these positions.